BIOLOGY
A CONTEMPORARY APPROACH

Sebastian Haskel

David Sygoda

AMSCO

AMSCO SCHOOL PUBLICATIONS, INC.
315 Hudson Street / New York, N.Y. 10013

Sebastian Haskel and David Sygoda are the authors of *Fundamental Concepts of Modern Biology* and *Biology Investigations.*

Cover photo: Jaguar in a tropical rain forest, by Jeanne Drake, Tony Stone Images. Jaguars live in the rain forests of Mexico, Central America, and South America.

Please visit our Web site at:

www.amscopub.com

When ordering this book, please specify:
R 628 S *or* BIOLOGY: A CONTEMPORARY APPROACH, SOFTBOUND
or
R 628 H *or* BIOLOGY: A CONTEMPORARY APPROACH, HARDBOUND

ISBN 0-87720-052-1 (Softbound edition)
ISBN 0-87720-060-2 (Hardbound edition)

Printed in the United States of America

5 6 7 8 9 10 02 01 00

PREFACE

Biology touches all aspects of life. This edition of *Biology: A Contemporary Approach* updates and discusses important physiological, genetic, ecological, and evolutionary processes and relationships. The subject matter of the book reflects current research in the biological sciences. As a result, many problems regarding the complexities of life are better explained and thus can be better understood by students. Certain social and environmental issues are discussed, enabling students to make connections with the community of life on Earth.

Biology: A Contemporary Approach invites scientific thinking through problem solving—a process that encourages inquiry. For example, the discussion of photosynthesis includes investigations that explain, clarify, and reinforce facts and ideas regarding this food-making process. The reading level and conversational style of the book cater to the needs of high school students, thus fostering interest and comprehension. *Biology: A Contemporary Approach* introduces basic concepts early and repeats them in relevant chapters, and uses headings, subheadings, outlines, and illustrations to highlight important ideas. Scientific terms that may be new to the student are italicized and defined in context. In addition, a glossary of important terms precedes the index.

Each chapter resembles a classroom lesson that opens with a motivating introduction followed by a logical presentation and development. The clear illustrations and summary tables are valuable aids in organizing, learning, and remembering essential information. Because of these features, *Biology: A Contemporary Approach* may be used effectively by all students to learn modern biological topics in an interesting and enjoyable manner.

Learning Features

Biology: A Contemporary Approach is designed to make learning easier. Special features that stimulate interest, enrich understanding, and enable students to evaluate their progress and review important concepts are provided. The three different types of question sets that are offered in each chapter may be used for review, testing, and/or homework assignments. The features of this book include:

1. ***Chapter Learning Objectives.*** Each chapter begins with a Learning Objectives section that provides guidelines for focusing learning on important topics in the chapter. The objectives will also be helpful as a way of reviewing each chapter's material.
2. ***Chapter Overview.*** The introductory Overview section is a concise description of the chapter's purpose and contents. In conjunction with the Learning Objectives, this section will help direct the students to the aim of their study in each chapter.
3. ***Section Quizzes.*** These short-answer question sets are provided to help students test themselves on discrete portions of each chapter.
4. ***Chapter Review Questions.*** These more extensive question sets are provided at the end of each chapter. They consist of multiple-choice questions that test the student's achievement in recall and understanding of the textual material.
5. ***Biology Challenge Questions.*** These question sets, placed after the Chapter Review questions, are varied in format and graded in difficulty. The answers to these questions involve the students in making associations, inferences, interpretations, and hypotheses, and may require further research. The questions stimulate and elicit higher-level thinking skills. In addition, they provide good preparation for SAT-II examinations.

With so many advances and discoveries—relevant to the life of each student—being made in the biological sciences today, the study of biology can be very exciting and rewarding. *Biology: A Contemporary Approach* aims to enhance the student's enjoyment of this interesting field.

CONTENTS

1

Modern Concept of Life

Learning Objectives

When you have completed this chapter, you should be able to:

- **Develop** a working definition for the word "life."
- **Classify** organic and inorganic compounds on the basis of elements they contain.
- **Identify** life processes and explain how they contribute to an organism's survival.
- **List** the special characteristics of living things.

OVERVIEW

Biology is the study of living things, or organisms. With the help of scientists in other fields, biologists try to answer the question, "What is life?" This question remains not fully answered because there are many things we still do not know about the complex nature of life. In this chapter, you will learn some of the things we do know about life. We hope this knowledge will encourage you to study further. Perhaps in the future, you will contribute to a better understanding of what life is.

ORGANIC SUBSTANCES

Living things or organisms that were once alive are called *organic*. Things such as table salt ($NaCl$) that are neither living nor were ever alive are called *inorganic*. To a chemist, an organic substance is a substance that contains the element carbon in complex combinations with hydrogen and other elements. For example, table sugar ($C_{12}H_{22}O_{11}$) is an organic substance. Both living matter and dead matter contain carbon, hydrogen, and other elements in many different combinations. As a result, both living and dead matter fit the definition of organic substances.

LIFE PROCESSES

What does an organism do to survive? All organisms carry on similar activities called *life processes*, such as movement, ingestion, digestion, transport, respiration, synthesis, assimilation, growth, excretion, sensitivity, and reproduction.

Movement

The movement of organisms helps them survive. Animals can move from place to place in search of food and shelter. Although rooted plants such as trees and shrubs do not move from place to place, they show a kind of movement in their ability to grow upward, downward, and in width. Some plants, such as the sunflower, follow the sun by bending toward it. Some algae and protozoans use whiplike structures called *flagella* (singular, flagellum) for swimming.

Ingestion

All organisms must have food to survive. The taking in of food by an organism is called *ingestion*. The food ingested by animals is usually organic—that is, it either is alive or was once alive. The food of nongreen plants, such as Indian pipe (Figure 1–1), and of plantlike organisms, such as mushrooms, is also organic. Nongreen plants and plantlike organisms are similar to animals in that they also use organic substances for food.

Figure 1–1 A nongreen plant (Indian Pipe)

Green plants make their own food using solar energy, carbon dioxide (CO_2), water (H_2O), and other inorganic substances. Green plants change inorganic substances into organic substances such as sugar, starch, protein, and oil. The organic substances that plants make and use for energy, growth, and repair are the same substances that animals use for food.

Organisms have different adaptations for ingestion. For example, the ameba, a one-celled organism, ingests food by engulfing it. The paramecium, another one-celled organism, sweeps food into its body by using tiny hairs called *cilia*. Bacteria take in inorganic and organic substances directly from their environment. Most animals possess a mouth opening through which food enters the body.

Digestion

The chemical process that changes complex food molecules to simpler molecules is called *digestion*. Digestion makes many chemicals available to an organism and enables the organism to release energy, grow, and repair itself. The digestive process is essentially the same in all organisms. For example, both animals and plants digest starch to a simple sugar called *glucose* that is used by most organisms as a source of energy.

Transport

All of the processes that move usable substances and waste substances to and from cells as well as into and out of cells are called *transport*. Transport is a continuous movement of substances. If transport stops, cells cannot function properly, disorder (lack of chemical balance) occurs within them, and the cells die.

Respiration

The process that releases, transforms, and stores the energy living things need to carry on their activities is called *respiration*. During respiration, oxidation reactions within cells change glucose to simpler substances and release energy. All organisms carry on *anaerobic respiration*, or reactions that do not require oxygen. Simple organisms, such as certain bacteria, use this process exclusively. Complex organisms, such as multicellular plants and animals, carry on both anaerobic and aerobic respiration. *Aerobic respiration*, a process that requires oxygen, releases much more energy than does anaerobic respiration.

Synthesis

All organisms combine, or *synthesize*, simple chemical substances to form more complex substances. Some of the energy released by respiration is used for this purpose. For example, a plant may synthesize starch from sugar, and a human may synthesize a starchlike product (glycogen) or fat from the same sugar.

Secretion

Some cells synthesize special substances that are used in other life processes. The formation of these useful substances is called *secretion*. For example, certain cells of human salivary glands produce *ptyalin*, an enzyme that helps digest starch in the mouth. Other cells secrete *hormones*, such as growth hormone, which stimulates growth of bones and muscles.

Assimilation

The process of digestion produces useful substances that are transported to all of the cells. Each substance has a different use in the body. After entering a cell, these substances may be synthesized into the different substances that make up living matter. The changing of chemical substances into living matter is called *assimilation*. As a result of synthesis and assimilation, organisms grow and repair themselves.

Growth

All living things grow. The process of assimilation increases the amount of protoplasm in an organism. Growth usually involves the enlargement of cells and the production of more cells.

Excretion

The reactions that occur in cells are called *metabolic activities*, or *metabolism*. Constructive metabolic activities, such as assimilation, which tend to build up cells, are called *anabolic*. The process is called *anabolism*. Destructive metabolic activities, such as movement, which expend energy and use up materials, are called *catabolic*. This process is called *catabolism*. Both types of activities produce waste products such as carbon dioxide and water. If not removed, these wastes disturb the chemical balance within cells, creating a disorder that prevents the normal functioning of the cells. The process that removes metabolic wastes from cells and from body fluids is called *excretion*. Thus, by maintaining a proper chemical balance between the interior and the environment (surroundings) of each cell of an organism, excretion enables the organism to survive.

Sensitivity (Regulation)

The ability of an organism to respond to changes in its environment is called *sensitivity*. For example, an ameba responds to a lack of oxygen within its body by taking in oxygen from its watery environment. The same ameba responds to a lack of water by forming a hard wall around itself. In this way, the ameba retains water and is able to survive until more water is available in its environment. Most animals react to environmental changes by means of a nervous system, a system of glands, or both. Plants lack a nervous system but possess groups of cells that function as glands. Both systems influence or regulate the responses an organism makes to its environment by sending out chemical messages to cells and organs.

Reproduction

The life activity by which organisms produce offspring is called *reproduction*. *Asexual reproduction*, which involves only one parent, does not require sex cells. For example, a bacterium produces another bacterium exactly like itself by splitting in two. This process is called *cell division*. Sexual reproduction requires two parents. Each parent produces sex cells—either egg cells or sperm cells. When these sex cells unite, *fertilization* occurs.

The cells of organisms contain tiny structures called *organelles*. When a cell divides, the organelles divide. Complex molecules within cells also divide when the cell divides. For example, DNA (*deoxyribonucleic acid*) molecules in the nucleus of a cell, which control the hereditary traits of an organism, produce copies of themselves during cell division (*replication*). As a result, the nucleus of each new cell receives a complete set of identical DNA molecules. Thus, reproduction occurs at the chemical level, the organelle level, and the individual level.

Section Quiz

1. Which life process must occur before table sugar ($C_{12}H_{22}O_{11}$) can be used by the cells for energy? (*a*) excretion (*b*) synthesis (*c*) digestion (*d*) respiration.
2. Aerobic respiration is a process that (*a*) is a characteristic of multicellular plants and animals (*b*) does not require oxygen (*c*) is used exclusively by simple organisms (*d*) does not release as much energy as anaerobic respiration.
3. Organisms grow and repair themselves mainly by (*a*) ingestion and transport (*b*) respiration and excretion (*c*) synthesis and assimilation (*d*) regulation and sensitivity.
4. Reproduction in an organism occurs (*a*) only at the chemical level (*b*) at the chemical level and the organelle level (*c*) only at the organelle level (*d*) at the chemical level, organelle level, and individual level.

SPECIAL CHARACTERISTICS OF LIFE

To carry out life processes, living things:

1. Release, transform, and store energy.
2. Respond to stimuli from their internal and external environments.
3. Maintain a stable internal environment.
4. Vary genetically and pass on adaptations.

Need for Energy

Living things require a constant source of energy to move, to manufacture organic substances, to change digested food into protoplasm, and to carry on other metabolic activities. Food molecules are the usual source of this energy. The reactions of cellular respiration release, transform, and store energy. Inorganic and organic compounds, obtained from nutrients in food, supply the reactants for biochemical reactions.

Response to Stimuli

All living things respond to changes, or *stimuli,* in both their internal and external environments. Responses to stimuli aid the survival of an organism in various ways. For example, the internal environment of an animal lets it know that it needs energy, or nutrients, so the animal seeks food. Likewise, the external environment may present a threat in the form of another animal that also is hungry, so the first animal responds by fleeing. Responses to stimuli range from the simple to the complex, depending upon the stimuli and the organisms responding to them.

Stability of Organization

Like nonliving things, living things are organized systems that tend to break down and eventually become disorganized. The constant use of energy by an organism tends to prevent the breakdown of its organization. In addition, organisms usually possess feedback systems that help them maintain a state of balance, or a *steady state*. The automatic maintenance of a steady state by an organism is called *homeostasis.*

A steady state helps an organism maintain a balanced internal environment. An unbalanced internal environment usually causes disorganization and abnormal functioning. For example, hot weather makes you perspire more than usual. As the perspiration on your skin evaporates, your body temperature drops when the temperature of the environment rises (Figure 1–2). Thus, perspiring helps to maintain a normal body temperature. If you do not perspire, your body temperature will rise beyond its normal temperature. This can interfere with the chemistry of the body and cause serious illness or even death.

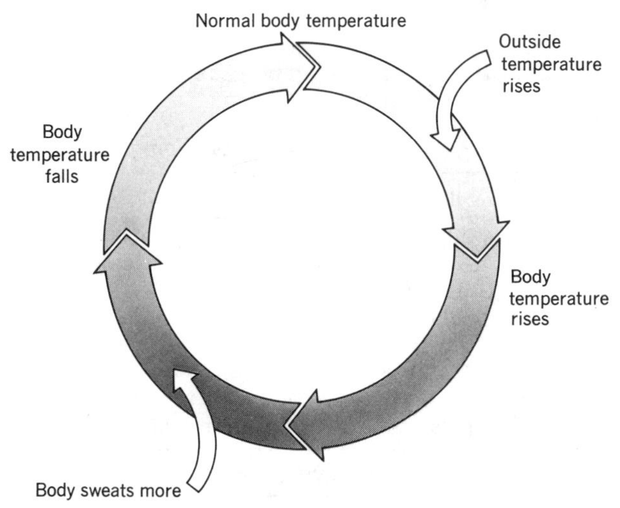

Figure 1–2 Maintenance of a steady state

Genetic Variation

Genetic variations that help an organism survive are called *advantageous*. These variations may result in specialized structural or chemical characteristics, such as the webbed toes of a frog or the poisonous toxins secreted by bacteria. Genetic variations that do not help an organism adjust to its environment usually lead to its death and extinction of its species. During the course of evolution, new species arose by passing on advantageous genetic variations by the process of reproduction.

Adaptation

The special characteristics of an organism that enable it to survive in its environment are called *adaptations.* For example, the highly complex human brain has helped people adjust to changing environments; the spreading branches of a tree help the leaves obtain as much light as possible for the manufacture of food. The gradual accumulation of genetic variations that enable living things to survive as their environment changes is called *evolution*. As a result of genetic variations, new adaptations are always arising. Consequently, evolution is a continual process.

WHAT IS LIFE?

Many more features of living things will be treated in the following chapters. In general, however, all living things are distinguished from nonliving things by being able to carry on similar life processes (Table 1–1).

Table 1–1 Summary of Life Processes and Characteristics

Process or Characteristic	*Explanation*
Movement	Ability of entire organism or a part to move
Ingestion	Taking-in of food
Digestion	Breakdown of food from complex to simpler chemical form
Transport	Movement of needed substances to cells and into cells, and movement of wastes and other cell products out of and away from cells
Respiration	Release of energy resulting from oxidation of food
Synthesis	Formation of complex substances from simpler substances
Assimilation	Change of digested substances into living matter, making possible self-repair and growth
Growth	Enlargement that results from synthesis, assimilation, and increase in number and size of cells
Excretion	Removal of wastes of metabolism
Secretion	Formation of substances used by the organism
Sensitivity	Response to environmental changes, thereby maintaining a steady state (homeostasis)
Reproduction	Giving rise to offspring of the same kind
Homeostasis	Automatic maintenance of a steady state by an organism
Variations	Differences between closely related organisms
Adaptation	Special feature of an organism that enables it to fit into its environment and to survive

Life is the use of energy to maintain a steady state and an organization that does not readily break down. Life also involves the ability to produce offspring and genetic variations, some of which enable organisms with new adaptations to survive in a changed environment. The ability to reproduce and adapt to environmental changes depends on living cells. The next chapter describes how biologists study cells and what their research has revealed about the structure and functions of cells.

Chapter Review Questions

The following questions will help you check your understanding of the material presented in the chapter.

1. All organic compounds contain (*a*) carbon (*b*) nitrogen (*c*) water (*d*) oxygen.
2. An inorganic compound that plays a major role in chemical reactions within living cells is (*a*) glucose (*b*) amino acid (*c*) water (*d*) fatty acid.
3. Carbon and hydrogen usually are not found together in molecules of (*a*) carbohydrates (*b*) proteins (*c*) nitrates (*d*) amino acids.
4. A molecule that can duplicate itself is (*a*) DNA (*b*) water (*c*) protein (*d*) glucose.
5. The total of biochemical activities that occur in an organism is called (*a*) replication (*b*) transformation (*c*) metabolism (*d*) respiration.
6. The process by which animals take in materials for nutrition is called (*a*) digestion (*b*) ingestion (*c*) egestion (*d*) circulation.
7. Aerobic respiration requires (*a*) oxygen (*b*) carbon dioxide (*c*) starch (*d*) protein.
8. An organism that makes its own food from carbon dioxide and water is a (*a*) mold (*b*) plant (*c*) mushroom (*d*) yeast.
9. Regulation in complex plants and animals is similar because both (*a*) have a nervous system (*b*) have mechanisms for homeostasis (*c*) contain auxins (*d*) have growth responses called tropisms.
10. Which life activity is not essential for the maintenance of an individual organism? (*a*) reproduction (*b*) synthesis (*c*) excretion (*d*) regulation.

11. Which term includes the other three? (*a*) digestion (*b*) circulation (*c*) metabolism (*d*) synthesis.
12. Asexual reproduction always involves (*a*) two parents (*b*) one parent (*c*) two sex cells (*d*) one sex cell.
13. In humans, the mechanism by which the level of blood sugar is kept relatively constant is an example of (*a*) active transport (*b*) cyclosis (*c*) homeostasis (*d*) digestion.
14. Evolution is a process that usually results from (*a*) genetic variations (*b*) catabolism (*c*) replication (*d*) transport.

Biology Challenge

The following questions will provide practice in answering SAT II-type questions.

Part I

Most of the useful facts and ideas found in science texts were discovered through a structured procedure called the *scientific method*. Scientists use the scientific method to investigate natural phenomena by making observations, collecting data, analyzing and evaluating data, doing controlled experiments, making hypotheses, and forming conclusions. An experiment or investigation must be repeated by many other scientists before a particular hypothesis is accepted by the international scientific community.

The paragraphs below describe a scientific investigation designed to find out what happens to fat consumed by a mammal. Answer questions *1 through 6*, which follow the data table.

After natural fats are digested, they are absorbed as fatty acid molecules into the blood. It is difficult, however, to keep track of fatty acid molecules because they mix with similar substances already present in the blood.

If fatty foods containing tritium (a radioactive isotope of hydrogen) are fed to an animal, the resulting fatty acid molecules can be detected. A near-starvation diet containing radioactive, or "tagged," fat was fed to a group of mice. Because the mice would be extremely underfed, the researchers assumed that the mice would use all the fat they ingested for energy release. After ten days, the researchers made careful measurements that revealed the following data.

Data

Total amount of radioactive fat consumed by the mice	35 grams
Total amount of fat found stored in the bodies of the mice after ten days	12 grams
Percentage of stored, radioactive ("tagged") fat in the mice	50 percent

1. Which one of the following hypotheses best relates to this experiment? (*a*) Underfed mammals quickly use all the fat in their diets. (*b*) Fat storage is a continual process in mice. (*c*) Radioactive substances kill animals. (*d*) Mice digest radioactive fat in a different way than they digest nonradioactive fat. (*e*) The metabolism of an animal on a diet containing tritium is greatly changed.
2. Which one of the following assumptions is a necessary part of this experiment? (*a*) Mice use radioactive fat in the same way they use nonradioactive fat. (*b*) Fat "tagged" with tritium is identical to "untagged" fat. (*c*) Mice can tolerate tritium better than other mammals. (*d*) Substances "tagged" with tritium separate from other substances. (*e*) All radioactive substances are harmless.
3. The radioactive fat stored in the mice after ten days was (*a*) 6 grams (*b*) 17.5 grams (*c*) 23 grams (*d*) 29 grams (*e*) 35 grams.
4. From the data, a researcher can determine the amount of radioactive fat used for energy during the experiment if (*a*) more radioactive fat than nonradioactive fat is used for energy release (*b*) fat that is not stored is used for energy release (*c*) the mice lost weight at the end of the experiment (*d*) one gram of fat yields 186 calories (*e*) there is no relationship between the loss of fat and energy release.
5. How much radioactive fat was used by the mice during the experiment? (*a*) 6 grams (*b*) 17.5 grams (*c*) 23 grams (*d*) 29 grams (*e*) 47 grams.
6. A logical conclusion that can be made from this experiment is that mice on near-starvation diets (*a*) can survive if fed radioactive substances (*b*) use half and store half of the fat in their diets (*c*) store some of the fat in their diets (*d*) use all the fat in their diets for energy release (*e*) store only the "tagged" fat in their diets.

Part II

Select the letter of the statement that best completes the sentence.

1. The cells of an apple tree, earthworm, fish, and chimpanzee are similar in that each (*a*) stores, transforms, and utilizes energy (*b*) metabolizes inorganic substances (*c*) excretes urea (*d*) has a nervous system (*e*) reproduces by asexual cells.
2. Of the following, which substance is organic? (*a*) NaCl (*b*) H_2O (*c*) C_2H_5OH (*d*) $CaCO_3$ (*e*) H_2SO_4.
3. A bird that ordinarily lays white eggs was placed on the endangered species list. This bird builds nests made of twigs on the bare branches of locust trees. Recently, the bird was removed from the endangered species list. A reasonable explanation for the removal of the bird is (*a*) their predators migrated, seeking other prey (*b*) some of them laid brown eggs with hard, unbreakable shells (*c*) the endangered species became extinct (*d*) some of the birds laid eggs on the ground (*e*) the environment changed as the twigs making up nests sprouted leaves.
4. Biological information is passed on from one generation to the next generation by (*a*) homeostasis (*b*) reproduction (*c*) metabolism (*d*) mutations (*e*) synthesis.
5. Of the following life processes, the one that does not belong with the others is (*a*) digestion (*b*) secretion (*c*) metabolism (*d*) ingestion (*e*) replication.

2

Cell Structure and Function

Learning Objectives

When you have completed this chapter, you should be able to:

- **Describe** the development of the cell theory.
- **List** the major ideas of the modern cell theory.
- **Relate** the development of modern tools of the biologist to the study of the cell.
- **Apply** some techniques used by biologists to study cells.
- **Classify** the structure and function of organelles present in a cell.
- **Relate** the structure of a cell's membrane to its ability to let certain substances enter and exit a cell.
- **Describe** the structure and function of the different types of tissues found in plants and animals.

OVERVIEW

Scientists are learning more about cell structure and function by using powerful instruments and techniques, such as biochemical analysis, electron microscopy, chromatography, and radioactive methods of analysis.

Current knowledge about cells includes these concepts:

- Cells are highly organized and complex.
- Organisms are cells or combinations of cells.
- All organisms evolved from ancient, primitive cells.

Discoveries about cells are being made daily, and more will be made in the future. Will you decide to explore the exciting field of cell studies?

THE CELL THEORY

In the mid-17th century, *Anton van Leeuwenhoek* constructed a microscope that magnified objects about 250 times (Figure 2–1). This simple microscope consisted of a single convex (curved outward) lens. With his microscope, Leeuwenhoek discovered tiny organisms such as bacteria, parameciums, and microscopic worms. Before Leeuwenhoek developed his microscope, the smallest organisms known were those large enough to be seen with an ordinary magnifying glass. Leeuwenhoek's microscopic observations led the way to the development of the cell theory, which includes the idea that living things are made up of cells. A summary of how the cell theory developed follows.

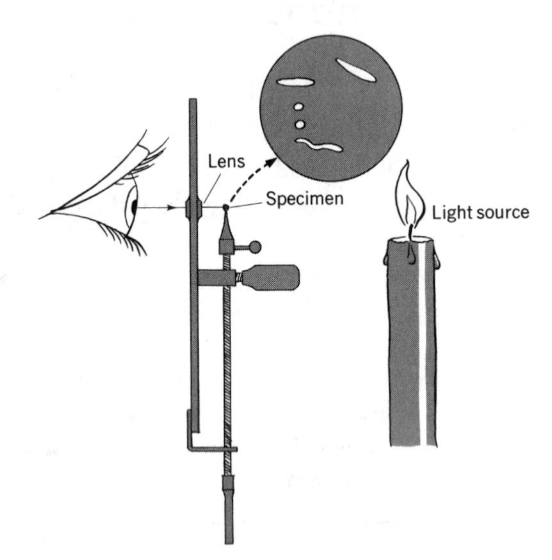

Figure 2–1 One of Leeuwenhoek's microscopes

Brief History of the Cell Theory

In 1665, *Robert Hooke* discovered the cell when he examined thin slices of cork under the microscope (Figure 2–2). Hooke used the word *cell* to describe the boxlike areas he saw in the cork. What he really saw were the cell walls of cells from which the living matter had dried up and disappeared.

In 1838, *Matthias Schleiden*, a *botanist* (plant specialist), and *Theodor Schwann*, a *zoologist* (animal specialist), agreed that the plant cells Schleiden was studying and the animal cells Schwann was studying had many similarities. In 1847, Schwann postulated that all animal tissue, including bone, blood, muscle, and glands, is composed of cells. Schleiden postulated the same for plant tissue.

In 1859, *Rudolf Virchow* proposed that all cells originate from other cells. This idea was expressed in Latin as *Omnis cellula e cellula:* All cells come from other cells.

The cell theory of Schleiden and Schwann and Virchow's idea have been extended by other scientists. In the early 20th century, *Edmund B. Wilson* made important contributions in the fields of cell division and heredity. And *James D. Watson* and *Francis H. Crick* in the 1950s discovered the chemical structure of DNA. Now scientists understand that cells are not only the basic building units of living things, but also the basic units of function.

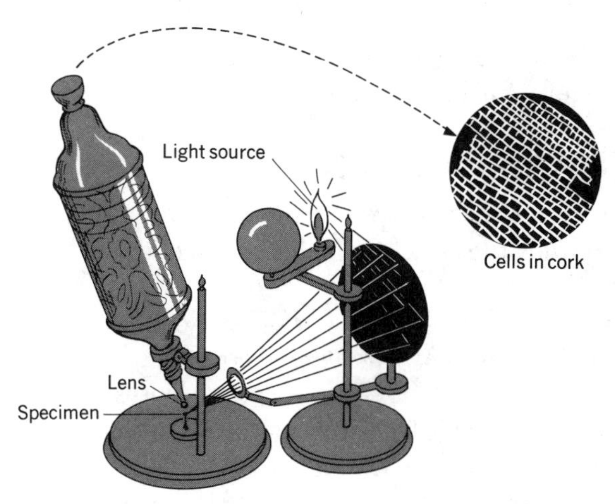

Figure 2–2 Hooke's microscope

Major Ideas of the Modern Cell Theory

The major ideas of the modern cell theory include the following:

1. The cell is the unit of structure and the unit of function of most living things; life processes go on because of the chemical reactions that occur within cells.

2. Cells arise from preexisting cells.
3. The chemical activities that occur within cells depend upon the presence of tiny cell structures called *organelles*.
4. The cells of all living things carry on similar chemical activities.

Exceptions to the Cell Theory

1. The cell theory does not answer the question, "Where did the *first* cell come from?" Obviously, the first cell could not have come from a preexisting cell.
2. Simple organisms, such as some algae and fungi, lack the separations that usually divide the bodies of organisms into cells. Is the cell, then, truly the unit of structure of living things?
3. Viruses, which consist mainly of a protein covering over a core of either DNA or RNA (ribonucleic acid), lack the organization found in cells.
4. Some organelles divide when the cell divides. Among these are chloroplasts and mitochondria (singular, mitochondrion). Organelles, however, are not cells but only cell structures.

These exceptions do not disprove the cell theory, but introduce challenges that require further investigation.

TOOLS OF THE BIOLOGIST

Biologists use different types of microscopes and various tools and techniques to study the structure and function of cells.

The Compound (Light) Microscope

The *compound*, or *light*, *microscope* consists of two or more sets of convex lenses (Figure 2–3). One set of convex lenses makes up an *objective*. Microscopes usually have two or more objectives. Light passes through a specimen and then through the first objective, which forms an enlarged image of the specimen. This image is then viewed with a second set of lenses, called the *eyepiece*, or *ocular*, which further enlarges the image. The eye then views the final enlargement.

A compound microscope consists of three main systems: the optical system, the mechanical system, and the light system.

Optical system. The optical system usually consists of two objectives and an eyepiece. The magnifying power of each objective usually is marked on the objective. The *low-power* objective usually is marked 10×, which means that this set of lenses magnifies the image ten times. The *high-*

power objective is usually marked 43×, which means that this set of lenses magnifies the image 43 times. The magnifying power of the eyepiece is usually 10×.

Occasionally it may be necessary to use an *oil-immersion objective*, which magnifies about 95 times. This objective is designed to use a layer of oil (such as mineral oil) between the object to be viewed and the objective. The oil layer provides greater resolution (ability to see details).

You can calculate the total magnification of a microscope by multiplying the magnifying power of the eyepiece by the magnifying power of the objective you are using. For example, the total magnification obtained by using a combination of a 10× eyepiece and a 10× objective is 100 times. When a 10× eyepiece is combined with a 43× objective, the total magnification is 430 times. Under low power, the field of vision is large and reveals many cells with few details; under high power, the field of vision is small and reveals fewer cells with greater detail.

Mechanical system. The mechanical system consists mainly of the coarse adjustment and the fine adjustment.

Coarse adjustment. The optical system can be moved up and down through relatively large distances by turning the coarse-adjustment knob. This movement provides a fairly good focus under low power. The coarse adjustment should never be used for high-power viewing.

Fine adjustment. Turning the fine-adjustment knob moves the optical system up and down through relatively small distances. This movement provides a sharp focus under either low or high power.

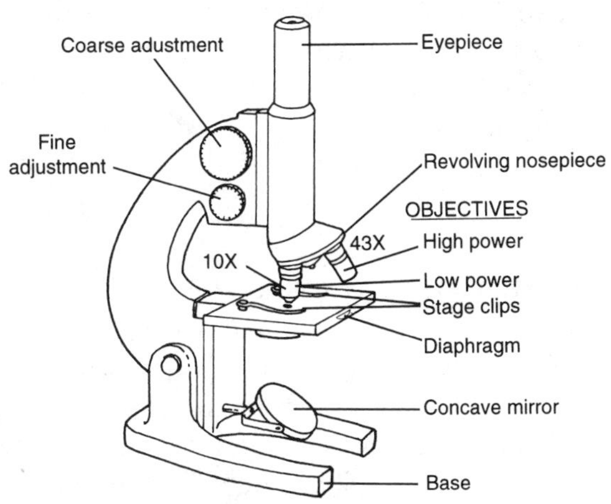

Figure 2–3 Modern compound microscope

Light system. Light must pass through the object that is being observed under the microscope. The light system of a microscope consists of a concave mirror (or a built-in lamp), a diaphragm, and condenser lenses.

Concave mirror. A concave (curved inward) mirror gathers light and reflects it to the opening in the stage.

Diaphragm. The diaphragm, which is usually attached underneath the stage, controls the width of a light beam before it passes through the specimen and enters the objective.

Condenser lenses. Condenser lenses are located either in the opening of the stage or in a special device attached under the diaphragm. Condenser lenses concentrate the light before it reaches the specimen.

The image. The lenses of a compound microscope change the direction of light rays. This causes an inverted image (upside down). The image also is reversed (changed in position from left to right). Thus, if a slide is moved toward or away from the viewer, or moved to the left or right, the image moves in the opposite direction.

The Electron Microscope

An *electron microscope* uses a beam of electrons instead of light rays to produce an image. The image produced by an electron microscope is formed on a fluorescent screen or photographic plate, both of which are sensitive to electron beams. Scientists use electron microscopes to observe *ultramicroscopic objects* or objects that are too small to be seen with a compound microscope. Electron microscopes provide magnifications greater than 250,000 times, which has enabled scientists to view virus particles and even large chemical molecules such as proteins. The main disadvantage of using an electron microscope is that it cannot be used with live specimens. Living material is destroyed by either the electron beam or the vacuum in the chamber in which the specimen is placed.

The Phase-Contrast Microscope

A *phase-contrast microscope* is used to view living material without the use of stains, which may kill or distort living material. A compound microscope can be converted into a phase-contrast microscope by inserting a device that causes light to reach the specimen at an oblique angle. Because certain structures in a cell are affected differently by oblique light rays, these structures stand out in sharp contrast to other structures.

SOME TECHNIQUES USED BY A BIOLOGIST

Slide Preparation

Specimens viewed with a compound microscope are mounted on glass slides that can be prepared for temporary or permanent use.

Temporary slides. A specimen that is used as a temporary slide is mounted in water and observed for a short time. Iodine solution and methylene blue are two dyes that may be used to make certain cell structures more visible. Figure 2–4 shows how a temporary slide is made.

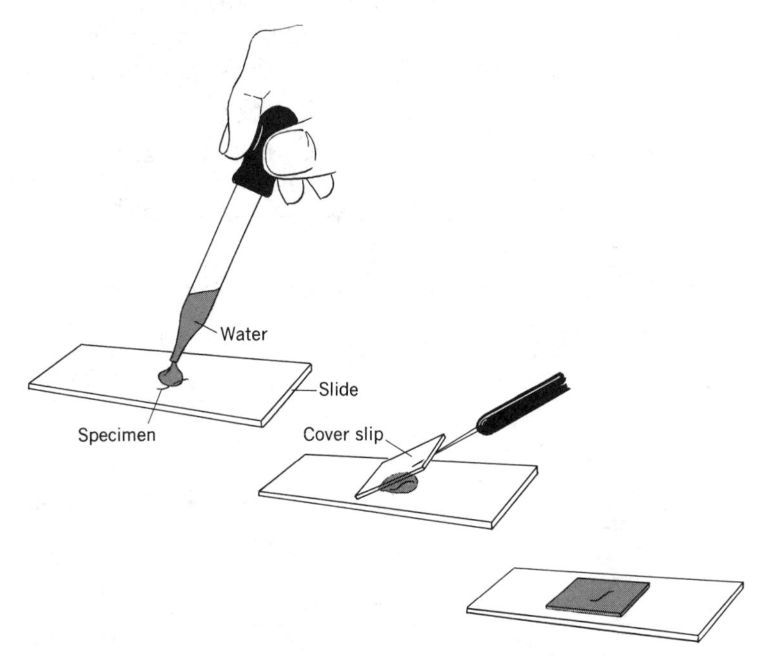

Figure 2–4 Making a temporary slide

Permanent slides. Specimens to be kept as permanent slides are first killed in preserving solutions that harden them. A special slicing machine, called a *microtome*, is then used to cut thin sections of the hardened material. Then, the thin sections are cemented to a slide and dyed. Finally, a clear cement is used to fasten a coverslip to the slide.

Measuring microscopic objects

Microscopic specimens are too small to be measured in ordinary units of length. The unit used for measuring microscopic objects is the micrometer.

The micrometer. A micrometer (formerly micron) is 1/1,000 (0.001) mm, or about 1/25,400 inch. The symbol for a micrometer is the Greek letter μ (pronounced "mew"). Table 2–1 lists major units of scientific measurement and their equivalents.

Estimating size with a microscope. The field of vision under low power usually has a diameter of 1.5 mm, or 1,500μ. If the specimen's length appears to be one-half the diameter of the field of vision, the estimated size of the object is then one-half of 1,500μ, or 750μ.

The field of vision under high power is much less than that under low power. Under high power, the field of vision usually is 0.3 mm, or 300μ. If the specimen's length appears to be one-half the diameter of the field of vision,the estimated size of the object is then one-half of 300μ, or 150μ.

Table 2–1 Abbreviated Table of Measurements

SI Unit		*Equivalent*
	Length	
1 meter (m)	=	100 centimeters (cm)
	=	1,000 millimeters (mm)
1 centimeter (cm)	=	1/100 m or 0.01 m
1 millimeter (mm)	=	1/1,000 m or 0.001 m
	=	1,000 micrometers (μ)
1 micrometer (μ)	=	1/1,000 mm or 0.001 mm
	Area	
1 square centimeter (cm^2)	=	0.0001 m^2

Microtools

A *micromanipulator* is used for precise handling of objects under magnification. The micromanipulator, which is attached to the microscope stage, introduces microknives, microneedles, microelectrodes (tiny electrical wires), or micropipets (tiny droppers) into cells. With these micro-tools, cell parts are removed, parts are transplanted, chemical substances are injected, and electricity within cells is measured.

Chromatography

Many complex substances in living material are studied by chromatography, which separates the different components of a solution or other

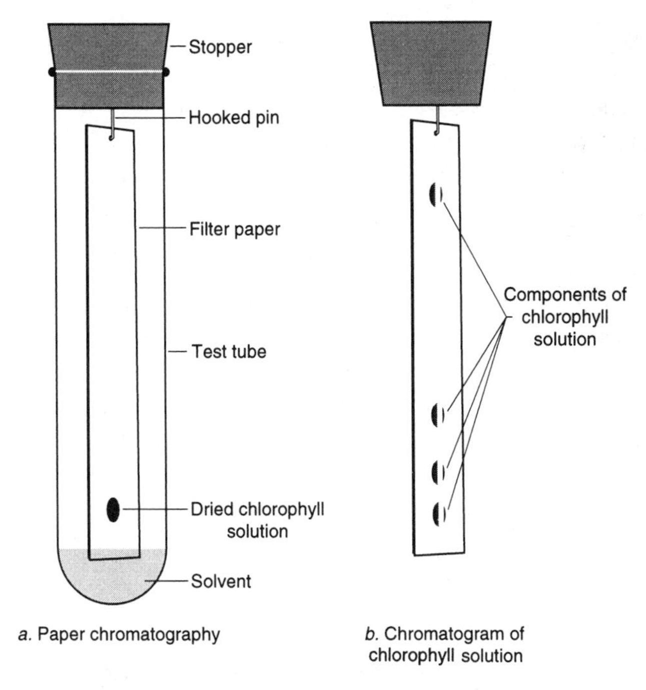

Figure 2–5 Paper chromatography

liquid mixture by allowing the mixture to flow through a special material called an *adsorbent*. An adsorbent is a substance on whose surface certain molecules may stick while other molecules are carried away by a fluid. *Paper chromatography* uses special filter paper as the adsorbent. *Thin-layer chromatography* uses a thin film of silica gel on a slide as adsorbent. A *chromatogram*, or picture of the components of chlorophyll, is produced by using paper chromatography to separate the components of a chlorophyll solution (see Figure 2–5).

The Centrifuge

A *centrifuge* is a device that separates materials of different densities suspended in a liquid. As the mixture is whirled rapidly in a tube, the denser materials are forced to the bottom of the tube. Tiny cell structures, such as nuclei

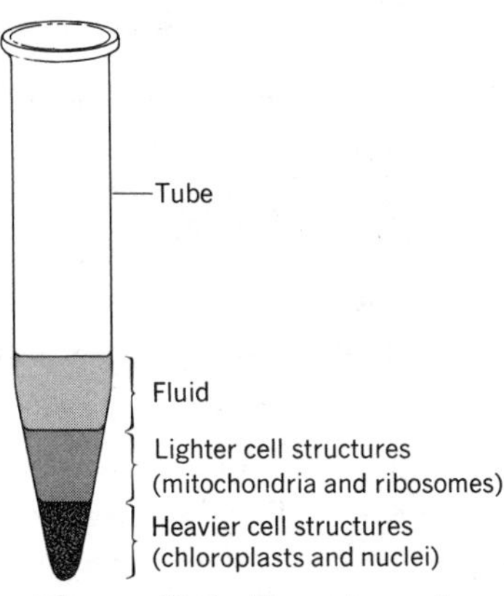

Figure 2–6 Fractionation

and chloroplasts, can be separated from one another, or *fractionated* (Figure 2–6), by using a high-speed centrifuge called an *ultracentrifuge*. Then the cell structures can be studied with an electron microscope and by biochemical methods. Keep in mind that it is the different densities of the cell structures that make the centrifuge a useful tool in the laboratory.

Radioactivity

The radioactive forms of elements such as iodine and carbon emit rays that can be detected by photographic paper (*radioautography*) or by instruments such as a *Geiger counter*. By using these methods, scientists can detect radioactive elements taken into an organism as they are used in the organism. Radioactive iodine and carbon are used as *tracer elements* to unlock the secrets of chemical reactions within living cells.

Section Quiz

1. Which represent an exception to the cell theory? (*a*) protozoans (*b*) yeasts (*c*) viruses (*d*) green algae.
2. Which concept is not part of the cell theory? (*a*) Cells are the unit of structure of living things. (*b*) All cells are genetically identical. (*c*) All cells arise from preexisting cells. (*d*) Cells are the unit of function of living things.
3. Using a microscope, a student observes a cell whose diameter just fills the field of vision. If the diameter of the field is 0.5 millimeter, the diameter of the cell is (*a*) 0.5 micrometer (*b*) 22.5 micrometers (*c*) 450 micrometers (*d*) 500 micrometers.
4. A student observes a section of cork under a compound microscope. As the student turns the fine adjustment knob, different areas of the cork come into focus, while other areas become blurred. Which of the following is the most probable reason for these observations? (*a*) There are varying thicknesses in the cork sample. (*b*) The fine adjustment knob was turned too rapidly. (*c*) There are variations in the light source. (*d*) The microscope has a defective objective lens.
5. In a compound microscope, which combination of eyepiece and objective will give the least detail and the largest field of vision? (*a*) 5× eyepiece, 10× objective (*b*) 5× eyepiece, 44× objective (*c*) 6× eyepiece, 10× objective (*d*) 6× eyepiece, 40× objective.

ORGANELLES PRESENT IN CELLS

The basic building block of most living things is the cell, which is made up of organelles (Figure 2–7). Each *organelle* has a definite structure and function.

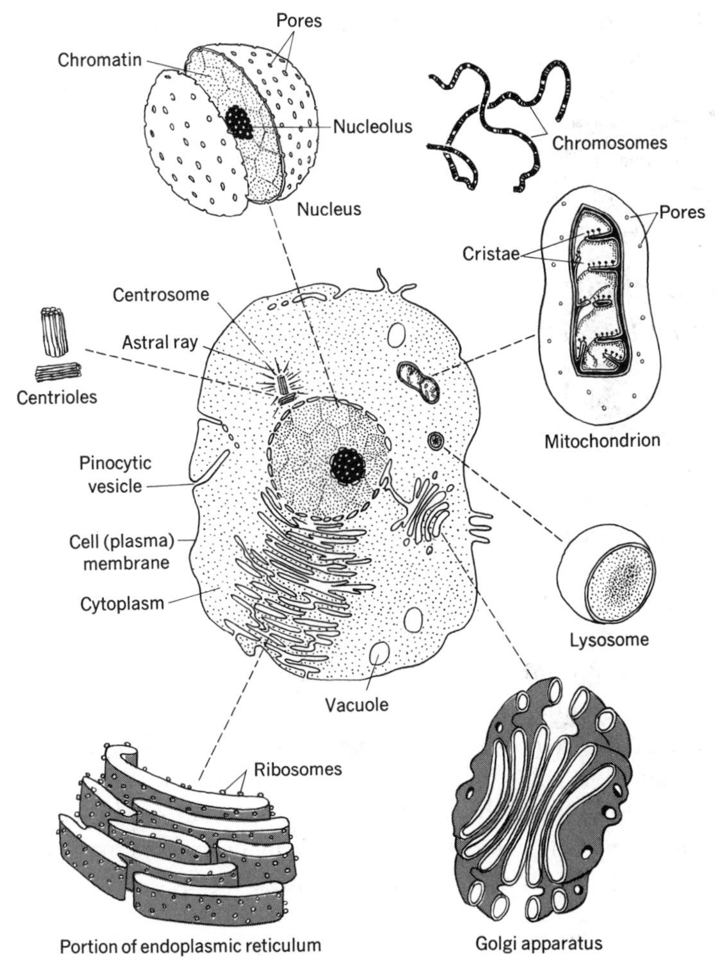

Figure 2–7 Cell organelles (electron-microscope views)

Prokaryotic and Eukaryotic Cells

An organism that lacks a definite membrane enclosing its genetic material is called *prokaryotic.* All bacteria are prokaryotic. Bacteria possess a coiled mass of DNA called a *nucleoid*, which is considered an organelle.

In fact, any cell structure is called an organelle. Bacteria possess other organelles, such as a rigid cell wall, an inner plasma membrane, and cytoplasm.

An organism that has a definite membrane enclosing its genetic material is called *eukaryotic*. Plants and animals are composed of eukaryotic cells. There are several organelles that are present in the cells of both plants and animals. However, certain organelles are found in plant cells only, and other organelles are found only in animal cells. Organelles found in eukaryotic cells include the nucleus, endoplasmic reticulum, Golgi bodies, ribosomes, lysosomes, plasma (cell) membrane, cytoskeleton, peroxisomes, vesicles, mitochondria, and chloroplasts. These organelles will be discussed in the following pages.

The Nucleus

The *nucleus* of the cell is the densest, most easily seen organelle. In 1831, Robert Brown discovered the nucleus while working with plant cells. He was able to observe the nuclei of orchid cells using a simple light microscope. However, it was not until the early part of the 20th century that a connection was made between the nucleus and the chromosomes and the transmission of hereditary traits. The nucleus carries hereditary instructions (DNA), which are used in controlled ways, such as the synthesis of proteins.

Surgical removal of a nucleus stops cell growth and is followed shortly by death of the cell. When a cell grows to its physical limit, DNA is triggered to replicate and the cell divides into two daughter cells. Each new cell possesses identical DNA codes for its life processes.

DNA is contained in *chromosomes*, which are visible only during cell division (see Figure 2–8). The body cells of each species of organism possess a characteristic number of chromosomes. For example, a human has 46 chromosomes in each body cell; a fruit fly has eight chromosomes in each body cell. Chromosomes are composed of *genes*, which occupy set

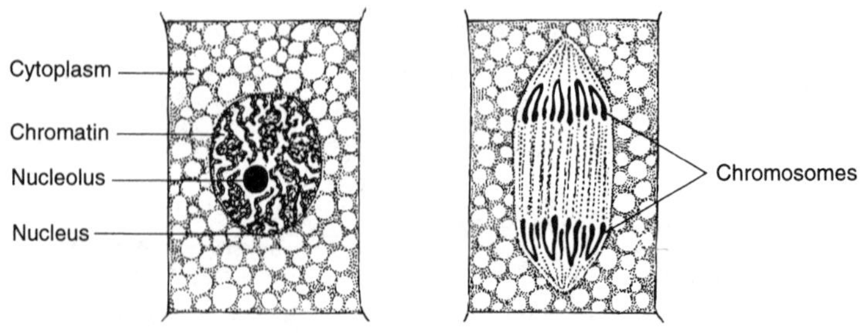

Figure 2–8 Cells of onion root tip

locations on the chromosomes. Genes are molecules of DNA that determine the physical and chemical traits of an organism.

The nucleus is a complex organelle because it is composed of the following structures:

1. *Chromosomes*, which contain genes (molecules of DNA).
2. A *nuclear membrane*, which is a double-layered structure with many openings, or pores. The pores function as pathways between the nucleus and the cytoplasm. Ribosomes, which manufacture proteins, are attached to the outer surface of the nuclear membrane facing the cytoplasm.
3. A spherical *nucleolus*, which is composed mostly of an RNA called *ribosomal RNA (rRNA)*. During cell division, rRNA travels through the cytoplasm to the ribosomes and becomes part of ribosomal structure. Ribosomes manufacture proteins.

   A nucleus may contain more than one nucleolus. Human cells usually have four nucleoli; occasionally, two are joined, which give the appearance of one large nucleolus. Recently, the nucleolus has been reevaluated by biologists. Is the nucleolus actually an organelle? Today many biologists do *not* consider the nucleolus to be an organelle. They think it is a temporary collection of rRNA molecules visible in nondividing cells, then disappearing when cells divide.
4. *Nucleoplasm*, which is the fluid portion of a nucleus.

The Plasma (Cell) Membrane

The *plasma membrane* is the outer limit of a eukaryotic cell. The plasma membrane is a layer about nine nanometers (nine billionths of a meter) thick, covered with pores, proteins, and twiglike carbohydrate groups that project from some of the proteins. Materials are exchanged between a cell's internal and external chemical environments through the plasma membrane.

Membrane structure. The plasma membrane is constructed of molecules called *phospholipids*, whose chemical structure forms two layers. Each layer of a phospholipid is oriented so that the lipid (fatty) "tails" extend toward each other and phosphate "heads" are located outside (Figure 2–9 on the next page). Thus, if a needle is thrust into a plasma membrane, the cell contents do not spill out. Instead, the membrane flows over itself, similar to the way skin heals.

Protein molecules are interspersed between phospholipid molecules; some are on the outer surface. These proteins are mainly responsible for

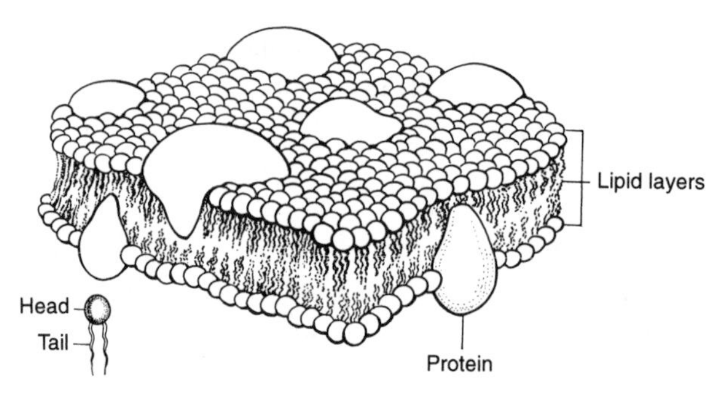

Figure 2–9 Cell membrane model

the many functions of the plasma membrane. *Cytologists* (cell researchers) think that some proteins control the passage of water-soluble substances into and out of cells; other proteins form enzymes that participate in metabolic reactions.

Twiglike groups of carbohydrate (polysaccharide) molecules project from the surface protein molecules. Research indicates that the twiglike carbohydrate molecules function as *surface receptors* that gather chemical messages; as *markers* that help cells identify each other; and as *control points* that regulate cell growth.

A plasma membrane is *selectively permeable*, or allows the passage of certain substances but not others. In response to a solute (dissolved substance), water molecules move from the region of greater concentration to a region of lesser concentration. This movement of water molecules is called *osmosis*.

Osmosis and Diffusion. Osmosis is a special case of diffusion, which only describes the movement of water molecules through a membrane. In Figure 2–10, the bulb of a thistle tube represents a cell and the cellophane membrane, which contains very tiny pores, is a portion of its plasma membrane. Only the water molecules from the beaker can enter the bulb through the membrane. From inside the bulb, however, both the water molecules and the larger glucose molecules can enter the beaker. As a result of molecular collisions, water molecules diffuse from the region where they are present in greater concentration (in the beaker) to where they are present in lesser concentration (in the bulb). As water moves into the bulb, the level of solution rises inside the stem of the thistle tube.

If the water in the beaker is tested for glucose, the results will show that glucose is present. We may therefore conclude that some glucose molecules move from a region where they are more concentrated (inside

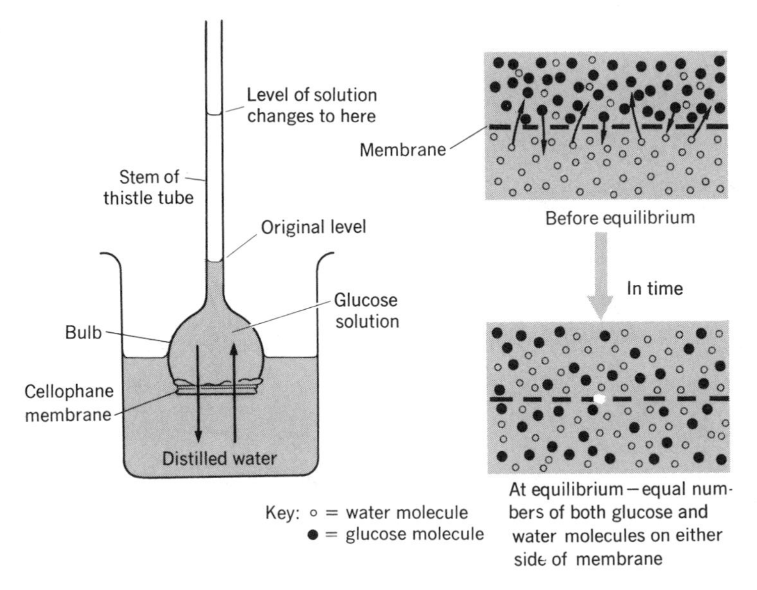

Figure 2–10 Osmosis and diffusion

the bulb) to a region where they are less concentrated (in the beaker). Since the water diffuses into the tube more rapidly than the glucose diffuses out of it, water rises in the stem of the thistle tube.

During the next several days, glucose will continue to leave the bulb and water will continue to enter the bulb until a state of equilibrium, or balance, is reached. At equilibrium, the rate at which water and glucose molecules enter the bulb and leave are equal. Then, there will be no further change in the level of liquid in the stem.

The plasma membrane of a cell functions like the cellophane membrane of the thistle tube. In living cells, however, equilibrium never occurs because concentration of substances within the cells and the substances outside the cells are changing continuously.

The process by which most substances move in and out of cells is called *simple diffusion*. This process is characterized by the random movement of solute particles (or ions) in response to a difference in concentration or *concentration gradient*. A particular solute diffuses according to its concentration gradient, independent of the concentration gradients of other solutes. Suppose drops of red dye and blue dye are added to one end of a large pan of water. In a short time, red and blue dye will spread throughout the water. You will notice that the red molecules and blue molecules move at different speeds depending on the physical properties and components of both dyes, as well as concentration gradients.

Passive and Active Transport. The simple passage of oxygen, water, carbon dioxide, and organic solvents (ethyl alcohol) across the lipid bilayer of a plasma membrane is called *passive transport*. Because they are electrically neutral, these and similar substances pass through the protein molecules of a plasma membrane without an energy boost. Large molecules, such as glucose, and electrically charged particles (ions) cannot diffuse through a lipid bilayer. Special proteins, called *channel* proteins or *carrier* proteins, transport these large molecules and charged particles across the plasma membrane.

Some plasma membrane proteins carry out *active transport* aided by an energy source called *ATP* (adenosine triphosphate). An energy boost from ATP helps the active transport proteins *pump* substances into and out of cells against their concentration gradients.

In the unicellular paramecium, water enters by passive transport (osmosis) and is expelled by active transport. This occurs because there is a higher concentration of mineral salts in the paramecium than in its watery environment.

Water continuously enters a paramecium. But if water flowed only into a paramecium, pressure within the cell would increase and the paramecium would burst. However, this rarely occurs because excess water is actively pumped out by the contractile vacuoles, pushed outward through the plasma membrane against its concentration gradient.

In human cells, a *sodium-potassium pump* uses energy to actively transport sodium ions (Na^+) and potassium ions (K^+). In relation to their chemical environment, animal cells normally contain a low concentration of Na^+ ions and a high concentration of K^+ ions. To maintain their different gradients, special protein carriers pump Na^+ and K^+ ions into and out of cells with the aid of energy boosts from ATP molecules. The sodium-potassium pump transports ions from regions of *low* concentration to regions of *high* concentration—against a gradient.

Exocytosis is a process by which certain substances are moved *out* of cells. Cytoplasmic vesicles containing substances such as glandular secretions travel to the plasma membrane. After merging with the membrane, the vesicle forms a pocket from which substances are ejected.

Endocytosis (*pinocytosis*) is a process by which substances are moved *into* cells. A pinched-in plasma membrane forms a pocket in which substances are engulfed. The pinched-in plasma membranes containing solids or fluids travel as vesicles through the cytoplasm (see Figure 2–11). Because they require energy, exocytosis and endocytosis are forms of active transport.

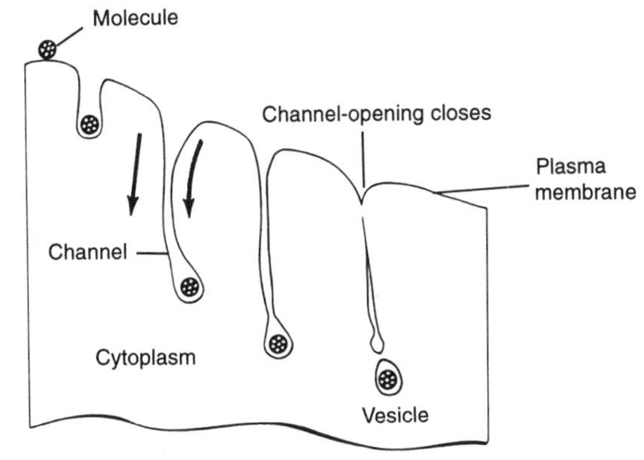

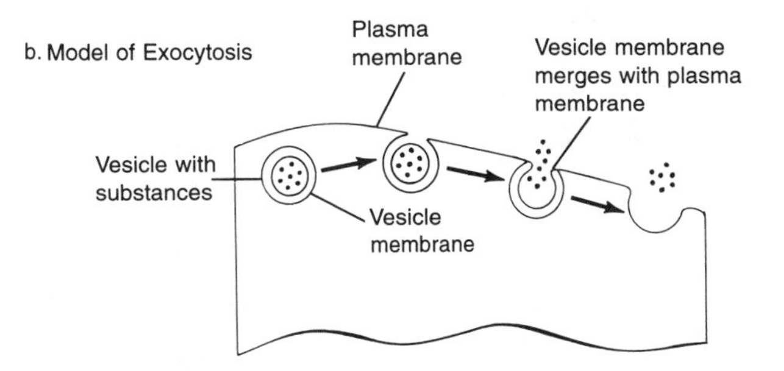

Figure 2–11 Endocytosis and exocytosis

Section Quiz

1. A prokaryotic cell is characterized by a (*a*) lack of cytoplasm (*b*) lack of ribosomes (*c*) plasma membrane (*d*) nucleoid.
2. The organelle that is visible in nondividing cells but absent in dividing cells is (*a*) a nucleus (*b*) chromatin material (*c*) a nucleolus (*d*) a centrosome.
3. The basic structure of a plasma membrane can be described as a (*a*) bilayer of phospholipid molecules (*b*) bilayer of phosphates and specialized proteins (*c*) nonliving molecular layer of charged particles (ions) (*d*) single layer of complex carbohydrate molecules with carrier proteins.
4. Active transport usually involves (*a*) phospholipids and carrier proteins (*b*) carrier proteins and ATP (*c*) the absence of a concentration gradient (*d*) the movement of metallic ions only.

5. Which organelle is involved in endocytosis? (*a*) Golgi complex (*b*) centriole (*c*) plasma membrane (*d*) lysosome.

Cytoplasm

The term *cytoplasm* refers to all substances inside the plasma membrane and outside the nucleus. The semifluid portion of cytoplasm, or *cytosol*, carries suspended particles, minute tubules, and filaments that form a structural network called a cytoskeleton. Cytosol contains about 80 percent water, and dissolved organic compounds, such as glucose, amino acids, and lipids.

Readily soluble inorganic compounds, such as mineral salts, are also part of an abundant collection of the chemicals of life, which represent reactants that are involved in about 3,000 metabolic reactions that occur in living things. An outstanding characteristic of all forms of life is the release, transformation, and storage of energy.

Cytoskeleton

Microfilaments, microtubules, and intermediate filaments make up most of a cell's structural network, or *cytoskeleton*. The cytoskeleton, composed of filamentous proteins, enables an entire cell to move. For example, a *phagocyte* is a white blood cell that moves and preys on bacteria and foreign proteins. The cytoskeleton also enables movement of organelles and chemicals inside a cell.

Rodlike structures composed mainly of a protein called *actin* are called *microfilaments*. Actin (and another protein microfilament called *myosin*) makes muscle contraction possible. In other body cells, actin filaments either give shape and support or they assist movements within cells, such as exocytosis and endocytosis.

Microtubules are relatively large, cylindrical structures, composed mainly of a protein called *tubulin*. Microtubules and microfilaments give shape and support to cells. The tubules also help distribute substances through the cytosol and chromosomes during cell division. Microtubules are found in the flagella and cilia of many unicellular organisms. Protozoans such as paramecium and euglena use *cilia* and *flagella* to move and eat. In response to specific needs, the microfilaments and microtubules of the cytoskeleton can assemble (as a network) and rapidly disassemble.

Intermediate filaments are the most stable fibers that compose a cell's cytoskeleton. In size, these filaments fall between microfilaments and microtubules. Like the microfilaments and microtubules, intermediate filaments also are composed of protein, and are more abundant in cells undergoing mechanical stress, such as skin cells. Current research shows that Alzheimer's and Parkinson's diseases are accompanied by disturbances in the arrangement of fibers that compose the intermediate filaments. Some medical researchers are studying the possible relationship between gene action and intermediate fibers.

The Ribosomes

Ribosomes are tiny organelles (about 30 nanometers in diameter) that serve as a cell's protein factories. Ribosomes are scattered throughout the cytoplasm or are attached to cell membranes. The free ribosomes release their manufactured proteins to float in the surrounding cytoplasm. The attached ribosomes transfer their proteins into a mazelike organelle called the endoplasmic reticulum. A ribosome is made of two unequally sized subunits composed of about 50 different proteins and ribosomal RNA (rRNA). Later (see Chapter 14), you will learn the details of translating a DNA message into the synthesis of a specific protein.

The Endoplasmic Reticulum (ER)

All eukaryotic cells feature a membranous organelle called *endoplasmic reticulum* (ER), which forms an extensive system of internal compartments in the cytoplasm. Prokaryotic cells lack an endoplasmic reticulum. Like the plasma membrane, the endoplasmic reticulum is a double layer of lipid molecules with attached enzyme molecules. Observation by an electron microscope shows that the endoplasmic reticulum is composed of a series of channels with many interconnections between its membranes and vesicles enclosed by the membranes.

Some parts of the endoplasmic reticulum are heavily dotted with ribosomes. As a result of their appearance, these sections are called *rough ER*; other portions with relatively fewer ribosomes are called *smooth ER*. Rough ER manufactures proteins, which are then sent to the Golgi complex for refining. The amount of rough ER corresponds closely to the amount of protein synthesized by a cell. For example, white blood cells that produce a variety of disease-fighting antibodies (proteins) possess highly developed rough ERs. Smooth ER is involved with the synthesis of carbohydrates and lipids. The smooth ER of detoxification cells in the liver is heavily laden with enzymes, which detoxify drugs such as codeine and morphine.

The Golgi Complex (Golgi Bodies)

Golgi bodies are named after the 19th-century Italian physician *Camillo Golgi*, who first observed these organelles in the nerve cells of a cat. Golgi hypothesized that this newly discovered organelle was involved in protein secretion. His hypothesis was finally supported in 1965. Research biologists using radioactive tags, staining, and electron microscopy showed that proteins manufactured by pancreatic cells move from rough ER, through the Golgi complex, into vesicles, and then to different areas of the cytoplasm. The Golgi complex modifies and refines proteins according to their reaction sites. For example, proteins bound for the plasma membrane are modified in one way and proteins bound for the nucleus are modified in another way.

Lysosomes and Peroxisomes

Some proteins that leave the Golgi complex "refinery" are incorporated into *lysosomes* and *peroxisomes*. Since these structures use enzymes to break down potential nutrients (energy sources), they are often called a cell's "digestive system." For example, when a phagocyte ingests bacteria, a lysosome fuses with the vesicle containing the bacteria and digests its contents. End products of digestion then diffuse out of a combined vesicle and are used by the cell. The used-up lysosome combined with the indigestible material, called a *residual body*, travels to the plasma membrane, where it is ejected by exocytosis.

Peroxisomes, which break down hydrogen peroxide (H_2O_2), are organelles mainly found in mammalian liver cells. Hydrogen peroxide, a toxic product, is the result of some metabolic reactions, such as the breakdown of alcohol. First, the liver cells detoxify alcohol by breaking it down to H_2O_2. Then using the enzyme *catalase*, the liver cells break down H_2O_2 to water and oxygen, which are eliminated as wastes. A rare genetic disease in humans (*Zellweger's syndrome*) is characterized by malfunctioning peroxisomes, which establishes a causal connection between proper peroxisome functioning and the health of a cell. In plant cells, there are enzyme-secreting organelles called *glyoxisomes*. These organelles change oils to carbohydrates, especially when seeds germinate.

Section Quiz

1. A cell's protein "refinery" is the (*a*) rough ER (*b*) smooth ER (*c*) ribosomes (*d*) Golgi complex.

2. Inorganic salts dissolve readily in (*a*) water (*b*) ethyl alcohol (*c*) oil (*d*) organic solvents.

3. Microtubules are structures that compose a cell's (*a*) endoplasmic reticulum (*b*) cytoskeleton (*c*) Golgi complex (*d*) plasma membrane.

4. In liver cells, smooth ER synthesizes substances that (*a*) add color to bile (*b*) prevent blood from clotting (*c*) detoxify drugs (*d*) change glucose molecules to lipid molecules.

5. "Digestive" enzymes are usually present in a cell's (*a*) endocytic vesicles (*b*) lysosomes (*c*) vacuoles (*d*) centrosome.

The Mitochondria

Mitochondria are the largest organelles in an animal cell, excluding the nucleus. Yet, they are extremely small, barely visible when magnified about 950× under an oil-immersion microscope. Mitochondria (singular, mitochondrion) vary from 0.5 to 1.0 micrometer in diameter and about 7.0 micrometers long. Mitochondria are usually oval-shaped in a dead cell, but readily change shape in living cells. They also enlarge or contract in response to ATP, drugs, and hormones. The enlargement of mitochondria is caused by water intake and is pronounced in living kidney cells, through which about 180 liters of water are filtered each day.

A mitochondrion has a folded inner membrane enclosed within an unfolded outer membrane. The inner folds, which extend into the cavity of the mitochrondrion similar to the shelves of a bookcase, are called *cristae,* (singular, crista). The outer membrane has pores that permit substances to enter and leave the mitochondrion.

Mitochondria normally move about but tend to gather near a cell's nucleus. Active cells contain more mitochrondria than do less active cells. For example, mitochrondria are more plentiful in muscle cells than in mature red blood cells. As a cell ages, mitochondria become less numerous and often disappear. In mature blood cells, mitochondria may be absent. Like nuclei, mitochondria contain DNA and divide when the cell divides. Consequently, about equal numbers of mitochondria are distributed to each new cell.

Many biologists now think that mitochrondria originated as *aerobic* bacteria (oxygen-requiring) that lived inside cells in a mutually beneficial relationship (a type of symbiosis called mutualism). The aerobic mitochondria provide oxygen for a cell's energy reactions. In return, they receive food, water, and a sheltered environment from the cell. Plant and animal cells possess mitochondria.

All of the mitochondria inside a eukaryotic cell are descendants of previous mitochondria. That is because mitochondria divide by simple fission when a cell divides. Mitochondria often are called "powerhouses" because they provide most of a cell's energy. In plant cells, however, chloroplasts are the main energy generators. The energy-generating reactions within mitochrondria are discussed in Chapter 7.

Centrioles and Centrosome

Most animal and protist (protozoans and algae) cells possess a pair of *centrioles*. (Plant cells lack centrioles.) Centrioles are made up of two sets of short microtubules. Each set consists of nine microtubule triplets (27 microtubules), which collectively have a cylindrical shape (Figure 2–12). The centrioles are situated at right angles to each other, and are located near the nuclear membrane. During cell division, centrioles help form the poles of mitotic spindles. This helps move the duplicated chromosomes into paired daughter cells. Like mitochondria, centrioles replicate during cell division. This observation provides support to the hypothesis that certain organelles originated as symbiotic organisms. Recently, DNA was discovered in centrioles, which also strengthens the case for the symbiosis theory of organelle origin.

A *centrosome* is the dense area of cytoplasmic material (located near the nucleus) that contains the centrioles. In nondividing cells, the

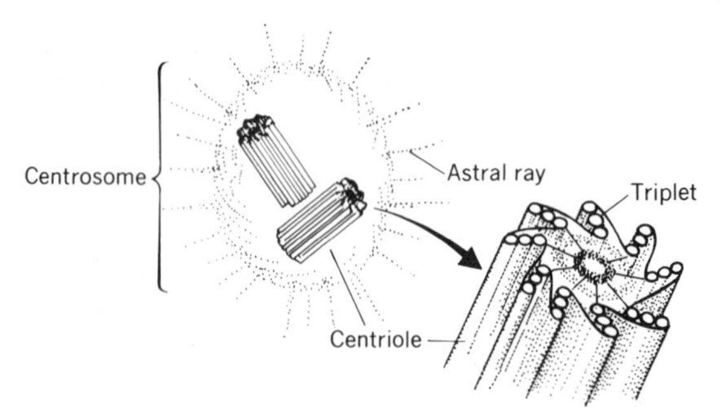

Figure 2–12 Structure of a centrosome

centrosomes organize microtubules, arranging them as centrioles. During cell division, centrosomes help form mitotic spindles (see Chapter 12, Reproduction). Plant cells lack a centrosome.

Cilia and Flagella

Cilia (singular, cilium) are short, and *flagella* (singular, flagellum) are long hairlike structures that project from cells and move with either a wavelike or whiplike motion. Protozoans, such as paramecium and blepharisma, are covered with rows of cilia, which they use for locomotion and to sweep food particles into their gullets. The flatworm planaria, commonly found in ponds, also travels by means of a cilia-covered body. In humans, the lining of the respiratory passages is made up of ciliated cells that sweep mucus and dust particles up toward the throat (to be coughed out). In contrast, sperm cells move by means of a single long flagellum.

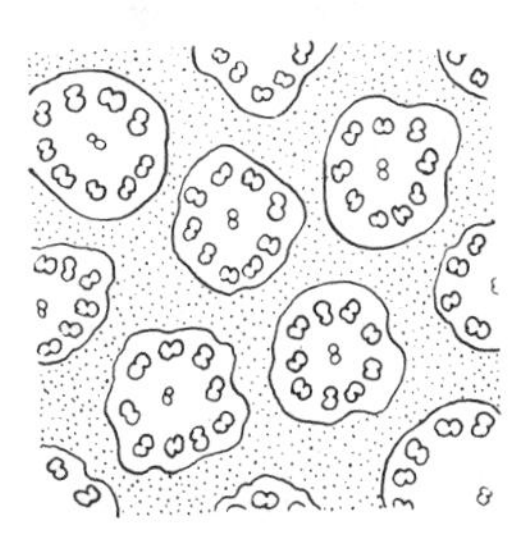

Figure 2–13
Structure of cilia (cross section)

Like centrioles, cilia and flagella are composed of microtubules. However, cilia and flagella are formed by nine pairs of microtubules (rather than triplets) surrounding a tenth pair (see Figure 2–13).

The Chloroplast

Chloroplasts are important membraned organelles in plant cells. They may be disk-shaped, cup-shaped, spiral, or spherical. Chloroplasts also vary in size. Some appear as dots in a plant cell, others almost fill up an entire cell.

Chloroplasts are green because they contain the pigment *chlorophyll*. Orange-yellow pigments, called *carotenoids*, also are present in plant cells. These pigments capture light energy, changing it into chemical energy. This energy drives the complex reactions of *photosynthesis*, which culminates in the production of food substances and the generation of oxygen. Details of the foodmaking process are discussed in Chapter 11.

The interior of a chloroplast contains many cylindrical structures called *grana* (singular, granum) that are connected by a system of flattened channels. Numerous openings in the channels communicate with semifluid material bathing the grana. A single granum resembles a pile of stacked coins; each coin is called a *thylakoid*. Light absorption occurs in a thylakoid (see Figure 2–14 on the following page).

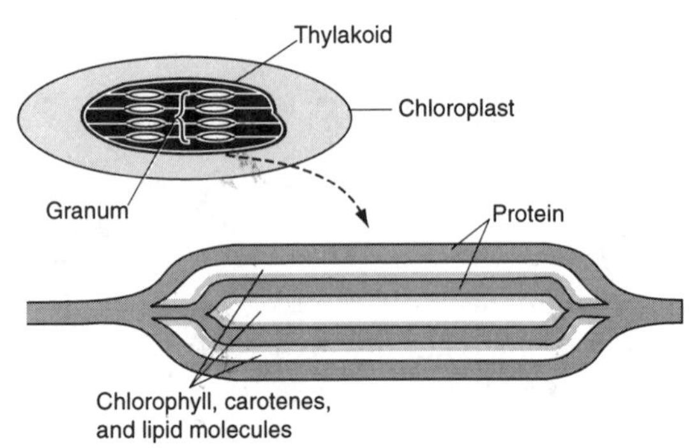

Figure 2–14 Chloroplast with granum

The Cell Wall

The *cell wall* is the rigid, outer layer of plant and fungi cells, and some bacteria and protists, that encloses the cell's plasma membrane. All plant cell walls are composed of cellulose molecules in the form of very thin *microfibrils*. The spaces between microfibrils are filled with *pectin*, a complex sugar. The spaces permit water, gases, and dissolved substances to pass through the cell wall.

The first portion of the cell wall laid down by a developing cell is called the *primary wall*. An intercellular layer, called the *middle lamella*, cements the cell walls of two adjacent cells together. Plant cells that form hard, woody parts add a *secondary wall*, located between the primary wall and the plasma membrane. Secondary walls contain cellulose microfibrils and *lignin*, which both strengthen the cell wall. Extracted from wood, lignin is used to manufacture adhesives, resins, drugs, and food. Cell walls usually have tiny holes, called *plasmodesma*, through which delicate cytoplasmic connections provide chemical continuity between adjacent cells. Table 2–2 lists major cell organelles and their functions.

Section Quiz

1. A plant cell's food-making system includes (*a*) chloroplasts (*b*) peroxisomes (*c*) centrosomes (*d*) cytosomes.
2. The mitochondria in your cells are responsible for (*a*) packaging proteins (*b*) breaking down proteins (*c*) providing energy for reactions (*d*) absorbing energy from light.
3. Because mitochondria and centrioles replicate during cell division, many biologists think that (*a*) some organelles originated as symbiotic organisms (*b*) cell division cannot occur without the presence of

Table 2–2 Cell Organelles and Their Functions

Organelle	*Functions*
Cell wall	Protects and supports cell
Centrosome and centrioles	Aid in cell division
Chloroplasts	Carry on photosynthesis
Chromosomes	Contain the genetic material (DNA) of the cell
Cilia and flagella	Enable locomotion, movement of fluids
Cytoplasm	Stores chemicals; carries on anaerobic respiration
Cytoskeleton	Structural network enables movement of cell and organelles within the cell
Endoplasmic reticulum	Transports substances within cell; protein synthesis
Golgi apparatus	Stores secretions; modifies and refines proteins
Lysosome	Digests materials in cell
Microfilaments	Provide shape and support; assist movements within cells
Microtubules	Provide shape and support; help move chromosomes during cell division
Mitochondria	Release energy
Nuclear membrane	Regulates passage of substances into and out of nucleus
Nucleolus	Stores RNA
Nucleus	Controls all cell activities, including reproduction
Peroxisome	Breaks down hydrogen peroxide in liver cells
Pinocytic vesicles	Transport large particles into cell
Plasma membrane	Regulates passage of water and dissolved substances into and out of cell
Ribosomes	Synthesize proteins
Vacuoles	Store water and dissolved substances

both organelles (*c*) mitochondria and centrioles are chemically related organelles (*d*) both organelles are composed of microtubules.

4. Thylakoids are mainly involved in (*a*) storing sugars and water (*b*) absorbing light (*c*) harnessing chemical energy (*d*) changing chlorophyll *b* to chlorophyll *a*.
5. A plant substance that helps strengthen cell walls is (*a*) pectin (*b*) suberin (*c*) lignin (*d*) tubulin.

TISSUES AND ORGANS

A group of similar cells and intercellular substances carrying on a specialized activity is called a *tissue*. Several tissues working together make up an *organ*. An organ usually is composed of a predominant kind of tissue. For example, your heart is predominantly cardiac muscle working in association with nerve, connective, and epithelial tissues. Several organs working together comprise an *organ system*; the heart, arteries, veins, capillaries, and lymph vessels are organs of the circulatory system.

In living things, *specialization*, or division of labor, is an evolutionary adaptation for survival of multicellular plants and animals. Specialization prevents one tissue or organ from functioning as another tissue. For example, brain tissue (composed of nerves) cannot function as muscle tissue. In plants, leaf, or epidermal, tissue cannot function as vascular tissue in the roots or stems.

Plant Tissues

The three main tissues that make up most of a plant's body are the parenchyma, collenchyma, and sclerenchyma. Other plant tissues include the vascular tissue (xylem and phloem), dermal tissue, chlorenchyma tissue, and meristem tissue.

Parenchyma tissue is found in most plant structures, including roots, stems, leaves, and fruits. Parenchyma cells are thin-walled, loosely packed, and have a large vacuole surrounded by a thin layer of cytoplasm. In leaves, most chloroplasts are found in the cells of parenchyma tissue. In stems and roots, parenchyma functions as storage centers for food and water. When *turgid*, or filled with water, parenchyma gives support and shape to plant organs.

Collenchyma tissue functions as supporting tissue in young plants, in the stems of *herbaceous* (nonwoody) older plants, and in leaves. The walls of collenchyma cells are thickened where cells meet. Otherwise, the collenchyma cells are similar to the parenchyma cells.

Sclerenchyma strengthens and supports mature plants. At maturity, most sclerenchyma cells die, leaving thick cell walls fortified with lignin. At times, the walls are so thick that the cell interiors are completely filled. Commercial textiles of flax and hemp are made from sclerenchyma fibers. The gritty inside of a pear is the result of *stone cells*, or masses of sclerenchyma.

There are two types of *vascular* (conducting) *tissue:* xylem and phloem. *Xylem* is vascular tissue that conducts water and dissolved substances *upward* from the roots to the leaves. Xylem forms a continuous system of tubes that runs through the roots, stems, and leaves. The main conducting vessels of the xylem are tracheids and vessel elements, which consist of dead cells. The walls of xylem vessels are composed mainly of cellulose, lignin, and other rigid materials.

Tracheids are long, tapering cells with secondary cell walls composed of lignin. Tracheids that form in a young plant stretch out as they mature,

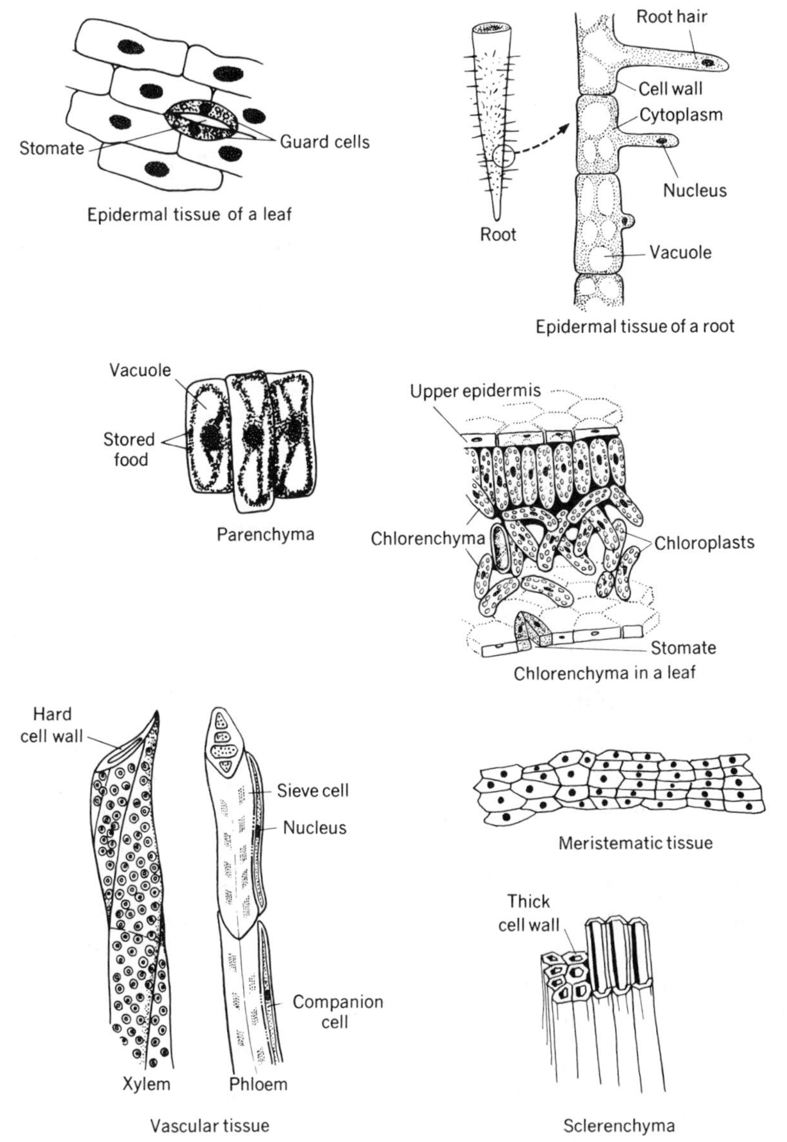

Figure 2–15 Plant tissues

acquiring secondary cell walls in the form of rings or spirals. Later, after lengthwise growth ceases, tracheids acquire many *bordered pits*. Water and dissolved substances move from one tracheid into another through pit-pairs concentrated at overlapping regions.

Vessel cells are more specialized conducting tissue than tracheids. The vessel cells are found only in flowering plants. Vessel cells are shorter and wider than tracheids; they also have numerous bordered pits through which water and dissolved substances move vertically from cell to cell (Figure 2–15). In addition to tracheids and vessel cells, xylem contains water-storing cells and long, thick-walled fiber cells, which help support the plant.

Phloem is composed mainly of *sieve elements*, or cells that form *sieve tubes*. The sieve tubes consist of elongated cells with *sieve plates* at their end walls (Figure 2–15). A sieve plate has many perforations, or holes, through which strands of cytoplasm form interconnections between cells. Sieve elements retain their cytoplasm, but lose their nucleus. In flowering plants, sieve tubes are accompanied by *companion cells*, which possess a nucleus. Companion cells move sugars from regions of photosynthesis to sieve tubes, and help direct activities in sieve elements.

Dermal tissue consists mainly of *epidermis*, a continuous layer of packed cells that covers the entire plant. The epidermis of stems and leaves may be covered with wax and cutin, forming an outer, protective layer called a *cuticle*. The cuticle reduces water loss by evaporation and protects the stems and leaves from insect and fungal attack. Some epidermal cells form long, extensive *root hairs* that probe the soil for water and dissolved minerals.

Chlorenchyma tissue consists of parenchyma cells that contain chloroplasts. Thus, photosynthesis occurs mainly in the chlorenchyma of leaves. In some plants, however, both leaves and stems carry out photosynthesis. Chlorenchyma tissue provides a plant with nutrients, including carbohydrates, lipids, and proteins. The process of photosynthesis (foodmaking) is discussed in Chapter 11.

Embryonic unspecialized cells capable of cell division compose *meristem tissues*. Young plants undergo many cell divisions, forming specialized tissues and organs. Unspecialized tissues are made up of meristem cells, usually small, thin-walled, and rich in cytoplasm. The growing tips of stems and roots that possess meristem tissue, called *apical meristem*, is responsible for an increase in plant length. Some roots and stems possess meristem tissue, called *lateral meristem*, that enables roots and stems to widen. Lateral meristem also promotes the development of the *vascular*

Table 2–3 Summary of Plant Tissues

Tissue	*Location*	*Functions*	*Adaptations*
Epidermal	Root	Protection	Cells close together
		Increases absorption area	Root hairs
	Stem	Protection	Cells close together
		Reduces loss of water	Waxy covering
	Leaf	Protection	Cells close together
		Reduces loss of water	Waxy covering
		Regulates exchange of gases	Guard cells
Parenchyma	Root, stem, and leaf	Storage of food and water	Boxlike cells; vacuoles
Chlorenchyma	Leaf and young stems	Photosynthesis	Chloroplasts
Vascular	Root, stem, and leaf	Upward conduction of fluids	Xylem tubes
		Downward conduction of fluids	Phloem tubes
Meristematic	Root and stem	Growth; formation of xylem, phloem, and other tissues	Unspecialized, dividing cells
Sclerenchyma	Stem and leaf	Support	Thick, hard cell walls

cambium which, in turn, promotes the development of new xylem and phloem after the primary plant body has formed. Lateral meristem also is the source of *cork cambium*, which produces the *periderm*, a protective cover that replaces epidermal tissue. The tissue organization of leaves, stems, roots, and their functions are discussed in Chapter 11 (see Table 2–3 above).

Section Quiz

1. Plant storage tissue is called (*a*) sclerenchyma (*b*) parenchyma (*c*) collenchyma (*d*) chlorenchyma.

2. A tissue that strengthens plants is (*a*) sclerenchyma (*b*) parenchyma (*c*) collenchyma (*d*) chlorenchyma.

3. In tracheids of xylem, bordered pits (*a*) bind cells together forming tubes (*b*) are composed of strengthening fibers (*c*) transfer water and dissolved substances between cells (*d*) mature as sieve plates in older xylem.

4. Unspecialized plant tissue capable of cell division is called (*a*) dermal tissue (*b*) collenchyma (*c*) periderm (*d*) meristem.

5. Chlorenchyma is a type of plant tissue that contains (*a*) tracheids (*b*) vessel cells (*c*) chloroplasts (*d*) meristem.

Animal Tissues

In the animal kingdom, there are four main types of tissues that make up body parts: epithelial, connective, muscle, and nervous. All four types are not found in every animal. For example, sponges are composed of cells that are not organized as tissues; hydra and its relatives (jellyfish, Portuguese man-of-war) are made up of two layers of tissue. Some simple animals, such as hydra, possess a nerve net and contractile fibers. These adaptations, however, are not considered tissues. All animals organized on an organ-system level have the four main types of tissue. The amount of a particular type of tissue in an organ depends on the organ's function. For example, the brain is composed mainly of nervous tissue and the skin is mainly epithelial tissue (see Figure 2–16 on page 45).

1. *Epithelial tissue*, or *epithelium*, covers body surfaces, lines internal ducts (tubes) and body cavities, and composes the secretory region of glands. The following are distinguishing features of epithelial tissue:

- Cells are tightly packed with little cementing material between them and no intercellular spaces.
- Cells are arranged in continuous sheets from one to several layers.

- Cells have a *basal surface* that is attached to other underlying tissue and a *free surface* that faces a body cavity, lining of a duct, or the exterior of the body.
- *Cell junctions*, or points of contact between cells, provide functional attachments between cells.
- Epithelial tissue lacks a direct blood supply; underlying tissue cells supply materials and remove wastes by diffusion.
- The functions of epithelial tissue include protection, secretion, digestion, absorption, excretion, and reproduction. The cells of epithelium are specialized; different types of epithelial tissue cells perform specialized functions. The following are types of epithelial tissues:
- *Simple squamous epithelium,* composed of a single layer of flat, platelike cells, is found lining air sacs in the lungs and in part of a *nephron* (filtration unit) of kidneys. The main functions of squamous epithelium are secretion, osmosis, diffusion, and filtration.
- *Simple cuboidal epithelium*, composed of a single layer of cube-shaped cells, is found covering ovaries, lining kidney tubules, and forming the pigmented epithelium of eyes. The main functions of this tissue are secretion and absorption.
- *Simple columnar epithelium* without cilia is composed of a single layer of column-shaped cells. Because many of these cells are partly filled with secretions that enter a gland's duct or cavity, they are called *goblet cells*. Simple columnar epithelium lines the alimentary canal between the stomach and anus, as well as the ducts of glands. The main functions of this tissue are secretion and absorption.
- *Simple columnar epithelium* composed of a single layer of ciliated column-shaped cells lines the upper respiratory passages and the Fallopian tubes (oviducts). Using cilia, the main function of simple columnar epithelium is to move fluids and particles along a passageway.
- *Glandular epithelium* is composed of large, irregular cells that line secretory glands, including sweat glands, salivary glands, oil glands, and the pancreas. The main functions of glandular epithelium are secretion of sweat, saliva, digestive juices, and mucus. Other glandular epithelium, composed of smaller, boxlike cells, are found in ductless *endocrine glands*, such as the adrenals, thyroid, and the pituitary. Secretions from these glands are called *hormones*, which pass directly into the blood.

2. *Connective tissue* cells, unlike epithelial cells, do not touch one another; cells are usually separated by intercellular material called a *matrix*. The following are distinguishing characteristics of connective tissue:

- Does not occur on the free surfaces of a body cavity or on the external surface of a body.
- Contains a rich supply of blood and nerves.
- Has a fluid, gelatinous, fibrous, or calcified matrix.

Most mature connective tissue cells do not divide readily, but are responsible for maintenance of the matrix. The matrix, or intercellular fluid, binds and supports cells and helps exchange substances between the connective tissue cells and the blood cells. *Collagen* fibers and *elastic* fibers usually are found in the matrix. Collagen fibers, which are abundant in bones, tendons, ligaments, and cartilage, add strength and flexibility. Elastic fibers, numerous in skin, lungs, and blood vessels, also provide strength and flexibility.

The following are types of connective tissues:

- *Blood* or *vascular tissue* is a connective tissue with a fluid matrix called *plasma*. Plasma consists mainly of water and dissolved substances. The solid, or cellular, part of blood includes red blood cells, white blood cells, and platelets. A detailed discussion of the composition and functions of blood is found in Chapter 5.
- *Areolar connective tissue* consists of cells and fibers (collagen and elastic) within a semifluid matrix. This tissue is widely distributed in the subcutaneous layer of the skin, blood vessels, and mucous membranes. The main functions of areolar connective tissue are elasticity and strength.
- *Adipose (fat) tissue* stores neutral fats. As a result, the nuclei of fat cells and a thin layer of cytoplasm are pushed outward to the plasma membrane. Adipose tissue is commonly found surrounding internal organs, the subcutaneous layer of skin, around bone joints, and behind the eyeballs. The main functions of adipose tissue include protection, support, energy, and temperature maintenance.
- *Dense connective tissue* consists mainly of collagen fibers arranged in bundles. The cells are located between bundles. *Tendons*, which attach muscles to bones, and *ligaments*, which attach bones to bones, are composed mainly of dense connective tissue.
- *Elastic connective tissue* consists mainly of elastic fibers and cells scattered among the fibers. Body parts that undergo stretching are

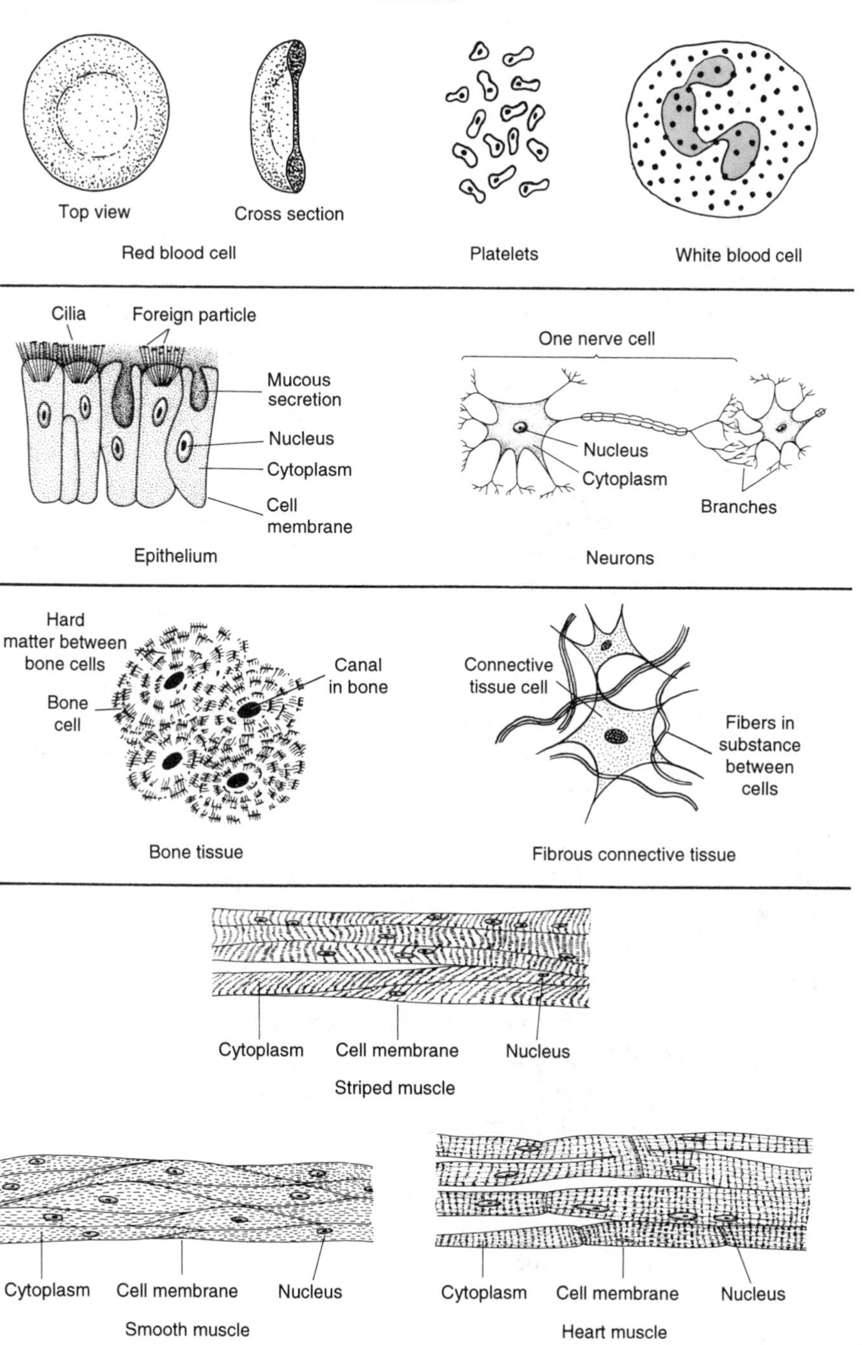

Figure 2–16 Animal tissues

rich in elastic connective tissue. This tissue is found in the lungs, artery wall, trachea (windpipe), bronchial tubes, and the vocal cords.

- *Cartilage* consists of a glossy, bluish-white matrix, collagen fibers, and cells. Cartilage is located at the ends of long bones, in ribs, nose, trachea, larynx (voice box), and bronchial tubes. The main functions of cartilage include smooth movement of bones at joints, support, and flexibility.
- *Bone tissue* is classified as either compact or spongy. The basic unit of *compact bone* is called an *osteon* or *Haversian system*. Each osteon consists of concentric rings of hard matrix called *lamellae*, composed mainly of a mineral called *hydroxyapatite* (a combination of calcium and phosphorus salts). Small spaces in the matrix, called *lacunae*, contain *osteocytes*, or bone cells. Lacunae are connected by a system of canals called *canaliculi*, which provide passageways for nutrients and waste removal. Blood vessels and nerves are located in a central Haversian canal. *Spongy bone* consists of thin plates, called *trabeculae*, which contain osteocytes, lamellae, lacunae, and canaliculi. Compact and spongy bone make up most of your skeletal system. A skeletal system supports soft tissues, protects internal organs, and in association with muscles enables movement and locomotion. Bone functions as a storehouse of calcium and phosphorus, produces blood cells in its red bone marrow, and stores lipids in its yellow bone marrow. Bone tissue and the skeletal system are discussed in detail in Chapter 9.

3. *Muscle tissue* is made up of cells that can expand and contract more than most other cells. Individual cells are elongated and bound together into sheets or bundles by connective tissue. Muscle tissue enables movement, maintains structure, and generates heat energy. The three main types of muscle tissue are skeletal, cardiac, and smooth.

- *Skeletal muscle*, which is closely connected with proper functioning of the skeletal system, is often called *striated muscle*. That is because certain elements within cells exhibit alternate light and dark bands called striations. Skeletal muscle also is called *voluntary muscle* because its action is consciously controlled. Muscle cells, or fibers, are cylindrical with many nuclei located at the periphery. The process and chemistry of muscle contraction are discussed in Chapter 9.
- *Cardiac muscle* comprises all of heart muscle. Cardiac muscle cells (fibers) are branched cylinders that contain one centrally located nucleus. Branching enables muscle fibers to attach to each other, which ensures synchronous contractions of all muscle fibers in the

Table 2–4 Summary of Animal Tissues

Tissue	*Location*	*Functions*	*Adaptation*
Epithelial	Covering of body	Protection	Cells fit closely
	Lining internal organs	Secretion	Cells arranged as sacs
Muscle Skeletal	Attached to bones of skeleton	Voluntary movement	Long, thin cells that contract
Smooth	Walls of internal organs	Involuntary movement	
Cardiac	Walls of heart	Pumping blood	
Nervous	Brain	Interpretation of impulses, mental activity	Long, thin, sensitive cells
	Spinal cord, nerves, and ganglions	Carrying impulses to and from all organs	
Connective Binding	Covering organs, and in tendons and ligaments	Holding tissues and organs together	Fibrous intercellular material
Adipose	Beneath skin and around internal organs	Fat storage, cushion, insulation	Cells have fat in their vacuoles
Cartilage	Ends of bones, part of nose and ears	Reduction of friction, support	Flexible intercellular material
Bone	Skeleton	Supporting framework, protection, movement	Hard intercellular substance
Blood	Blood vessels and heart	Carrying materials to and from cells	Intercellular materials in fluid
		Carrying oxygen	Red blood cells have hemoglobin
		Fighting germs	Some white blood cells have pseudopods and lysosomes
		Clotting	Platelets

heart chamber. Like skeletal muscle, cardiac muscle tissue is striated. However, this type of muscle is involuntary; heartbeats are not consciously controlled.

- *Smooth muscle* is found in the walls of blood vessels, air passages, urinary bladder, and the alimentary canal. Smooth muscle cells (fibers) are spindle-shaped and have a single nucleus. They lack striations, are mainly involuntary, and help move materials through the alimentary canal, as well as contract the urinary bladder, blood vessels, bronchial tubes, and air sacs in the lungs.

4. *Nervous tissue* is adapted to respond to internal and external stimuli. *Neurons* (nerve cells) convert stimuli into nerve impulses (electrochemical signals), which are then transmitted to other neurons, muscles, or glands. Neurons are the longest cells in the body. An individual neuron may be 1.0 meter to 1.5 meters long. A neuron consists of a *cyton*, or *cell body*, which contains a nucleus and other organelles; *dendrites*, or branched extensions of the cell body; and an *axon*, which is a single, long extension of the cell body that conducts impulses away from the cell body. A detailed discussion of neurons and nervous control is found in Chapter 8. (See also Table 2–4 on the previous page.)

The life processes that occur within cells are mainly chemical reactions. The following chapter, which discusses important facts and ideas of chemistry, will provide a better understanding of the chemical reactions that occur during these life processes.

Section Quiz

1. A type of connective tissue with a fluid matrix is (*a*) areolar (*b*) cartilage (*c*) adipose (*d*) blood.
2. Goblet cells are a characteristic of (*a*) connective tissue (*b*) ciliated columnar epithelium (*c*) the mucous membrane of the alimentary canal (*d*) the inner lining of endocrine glands.
3. The organ that moves fluids or solids by ciliary action is the (*a*) Fallopian tube (*b*) skin (*c*) esophagus (*d*) kidney tubule.
4. Collagen fibers and elastic fibers are numerous in (*a*) connective tissue (*b*) epithelial tissue (*c*) nervous tissue (*d*) muscle tissue.
5. The hard, intercellular matrix of bone tissue is composed mainly of (*a*) osteons (*b*) hydroxyapatite (*c*) calcium and magnesium salts (*d*) phosphorus and magnesium salts.

Chapter Review Questions

The following questions will help you check your understanding of the material presented in the chapter.

1. A microscope has a 10× low-power objective and a 40× high-power objective with a 10× ocular (eyepiece). A slide showing evenly distributed red blood cells is examined under low power and 320 cells are observed. If you turn to a high power lens, you should be able to observe about (*a*) 20 cells (*b*) 320 cells (*c*) 640 cells (*d*) 1,280 cells.
2. The diameter of the field of a microscope under low-power magnification is 1.2 millimeters. What is its diameter in microns? (*a*) 12 microns (*b*) 120 microns (*c*) 1,200 microns (*d*) 12,000 microns.
3. Microdissection instruments could best be used to (*a*) transplant cell nuclei (*b*) remove enzymes from the mitochondria (*c*) attach ribosomes onto the endoplasmic reticulum (*d*) rearrange the sequence of bases on DNA.
4. Radioactive carbon in a cell may be detected by means of a (*a*) silica gel film (*b*) microelectrode (*c*) Geiger counter (*d*) Celsius thermometer.
5. To view a virus, it is best to use (*a*) an oil-immersion objective (*b*) a condenser lens (*c*) methylene blue (*d*) an electron microscope.
6. Hereditary information is transferred from the cell's nucleus to the cytoplasm by means of (*a*) RNA (*b*) ATP (*c*) PGAL (*d*) ACTH.
7. Which organelle of an animal cell contains most of the enzymes involved in the release of energy for cell processes? (*a*) gene (*b*) centrosome (*c*) mitochondrion (*d*) nucleolus.
8. ATP is not essential during (*a*) active transport (*b*) passive transport (*c*) phagocytosis (*d*) pinocytosis.
9. Membranes within cells are structural components that are formed directly by (*a*) osmosis (*b*) excretion (*c*) hydrolysis (*d*) synthesis.
10. Which would most probably be found in a chemical analysis of a cell membrane? (*a*) proteins only (*b*) carbohydrates only (*c*) both proteins and carbohydrates (*d*) both proteins and lipids.
11. Ions and soluble molecules move from a region of high concentration to a region of lower concentration by (*a*) pinocytosis (*b*) digestion (*c*) active transport (*d*) diffusion.

12. Which substance would be least likely to pass through a cell membrane by diffusion? (*a*) water (*b*) starch (*c*) glucose (*d*) fatty acid.
13. Some algae accumulate iodine salts in concentrations 100 times greater than the concentration in the surrounding seawater. The process responsible for this phenomenon is (*a*) active transport (*b*) osmosis (*c*) passive adsorption (*d*) diffusion.
14. A certain type of poison affects living organisms by interfering with protein formation. The cell structures that might be affected by this poison are (*a*) ribosomes (*b*) centrosomes (*c*) chloroplasts (*d*) vacuoles.
15. Genetic control of cellular activities is least likely to involve (*a*) DNA (*b*) RNA (*c*) amino acids (*d*) fatty acids.
16. In the cytoplasm, the structures that contain the greatest amount of RNA are the (*a*) mitochondria (*b*) ribosomes (*c*) vacuoles (*d*) chloroplasts.
17. The process that requires the use of metabolic energy by a cell is (*a*) diffusion (*b*) active transport (*c*) plasmolysis (*d*) turgor.
18. The organelles usually found in the cells of both humans and maple trees are (*a*) chloroplasts (*b*) mitochondria (*c*) centrosomes (*d*) cell walls.
19. An organelle found in the cells of a grasshopper but not in the cells of a corn plant is the (*a*) cell membrane (*b*) chloroplast (*c*) centrosome (*d*) cell wall.
20. One way to classify an organism as a plant or an animal is to determine whether its cells (*a*) have vacuoles (*b*) have thick cellulose walls (*c*) reproduce (*d*) carry on cyclosis.
21. Flagella and cilia are structures responsible for locomotion of (*a*) protozoans (*b*) hydras (*c*) earthworms (*d*) grasshoppers.
22. Within an animal cell the self-replicating structures are (*a*) mitochondria and chromosomes (*b*) chromosomes and chloroplasts (*c*) ribosomes and vacuoles (*d*) chloroplasts and mitochondria.
23. *Escherichia coli*, a bacterium that inhabits the human colon, has a cell body that contains enough ribosomes to account for about 25 percent of its weight. Thus, its major products of synthesis are (*a*) carbohydrates (*b*) lipids (*c*) proteins (*d*) phospholipids.

Biology Challenge

The following questions will provide practice in answering SAT II-type questions.

Part I

Each of the following questions consists of a *statement* in the left column and a *reason* in the right column. Select the correct letter based on the following conditions:

A — The statement is true, but the reason is false.
B — The statement is false, but the reason is true (as a statement of fact).
C — Both statement and reason are false.

Statement	*Reason*
1. Genes are DNA molecules that encode instructions for cell activities.	1. Encoded instructions are used only for replication.
2. Ribosomes are often located attached to mitochondria.	2. Ribosomes are important generators of energy.
3. A plasma (cell) membrane is permeable to most organic and inorganic substances.	3. Permeability of a membrane permits diffusion of selected small molecules, ions, and particles into and out of a cell.
4. Macrophages (phagocytes) are the body's scavengers.	4. These "eaters" ingest debris of all kinds, except worn-out red blood cells.
5. Spindle fibers are part of a cell's cytoskeleton.	5. The cytoskeleton is involved in the movement of chromosomes during cell division.

Part II

Base your answers to questions *1 through 5* on the following diagrams, which show normal human red blood cells and their reactions to solutions of different salt concentrations.

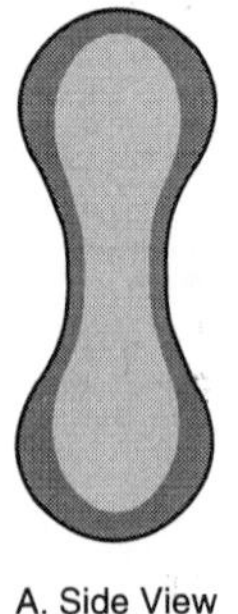

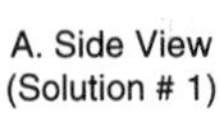

A. Side View
(Solution # 1)

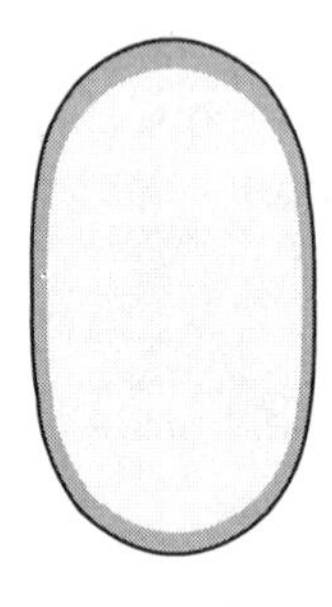

B. Side View
(Solution # 2)

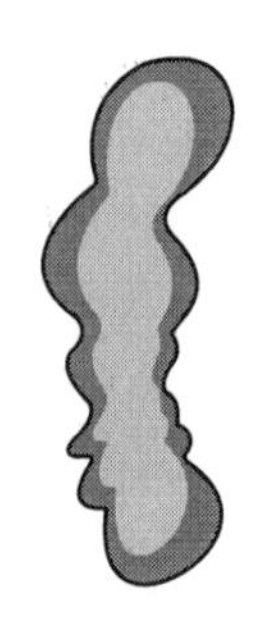

C. Side View
(Solution # 3)

1. The shading in red blood cells A, B, and C most likely represents (*a*) chromatin material (*b*) hemoglobin (*c*) salt solution (*d*) cytoskeleton (*e*) endoplasmic reticulum.
2. Cell A does not swell or shrink when immersed in a solution that is (*a*) isotonic (*b*) hypotonic (*c*) hypertonic (*d*) mainly distilled water (*e*) similar to sea water.
3. Cell C shrinks and assumes a wrinkled shape in solution 3 because (*a*) its plasma membrane lacks phospholipids (*b*) protein molecules coagulate, thus reducing the cell's volume (*c*) the solution is hypotonic and it loses water (*d*) the solution is hypertonic and it loses water (*e*) its nucleus is destroyed.
4. The sodium chloride concentration of normal human blood is about 0.9 percent. What is the approximate salt concentration of solution 2? (*a*) 1.5 percent (*b*) 1.4 percent (*c*) 1.0 percent (*d*) 0.9 percent (*e*) 0.7 percent.
5. Cell B is lighter in shading than either A or C. It may be inferred that (*a*) lighter shading indicates a loss of hemoglobin (*b*) vacuoles have filled with fluid (*c*) hemoglobin molecules are more widely dispersed in a fluid medium (*d*) more light is reflected from the cell's surface (*e*) solution 2 is hypertonic.

3

Chemistry of Life

Learning Objectives

When you have completed this chapter, you should be able to:

- **Differentiate** between elements, compounds, and mixtures.
- **Explain** some physical and chemical properties of matter.
- **Relate** potential to kinetic energy.
- **Describe** atomic mass and atomic number.
- **Identify** different types of chemical bonds.
- **Distinguish** different properties of acids and bases.
- **Identify** factors that affect the rates of chemical reactions.
- **Describe** the structure of nucleic acids.

OVERVIEW

Most living things are composed of the same chemical elements and perform similar biochemical reactions. Although there are about thirty elements found in living things, carbon, oxygen, hydrogen, and nitrogen are the most abundant. Molecules and compounds of these elements, and the reactions they undergo, enable organisms to carry out their life processes and exhibit the characteristics of life. This chapter covers the facts and concepts of chemistry that relate to the structure and function of living things.

ELEMENTARY CHEMISTRY

There are two main factors that determine whether or not matter will undergo change. One factor is the properties of matter and the other is the availability of energy.

Matter

Matter is anything that has mass and occupies space. All matter is composed of one or more substances that can change under the proper conditions. Understanding the changes that take place in living matter requires some knowledge of elementary chemistry and biochemistry.

Composition of matter. All matter, both living or nonliving, is composed of elements, compounds, and mixtures.

Table 3–1 Elements in Living Matter

Symbol	*Element*	*Atomic Number*	*Percentage of Earth's Crust by Weight*	*Percentage of Human Body by Weight*
B	Boron	5	Trace	Trace
Ca	Calcium	20	3.6	1.5
C	Carbon	6	0.03	18.5
Cl	Chlorine	17	0.01	0.2
Cr	Chromium	24	0.01	Trace
Co	Cobalt	27	Trace	Trace
Cu	Copper	29	0.01	Trace
F	Fluorine	9	0.07	Trace
H	Hydrogen	1	0.14	9.5
I	Iodine	53	Trace	Trace
Fe	Iron	26	5.0	Trace
Mg	Magnesium	12	2.1	0.1
Mn	Manganese	25	0.1	Trace
Mo	Molybdenum	42	Trace	Trace
Ni	Nickel	28	Trace	Trace
N	Nitrogen	7	Trace	3.3
O	Oxygen	8	46.6	65.0
P	Phosphorus	15	0.07	1.0
K	Potassium	19	2.6	0.4
Se	Selenium	34	Trace	Trace
Si	Silicon	14	27.7	Trace
Na	Sodium	11	2.8	0.2
S	Sulfur	16	0.03	0.3
Sn	Tin	50	Trace	Trace
V	Vanadium	23	0.01	Trace
Zn	Zinc	30	Trace	Trace

Elements. A substance that cannot be changed into a simpler substance by ordinary means is called an *element.*

Table 3–1 shows some familiar examples of elements and their chemical symbols.

Living things are composed mainly of the elements carbon, hydrogen, oxygen, and nitrogen. In addition, living things usually contain smaller quantities of sodium, chlorine, iron, sulfur, magnesium, phosphorus, calcium, potassium, cobalt, and other elements necessary for life.

Compounds. A combination of chemical elements in definite proportions by weight is called a *compound.* A chemical formula shows the kinds of elements in a compound and the proportions of each element by weight. For example, the formula for glucose is $C_6H_{12}O_6$, which indicates that the compound is composed of 24 atoms. It has six atoms of carbon each weighing 12 atomic mass units, 12 atoms of hydrogen each weighing 1 atomic mass unit, and six atoms of oxygen each weighing 16 atomic mass units. Atomic weights will be discussed later. A compound can be decomposed by chemical means, such as heating, digestion, or cellular respiration.

Mixtures. The components of a *mixture* are not fixed or chemically combined. Thus, a mixture has no chemical formula. Any quantity of sand (SiO_2) and table salt (NaCl) may be combined to form a mixture of sand and salt. Unlike a compound, this mixture has no special properties; the properties of a mixture are the properties of each component. Each component also can be separated from the other by simple means. Living matter is a continually changing mixture.

A mixture in which one substance dissolves in another is called a *solution.* The substance that dissolves is called a *solute* and the substance in which the solute dissolves is called a *solvent.* In a solution of table salt and water, the solute is table salt and the solvent is water.

Equal quantities of a mixture have the same composition. Thus, solutions are *homogeneous.* Equal amounts of a solution of table salt and water have the same quantity of salt and water.

Change in matter. Both the phase and the composition of matter may change under different conditions.

Changes in phase. At a given temperature and pressure, all matter exists in one of three *phases*—solid, liquid, or gas. When the temperature or pressure changes, matter may undergo a change in phase. For example, at sea level, water freezes (becomes a solid) at 0°C. Water boils (becomes a gas) at 100°C. Importantly for life on Earth, water is a liquid between these temperatures.

Although living matter appears solid, it usually includes liquids and gases. For example, in a fish we find solid bone, liquid body fluid, and the gaseous contents of the swim bladder.

When matter changes phase, a *physical change* takes place, in which matter retains its original composition. The change from solid ice to liquid water and the change from liquid water to gaseous steam are examples of physical changes. In these examples, the phase of matter has changed but the composition of each form of water (H_2O) is unchanged.

Changes in composition. When matter undergoes a change in composition, a *chemical change* takes place. For example, when potassium chlorate is heated, it decomposes (breaks down) and changes in *composition.* As a result of decomposition, new substances, called *products*—potassium chloride and oxygen—are formed. Similarly, when excess protein in the body is acted upon by the liver, a chemical change occurs producing water and urea.

When a cat tears a piece of meat into smaller pieces, the meat only undergoes a physical change. After the cat eats the meat, digestion changes the meat into new and simpler substances—a chemical change. Another example of a chemical change occurs when these simpler substances are assimilated into living matter.

Changes in properties. When matter undergoes change, its *properties* (characteristics) also change. All matter possesses *physical properties* such as phase, color, odor, and solubility. Matter also possesses *chemical properties*, such as the ability to burn or to support burning. Hydrogen is a colorless and odorless gas that burns in the presence of oxygen to form liquid water. And oxygen is a colorless and odorless gas that supports the burning of hydrogen to form liquid water. Biology studies the physical and chemical properties of substances present in living things.

Energy

Physical scientists define *energy* as the ability to do work. Because living things require a constant supply of energy to survive, it is important to understand some terms used to describe energy.

Kinds of energy. The two kinds of energy are potential energy and kinetic energy.

Potential energy is the energy matter possesses, or stores, because of position or condition. A large rock perched on a hilltop possesses potential energy. Substances, such as sugar, also possess potential energy in the form of stored chemical energy.

Kinetic energy is the energy of motion. As a rock rolls down a hill, the potential energy that the rock had at the top of the hill is transformed into kinetic energy. A similar energy transformation occurs when sugar is oxidized in a chemical reaction. In the oxidation of sugar, the reacting substances, or *reactants*, are changed. At the same time, the potential energy of the sugar is changed into kinetic energy in the form of light and heat. When sugar is oxidized in organisms, some potential energy is changed to heat energy and some potential energy is transferred to molecules of other substances. In a living thing, this energy is used to do work as an organism carries out metabolic activities.

Energy and chemical reactions. Chemical reactions usually require energy to occur. This energy, called *activation energy*, causes molecules of the reactants to move faster, thus increasing the chance of molecular collisions.

To begin the reaction in which potassium chlorate decomposes to potassium chloride and oxygen, requires an input of heat energy. A continuous input of heat energy is required to keep this reaction going. In living things, the foodmaking process (photosynthesis) of plants is an example of a reaction that requires a continuous input of light energy.

Other chemical reactions also release energy. For example, when a substance burns in oxygen, energy is released as heat and light. Similarly, when oxidation occurs in living cells, some energy is transferred as chemical energy to other molecules and some is released as body heat.

Measuring energy. All kinds of energy can be changed into heat energy. The amount of heat energy that a substance absorbs or releases during a chemical reaction is measured in units called calories. A *calorie* is the amount of heat required to raise the temperature of 1 gram of water 1°C. When a peanut burns, the peanut's stored chemical energy is changed into heat and light. The calories released by the burning peanut can be determined by allowing the flame to heat a given amount of water and determining the difference between the water's original temperature and its final temperature. If the temperature of 10 grams of water is raised 2°C by the burning peanut, the number of calories is calculated by multiplying 10 by 2 (20 calories). Respiration, which is another type of "burning," releases chemical energy and heat energy.

ELEMENTS AND ATOMS

The smallest particle of a particular element that can combine with other elements is called an atom. Each element has a different atomic structure.

Structure of Atoms

Figure 3–1 shows models of the atomic structure of some common elements.

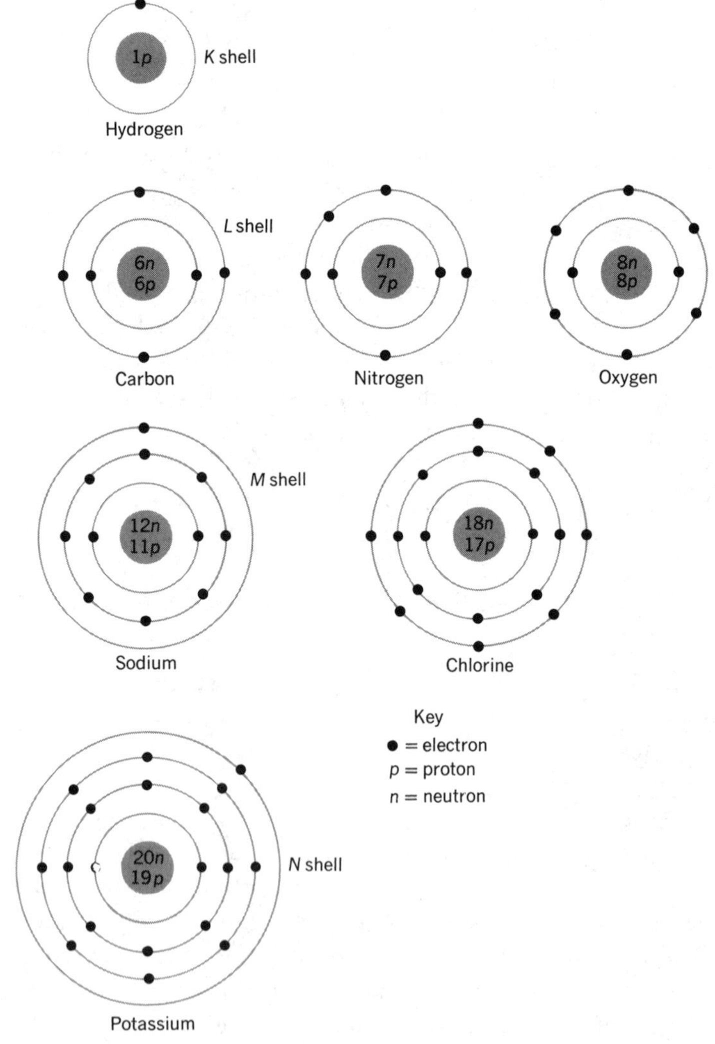

Figure 3–1 Atomic structure of some common elements

Atomic nucleus. Most of an atom's mass is found in the nucleus, which is located in the center of an atom. Excluding hydrogen, the nucleus of each kind of atom contains *protons* (p) and *neutrons* (n). The nucleus of an atom of ordinary hydrogen has a single proton but lacks a neutron. Each proton has a positive (+) electrical charge. A neutron has no electrical charge.

Electrons. Electrons are negatively charged (-) particles located outside the nucleus. Because the number of electrons in an atom equals the number of protons, atoms are electrically neutral. Electrons are arranged in patterns, called *shells,* or *energy levels,* that are identified by certain letters. The shell closest to the nucleus is called the K shell; the second shell from the nucleus is called the L shell; the third, the M shell, and so on. The K shell can hold a maximum of two electrons; the L shell a maximum of eight; and the M shell 18 electrons. Atoms differ because the numbers of protons and neutrons in their nuclei differ. In addition, the number of shells and arrangement of electrons in the shells of one atom differ from the number and arrangement of electrons in the shells of other atoms.

Electrons may jump from one energy level to another. When an electron acquires energy from its environment, the electron jumps to a higher energy level. When the electron returns to a lower energy level, energy is released that is precisely equal to the amount of energy used when the electrons jumped to the higher shell. The energy may be heat, light, electricity, or radiation. The *chemical activity* of elements depends upon (a) the number and arrangement of electrons in their atoms, and (b) the energy levels the electrons reach as the result of gaining or losing energy.

Atomic Mass and Atomic Number

The combined mass of the protons and neutrons in an atom's nucleus is called the *atomic mass*. Each proton and neutron weighs one unit. Since the mass of an electron is only about 1/1,837 the mass of a proton, the mass of an electron is not included in the atomic mass. Thus, the atomic mass of hydrogen is 1 (1 proton); carbon is 12 (6 protons + 6 neutrons); oxygen is 16 (8 protons + 8 neutrons); and potassium is 39 (19 protons + 20 neutrons). Atomic mass is designated by placing the combined number of protons and neutrons to the upper left of the chemical symbol of the element.

The *atomic number*, or the number of protons, is designated by placing the appropriate number to the lower left of the element's symbol. Thus, ${}^{1}_{1}H$ shows the atomic mass and the atomic number of hydrogen, ${}^{12}_{6}C$

shows the atomic mass and the atomic number of carbon, and $^{16}_{8}O$ shows the atomic weight and the atomic number of oxygen.

Isotopes

All atoms of a particular element do not have the same mass. An atom of an element that does not have the same mass as other atoms of the same element is called an *isotope*. Figure 3–2 shows the isotopes of hydrogen.

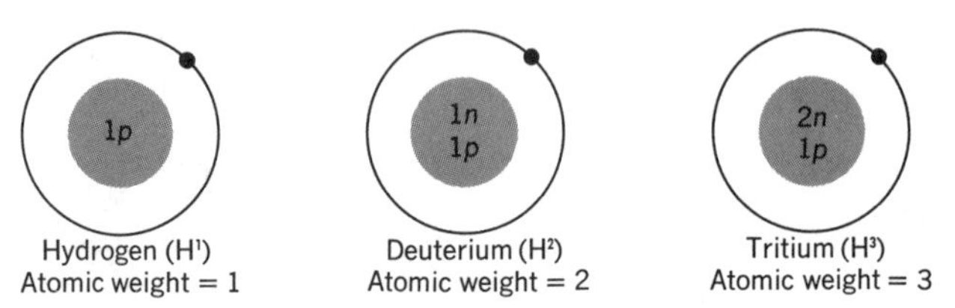

Figure 3–2 Isotopes of hydrogen

Many isotopes are *radioactive*. They emit radiations such as alpha rays (helium nuclei), beta rays (high-speed electrons), and gamma rays (which are similar to X rays). Radioactive isotopes of hydrogen, carbon, phosphorus, iodine, and cobalt are among many radioactive isotopes used in research and medicine. Figure 3–3 shows two common isotopes of oxygen.

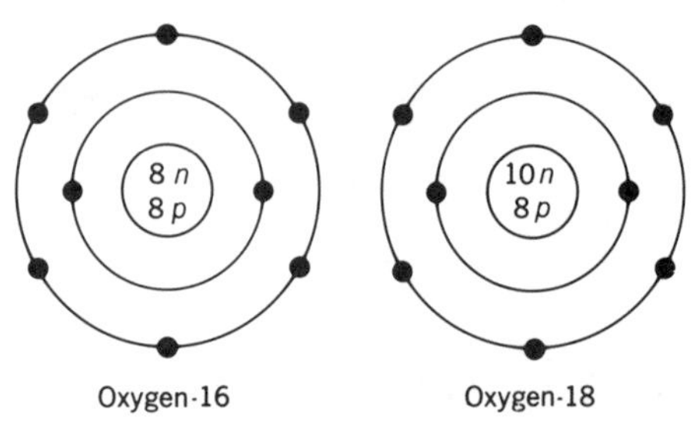

Figure 3–3 Two isotopes of oxygen

TYPES OF COMPOUNDS

All compounds are classified as either inorganic or organic.

Inorganic Compounds

In general, an *inorganic compound* does not contain the element carbon. For example, hydrochloric acid (HCl) and sodium chloride (NaCl) are inorganic compounds. If carbon is present in an inorganic compound, the carbon is usually combined with oxygen, as in calcium carbonate ($CaCO_3$) and carbon dioxide (CO_2).

Organic Compounds

An *organic compound* always contains carbon in complex combinations with hydrogen, oxygen, and other elements. Most organic compounds are associated with living things and their products. Proteins, carbohydrates, lipids, and nucleic acids are organic compounds found in all living things.

CHEMICAL BONDS

The attractive force that binds atoms in a compound is called a *chemical bond*. The two general types of chemical bonds are ionic and covalent.

Ionic Bonds

Atoms and compounds are electrically neutral because they contain equal numbers of positive and negative electrical charges. When a metallic atom loses one or more electrons, it becomes positively charged (+) because the balance of positive and negative charges is upset. For example, the reaction of $Na^{\circ} \rightarrow Na^{+} + e^{-}$ (the superscript ^{o} indicates electrical neutrality) shows a neutral sodium atom changing into a sodium ion by losing an electron from its outermost energy level. When a nonmetallic atom of chlorine (Cl°) gains an electron, the chlorine atom becomes a chloride ion ($Cl^{\circ} + e^{-} \rightarrow Cl^{-}$).

When a fragment of sodium metal is dropped into a container filled with chlorine gas, a vigorous reaction occurs in which sodium atoms lose electrons and chlorine atoms gain electrons. As a result, sodium ions (Na^{+}) and chloride ions (Cl^{-}) are formed. The equation for this chemical reaction is $2Na + Cl_2 \rightarrow 2NaCl$ which shows that two atoms of sodium have reacted with two atoms of chlorine to yield two particles of the compound sodium chloride. In this reaction, each sodium atom donates one electron to each chlorine atom. The resulting ions attract each other and unite, forming an electrostatic bond, called an *ionic bond*. Thus, the compound NaCl can be represented as $Na^{+}\ Cl^{-}$.

In the synthesis of sodium chloride, four reactant atoms yield four product atoms. This conforms to the *First Law of Thermodynamics*, which states that the quantity of mass/energy in the universe is constant. Figure 3–4 shows how sodium chloride is formed from the union of a positively charged sodium ion and a negatively charged chloride ion. Ionic compounds, which are usually soluble in water, consist of metallic and nonmetallic ions.

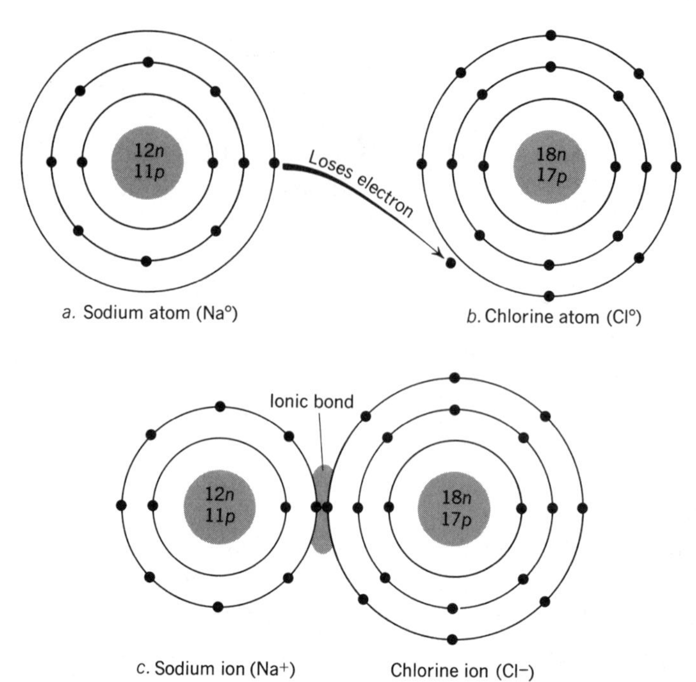

Figure 3–4 Formation of sodium chloride

Covalent Bonds

A *molecule* is made up of two or more atoms joined by covalent bonds. A *covalent bond* is formed when atoms share electrons. The atoms that take part in covalent bonding are usually nonmetals. The overlapping outermost energy levels are involved in covalent bonding. Each atom contributes one or more electrons to each pair that is shared. Sharing involves electrons that occupy the outer energy levels of both atoms. Thus, shared electrons actually "belong" to both atoms. For example, the outermost energy level of an oxygen atom holds six electrons, called *valence electrons*. Valence electrons are the only electrons that take part in chemical reactions. A hydrogen atom has one valence electron and an oxygen atom has six valence electrons. Figure 3–5 shows the sharing of electrons in a water molecule, in which each hydrogen atom has two electrons and the oxygen atom has eight electrons, forming an *octet*.

Water as a solvent. Salts, including sodium chloride (NaCl) and silver nitrate ($AgNO_3$), are composed of ions joined by an ionic bond. For example, NaCl (common table salt) has oppositely charged ions and is represented as Na^+Cl^-. AgNO is represented as Ag^+ NO_3^-. The sum of the charges of each compound must equal zero to make the compound electrically neutral.

When ionic compounds are mixed with water, the compounds are called *solutes*, and the water is the *solvent*. Water is a good solvent for ionic compounds and some organic molecules because of the polar nature of water molecules. When ionic compounds such as NaCl and $AgNO_3$ are mixed with water, their ions fall apart. The negatively charged ends of

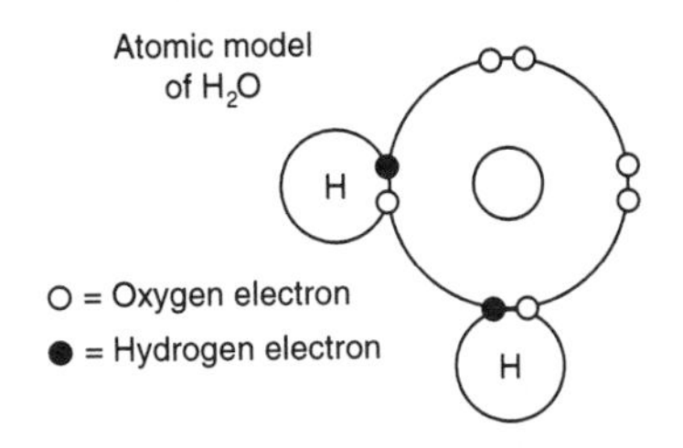

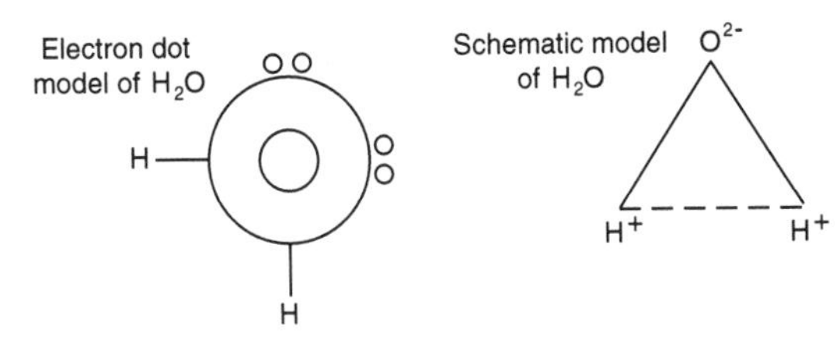

Figure 3–5 Models of water molecule

some water molecules are attracted to and cluster around the positively charged ends of these compounds (Na^+ and Ag^+). The positively charged ends of other water molecules are attracted to and cluster around the negatively charged ends of these compounds (Cl^- and NO_3^-). The result is a tug of war that breaks ionic bonds and releases ions in the solvent. The released ions can unite and reform the original reactants NaCl and $AgNO_3$, or unite in another way to form sodium nitrate ($NaNO_3$) and silver chloride (AgCl). The latter reaction takes place more frequently, eventually involving all available ions. The final outcome is summarized by the following equations:

$$NaCl + AgNO_3 \rightarrow NaNO_3 + AgCl$$

or

$$Na^+ Cl^- + Ag^+ + NO_3^- \rightarrow Na^+ + NO_3^- + AgCl$$

Because silver chloride (AgCl) is practically insoluble in water and does not form ions, the reaction can go to completion. Many different metallic and nonmetallic ions as well as organic polar compounds are found in the cytosol of cells. These ions and organic compounds are the reactants involved in many of the 3,000 or more reactions that occur in cells.

Bonds and Chemical Reactions

When a chemical reaction occurs, old bonds are broken and new bonds are formed. Energy (activation energy) is necessary. New bonds are formed in the products of the reaction. In many reactions, the formation of new bonds is accompanied by the release of some energy. The chemical reactions are called *exothermic reactions*.

Section Quiz

1. Water is a solvent for a great number of substances because of its (*a*) covalent bonds (*b*) cohesion (*c*) polarity (*d*) linear structure.
2. Covalent bonds are usually formed between (*a*) nonmetallic atoms (*b*) metallic atoms (*c*) nonmetallic and metallic atoms (*d*) atoms of gases and metals.
3. An atom that gains an electron (*a*) is neutral (*b*) is negatively charged (*c*) is positively charged (*d*) always forms an octet.
4. Atoms of elements are chemically active mainly because of differences in (*a*) atomic mass (*b*) atomic number (*c*) number of electrons in their outer energy levels (*d*) number of electrons in the energy level closest to the nucleus
5. Organic compounds always contain one or more atoms of (*a*) nitrogen (*b*) oxygen (*c*) sulfur (*d*) carbon

Acids, Bases, and Neutralization

Most inorganic and organic compounds in living things are classified as acids, bases, or salts.

Acids. A water solution of a substance that ionizes into positively charged hydrogen ions (H^+) is called an *acid*. For example, hydrochloric acid ionizes into a hydrogen ion and a chloride ion: $HCl \rightarrow H^+ + Cl^-$.

Hydrogen ions give acids their characteristic properties. Acids taste sour and change the color of blue litmus (an *indicator*) to red. Acids are classified as strong or weak, based on the number of hydrogen ions formed. Inorganic acids such as hydrochloric acid, sulfuric acid, and nitric acid are strong acids because they readily ionize in water (many H^+ ions are formed). Organic acids such as acetic acid (in vinegar), citric acid (in citrus fruits), and lactic acid (sour milk) are weak acids because they do not readily ionize in water (few H^+ ions are formed).

Bases. A water solution of a substance that ionizes into negatively charged hydroxide ions (OH^-) is called a *base.* For example, sodium hydroxide ionizes into a sodium ion and a hydroxide ion: $NaOH \rightarrow Na^+ + OH^-$. The hydroxide ions give bases their characteristic properties. Bases taste bitter, feel slippery, and change the color of red litmus to blue. Organic compounds that contain the OH^- ion are called *alcohols,* which include glycerin.

Neutralization. The reaction between an acid and a base that yields a salt and water is called *neutralization*. The neutralization reaction between hydrochloric acid and sodium hydroxide is represented by the following equations:

$$HCl + NaOH \rightarrow NaCl + H_2O$$
$$\text{or } H^+ + Cl^- + Na^+ + OH^- \rightarrow NaCl + H_2O$$

The equations show that a *salt* is a compound with a positive ion (released by a base) and a negative ion (released by an acid). Inorganic salts are usually called *mineral salts* because they are found independently in rocks and minerals. Sodium citrate and sodium stearate are examples of organic salts.

The pH Scale

A *pH scale* (Figure 3–6) is used to indicate the degree of acidity (concentration of hydrogen ions) or the degree of alkalinity (concentration of hydroxide ions) of a particular solution. This scale, which ranges from 0 to 14, quantifies the hydrogen ion concentration in a given quantity of solution. A pH of 7 indicates a neutral point between acidity and alkalinity. At a pH of 7, there are equal concentrations of hydroxide ions and hydrogen ions in a solution. A pH between 14.0 and 7.0 indicates a basic (or alkaline)

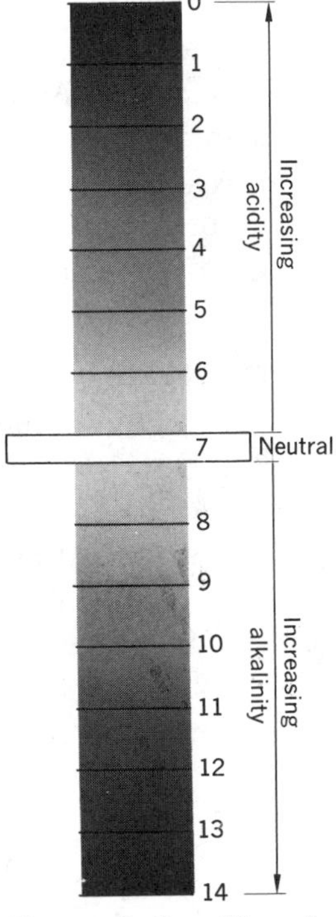

Figure 3–6 pH scale

solution. A pH between 7.0 and 0 indicates an acidic solution. In effect, the greater the acidity, the lower the pH number. The greater the alkalinity, the higher the pH number. Table 3–2 shows the approximate pH of some human body fluids.

Table 3–2 Approximate pH of Some Human Body Fluids

Fluid	*pH*	*Fluid*	*pH*
Gastric juice	1.5	Milk	7.2
Urine	6.0	Tears	7.3
Saliva	6.8	Blood	7.4
Water	7.0	Pancreatic juice	8.0

TYPES OF CHEMICAL REACTIONS IN LIVING THINGS

Small molecules combine to form larger and more complex molecules in a *dehydration synthesis reaction*. During dehydration synthesis, hydrogen ions and hydroxide ions are removed from the reactant molecules and water is formed. At the same time, the remainder of the reactant molecules are joined by chemical bonds, forming a long, complex molecule. Figure 3–7 shows how glucose molecules join to form starch, how amino acids join to form protein, and how fatty acid molecules and glycerin molecules join to form a lipid (fat or oil).

Hydrolysis

Large complex molecules are broken down to smaller ones with the aid of water in a *hydrolysis reaction*. The reaction replaces water (H^+ and OH^- ions) where water was originally removed during dehydration synthesis. During hydrolysis, the bonds holding together the units of a large molecule are broken, and the individual units are released. Figure 3–8 on page 68 shows how the hydrolysis of starch yields individual glucose molecules. Note that hydrolysis, which restores water, is the opposite of dehydration synthesis, which removes water.

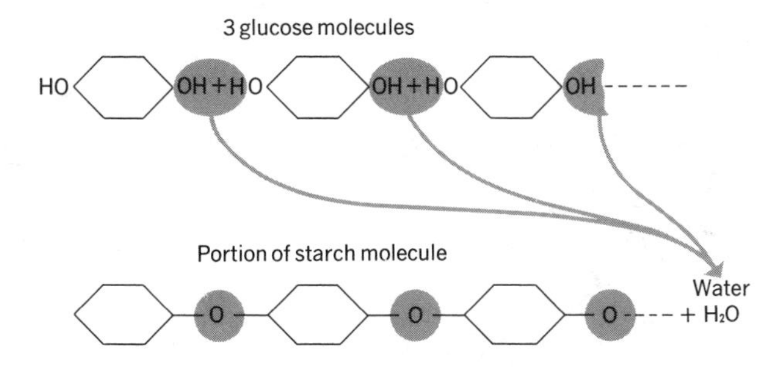

a. Scheme of dehydration synthesis of starch

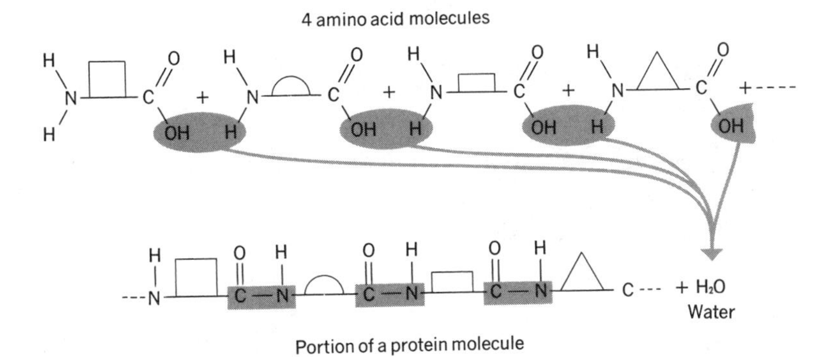

b. Scheme of dehydration synthesis of protein

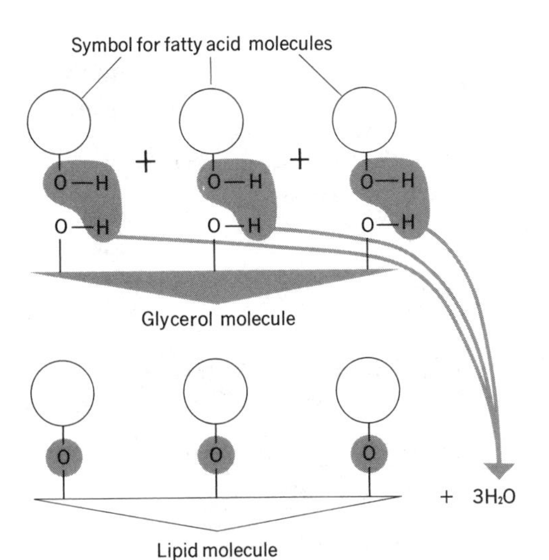

c. Scheme of dehydration synthesis of lipid

Figure 3–7 Dehydration synthesis

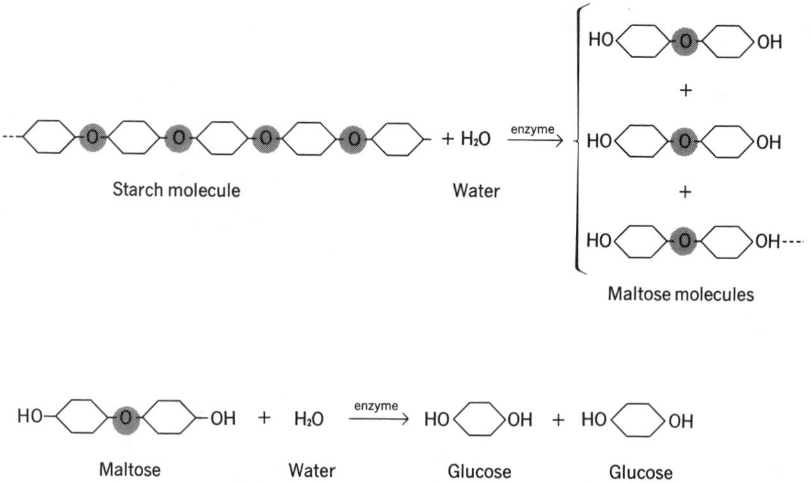

Starch molecule + H_2O (Water) $\xrightarrow{\text{enzyme}}$ Maltose molecules

Maltose + H_2O (Water) $\xrightarrow{\text{enzyme}}$ Glucose + Glucose

Figure 3–8 Scheme of hydrolysis of starch

Replacement

One or more elements in a compound replace one or more elements in another compound during a *replacement reaction*.

For example, iron reacts with copper chloride to yield copper and iron chloride. In the lungs, oxygen combines with hemoglobin that carries carbon dioxide (carbon dioxide–hemoglobin) to yield oxyhemoglobin and to release carbon dioxide:

oxygen (inhaled) + carbon dioxide–hemoglobin →
oxyhemoglobin + carbon dioxide (exhaled)

When the concentration of carbon dioxide is relatively high in tissues, carbon dioxide replaces the oxygen of oxyhemoglobin to yield carbon dioxide hemoglobin:

oxyhemoglobin + carbon dioxide (from cells) →
oxygen (to cells) + carbon dioxide–hemoglobin

The reactions are repeated in the next breathing cycle.

FACTORS AFFECTING RATES OF CHEMICAL REACTIONS

The rate of a chemical reaction depends upon the nature and concentrations of the reactants, sizes of the reactant particles, temperature, and the presence of catalysts.

Nature of Reactants

Different substances react at different rates. For example, a match burns faster than a candle, and a stick of dynamite, when ignited, burns so rapidly that it produces a violent explosion. Energy in the form of heat (and often light) is released in all cases of burning. When glucose is burned (oxidized) in the cells of living things, energy is also released. In cells, however, energy is released slowly.

Concentration of Reactants

A chemical reaction occurs because of collisions between reactant molecules. Thus, increasing the quantity, or concentration, of reactants in a given reaction increases the rate of reaction. More reactant particles increases the probability of collisions. Fewer reactant particles decreases the probability of collisions between particles. Thus, when dilute (weak) solutions of sodium carbonate (Na_2CO_3) and hydrochloric acid (HCl) are mixed, carbon dioxide (CO_2) gas is released slowly. If larger quantities of reactants (one or both) are used, the gas is released rapidly.

A given volume of blood contains red blood cells and plasma, which contains dissolved oxygen. It takes time for the hemoglobin in the red blood cells to form the maximum amount of oxyhemoglobin possible. When the amount of dissolved oxygen in the plasma is doubled, the maximum amount of oxyhemoglobin is formed in about half the original time.

If the number of red blood cells and the quantity of oxygen in the plasma are doubled, the rate of oxyhemoglobin formation will increase accordingly.

Size of Reactant Particles

The smaller the sizes of reactant particles, the more rapid is the reaction. For example, when dry sodium chloride and dry silver nitrate are mixed, they do not react readily. However, when both substances are dissolved in water, they react readily to form white silver chloride. When the substances are dry, their crystals are so large that the chances of contact between reacting particles is reduced. When sodium chloride and silver nitrate are dissolved in water, both substances form ions. Ions are tiny and move about freely, increasing the chances of contact between reacting particles.

In living things, ingested salts of sodium, potassium, and other elements dissolve in water to form ions. Ions in a solution react readily within cells. Large organic molecules, such as starch, are broken down into smaller molecules, such as glucose, which can enter the cells and take part in energy-releasing reactions.

Temperature

An increase in temperature usually increases the rate of reaction. That is because the additional heat energy causes the reactant molecules to move faster. Rapid motion increases the number of collisions between the reactant molecules. If the temperature is lowered, the reactant molecules move more slowly, the number of collisions decreases, and the rate of reaction decreases.

The chemical test for starch clearly illustrates the effect of temperature upon the rate of a reaction. In the chemical test for starch, a blue-black color is produced when iodine reacts with starch. When a dilute solution of potassium iodate (KIO_3) is added to a mixture of starch, water, and sulfurous acid (H_2SO_3) and then placed in a hot-water bath at 90°C (194°F), iodine is released rapidly from the KIO_3, which turns the starch blue-black almost immediately. When the same mixture is placed in a water bath at 22°C (72°F), iodine is released much more slowly, and several seconds elapse before the starch turns blue-black.

The effect of temperature upon reactions in living things may be readily observed. A tropical fish becomes sluggish when placed in cold water. The low temperature reduces the rate of many metabolic reactions in the fish. As a result, its life processes slow down, and unless the fish is soon warmed, it may die. The opposite effect, an increase in the rate of metabolic reactions, is readily observed in the increased activity of insects, fish, and other animals on hot days.

Catalysts

A substance that speeds up the rate of reaction without being changed itself is called a *catalyst*. The catalyst does this by lowering the activation energy required for the reaction.

Catalysts may be inorganic or organic. When potassium chlorate ($KClO_3$) is heated to a temperature of about 360°C (680°F), the reaction is slow and oxygen is released slowly. If $KClO_3$ is mixed with a small quantity of an inorganic catalyst, such as manganese dioxide (MnO_2), and then heated to about 200°C (392°F), the reaction rate increases noticeably, and oxygen is released rapidly.

Organic catalysts, called *enzymes*, are found in living things. Intense heat is required to oxidize glucose in a test tube. However, in a human cell in which the appropriate enzymes are present, glucose is oxidized at the relatively low body temperature of 37°C (98.6°F).

BIOCHEMISTRY

Living things are a complex mixture of water, mineral salts, carbohydrates, proteins, lipids, nucleic acids, nucleotides, and vitamins. Of these compounds, only water and mineral salts are inorganic; the other compounds are organic. A basic knowledge of biochemistry, or the chemistry of organic compounds found in living things, is necessary to learn more about life.

Carbohydrates

Carbohydrates contain carbon, hydrogen, and oxygen. The ratio of hydrogen atoms to oxygen atoms in a carbohydrate is always the same as it is in water—2:1. Carbohydrates include many types of sugars, starches, glycogen, and cellulose.

Single sugars (monosaccharides). A *single sugar* cannot be broken down into a simpler sugar. Single sugars include the *hexoses* (six-carbon sugars)—glucose, fructose, and galactose; all are isomers with the formula $C_6H_{12}O_6$. *Isomers* are compounds that have the same molecular formula (composition) but different structural formulas. Figure 3–9 shows a molecule of fructose and a molecule of galactose. In these molecules, the atoms and groups of atoms are arranged differently. Thus, the structural formulas of fructose and galactose are different. The structural differences give each isomer different properties. For example, fructose is sweeter than glucose, and galactose passes into a cell more readily than does either glucose or fructose.

Glucose (also called *dextrose* or *grape sugar*) is a very important single sugar in organisms because it is easily used in exothermic reactions. Other single sugars include the *pentoses* (five-carbon sugars); *deoxyribose*, one of the components of DNA; and *ribose*, one of the components of RNA.

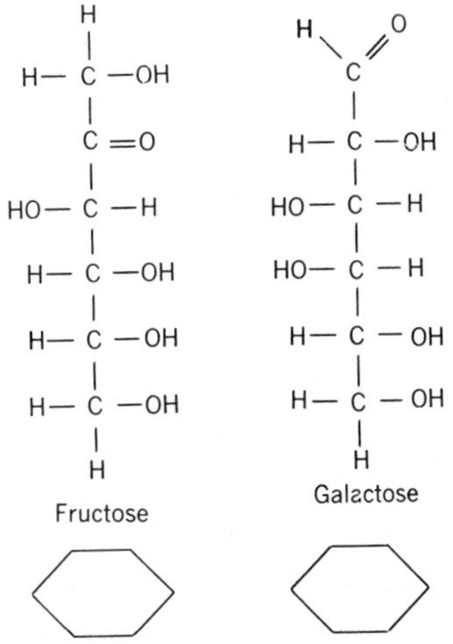

Fructose

Galactose

Figure 3–9 Isomers of six-carbon sugars

Double sugars (disaccharides). A *double sugar* consists of two single sugars joined by dehydration synthesis (Figure 3–10). Common double sugars include sucrose (table sugar), maltose (found in many seeds), and lactose (found in milk). All double sugars are isomers with the formula $C_{12}H_{22}O_{11}$. In some cells, sucrose is synthesized from a glucose molecule and fructose molecule; maltose is synthesized from two glucose molecules; and lactose is synthesized from a glucose molecule and a galactose molecule.

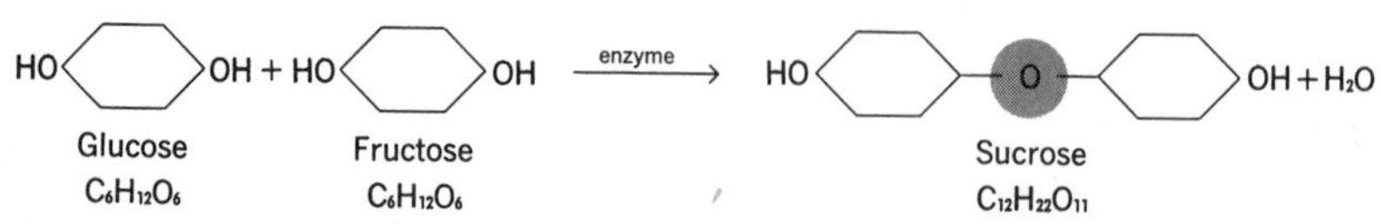

Figure 3–10 Double-sugar formation by dehydration synthesis

Multiple sugars (polysaccharides). A *multiple sugar* is a complex molecule formed by the joining of hundreds of glucose molecules (Figure 3–7a). A multiple sugar may consist of a long chain of 300 to 2,500 glucose molecules joined by dehydration synthesis.

Examples of multiple sugars include starch, glycogen, and cellulose. *Glycogen* is a storage form of glucose found in the liver, and *cellulose* makes up the cell walls of plant cells. All multiple sugars have the general formula $C_6H_{10}O_5$. However, each one has a different number of $C_6H_{10}O_5$ units. Thus, the formula for all multiple sugars is $(C_6H_{10}O_5)_n$, where *n* means more than one unit. The exact number of units depends on the nature of the multiple sugar.

Proteins

Most *proteins* are composed of the elements carbon, hydrogen, oxygen, nitrogen, phosphorus, and sulfur. Protein molecules are enormous; the more complex protein molecules are composed of thousands of atoms arranged in complex patterns. The patterns of many proteins are so complex that their molecular and structural formulas are still undetermined.

Proteins are found throughout the cell and play a significant role in many chemical reactions because the enzymes that catalyze these reactions also are proteins. In addition, proteins are the building and binding materials of living organisms.

Protein formation from amino acids. Proteins are formed by joining smaller molecules called amino acids. An *amino acid* is an organic acid containing the *amino group* (NH_2) and the *carboxyl*, or *acid, group* (COOH). Figure 3–11 shows the general structure of an amino acid and the structural formulas of several amino acids. There are 20 different amino acids. Various combinations of amino acids make it possible to have tremendously large numbers of different proteins. The number of amino acid molecules in a protein varies from about 300 to several thousand.

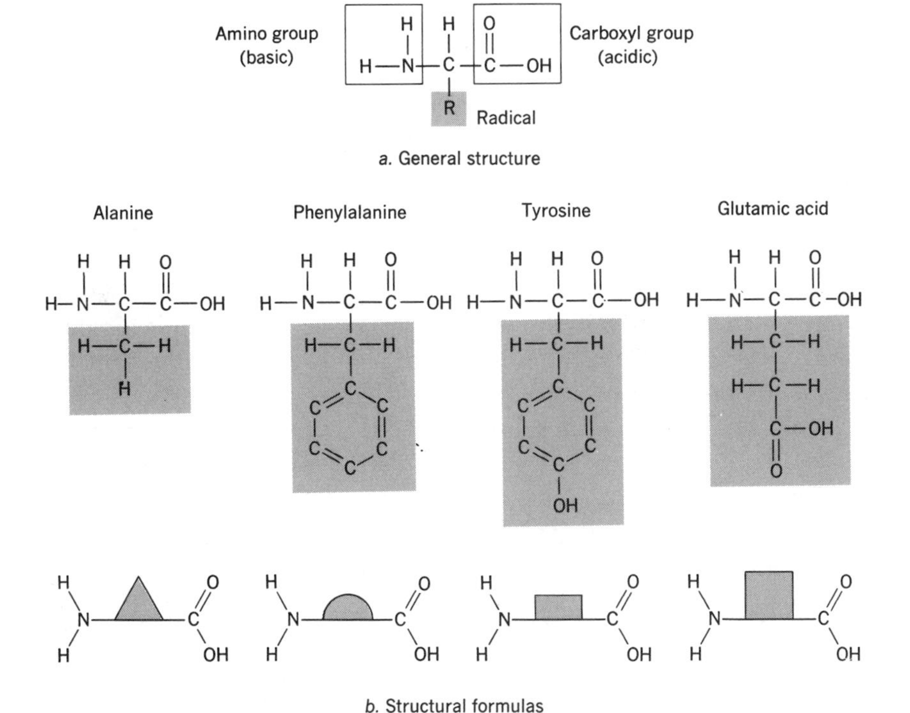

Figure 3–11 Some typical amino acids

Like the synthesis of complex carbohydrate molecules, proteins are synthesized from amino acids by dehydration synthesis. Figure 3–12 shows how a *peptide bond* is formed between amino acids after water comes out. The N from the NH_2 group of one amino acid connects with the C from the COOH group of the adjacent amino acid.

Lipids

All *lipids* (fats and oils) are composed of carbon, hydrogen, and oxygen. Unlike the 2:1 ratio found in carbohydrates, the ratio of hydrogen atoms to oxygen atoms varies in lipids. For example, the formula for the major lipid in castor oil is $C_{18}H_{34}O_3$. In animals, lipids are usually found as fats; in plants, lipids are usually found as oils.

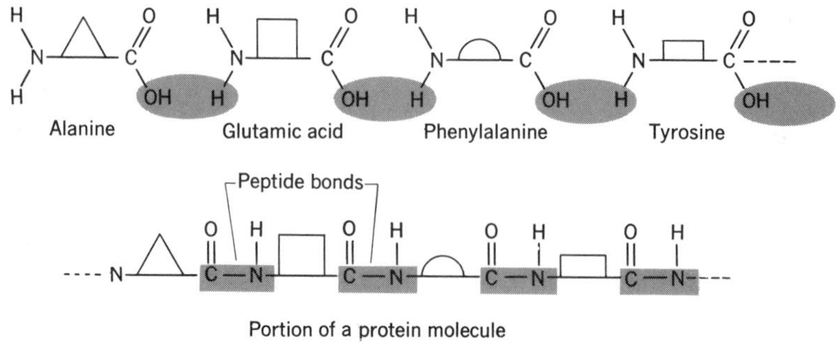

Figure 3–12 Formation of peptide bonds

Lipid formation from fatty acids and glycerin. Like carbohydrates and proteins, dehydration synthesis combines small molecules, such as a fatty acid and glycerin, to form a complex lipid molecule (Figure 3–7c). Some common fatty acids are stearic acid ($C_{17}H_{35}COOH$), palmitic acid ($C_{15}H_{31}$ COOH), and butyric acid (C_3H_7COOH). The formula for glycerin is $C_3H_5(OH)_3$. The synthesis of lipids involves three fatty acid molecules that supply three carboxyl (COOH) groups, and one glycerin molecule that supplies three hydroxyl (OH) groups. Each OH group bonds with an H atom to form a water molecule. In this example, three water molecules and three chemical bonds are formed.

Enzymes

Organic catalysts, or *enzymes,* control the chemical activity in living cells. Each reaction proceeds at the appropriate speed at just the right time with the precision of a finely programmed computer.

Composition of enzymes. All enzymes are composed of single proteins or proteins attached to other molecules. The protein part of an enzyme is called an *apoenzyme*; the nonprotein part is called a *coenzyme*. Specific vitamins and minerals usually are the coenzymes in one or more enzymes that catalyze cellular reactions. For example, vitamin C (ascorbic acid) is the coenzyme of an enzyme that controls the synthesis of a cementlike substance that binds the cells of capillary walls. Without sufficient ascorbic acid, the cells of capillaries separate because they are not properly cemented together. This results in blood oozing out of capillaries and black and blue spots forming under the skin. These spots are characteristic of the vitamin C deficiency disease called scurvy.

Function of enzymes. The molecules upon which an enzyme acts are called *substrate* molecules. Figure 3–13 shows how reactants A and B

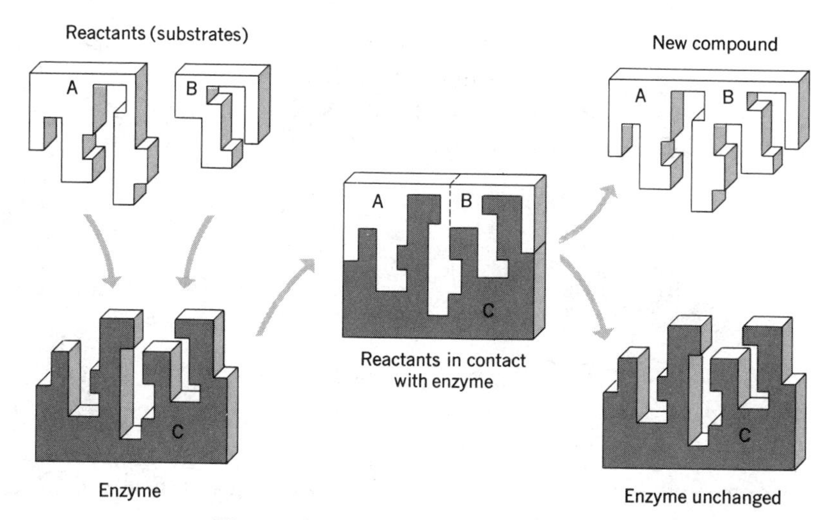

Figure 3–13 How a catalyst acts

form the new compound AB in the presence of catalyst C. The surface of the catalyst is shaped to admit A and B. Proper shape allows only one kind of substrate molecule to fit into the enzyme. In effect, the relation between an enzyme and its substrate is like the relation between a lock and a key. Only one key can fit a particular lock and open it.

An enzyme usually is named by adding the suffix *ase* to the root of its substrate name. For example, *sucrase* acts on *sucrose*, and *urease* acts on *urea*.

The reaction between an enzyme and its substrate occurs at *active sites* in the enzyme molecule. Substrate molecules lack active sites. The pattern of atoms making up the active sites of one enzyme differs from that of all other enzymes. Thus, enzymes are action *specific*, each acts on its own substrate. The shape of the active sites of the enzyme molecule in

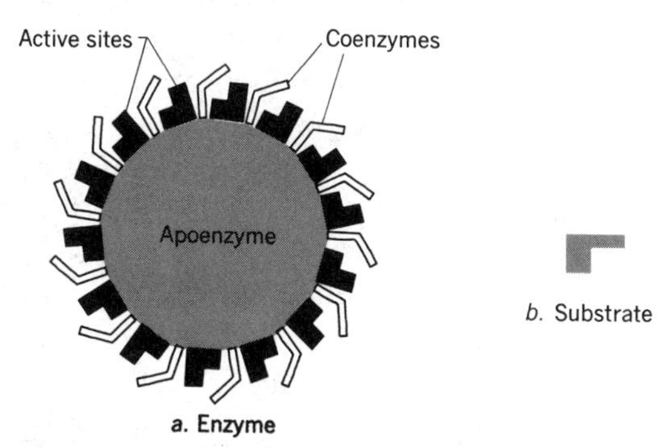

Figure 3–14 Structure of an enzyme and a substrate

Figure 3–14a and the shape of the substrate molecule in Figure 3–14b allow the molecules to fit together like pieces of a jigsaw puzzle. When matching enzyme and substrate molecules are concentrated, the molecules collide. Then, substrate molecules fit into each active site of the enzyme (Figure 3–15). The relation between an enzyme and its substrate is called an *enzyme-substrate complex*. This complex combination is highly reactive: old bonds are quickly broken and new bonds are quickly formed, producing new compounds. After releasing new products, the active sites are free and the unchanged enzyme is ready to react with more substrate molecules.

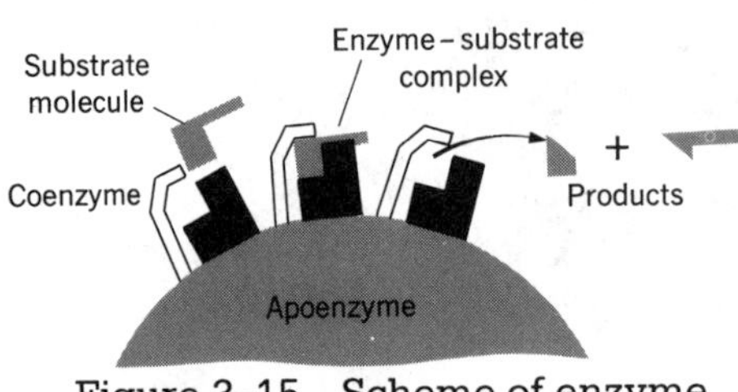

Figure 3–15 Scheme of enzyme action

Factors influencing the rate of enzyme action. The rate of an enzyme reaction is influenced by pH, temperature, and the relative amounts of enzyme and substrate molecules.

pH. Enzymes are sensitive to changes in acidity or alkalinity. The pH at which an enzyme reacts best with its substrate is called the *optimum pH*. The optimum pH of pepsin, a digestive enzyme in the stomach, is about 2.0. If the pH rises above 2, the hydrolysis of proteins in the stomach either slows down or stops. The graph in Figure 3–16a shows how differences in pH influence the rate of enzyme action. For most reactions in living things, the optimum pH is near 7. Low and high pH values tend to slow down enzyme activity.

Temperature. Like other chemical reactions, enzyme reactions are affected by temperature. The temperature at which an enzyme reacts best with its substrate is called the *optimum temperature*. Below optimum temperatures, an enzyme-controlled reaction proceeds as if the enzyme was not present. At lower than optimum temperature, the reactions may start slowly or not at all. As the temperature rises above the optimum temperature for an enzyme, the rate of reaction increases slightly and then decreases sharply. Biologists think that high temperatures may alter the protein and the active sites of an enzyme. All enzymes are destroyed at temperatures above 50°C (122°F). At the freezing point of water, 0°C (32°F), most enzymes are inactive, but are not destroyed. When warmed to their optimum temperatures, most enzymes become active again.

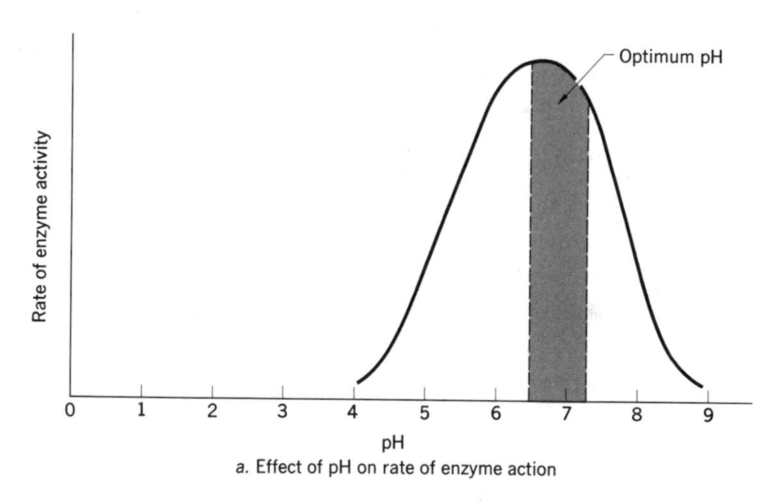

a. Effect of pH on rate of enzyme action

Optimum temperature
Rate of enzyme activity
0 10 20 30 40 50 60
Degrees Celsius

b. Effect of temperature upon rate of enzyme action

Figure 3–16 Some factors affecting the rate of enzyme activity

Enzymes of different organisms have different optimum temperatures. In humans, the optimum temperature for enzyme activity is body temperature 37°C (98.6°F). Plant enzymes function best at about 25°C (77°F). The graph in Figure 3–16b shows how temperature affects the rate of enzyme action. As you can see, the optimum temperatures for most enzymes is from 37°C to 50°C.

Concentrations of enzyme and substrate. When other factors are constant, the rate of an enzyme-controlled reaction is influenced by the concentration of enzyme molecules and substrate molecules. When there is a fixed amount of enzyme and an excess of substrate molecules, the rate of reaction rises to a certain point and then levels off (see Figure 3–17a). At the point where leveling off begins, all the active sites in the enzyme are in

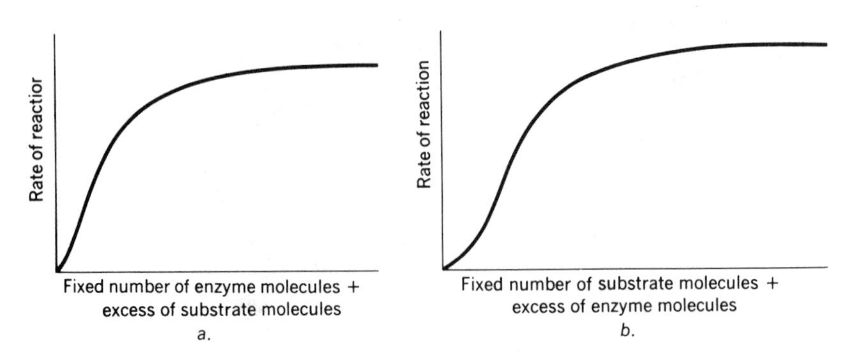

Figure 3–17 Effects of concentrations of enzyme and substrate on rate of reaction

use and no sites are available for excess substrate molecules. If the situation is reversed (Figure 3–17b)—if more enzyme molecules than substrate molecules are available—the rate of reaction also rises to a certain point and then levels off. At the point where leveling off begins, all the substrate molecules are reacting with enzyme molecules and there are excess enzyme molecules for which there are no substrate molecules. Thus, there are many available active sites but no substrate molecules to occupy them.

Section Quiz

1. The amino group (-NH_2) and the carboxyl group (-COOH) of an amino acid indicate that the compound (*a*) always functions as an acid (*b*) always functions as a base (*c*) may function as either an acid or base (*d*) may function as a neutral molecule.
2. The ratio of hydrogen atoms to oxygen atoms in lipids is (*a*) always 2:1 (*b*) always 1:2 (*c*) always 3:1 (*d*) variable.
3. Enzymes speed up the rate of a reaction by (*a*) decreasing the activation energy (*b*) increasing the activation energy (*c*) decreasing the potential energy of reactants (*d*) increasing the potential energy of reactants.
4. The compound $CuSO_4$ (copper sulfate) dissolves in water forming copper ions (Cu^{2+}) and sulfate (SO_4^{2-}) ions. $CuSO_4$ is a(an) (*a*) acid (*b*) salt (*c*) base (*d*) covalent molecule.

5. The pH of a carbonated soft drink is about 3.5 and that of coffee (without milk or cream) is about 5.0. It may be stated that (*a*) the soft drink is less acidic than coffee (*b*) coffee is less basic than the soft drink (*c*) the soft drink is more acidic than coffee (*d*) milk increases the acidity of coffee.

Nucleic Acids

The two most important *nucleic acids* found in cells are deoxyribonucleic acid (DNA) and ribonucleic acid (RNA). DNA and RNA influence heredity and are involved in the synthesis of cellular proteins.

Chromosomes and genes are composed mainly of DNA. Chromosomes are cell structures that carry hereditary traits; genes are the units that make up each chromosome. DNA also is found in other organelles, including the plasma membrane, mitochondria, and the chloroplasts. Large quantities of RNA are found in the nucleoli, the cytoplasm, and the ribosomes.

Structure of nucleic acids. DNA and RNA are composed of subunits (smaller molecules). Each subunit of a DNA and RNA molecule is called a nucleotide. A *nucleotide* is a complex molecule composed of three smaller units joined together (Figure 3–18a). These units consist of a *nitrogenous* (organic nitrogen-containing) *base*, a *pentose* (five-carbon sugar), and a *phosphate* (PO_4) *group*. Figure 3–19 shows that the nitrogenous base may be any one of the following five: *adenine* (*A*), *guanine* (*G*), *thymine* (*T*), *cytosine* (*C*), or *uracil* (*U*). In DNA, the nitrogenous bases A, G, C, and T are found in the nucleotides. The same bases are found in RNA, except that the base uracil (U) is present instead of the base thymine (T). Thus, the bases in RNA are A, G, C, and U. Some nucleotides and their bases are shown in Figure 3–18b.

DNA and RNA are composed of many linked nucleotides that form a long, chainlike strand. As a result, both DNA and RNA are called *polynucleotides*. The details of the structure and function of DNA and RNA are discussed in Chapter 13.

ADP and ATP

Adenosine diphosphate (ADP) and *adenosine triphosphate* (ATP) are special nucleotides that store energy. ADP is symbolized as A–P~P, and ATP is symbolized A–P~P~P (Figure 3–20). The curved-dash symbol (~) indicates the presence of a high-energy bond between the phosphate groups. A *high-energy bond* is a chemical bond that has more potential

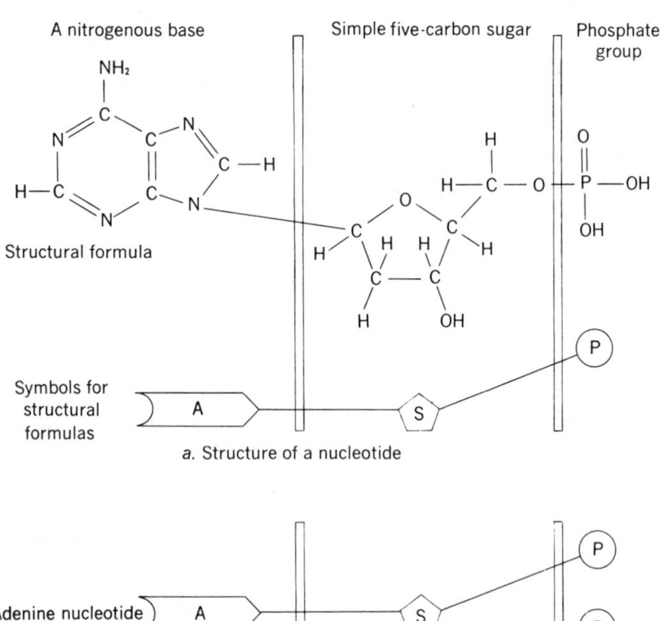

a. Structure of a nucleotide

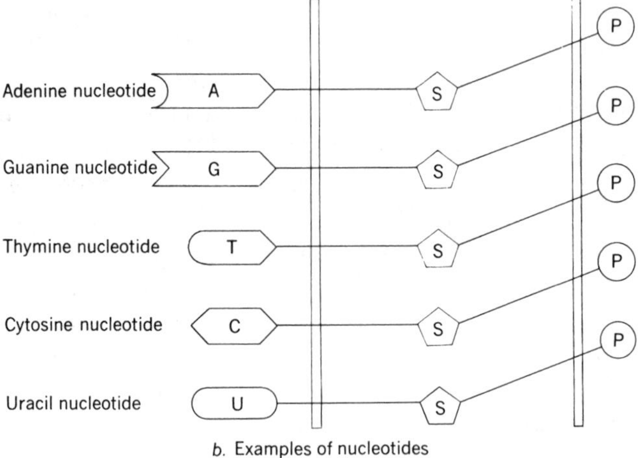

b. Examples of nucleotides

Key:
A, G, T, C, U = Nitrogenous bases
S = Sugar
P = Phosphate

Figure 3–18 Important nucleotides

energy than do other chemical bonds. ATP has two high-energy bonds. Thus, ATP has more potential energy than ADP, which only has one high-energy bond.

As the result of respiration in living cells, ADP is changed to ATP by the addition of a high-energy phosphate bond (~P). When ATP is used to provide energy for a reaction within a cell, the high-energy phosphate

Nitrogen base	Type	Structure	Symbol
Adenine	Purine	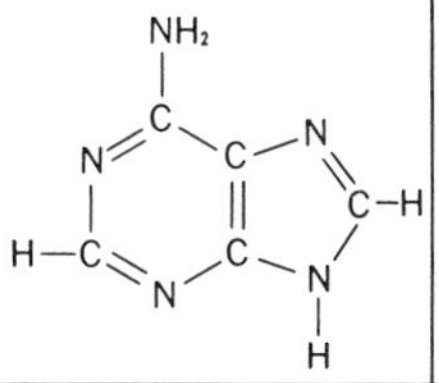 NH_2 C N C N C–H H–C N C N H	 A
Guanine	Purine	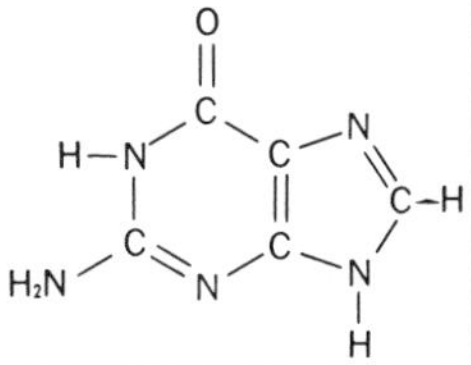 O C H–N C N C–H H_2N C N C N H	 G
Cytosine	Pyrimidine	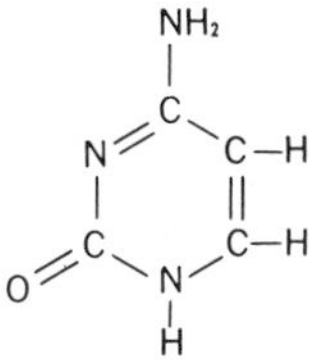 NH_2 C N C–H O C C–H N H	 C
Thymine	Pyrimidine	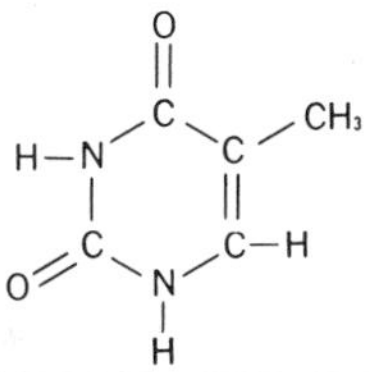 O C CH_3 H–N C O C C–H N H	 T
Uracil	Pyrimidine	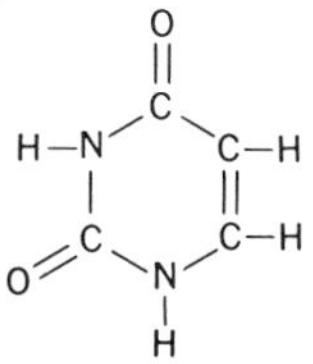 O C H–N C–H O C C–H N H	 U

Figure 3–19 Nitrogenous bases

bond is broken. This break is accompanied by a release of energy, and is changed back to ADP (Figure 3-21). The release of energy occurs slowly, with the aid of enzymes. This controlled release of energy prevents a cell from receiving too much energy at once. If this occurred, the cell would be seriously damaged or the homeostasis of the cell would be disturbed.

Like a fully charged storage battery, an ATP molecule is ready to give instant service in the form of energy. When it releases energy, a battery is

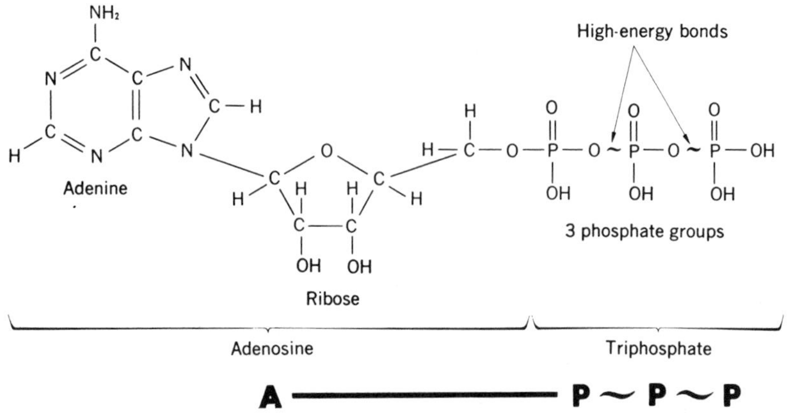

High-energy bonds
Adenine
Ribose
3 phosphate groups
Adenosine
Triphosphate
A —— P ~ P ~ P

Figure 3–20 ATP molecule

discharged. When it is connected to a source of electricity, the battery is recharged. Similarly, when the energy in ATP is discharged, ATP changes to ADP. When ADP is recharged by the energy supplied through oxidation (respiration), ADP is changed again to fully charged ATP.

Oxidation reactions in cells
Energy
ATP
−PO₄
+PO₄
Energy for life processes
ADP

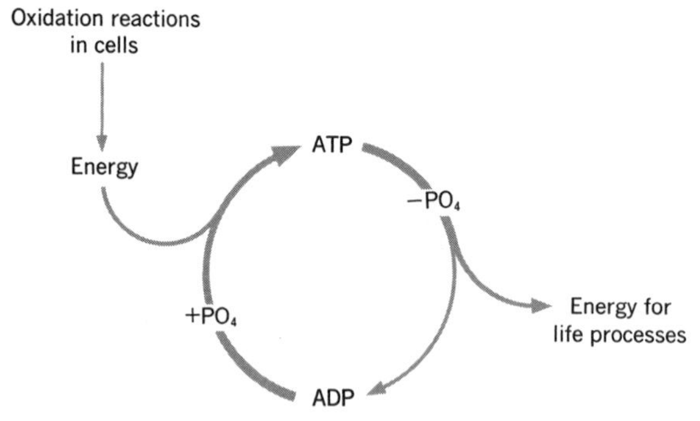

Figure 3–21 ADP-ATP cycle

Let us apply what we have learned about chemistry to the structure and function of organisms. Chapter 4 will discuss the nutrients in food, how food is taken into the body, and what happens to food after it enters the body.

Section Quiz

1. Genes are composed mainly of (*a*) RNA (*b*) DNA (*c*) ATP (*d*) ADP.
2. Each DNA nucleotide molecule is composed of a nitrogenous base, a pentose (sugar), and a (*a*) phosphate group (*b*) diphosphate group (*c*) triphosphate group (*d*) polynucleotide.
3. Which nitrogenous base would *not* be found in a DNA strand? (*a*) adenine (*b*) cytosine (*c*) uracil (*d*) guanine.
4. Special nucleotides that store energy are (*a*) DNA and ATP (*b*) ADP and ATP (*c*) ATP and RNA (*d*) RNA and ADP.
5. When the high-energy phosphate bond of an ATP nucleotide is broken, (*a*) energy is released (*b*) energy is stored (*c*) nucleotides are formed (*d*) polynucleotides are broken.

Chapter Review Questions

The following questions will help you check your understanding of the material presented in the chapter.

1. Which process uses energy to make complex molecules? (*a*) nutrition (*b*) transport (*c*) excretion (*d*) synthesis.
2. An enzyme is mainly a (*a*) lipid material (*b*) polysaccharide (*c*) protein (*d*) vitamin.
3. The structural formulas below represent two molecules. These molecules are normally joined by the removal of (*a*) H and OH (*b*) N (*c*) O (*d*) R and NH_2.

```
H    R    O          H    R    O
 \   |   //           \   |   //
  N--C--C              N--C--C
 /   |   \            /   |   \
H    H    OH         H    H    OH
```

4. Which process does not directly involve the synthesis of chemical compounds? (*a*) growth (*b*) cell division (*c*) reproduction (*d*) egestion.
5. A difference between organic compounds and inorganic compounds is that (*a*) inorganic compounds are never found in living

things (*b*) inorganic compounds always contain oxygen and hydrogen (*c*) organic compounds always contain carbon (*d*) organic compounds always dissolve in water.

6. Changing the pH value within the human digestive tract would have the greatest effect upon (*a*) swallowing (*b*) water absorption (*c*) chemical digestion (*d*) ingestion.
7. A neutral solution has a pH of (*a*) 1 (*b*) 7 (*c*) 9 (*d*) 14.
8. Which pH indicates the strongest basic solution? (*a*) 3.5 (*b*) 6.4 (*c*) 10.0 (*d*) 13.9.
9. Which statement best describes dehydration synthesis? (*a*) Water is a substance needed for the construction of molecules. (*b*) Molecules are constructed and water is released. (*c*) Molecules of the same type must always be involved. (*d*) Molecules of different types must always be involved.
10. Which statement may be applied to enzymes? (*a*) They are not specific. (*b*) They are unaffected by pH changes. (*c*) They become permanently attached to substrate molecules. (*d*) They may speed up or slow down a reaction.
11. The action of enzymes involved in hydrolysis always depends on the presence of (*a*) nitrogen (*b*) water (*c*) acids (*d*) bases.
12. In plant seeds, the organic catalysts that make digestion possible at relatively low temperatures are the (*a*) auxins (*b*) nitrogenous bases (*c*) lipids (*d*) enzymes.
13. At 20°C the optimum reaction rate of a certain enzyme occurs at a pH of 7. A faster reaction rate could be attained by increasing (*a*) the temperature to 30°C and keeping the pH at 7 (*b*) both the temperature and the pH (*c*) the pH and keeping the temperature at 20°C (*d*) the pH and decreasing the temperature.
14. A chain of hundreds of amino acid molecules forms a (*a*) polynucleotide (*b*) polysaccharide (*c*) polyploid (*d*) polypeptide.

For questions *15 through 19*, select the number of the general term from the list below that is most closely related to each item. A number may be used more than once or not at all.

General Terms

(1) carbohydrates (3) proteins
(2) lipids (4) nucleic acids

15. Fats, oils, and waxes
16. Peptide bonds
17. Glycerol molecules combined with three fatty acid molecules
18. Glucose, starch, and cellulose
19. DNA and RNA
20. Which process is indicated by the equation below? (*a*) hydrolysis (*b*) osmosis (*c*) decomposition (*d*) dehydration synthesis

$$C_6H_{12}O_6 + C_6H_{12}O_6 \xrightarrow{\text{enzymes}} C_{12}H_{22}O_{11} + H_2O$$

21. Which organic molecule is correctly paired with its simpler subunits? (*a*) lipid—nitrogenous bases (*b*) nucleic acid—inorganic bases (*c*) carbohydrate-fatty acids (*d*) protein—amino acids.
22. In humans, fats and carbohydrates are not directly converted into proteins because fats and carbohydrates lack (*a*) carbon (*b*) hydrogen (*c*) oxygen (*d*) nitrogen.
23. The formula $C_6H_{12}O_6$ represents a molecule that is a building block for many (*a*) proteins (*b*) lipids (*c*) carbohydrates (*d*) enzymes.
24. Although there are only twenty amino acids, a large number of proteins may be formed because (*a*) differences in numbers and kinds of amino acids result in the possibility of many different configurations (*b*) amino acids are not part of the structure of proteins (*c*) each protein is composed of one type of amino acid (*d*) the number of proteins produced by cells is very large.
25. Enzymes function primarily as (*a*) organic catalysts (*b*) inorganic catalysts (*c*) hormones (*d*) substrates.
26. The most abundant inorganic substance in a cell is (*a*) salt (*b*) sugar (*c*) amino acid (*d*) water.
27. Amino acids always contain (*a*) a long carbon chain and a carboxyl group (*b*) an amino group and twice as many atoms of hydrogen (*c*) an amino group and a long carbon chain (*d*) an amino group and a carboxyl group.

Biology Challenge

The following questions will provide practice in answering SAT II-type questions.

Part I

Select the letter of the statement that best completes the sentence or answers the question.

1. An organic molecule with the chemical formula $C_{18}H_{36}O_2$ is a (*a*) carbohydrate (*b*) protein (*c*) lipid (*d*) mineral (*e*) vitamin.

Base your answers to questions *2 through 5* on the chemical reaction below and on your knowledge of biology.

```
H   H  O          H  R  O                       H  |H   O | R  O
 \  |  //          \ |  //        enzyme         \ ||   //|  |  //
  N-C-C     +       N-C-C      ------------>      N-|C-C-N|-C-C     + H2O
 /  |  \           / |  \                        /  ||   /|  |  \
H   R  OH         H  H  OH                      H  |R___H| H  OH
   A                                                   B
```

2. The structural formula indicated by A represents a (an) (*a*) carbohydrate (*b*) protein (*c*) amino acid (*d*) complex fatty acid (*e*) simple sugar.
3. This chemical reaction is one of more than 3000 that occur in living cells. This type of reaction is called (*a*) dehydration synthesis (*b*) hydrolysis (*c*) anaerobic respiration (*d*) oxidation (*e*) glycolysis.
4. The grouping of atoms within area B and outlined in a box is called a(an) (*a*) peptide bond (*b*) radical (*c*) ionic bond (*d*) saturated hydrocarbon (*e*) valence group.
5. Which organisms carry out this chemical reaction? (*a*) only plants (*b*) only animals (*c*) plants and bacteria (*d*) plants, animals, and bacteria (*e*) animals and bacteria.

Part II

Base your answer to questions *1 through 6* on the information and diagrams below, library research, and on your knowledge of biology.

An investigation was performed to determine the ability of yeast to metabolize different carbohydrates. Four experimental tubes—A, B, C, and D—each containing 0.5 grams of yeast, were filled with water, as shown in the diagram. Glucose was added to tube A, maltose to tube C, and lactose to tube D in equal amounts. Then, the tubes were sealed. After 45 minutes at 37°C, the displacement of liquid in each tube was measured to determine the amount of gas collected. The results are shown in the diagrams below.

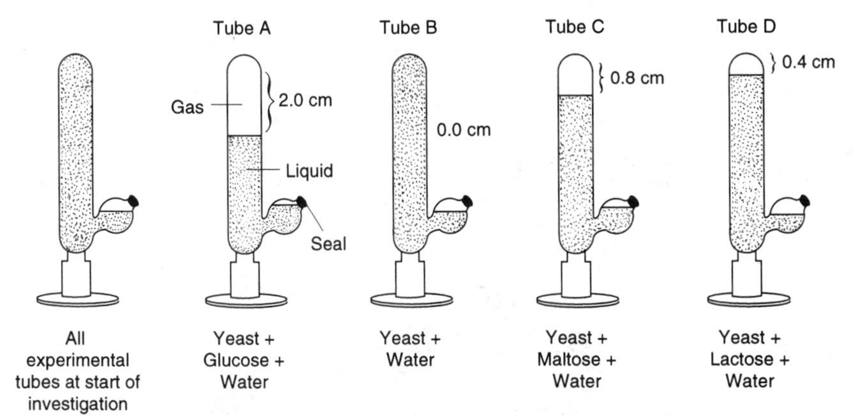

1. In this investigation, the gas collected and measured was most likely (*a*) hydrogen (*b*) oxygen (*c*) nitrogen (*d*) carbon dioxide (*e*) methane.

2. Based on the results of this investigation, which substrate did the yeast metabolize most efficiently? (*a*) glucose (*b*) water (*c*) lactose (*d*) maltose (*e*) air.

3. The metabolic process that produced the gas in the experimental tubes probably was (*a*) photosynthesis (*b*) dehydration synthesis (*c*) alcoholic fermentation (*d*) lactic acid fermentation (*e*) phosphorylation.

4. If the enzyme lactase is added to tube D at the beginning of the investigation, which prediction is probably correct? (*a*) tube D will resemble tube A (*b*) tube D will resemble tube B (*c*) tube D will resemble tube C (*d*) tube D will generate 4.0 cm gas (*e*) tube D will generate 1.6 cm gas.

5. In this investigation, the control is (*a*) tube A (*b*) tube B (*c*) tube C (*d*) tube D (*e*) not shown.

6. An intermediate substance formed during metabolism is (*a*) pyruvate (*b*) adenosine triphosphate (ATP) (*c*) phosphoglyceraldehyde (PGAL) (*d*) nicotinamide adenine dinucleotide (NAD) (*e*) all of the above.

4

Nutrition and Digestion

Learning Objectives

When you have completed this chapter, you should be able to:

- **Define** nutrients and identify the parts they play in living things.
- **Discuss** the importance of a balanced diet.
- **Explain** the term calorie and relate it to body needs.
- **Differentiate** between calorie and Calorie.
- **Distinguish** between intercellular and extracellular digestion.
- **Identify** the organs that make up the human digestive system.
- **Describe** ingestion and digestion in some other representative organisms.

OVERVIEW

The process of obtaining and using food is called nutrition. Organisms such as plants and some protists make food by carrying out photosynthesis. Other organisms, such as animals, fungi, bacteria, and some protists, obtain food that has been made by autotrophs, or from the bodies of other organisms. In this chapter, you will learn more about the nutrients organisms need to support life, and the ways in which they obtain and use these nutrients.

NUTRIENTS

Food contains usable organic compounds called *nutrients.* Examples of nutrients are carbohydrates, proteins, lipids, water, mineral salts, and vitamins. Table 4–1 summarizes the nutrients, their composition, uses in the body, sources, and chemical tests. The main functions of nutrients are the following:

- Provide energy for cells.
- Supply chemical compounds needed for growth and repair.
- Provide raw materials for the synthesis of enzymes, hormones, and other cell secretions.

VITAMINS

A *vitamin* is an organic compound needed in very small quantities for the proper functioning of an organism. Organisms must obtain vitamins from an outside source. An orange tree manufactures ascorbic acid (vitamin C), but because ascorbic acid is made by the tree, it is not considered a vitamin for the tree. In contrast, humans require vitamin C as the coenzyme of an important enzyme. Since it must be obtained from an outside source, ascorbic acid is considered a vitamin for humans.

Vitamins are either *fat-soluble* or *water-soluble*. Vitamins A, D, E, and K are fat-soluble. Vitamins of the B family (B complex) and vitamin C are water-soluble. Table 4–2 lists fat-soluble and water-soluble vitamins, their sources, and the results of vitamin deficiencies.

Tests for Vitamins

The two types of tests for vitamins are assay tests and chemical tests. An *assay test* is a controlled experiment conducted on two groups of animals of the same litter, sex, and age. The control group is fed a diet containing all of the vitamins required for good health. The experimental group is fed a diet that is deficient in a given vitamin. Figure 4–1 shows the

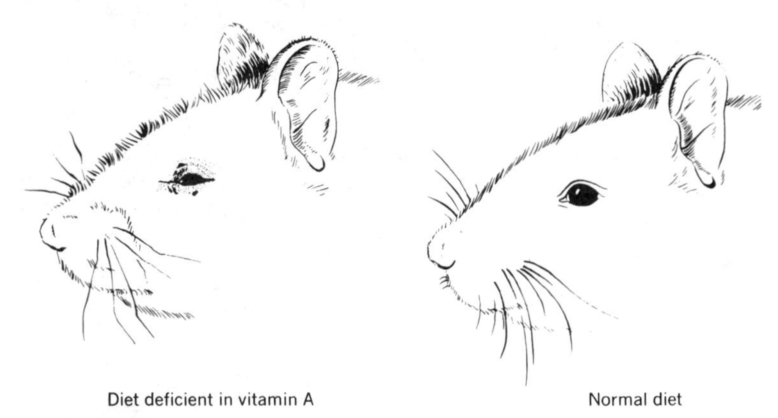

Figure 4–1 Vitamin assay test

Table 4–1 Nutrient Summary

Nutrient	*Composition*	*Uses in the Body*	*Rich Sources*	*Chemical Test*
Carbohydrates (sugars and starch)	Carbon, hydrogen, and oxygen; ratio of hydrogen atoms to oxygen atoms is 2 : 1	The chief source of energy; excess carbohydrates are stored for future use	Sugars: fruit, honey, candy, ice cream	Cover with Benedict's solution; heat strongly; orange-to-red color indicates glucose
			Starch: cereal grains, potatoes, bread	Add iodine solution; blue-black color indicates starch
Lipids (fats and oils)	Carbon, hydrogen, and oxygen; ratio of hydrogen atoms to oxygen atoms is not 2 : 1	A source of energy; excess fats and oils are stored for future use	Butter, lard, cream, bacon, meat, olive oil, nuts	Mix with the fat solvent perchloroethylene; pour liquid on unglazed paper; grease spot after liquid evaporates indicates fats
Proteins	Nitrogen in addition to carbon, hydrogen, and oxygen; some proteins also contain sulfur, iron, phosphorus, and other elements	Provide materials for assimilation and for growth and repair of all body cells; can supply energy	Milk, cheese, eggs, beef, liver, fish, peas, beans, nuts	Cover with dilute nitric acid; heat gently; yellow color develops; pour off acid; cover with ammonia water; orange color indicates protein

Nutrient	*Composition*	*Uses in the Body*	*Rich Sources*	*Chemical Test*
Water	Hydrogen and oxygen	Makes up more than 70 percent of protoplasm; it is the solvent in which the chemical reactions of protoplasm occur	Drinking water, milk, and other beverages; most foods, particularly fruits and vegetables	Heat food in test tube; condensed droplets on glass indicate water; verify with blue cobalt chloride paper, which turns pink in presence of water
Mineral salts	Chiefly calcium, phosphorus, iron, iodine, fluorine, sodium, and chlorine	Calcium and phosphorus make up hard parts of bones and teeth	Milk, eggs, cheese	Heat strongly in fireproof container until contents glow; allow to cool; white ash indicates minerals
		Iron makes up part of hemoglobin and cytochrome enzymes	Liver, eggs, beef, green vegetables	
		Iodine constitutes a large part of the secretion of the thyroid	Sea foods, iodized table salt	

(Continued on next page)

Table 4–1 *(Continued)*

Nutrient	*Composition*	*Uses in the Body*	*Rich Sources*	*Chemical Test*
Mineral salts (continued)	Chiefly calcium, phosphorus, iron, iodine, fluorine, sodium, and chlorine	Fluorine makes tooth enamel hard and resistant to decay	In some natural water, but often added	Heat strongly in fireproof container until contents glow; allow to cool; white ash indicates minerals
		Sodium takes part in the life functions of nerve and other cells	Table salt	
		Chlorine is a part of hydrochloric acid, necessary in digestion	Table salt	
Vitamins	Carbon, hydrogen, oxygen, nitrogen, and other elements	See Table 4–2		

results of this experiment on rats. *Chemical tests* involve color change. For example, a chemical test for vitamin A requires the mixing of food with a solution of antimony trichloride and chloroform. If vitamin A is present, the test solution will turn blue. Vitamin C can be detected by

Table 4–2 Vitamin Summary

FAT-SOLUBLE VITAMINS		
Vitamin	*Sources*	*Result of Deficiency*
Vitamin A (carotene is converted to vitamin A in the body)	Carrots and other yellow vegetables; whole milk, butter, eggs; leafy green vegetables, peas; fish-liver oils	Xerophthalmia, or "dry-eye" (an eye infection); night blindness (inability to see in dim light); increased susceptibility to infections of the nose, throat, and skin
Vitamin D (calcitriol)	Fish-liver oils; milk, eggs (Not commonly found in foods. Some foods are irradiated to increase their vitamin D content. If directly exposed to sunlight, the body is able to manufacture vitamin D. For this reason it is called "the sunshine vitamin.")	Rickets (bow-legs, knock-knees, swollen joints, especially wrists and ankles); badly formed teeth
Vitamin E (tocopherol)	Wheat germ; leafy green vegetables; meat; whole grain cereals	Sterility (inability to reproduce) in rats, and possibly in humans
Vitamin K	Leafy green vegetables	Hemorrhage (excessive bleeding, even from minor wounds; the body is unable to make prothrombin, a protein necessary for the normal clotting of blood)

(Continued on next page)

Table 4–2 *(Continued)*

	WATER-SOLUBLE VITAMINS		
	Vitamin	*Sources*	*Result of Deficiency*
B-complex	Vitamin B_1 (thiamin)	Whole grain cereals and enriched bread; beans and peas; yeast; milk; lean beef, liver, pork, poultry	Loss of appetite; limited growth; improper oxidation of foods, especially of carbohydrates; beriberi (exhaustion, paralysis, heart disease)
B-complex	Vitamin B_2 or G (riboflavin)	Yeast; lean beef, liver; milk, eggs; green vegetables; whole wheat; prunes	Inflamed lips; general weakness; eyes excessively sensitive to light (deficiency of riboflavin is often associated with pellagra)
B-complex	Niacin	Lean beef, liver; milk, eggs; leafy green vegetables, tomatoes; yeast	Pellagra (skin irritation, tongue inflammation, digestive and nervous disturbances)
B-complex	Vitamin B_{12} (cobalamin)	Liver, kidney; fish	Pernicious anemia; retarded growth; disorders of the nervous system
B-complex	Folic acid	Leafy green vegetables; yeast; meat	Some types of anemia
	Vitamin C (ascorbic acid)	Citrus fruits (oranges, grapefruits, lemons, limes); leafy green vegetables, tomatoes	Scurvy (soft and bleeding gums, loose teeth, swollen and painful joints, bleeding under the skin)

mixing food with blue indophenol. If the food contains vitamin C, blue indophenol will become colorless.

DIET

The food normally consumed by an organism is its *diet*. In general, an animal's diet provides the nutrients and fiber needed to maintain good health. Substances such as cellulose provide *fiber*, which cannot be digested. Fiber stimulates the muscles of the *alimentary canal*, or food tube. This muscular action keeps food moving through the food tube. In addition, the diet should be balanced and contain enough calories for optimum health of the organism.

Balanced Diet

A diet that includes a wide variety of foods containing sufficient amounts of the necessary nutrients is called a *balanced diet*. The food requirements of people vary with gender, age, size, occupation, and with the climate in which they live. During infancy, childhood, and adolescence, when growth is rapid, more proteins are required than in later life. People living in polar regions usually eat more fat than people in warmer areas. This provides them with a rich source of energy that helps them function in low temperatures.

Calories and Diet

A *calorie* is the amount of heat necessary to raise the temperature of one gram of water 1°C. The *kilocalorie* (1,000 calories), or Calorie, is often used instead of the calorie. A *Calorie* (spelled with an upper-case C) is the amount of heat required to raise the temperature of 1,000 grams (1 kilogram) of water 1°C. For example, it takes 5 Calories to raise the temperature of 1,000 grams of water from 15°C to 20°C.

The energy stored in food usually is expressed in Calories. This energy is released within cells when respiration, or oxidation reactions, break the carbon-carbon bonds and the carbon-hydrogen bonds in foods. A specific quantity of food is burned in a *bomb calorimeter* (Figure 4–2) to calculate the number of Calories. The food is placed in a chamber surrounded by a given quantity of water. Oxygen enters the chamber, and an electric spark

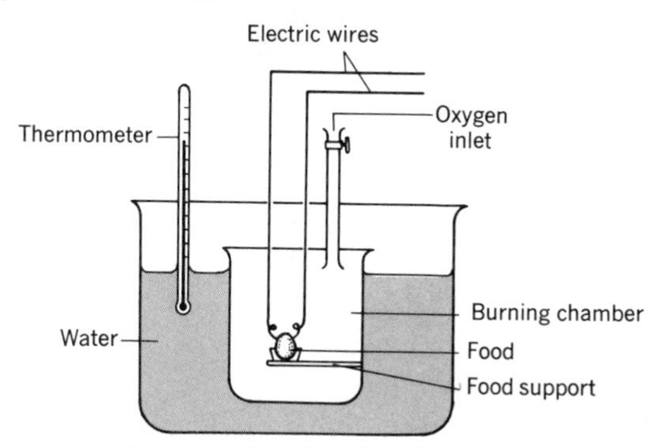

Figure 4–2 Bomb calorimeter

ignites the food. The water surrounding the chamber absorbs the heat given off as the food burns. The number of Calories released by the food can be calculated by measuring the increase in water temperature. A gram of fat releases about 9 Calories and a gram of glucose or protein releases about 4 Calories. Thus, fats and oils release more than twice as much energy as glucose or proteins.

INGESTION AND DIGESTION

Animals store, use, and transform energy. Organic chemicals, which are found in nutrients, are ingested and digested. *Digestion* decomposes, or breaks down, complex organic molecules into simpler molecules. This is accomplished by *hydrolysis*, which breaks chemical bonds by using water molecules (see Figure 3-8). Hydrolysis provides reactants for the thousands of chemical reactions that occur in each cell.

Types of Digestion

There are two types of digestion that occur in organisms. *Intracellular digestion*, or digestion inside a cell, occurs within a vacuole, as in protozoans (Figure 4–3*a*). *Extracellular digestion*, or digestion outside a cell, occurs within the cavity of a food tube, as in humans and other multicellular organisms (Figure 4–3*b*).

Phases of Digestion

In multicellular organisms, extracellular digestion occurs in two phases, a mechanical phase and a chemical phase.

Mechanical phase of digestion. The chewing and grinding actions of teeth and the muscular action of the walls of the alimentary canal break food down into smaller pieces in the *mechanical phase* of digestion. This is a physical change because the chemical composition of the food remains unchanged. Smaller pieces of food provide a larger total surface area, which aids chemical digestion.

Chemical phase of digestion. Large nutrient molecules are broken down to smaller molecules as a result of chemical reactions in the *chemical phase* of digestion. Hydrolysis, carried on by enzymes, breaks apart the chemical bonds holding together the units of large molecules. The units, or smaller molecules, can then pass into a cell through its plasma membrane. Minerals and vitamins do not require chemical digestion. Upon dissolving, mineral and vitamin particles are small enough to enter cells. The digestion of proteins, starches, and lipids is discussed below.

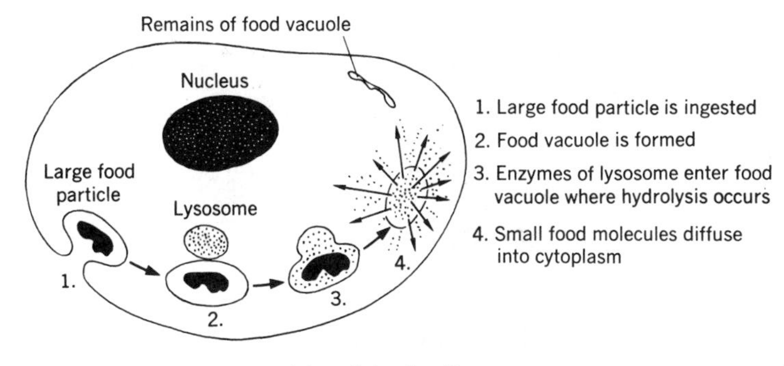

a. Intracellular digestion

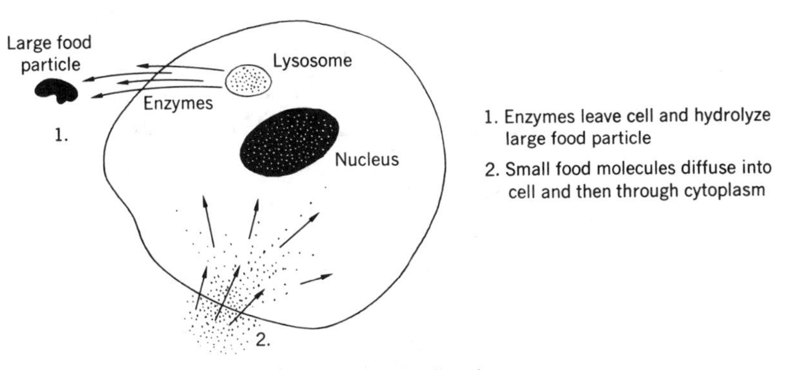

b. Extracellular digestion

Figure 4–3 Types of digestion

Digestion of proteins. Protein molecules, consisting of long chains of amino acids, are first hydrolyzed to smaller units called *polypeptides* by a *protease* (protein-digesting enzyme). Other enzymes then break down the polypeptides to amino acids. The digestion of proteins is summarized in the following equations:

$$\text{proteins} + \text{water} \xrightarrow{\text{protease}} \text{polypeptides}$$

$$\text{polypeptides} + \text{water} \xrightarrow{\text{enzymes}} \text{amino acids}$$

Digestion of starch. With the aid of an *amylase* (starch-digesting enzyme), starch molecules, consisting of long chains of glucose units, are first hydrolyzed to smaller units called *maltose* by an *amylase* (starch-digesting enzyme). Like sucrose and lactose, maltose is a double sugar and is too large to enter a cell. With the aid of *maltase* and similar enzymes, the double sugars are later broken down to the simple sugars—*glucose, fructose,* and *galactose*. The digestion of starch is summarized in the following equations:

$$\text{starch} + \text{water} \xrightarrow{\text{amylase}} \text{maltose}$$

$$\text{maltose} + \text{water} \xrightarrow{\text{maltase}} \text{glucose}$$

Digestion of lipids. Lipid molecules, consisting of three fatty acids bonded to a glycerin molecule, are hydrolyzed to form fatty acids and glycerin by a *lipase* (fat-digesting enzyme). The digestion of lipids is summarized in the following equation:

$$\text{lipids} + \text{water} \xrightarrow{\text{lipase}} \text{fatty acids} + \text{glycerin}$$

INGESTION AND DIGESTION IN HUMANS

Mechanical and chemical digestion begin in the *mouth*, or *oral cavity*. From the mouth, partially digested food particles pass through the organs of the alimentary canal. Figure 4–4 shows the organs of the alimentary canal and the glands that make up the *digestive system*. In the alimentary canal, the food is acted upon by various digestive juices secreted by different glands. Each digestive juice contains one or more enzymes. The organs of the alimentary canal into which digestive juices enter are the mouth, stomach, and small intestine.

Digestion in the Mouth

Teeth. Food is broken down into smaller pieces by four different types of teeth. The *incisor* teeth have sharp edges that are adapted for biting. The *canine* teeth have sharp points that are adapted for tearing. The *premolar* and *molar* teeth have broad, ridged surfaces that are adapted for crushing and grinding.

Tongue. Numerous sense organs, called *taste buds*, are located in different parts of the tongue. The taste buds are used to distinguish between sweet, salty, sour, and bitter substances. Stimulation of the taste buds usually results in the automatic secretion of saliva from the salivary glands. Saliva is a digestive juice that consists of water, mucus, and the enzyme *ptyalin* (an amylase). The water in saliva softens food and dissolves some minerals and vitamins. Mucus makes food slippery, which facilitates swallowing. Ptyalin hydrolyzes starch molecules to maltose.

Pharynx and Esophagus

Swallowing begins as a voluntary process. However, after food passes the *pharynx*, or upper part of the throat, swallowing becomes involuntary. As food moves downward from the pharynx to the *esophagus*, or gullet, the food passes over the *epiglottis*. This flaplike structure closes the *glottis*

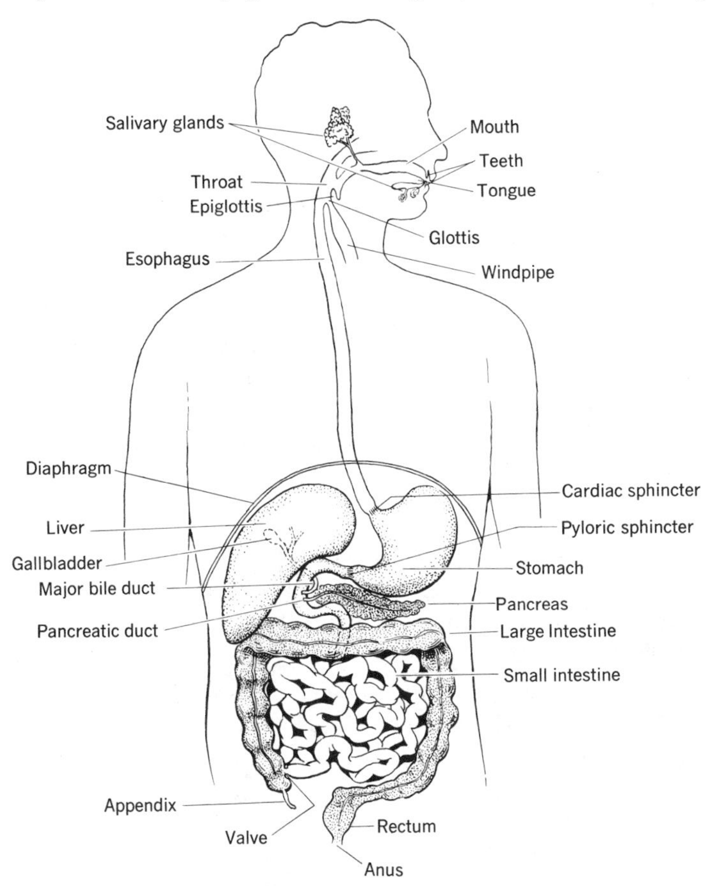

Figure 4–4 Human digestive system

(the entrance to the windpipe), preventing food from entering the windpipe. The action of the epiglottis directs food into the esophagus, where rhythmic, wavelike muscular contractions, called *peristalsis*, push food down the esophagus into the stomach.

Digestion in the Stomach

The stomach is a saclike, muscular organ that churns food and mixes it with the gastric juice. This juice is secreted by millions of tiny gastric glands in the stomach lining. Gastric juice consists of a mixture of water, mucus, hydrochloric acid, and the enzyme *pepsin* (a protease). Hydrochloric acid makes food strongly acidic (pH about 2.0).

Action of hydrochloric acid. As a result of chemical reactions, hydrochloric acid dissolves minerals in foods and changes the inactive form of pepsin, called *pepsinogen*, into an active enzyme. Hydrochloric acid also provides the optimum pH for the action of pepsin and kills many bacteria that often enter the stomach along with food.

Action of pepsin. Pepsin hydrolyzes proteins into polypeptides called *proteoses* and *peptones*. The souplike mixture of gastric juice, partially digested proteins, and other food substances in the stomach is called *chyme*.

Contractions of the stomach muscles (peristalsis) push food toward the *pyloric sphincter*, which is a valve located between the stomach and the small intestine. The texture of chyme regulates the opening and closing of the pyloric sphincter. When the chyme is sufficiently liquid, it passes through the valve into the small intestine; when solid particles of food are present, the valve remains closed.

Digestion in the Small Intestine

The small intestine is a long, narrow, folded tube that occupies most of the abdominal cavity. It is about 7.5 meters long and 2.5 centimeters wide. The small intestine is divided into three regions: the *duodenum* (about 25 centimeters long), the *jejunum* (about 3 meters long), and the *ileum* (about 4.5 meters long). When food passes from the stomach into the duodenum, certain cells lining the duodenum walls secrete a hormone called *secretin*. This hormone enters the blood and stimulates the liver and pancreas to secrete their digestive juices. The latter is an excellent example of homeostasis, or feedback control.

The liver. The liver is the largest gland in the body. It secretes *bile*, which enters the duodenum by way of a duct. Excess bile is stored in the gallbladder. Although it does not contain enzymes, bile is a digestive juice because it acts on drops of fat and oil by *emulsification*, a physical process that breaks large drops of fats and oils into tiny droplets. This results in an increased surface area upon which the enzyme lipase can act. Besides secreting bile, the liver has many other functions, including the decomposition of excess amino acids, the storage of excess glucose in the form of *glycogen*, and the breakdown of worn-out red blood cells.

The pancreas. The pancreas is a spearhead-shaped gland located near the duodenum. Most of the pancreatic tissue secretes pancreatic juice, which reaches the duodenum by way of the pancreatic duct. *Pancreatic juice* consists of a mixture of water, mucus, sodium bicarbonate, an amylase, a lipase, several proteases, and nucleases. The action of enzymes in pancreatic juice is best when the pH is about 8.0. This alkaline condition results from the presence of the sodium bicarbonate.

Action of the amylase. The amylase in pancreatic juice is called *amylopsin*, an enzyme that digests starch. Amylopsin completes the work begun by the ptyalin in saliva.

Action of the lipase. The lipase in pancreatic juice is called *steapsin*. This enzyme digests emulsified fats and oils by changing them into fatty acids and glycerin.

Action of the proteases. The main protease in pancreatic juice is called *trypsin*. Like pepsin, trypsin has an inactive form called *trypsinogen*, which is changed to the active form by the enzyme *enterokinase* found in intestinal juice. Trypsin and other proteases change proteins into polypeptides and simpler peptides. Proteoses and peptones (from the stomach) also are changed to simpler peptides. Many simple *peptides* consist of two joined amino acids (dipeptides).

Action of the nucleases. The nucleases act upon nucleic acids, which consist of double-stranded molecules of DNA and single-stranded molecules of RNA, and break them down into smaller nucleotide molecules.

Intestinal Glands

The lining of the small intestine contains numerous small glands that secrete *intestinal juice*. This juice consists of a mixture of water, mucus, and several enzymes that complete the digestion begun in other parts of the alimentary canal. The pH of intestinal juice is slightly alkaline (about 7.3). Among the enzymes in intestinal juice are *peptidase, maltase, sucrase,* and *lactase.*

Action of peptidase Peptidase changes polypeptides and dipeptides into individual amino acid molecules.

Action of maltase, sucrase, and lactase. Maltase changes maltose into glucose; sucrase changes sucrose into glucose and fructose; and lactase changes lactose into glucose and galactose.

Table 4–3 Action of Digestive Enzymes on Nutrients

Juice	*Gland*	*Place of Action*	*Enzymes*	*Substrates*	*Intermediate Products*	*End Products*
Saliva	Salivary	Mouth	Ptyalin (amylase)	Starch	Maltose	
Gastric*	Gastric	Stomach	Pepsin (protease)	Proteins	Polypeptides and peptides	
Bile	Liver	Small intestine	None	Lipids (fats and oils)	Emulsified lipids	
Pancreatic	Pancreas	Small intestine	Trypsin (protease)	Proteins	Polypeptides	
			Amylopsin (amylase)	Starch	Maltose	
			Steapsin (lipase)	Emulsified lipids		Fatty acids and glycerol
			Nucleases	Nucleic acids		Nucleotides
Intestinal	Intestinal	Small intestine	Peptidases	Polypeptides and dipeptides		Amino acids
			Maltase	Maltose		Glucose
			Sucrase	Sucrose		Glucose and fructose
			Lactase	Lactose		Glucose and galactose

*Hydrochloric acid (not an enzyme) is a component of gastric juice that helps dissolve minerals.

Chemical digestion, which starts in the mouth, is finally completed in the small intestine. Table 4–3 summarizes the action of digestive enzymes on nutrients.

The Large Intestine

The large intestine is about 6 centimeters wide and about 1.8 meters long. At the junction of the small intestine and the large intestine is a small fingerlike extension called the *appendix*, which plays no part in human digestion. However, the appendix can cause problems if it becomes infected; it must then be removed surgically.

The material entering the large intestine lacks digested nutrients because they were absorbed into the bloodstream from the small intestine. Most of the material in the large intestine consists of undigested food and indigestible substances such as cellulose and water. Most of the water in the large intestine is absorbed and passed into the bloodstream. The remainder of the water and the other materials are stored temporarily in the rectum, and then removed from the body through the *anus*.

In addition to cellulose, the large intestine harbors large numbers of bacteria that feed on undigested particles. The bacteria manufacture vitamins B_{12} and K, which are useful to the body.

Section Quiz

1. Which vitamin is fat-soluble? (*a*) K (*b*) C (*c*) B_{12} (*d*) B_2.
2. Hydrolyzed proteins include (*a*) polypeptides (*b*) proteoses (*c*) amino acids (*d*) all of these.
3. Excess bile is stored in the (*a*) gall bladder (*b*) duodenum (*c*) liver (*d*) glottis.
4. The enzyme steapsin acts on (*a*) carbohydrates (*b*) polypeptides (*c*) emulsified lipids (*d*) nucleic acids.
5. Which part of the human alimentary canal contains large numbers of helpful bacteria? (*a*) duodenum (*b*) ileum (*c*) large intestine (*d*) appendix.

INGESTION AND DIGESTION IN OTHER ORGANISMS

Protozoans

Different protozoans have developed different adaptations by which they ingest food. For example, an ameba forms pseudopods (false feet) that engulf a food particle. Then a food vacuole forms within the cytoplasm of the ameba (Figure 4–5). A paramecium, using cilia, sweeps food into its oral groove and then into its gullet (Figure 4–6). At the end of the gullet, a food vacuole is formed.

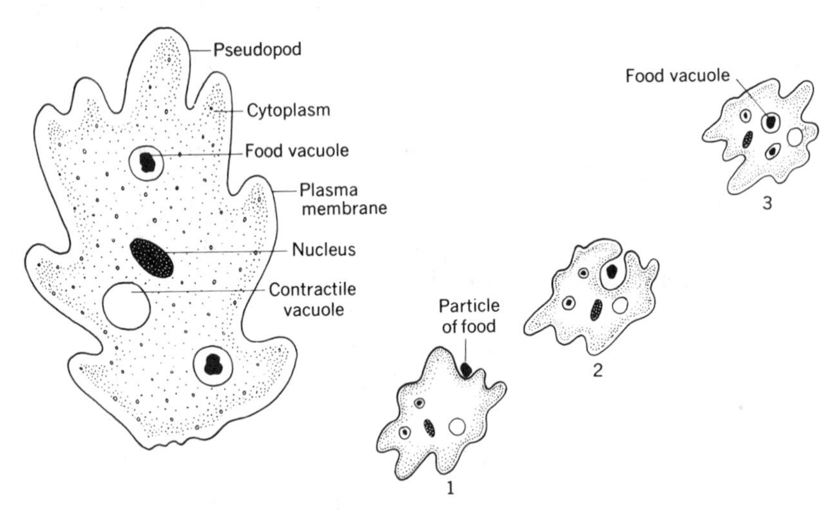

Figure 4–5 Ameba

Digestion is intracellular in protozoans. In protozoans, the food vacuole breaks away and is carried to different regions of the cytoplasm. Eventually, a lysosome unites with the food vacuole and the membranes between them dissolve. The enzymes of the lysosome enter the food vacuole, where they digest the food into usable substances.

Hydra

The mouth of a hydra is surrounded by tentacles that contain numerous stinging cells called *nematocysts* (Figure 4–7). When a hydra is stimulated by food, such as a tiny water flea, the nematocysts shoot out fine, hollow threads filled with a poison. The poison paralyzes the water flea, which is then pushed into the hydra's mouth by the tentacles. From the mouth, food enters the *digestive,* or *gastric, cavity*. There are several types

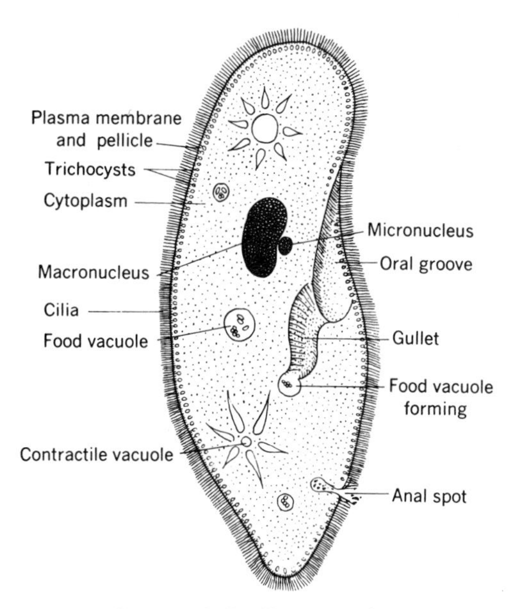

Figure 4–6 Paramecium

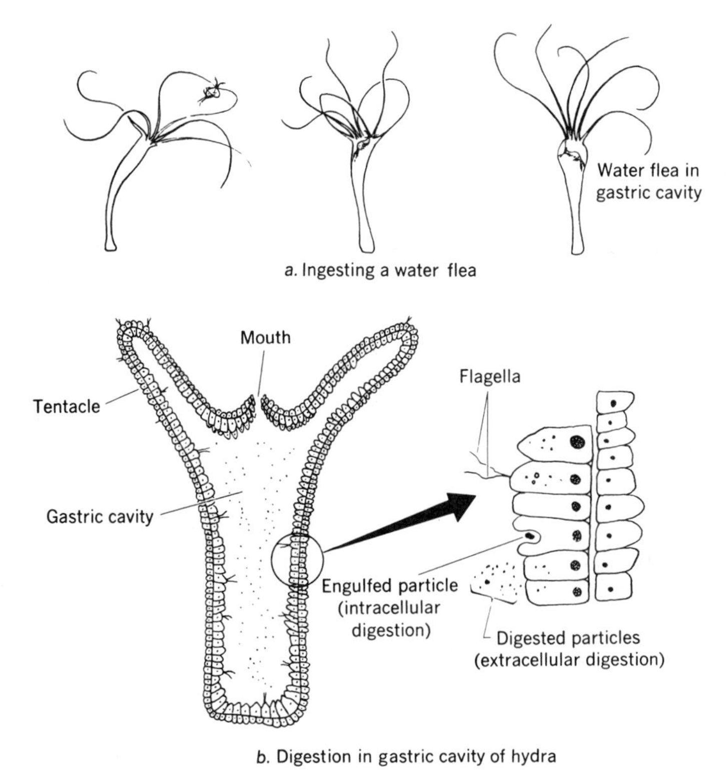

Figure 4–7 Ingestion and digestion in the hydra

of cells that line the gastric cavity: cells that form pseudopods and engulf particles of food (intracellular digestion); cells that empty enzymes into the gastric cavity (extracellular digestion); and cells bearing whiplike flagella that create water currents by moving back and forth. The current circulates food in the gastric cavity and helps eliminate undigested material through the mouth.

Earthworm

As an earthworm burrows through soil, the suctionlike action of its pharynx draws soil (containing organic matter and sand) into its mouth. From the mouth, material passes into the pharynx, the esophagus, and then the *crop*, which is a temporary storage organ (Figure 4–8). From the

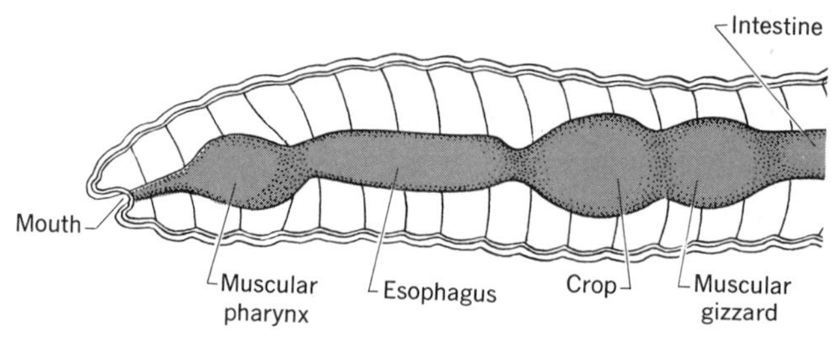

Figure 4–8 Digestive system of the earthworm

crop, the organic and inorganic material passes into the muscular *gizzard*, which grinds the food material into small bits aided by sand present in the soil. The material then passes into the intestine, where enzymes secreted from glands lining the intestine chemically digest the food particles (extracellular digestion). Undigested food and sand are eliminated through the anus. Peristalsis helps push the food and inorganic material through the earthworm's alimentary canal.

Grasshopper

Figure 4–9 shows the mouthparts of a grasshopper, which include the upper lip (*labrum*); the lower lip (*labium*); the jaws (*mandibles* and *maxillae*); and two pairs of taste organs (*palpi*). With the help of the mouthparts and the muscular action of the pharynx, pieces of food are drawn into the grasshopper's mouth and then into its pharynx (Figure 4–10). In the pharynx, food is mixed with saliva secreted by several salivary glands. Peristalsis moves the food from the pharynx into the esophagus and then into the *crop*. Food then moves into the

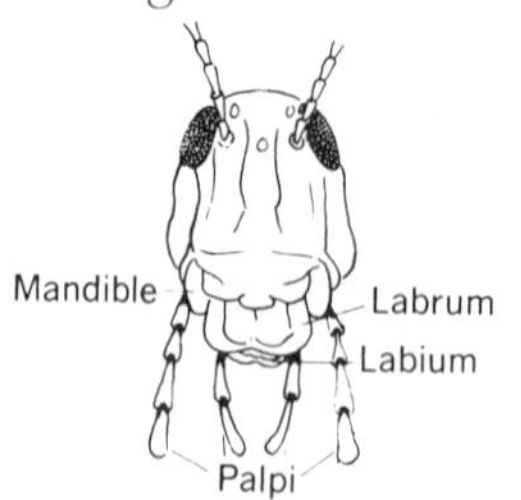

Figure 4–9 Mouthparts of the grasshopper

muscular *gizzard*, where the food is ground into smaller bits before entering the *stomach*. Digested food is absorbed in the stomach; undigested material passes into the *intestine* and *rectum*, and is expelled through the anus. As in the earthworm, digestion in a grasshopper is extracellular.

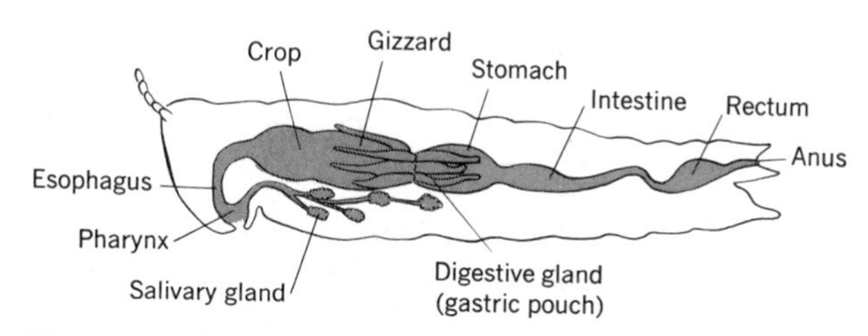

Figure 4–10 Digestive system of the grasshopper

Bird

The digestive system of birds, like that of all *vertebrates* (animals with backbones) and many higher *invertebrates* (animals lacking backbones), consists of an alimentary canal and associated digestive glands. The alimentary canal includes a *crop* and a *gizzard*, which physically digests food with the aid of small bits of ingested gravel. Most of the chemical digestion occurs in the small intestine. Undigested material and urine exit the body through a common passageway called a *cloaca* (see Figure 4–11).

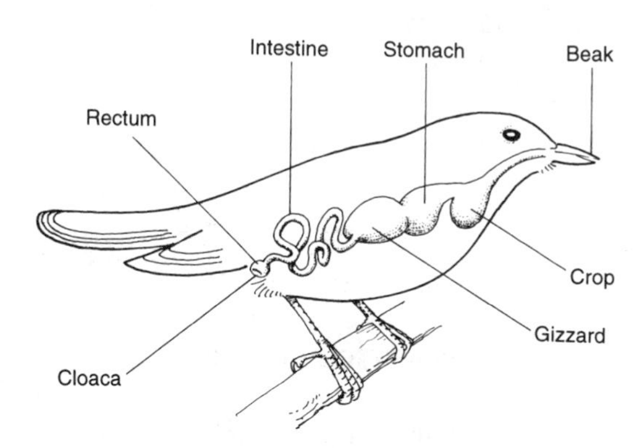

Figure 4–11 Digestive system of the bird

Table 4–4 on page 108 compares the adaptations for ingestion and digestion in the organisms discussed in this chapter. Note that in protozoans, digestion is only intracellular. In multicellular organisms, such as hydra, digestion is intracellular and extracellular. In animals that have a food tube, digestion is extracellular. Chapter 5 discusses the phase of nutrition in which the end products of digestion are transported to cells.

Table 4–4 Comparison of Adaptations for Ingestion and Digestion

Organism	*Adaptations for Ingestion*	*Type of Digestion*	*Adaptations for Digestion*
Protozoans			
Ameba	Pseudopods	Intracellular	Food vacuole
Paramecium	Cilia, oral groove	Intracellular	Food vacuole
Hydra	Tentacles with nematocysts, mouth	Extracellular and intracellular	Gastric cavity, gland cells, and pseudopod-forming cells
Earthworm	Mouth, pharynx	Extracellular	Digestive system (gizzard, intestine)
Grasshopper	Mouthparts, lips, mandibles, and maxillae	Extracellular	Digestive system (gizzard, intestine)
Bird	Mouth, beak, tongue, and crop	Extracellular	Digestive system (gizzard, intestine)
Human	Mouth, lips, tongue, and teeth	Extracellular	Digestive system

Section Quiz

1. A cloaca is part of the alimentary canal of a (*a*) bird (*b*) hydra (*c*) grasshopper (*d*) human.
2. A bird and an earthworm both possess a mouth, crop, gizzard, and anus. It is logical to state that (*a*) they are closely related (*b*) they both possess an alimentary canal (*c*) both animals use lysosomes to digest food (*d*) the gizzards have different functions.

3. Digestion in hydra is (*a*) intracellular (*b*) extracellular (*c*) intracellular and extracellular (*d*) intercellular.
4. Nematocysts help hydra (*a*) digest food (*b*) ingest food (*c*) capture food (*d*) assimilate food.
5. Palpi are the taste organs found in (*a*) birds (*b*) frogs (*c*) humans (*d*) grasshoppers.

Chapter Review Questions

The following questions will help you check your understanding of the material presented in the chapter.

1. Which statement best identifies heterotrophs? (*a*) They synthesize chlorophyll. (*b*) They extract minerals from the soil. (*c*) They consume organic compounds. (*d*) They make organic compounds from inorganic materials.
2. Humans can synthesize few (*a*) fats (*b*) proteins (*c*) hormones (*d*) vitamins.
3. In humans, an end product of lipid digestion is (*a*) maltose (*b*) glucose (*c*) glycerol (*d*) amino acids.
4. In animal cells, amino acids are end products resulting from the digestion of (*a*) lipids (*b*) starches (*c*) proteins (*d*) fatty acids.
5. Amino acids are required in the human diet mainly for the synthesis of (*a*) proteins (*b*) sugars (*c*) starches (*d*) lipids.
6. Fats are digested to fatty acids and glycerin by the action of (*a*) maltase (*b*) protease (*c*) lipase (*d*) amylase.
7. In the cells of muscle and liver tissues, sugars are converted to a storage product called (*a*) protein (*b*) glycogen (*c*) cellulose (*d*) glucose.
8. After a food particle has entered a cell by pinocytosis, the next process to occur would be (*a*) ingestion (*b*) digestion (*c*) respiration (*d*) synthesis.
9. An end product of carbohydrate digestion is (*a*) glycerin (*b*) a fatty acid (*c*) glucose (*d*) an amino acid.
10. A product formed by hydrolysis under the control of an amylase is (*a*) $C_3H_8O_3$ (*b*) $C_6H_{12}O_6$ (*c*) $C_3H_7O_2N$ (*d*) $C_{18}H_{36}O_2$.

For each body function given in questions *11 through 14*, select the number of the organ, chosen from the drawing at the right, that best performs that function.

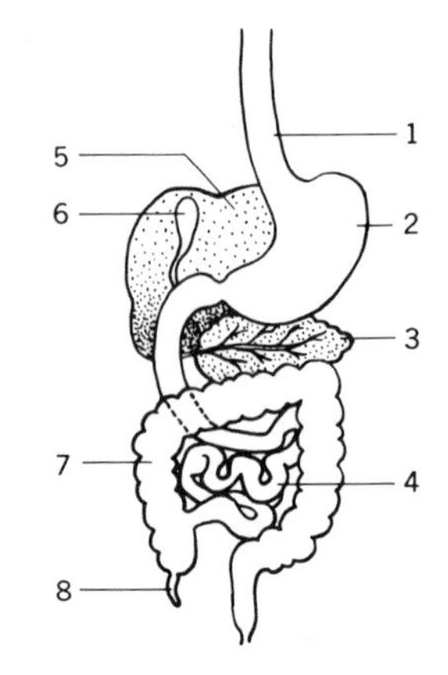

11. Initiates peristalsis.
12. Secretes pancreatic juice.
13. Begins protein digestion.
14. Absorbs most water.
15. As a result of normal liver function, there is no prolonged increase in blood sugar after meals. Thus, the liver helps maintain (*a*) homeostasis (*b*) extracellular digestion (*c*) assimilation (*d*) intracellular digestion.
16. In which part of the human alimentary canal does most water absorption occur? (*a*) mouth (*b*) stomach (*c*) large intestine (*d*) small intestine.
17. In protozoans, digestion occurs in the (*a*) food vacuole (*b*) stomach (*c*) contractile vacuole (*d*) ribosome.
18. In general, the digestive system of an earthworm is most similar to that of a (*a*) hydra (*b*) human (*c*) paramecium (*d*) mushroom.
19. In ameba, most hydrolysis takes place in (*a*) ribosomes (*b*) mitochondria (*c*) food vacuoles (*d*) contractile vacuoles.
20. Which organism does not carry on extracellular digestion? (*a*) paramecium (*b*) earthworm (*c*) hydra (*d*) grasshopper.
21. The structure of a grasshopper's digestive system is similar to a (*a*) sac (*b*) spider's web (*c*) tube (*d*) bunch of grapes.
22. In the section of an earthworm shown in the diagram below, the intestine is located at (*a*) A (*b*) B (*c*) C (*d*) D.

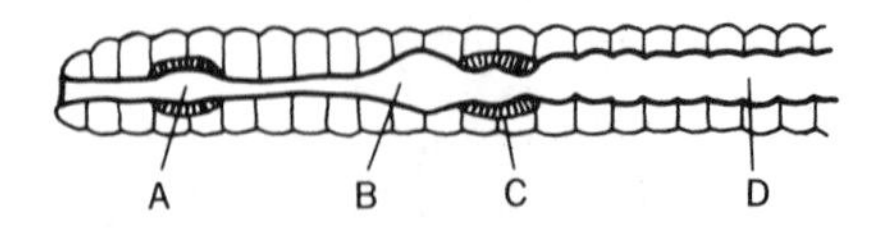

23. In the "tube within a tube" body plan of an earthworm, grasshopper, and human, the inner tube is called the (*a*) nerve tube (*b*) respiratory tube (*c*) excretory tube (*d*) digestive tube.

Biology Challenge

The following questions will provide practice in answering SAT II-type questions.

Part I

Select the letter of the statement that best completes the sentence.

1. The enzyme precursor, *pepsinogen*, changes to its active form *pepsin* when the pH of hydrochloric acid is about (*a*) 5.70 (*b*) 4.50 (*c*) 4.03 (*d*) 3.60 (*e*) 2.10.
2. Which substances released into the small intestine would aid chemical digestion? (*a*) bile, pancreatic juice, intestinal juice (*b*) hydrochloric acid, pancreatic juice, intestinal juice (*c*) salivary amylase, intestinal juice, pancreatic juice (*d*) bile, hydrochloric acid, salivary amylase (*e*) bile, secretin, pancreatic juice
3. What do the digestive processes of grasshoppers, earthworms, birds, and humans have in common? They each (*a*) rely on the liver to secrete hydrolytic enzymes (*b*) store partially digested food for digestion at a later time (*c*) use specialized organs for mechanical and chemical digestion (*d*) absorb end products of digestion through their caecum (*e*) use hydrolytic enzymes that work best at 37°C.
4. You can test for the presence of vitamin A in a food sample by adding a mixture of antimony trichloride and chloroform to the sample. The main reason for using chloroform is that (*a*) vitamin A is water soluble (*b*) vitamin A is fat soluble (*c*) antimony trichloride requires an organic solvent (*d*) antimony trichloride cannot react directly with vitamin A (*e*) other cell substances may interfere with the reaction.

Part II

The following two procedures are associated with possible objectives for each of two investigations. Select the letter of the statement that best describes the objective of each procedure.

1. At different times, 5 grams of olive oil, 5 grams of cane sugar, and 5 grams of lean chicken meat were placed in a calorimeter. Observations were recorded for each food sample to determine (*a*) that energy is stored in nutrients (*b*) that energy is released by nutrients

(*c*) the number of kilocalories released by nutrients (*d*) if olive oil releases the most kilocalories per gram (*e*) if cane sugar releases the most kilocalories per gram.

2. A test tube was partly filled with starch suspension. A small drop of iodine solution was added to the starch suspension and the contents of the test tube were mixed. The starch-iodine mixture was then divided into two test tubes. A dropperful of amylase solution was added to one test tube, and a dropperful of water was added to the other test tube. Each test tube was stoppered, labeled, and mixed. Both test tubes were placed in a hot-water bath at 37°C. Results should indicate that (*a*) a control is not required for certain investigations (*b*) this was a test for starch (*c*) amylase hydrolyzes starch to sugars (*d*) water interferes with digestion (*e*) an optimum temperature is required for starch digestion.

Part III

Questions *1 through* 5 are based on the graph below, which shows the relationship between blood glucose level and time of day.

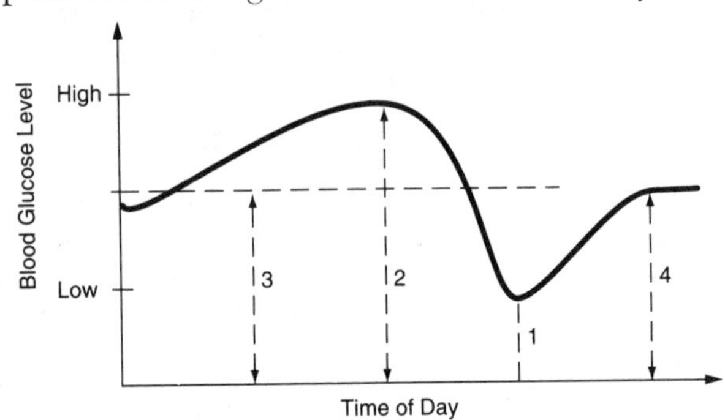

1. Which number on the graph represents a normal blood glucose level? (*a*) 1 (*b*) 2 (*c*) 3 (*d*) 4.
2. Which number on the graph represents a between-meal snack? (*a*) 1 (*b*) 2 (*c*) 3 (*d*) 4.
3. Which number on the graph represents blood glucose level after a full meal? (*a*) 1 (*b*) 2 (*c*) 3 (*d*) 4.
4. Which number on the graph represents the time between meals? (*a*) 1 (*b*) 2 (*c*) 3 (*d*) 4.
5. Which number on the graph represents a time when the liver is most active? (*a*) 1 (*b*) 2 (*c*) 3 (*d*) 4.

5

Circulatory System

Learning Objectives

When you have completed this chapter, you should be able to:

- **Describe** the process of absorption in humans.
- **Relate** the circulation of blood to the delivery of nutrients throughout the body.
- **Discuss** the structure of the heart and describe its functions.
- **Differentiate** between arteries, veins, capillaries and lymph vessels.
- **Explain** pulmonary and systemic circulation.
- **Identify** the different components of blood and their functions.
- **Compare** the circulatory systems of representative animals to that of humans.

OVERVIEW

An organism's survival depends upon the transport, or carrying, of useful materials (such as nutrients) to its cells and the transport of metabolic wastes away from its cells. Transport also includes the passage of materials into and out of cells by way of the plasma membrane. In this chapter, you will learn how an organism's circulatory system contributes to its survival.

ABSORPTION IN HUMANS

Absorption is the process by which glucose, amino acids, fatty acids, glycerol, mineral salts, water, vitamins, oxygen, and the other end products of digestion move from the small intestine into the blood by active and passive transport. *Circulation* involves the movement of blood in major pathways (arteries, veins, capillaries, and lymph vessels) and a pump (heart) to transport dissolved materials to and from all cells.

In humans, the end products of digestion are passed into the blood as it flows through the small intestine. The small intestine has several adaptations for absorption, including its length, narrow diameter, villi and microvilli, and the presence of smooth muscles in its walls.

Dimensions and Structure of the Small Intestine

The great length of the small intestine provides a large internal surface area for absorption. Its narrow diameter ensures close contact between the internal surface of the small intestine and the end products of digestion.

Villi

Villi (singular, villus) are microscopic, fingerlike projections from the internal surface of the small intestine (Figure 5–1). The epithelial cells of the villi are fringed with *microvilli*, which are threadlike extensions of the cell membrane. The microvilli increase the surface area and the rate of absorption.

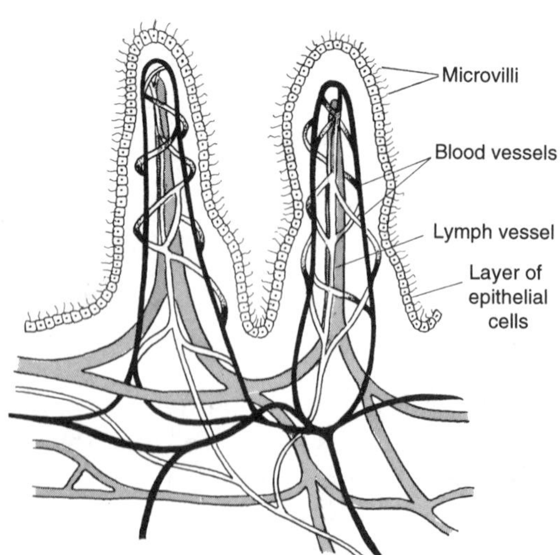

Figure 5–1 Villi

Absorption occurs in a villus by means of passive and active transport, in which the end products of digestion pass through the outer layer of epithelial cells of the villus and become part of the lymph within the villus. From the lymph, most end products (excluding some fatty acids) enter the blood within the *capillaries* (microscopic blood vessels) and then are carried to *venules* (tiny veins). Venules carry the blood with the dissolved end products to larger veins that eventually carry blood to the heart. Complex fatty acids enter the lymph within the *lacteals* (microscopic lymph vessels). Then the dissolved fatty acids travel to increasingly larger lymph vessels, which join certain veins in the neck region. The veins in the neck region join a larger vein that leads to the heart. The heart pumps blood rich in nutrients to all parts of the body.

Smooth Muscles

The absorption of the end products of digestion is aided by *peristalsis*. These contractions of the smooth muscles of the small intestine bring the end products into close contact with the absorbing surface of the small intestine. Peristalsis also moves the fluid food mass over the villi and forces the villi into the food mass.

CIRCULATION IN HUMANS

In 1628, *William Harvey* demonstrated that the blood circulates in a closed system of blood vessels to all parts of the body. This circulatory system consists of a heart, arteries, veins, capillaries, lymph vessels, blood, and lymph.

The Heart

Figure 5–2 shows the structure, function, and action of a human heart.

Structure of the heart. Most of the *heart* consists of cardiac muscle tissue, called the *myocardium*. The myocardium is covered by an elastic protective membrane called the *pericardium*. The inner, smooth surface of the heart, or the *endothelium*, is composed mainly of epithelial and connective tissues.

The four chambers of the heart are enclosed by the walls composed mainly of cardiac muscle tissue. The two upper chambers and the two lower chambers are separated from each other by a wall. Each upper chamber connects with a lower chamber through a passageway lined with a valve.

The thin-walled upper chambers, called *atria* (singular, atrium), receive blood, which they pump into the ventricles. The thick-walled lower

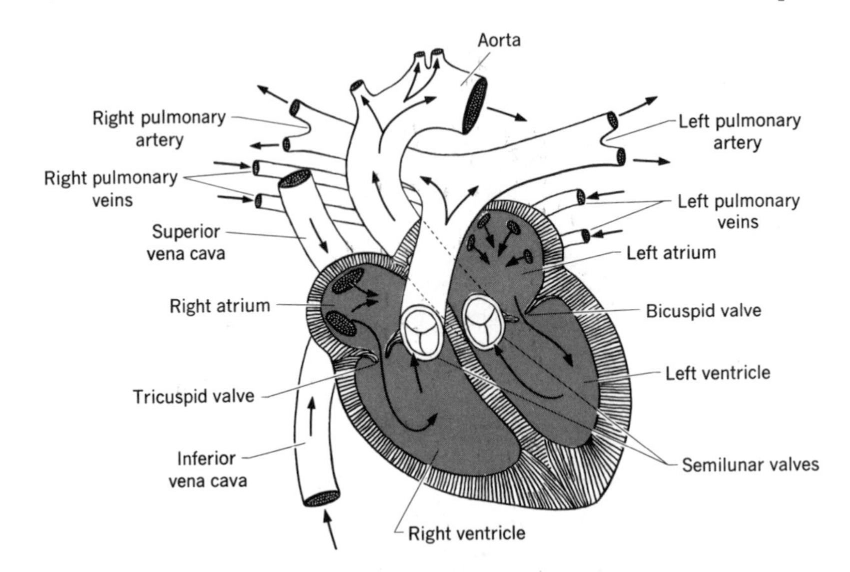

Figure 5–2 Structure of the heart

chambers, called *ventricles,* pump blood out of the heart. In the passageway between the left atrium and the left ventricle is the *bicuspid,* or *mitral, valve*; it prevents the backflow of blood into the atrium when the ventricle contracts. In the passageway between the right atrium and the right ventricle is the *tricuspid valve*; its function is similar to that of the bicuspid valve.

Function of the heart. The heart is a pump that distributes blood to all parts of the body. Its rhythmic beats are regulated by a *pacemaker*, groups of specialized cells in the right atrium and in the wall that separates the ventricles. Two pairs of nerves from the brain also regulate the heartbeat. When it is too rapid, a pair of *vagus nerves* slows down the heartbeat, and a pair of *accelerator nerves* speeds up the heartbeat when it is too slow.

Cardiac Cycle

A heartbeat is a cardiac cycle that features a period of contraction called *systole* and a period of relaxation called *diastole*. During diastole, the right atrium receives blood from the head and neck through a large vein called the *superior vena cava* and from the rest of the body through another large vein called the *inferior vena cava*. At the same time, the left atrium receives blood from the lungs through the pulmonary veins. As both atria receive blood, some blood begins to fill both ventricles.

During the first part of the systole, both atria contract simultaneously, which squeezes residual blood into the ventricles. During the second part of the systole, both ventricles, filled with blood, contract forcibly. The right ventricle pushes deoxygenated, or oxygen-poor, blood through the *pulmonary artery* (toward the lungs). At the same time, the left ventricle pushes oxygenated, or oxygen-rich, blood through the *aorta* toward all other parts of the body. After both atria and ventricles relax briefly, the cardiac cycle continues (see Figure 5–3).

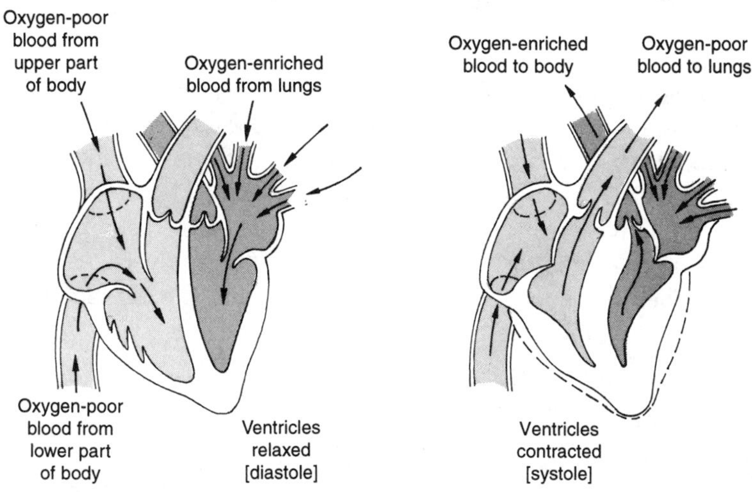

Figure 5–3 The heart during diastole and systole

The tricuspid valve and bicuspid valve close after diastole, which prevents blood from flowing backward into the atria. At the entrance to the aorta and the pulmonary artery, the *semilunar valves* prevent blood from flowing backward into ventricles following systole.

Blood Vessels and Lymph

Arteries. *Arteries* carry blood away from the heart and usually are located deep inside the body. Excluding the pulmonary artery, every artery carries oxygen-rich blood. Arteries possess a thick muscular layer (Figure 5–4) that contracts rhythmically, producing a *pulse*. The pulse helps blood circulate through arteries and causes blood to spurt out of a cut artery.

Veins. *Veins* carry blood to the heart and usually are closer to the body surface than arteries. Excluding the *pulmonary veins*, which lead away from the lungs, every vein carries oxygen-poor blood. Veins are made of

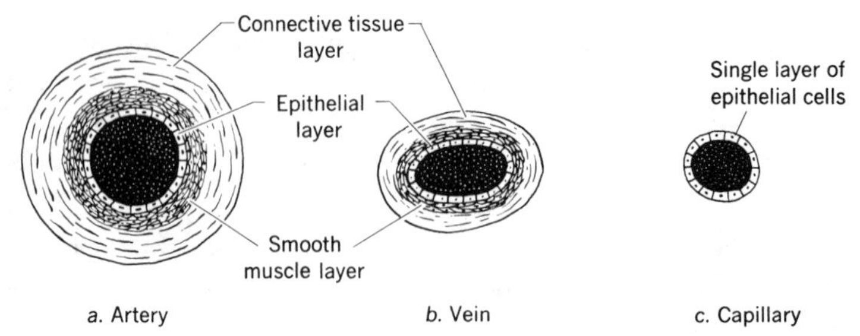

Figure 5–4 Comparison of the three types of blood vessels

less muscle tissue and are thinner than arteries (Figure 5–4). Unlike arteries, veins have valves that prevent backward flow (Figure 5–5).

Blood flows more smoothly in veins than in arteries because veins are farther from the pumping action of the heart. Blood in the lower part of the body and in the arms is moved through veins by the action of muscles attached to the skeleton. These muscles squeeze the veins as the body moves. When you inhale, the lungs squeeze veins, forcing blood toward the heart.

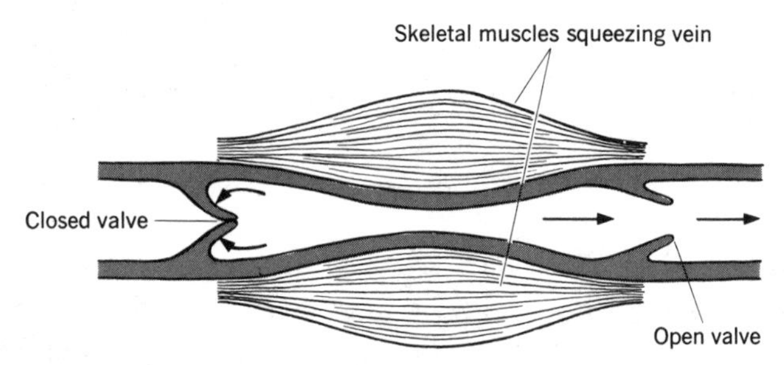

Figure 5–5 Blood flow in a vein

Arterioles and venules. Arteries branch out many times into smaller blood vessels called *arterioles*. These narrow blood vessels are made of rings of smooth muscle. Unlike arterial walls, which feature a continuous layer of smooth muscle, the rings of arteriole walls enable them to contract and recoil when stimulated by either the nervous system or endocrine (hormone) system. As a result, blood is forcibly propelled through the *capillaries*, or tiny vessels connecting arteries and veins.

Blood flows from the capillaries into *venules*, whose walls are thinner and less elastic than arterioles. Blood pressure in venules is less than in arterioles. Consequently, blood flows evenly from venules into veins and then back to the heart.

Capillaries. The microscopic *capillaries* are the most numerous blood vessels in the body. Capillaries thread their way through tissues, passing close to most cells. The walls of capillaries are composed of a single layer of flattened epithelial cells (Figure 5–4). The exchange of substances, including oxygen, carbon dioxide, and soluble nutrients, takes place between *interstitial* (intercellular) fluid and the circulatory system through the walls of capillaries (see Figure 5–6). Cellular metabolic wastes enter the interstitial fluid and then flow into capillaries before entering the general circulation. Passive and active transport mechanisms are involved in all exchanges.

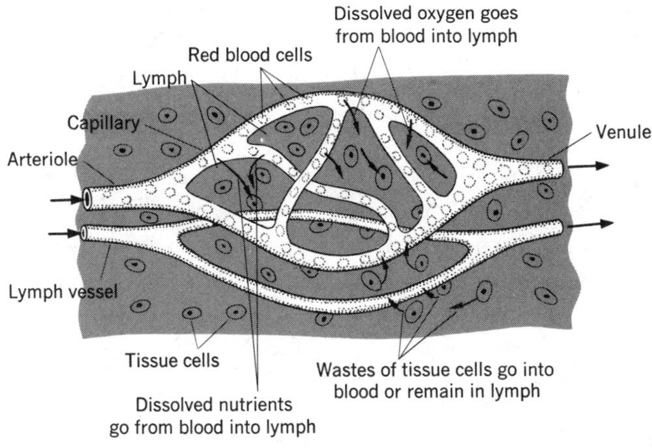

Figure 5–6 Exchange of materials between blood and tissue cells

Two opposing forces influence the movement of fluids between capillaries and tissue cells. One force is the pressure difference between the fluid in capillaries and the interstitial fluid. As a result of the higher pressure in capillaries, an outward flow of substances takes place from capillaries to interstitial fluid. A second force is the difference in water concentration between fluid in capillaries and in interstitial fluid. More water usually is found in interstitial fluid. Thus, water and dissolved wastes move from the interstitial fluid into the capillaries.

Lymph. The interstitial fluid that bathes every body cell is called *lymph*. Lymph is the fluid through which materials are exchanged between the cells of tissues and the blood. The composition of lymph is similar to blood except for a lack of red blood cells and certain complex protein molecules.

Lymph vessels accompany every artery, vein, and capillary, forming an extensive network throughout the body. All lymph vessels carry lymph in one direction from the body organs toward the heart. Tiny lymph vessels, similar to lacteals, carry lymph from tissue spaces to larger lymph vessels called *lymphatics*. For example, the *thoracic duct* is a lymphatic that carries lymph upward across the chest region and then empties into a large vein near the base of the neck. Here, lymph becomes part of the blood. Like veins, lymph vessels depend upon muscular movements and breathing movements for circulation. Lymph vessels also possess valves that prevent the backflow of fluid.

Pea-shaped structures found at intervals along the lymph vessels are called *lymph nodes*, or *lymph glands*. Lymph nodes help protect the body by filtering bacteria from lymph and by producing certain types of white blood cells. These white blood cells destroy the bacteria removed by the lymph nodes. The swollen glands that often accompany an illness are lymph glands that have enlarged as a result of fighting off a bacterial infection.

The *spleen*, which resembles a giant lymph node, is located near the stomach. In addition to removing bacteria from lymph, the spleen is a temporary storage area for red blood cells. When excessive bleeding occurs, the spleen contracts and puts the stored red cells into circulation.

Section Quiz

1. Lymph does not contain (*a*) plasma proteins (*b*) fatty acids (*c*) red blood cells (*d*) white blood cells.
2. Microvilli are useful because they (*a*) increase surface area for absorption (*b*) decrease capillary connections (*c*) replace worn-out villi (*d*) trap large molecules for pinocytosis.
3. Ringlike muscles in arteriole walls (*a*) add strength to resist systolic pressure (*b*) help propel blood into capillaries (*c*) prevent backflow of blood (*d*) are composed of cardiac muscle tissue.
4. The term that does *not* belong with the others is (*a*) cardiac cycle (*b*) systole (*c*) diastole (*d*) pulse.
5. Interstitial fluid is composed mainly of (*a*) wastes (*b*) water (*c*) plasma proteins (*d*) complex fatty acids.

Circulation of Blood: Circuits

The pulmonary artery and the aorta carry the blood into branch circuits, or circulation. Figure 5–7 shows the major blood circuits.

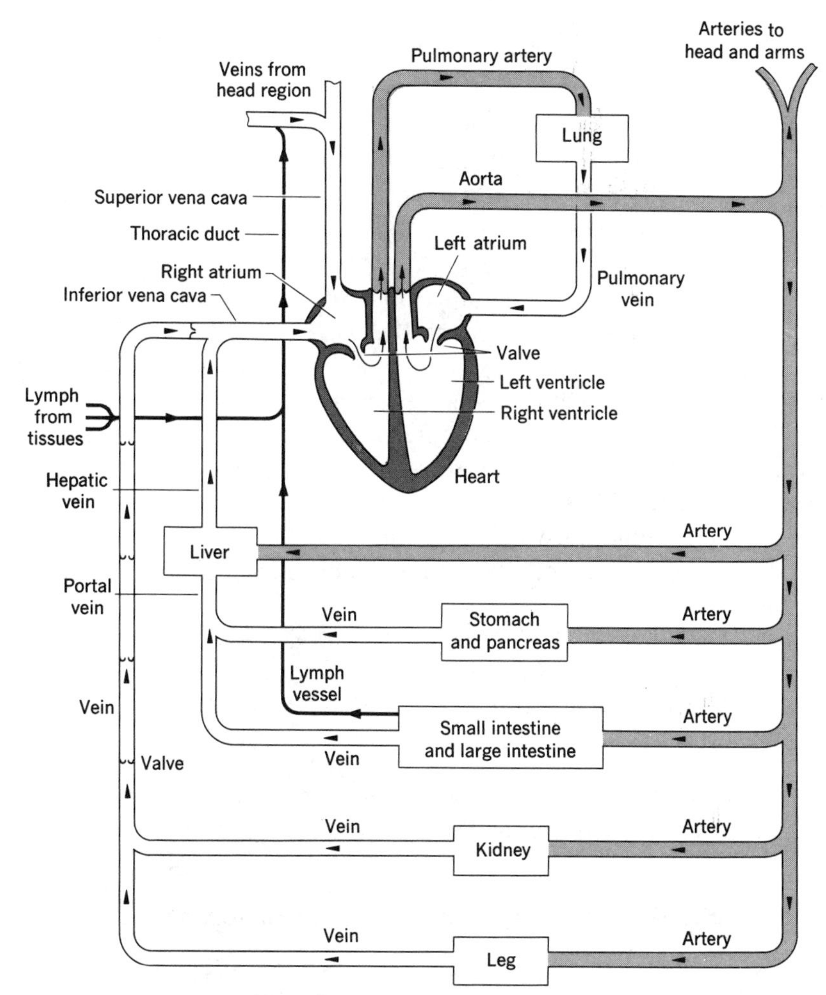

Figure 5–7 Scheme of major blood circuits

Pulmonary circulation. The pulmonary artery emerges from the right ventricle and divides into two branches, one leading to the right lung and the other to the left lung. Oxygenated blood from the lungs returns to the left atrium of the heart through the pulmonary veins. This blood circuit, from heart to lungs back to heart, is called the *pulmonary circulation*.

Systemic circulation. In the *systemic circulation,* blood moves from the left ventricle through the aorta, then circulates to all body systems except the lungs, before returning to the heart. Blood from all parts of the body returns to the right atrium through two large veins. The superior vena cava drains blood from the head and neck. The inferior vena cava drains blood from the lower regions of the body. The systemic circulation has three branches:

1. *Coronary circulation* consists of two *coronary arteries* that lead out of the aorta and branch into many arterioles and capillaries within the heart muscle. The capillaries form a network that leads into many venules, and then into larger veins. These finally drain into a single vein that enters the right atrium.
2. *Hepatic portal circulation* brings nutrient-rich blood from the small intestine to the liver through the *portal vein*. Liver cells remove worn-out red blood cells, excess glucose and amino acids, and certain drugs, if present, from the blood. In the liver, worn-out red blood cells are converted into bile; excess glucose is stored as glycogen; excess amino acids are changed into urea; and drugs are either neutralized or stored. If the blood glucose level is low, certain enzymes in the liver cells change the glycogen to glucose, which then enters the blood. From the liver, the blood flows into the *hepatic vein*, to the inferior vena cava, and then into the right atrium.
3. *Renal circulation.* The *renal circulation* passes through the kidneys. As blood passes through the kidneys, excess water, urea, and excess salts are removed from the blood. The renal circulation is explained in more detail in Chapter 6.

The composition of blood changes as it passes through the capillaries in different organs of the body. The major changes in the blood as it passes

Table 5–1 Changes in the Composition of Blood

Organs	*Blood Loses*	*Blood Gains*
Muscles	Glucose and oxygen	Lactic acid and carbon dioxide
Digestive glands	Raw materials used to synthesize digestive juices and enzymes	Carbon dioxide
Villi	Oxygen	End products of digestion
Lungs	Carbon dioxide and water	Oxygen
Liver	Excess glucose, amino acids, and old red blood cells	Released glucose, urea, and plasma proteins
Kidneys	Water, urea, and mineral salts	Carbon dioxide

through certain specialized structures and organs are summarized in Table 5–1. Blood provides the cells of all organs with digested nutrients, oxygen, mineral salts, water, and other substances. At the same time, blood removes waste products such as water, carbon dioxide, urea, and excess mineral salts from the cells.

COMPOSITION OF BLOOD

Blood is a fluid tissue that performs the following functions:

- Transports useful substances to cells and removes waste products from cells.
- Regulates body temperature by distributing the heat that cells release during oxidation.
- Protects the body by producing *antibodies* and certain white blood cells which combat foreign substances such as bacteria.

An adult human possesses about five liters of blood. About 55 percent of the total volume of blood is plasma; the remaining 45 percent is composed of different types of blood cells and cell fragments (Figure 5–8).

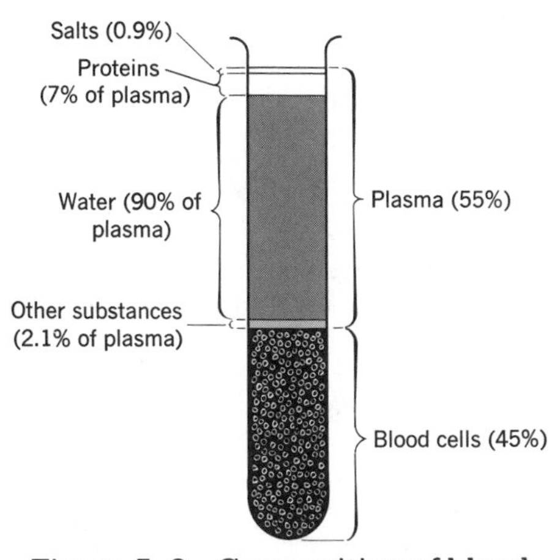

Figure 5–8 Composition of blood

Plasma

The liquid part of the blood, or *plasma*, is composed of about 90 percent water and 7 percent plasma proteins such as fibrinogen and globulins. About 2 percent of the plasma consists of hormones, antibodies, urea, end products of digestion, oxygen, and carbon dioxide. The remaining 1 percent is made up of inorganic compounds, such as sodium chloride.

Plasma helps maintain homeostasis by providing a proper balance between substances needed by the cells to carry out life processes and cellular wastes.

Blood Cells

The cellular part of blood is composed of erythrocytes, or red blood cells; leukocytes, or white blood cells; and thrombocytes, or platelets (see Figure 5–9).

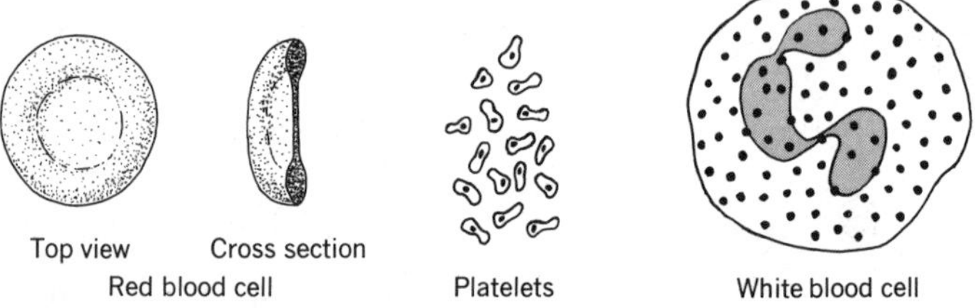

Figure 5–9 The three types of blood cells

***Erythrocytes* (*red blood cells*).** One microliter of blood (a very small drop) contains about five million red blood cells in a normal adult male and about four and one-half million red blood cells in a normal adult female. The *erythrocytes* are biconcave disks that lack a nucleus and contain hemoglobin.

Hemoglobin is a complex compound that contains the element iron. A hemoglobin molecule chemically resembles a chlorophyll molecule. This provides evidence that animals and plants may have evolved from a common ancestor.

In the lungs, where oxygen is plentiful, hemoglobin changes to *oxyhemoglobin*, which is bright red. Oxyhemoglobin transports oxygen to all parts of the body. Upon reaching the cells, oxygen is released. Carbon dioxide, released by the cells, changes hemoglobin to *carbon dioxide–hemoglobin*, which is dull red. In the lungs, the carbon dioxide is released and hemoglobin is again free to combine with oxygen. Thus, red blood cells transport oxygen to the cells and remove some carbon dioxide from them.

Red blood cells are formed mainly in the red marrow of long bones, such as the thigh bone, and in certain flat bones, such as the breastbone. Newly formed red blood cells possess a nucleus that disappears as the cells mature. The life span of a red blood cell is about 120 days. Old red blood cells are destroyed either in the spleen or liver. In the liver, hemoglobin is changed to bile, and the element iron is released and reused by the body.

If the blood does not carry enough oxygen, a condition called *anemia* develops. This condition results from blood loss due to hemorrhage, from too few red blood cells, or from inadequate amounts of hemoglobin. A proper diet including vitamin B_{12}, folic acid, iron salts, and proteins cures certain types of anemia.

Leukocytes (white blood cells). *Leukocytes* defend the body against foreign invaders, such as bacteria, viruses, and other substances. Thus, they play an important part in the body's immune system.

A drop of blood that contains about five million red blood cells also contains about 7,000 to 10,000 white blood cells. White blood cells are much larger than red blood cells, lack hemoglobin, and possess a nucleus at maturity. Most types of white blood cells are produced by the red marrow of bones; some white blood cells form in the lymph nodes.

There are five types of white blood cells: neutrophils, eosinophils, basophils, monocytes, and lymphocytes. When stained and observed under the microscope, each type can be identified by the shape of its cell and nuclei, size, and color of granules in the cytoplasm. White blood cells with granules (neutrophils, eosinophils, and basophils) are called *granulocytes*. White blood cells lacking granules (monocytes and lymphocytes) are called *agranulocytes*.

Granulocytes, also called *phagocytes*, defend the body against disease by engulfing and destroying bacteria and viruses similar to the way an ameba ingests food. Phagocytes travel through the blood and can also leave the blood by squeezing between the cells of capillaries. In body tissues, the phagocytes seek and destroy foreign invaders. For that reason, phagocytes are often called the body's police force.

Lymphocytes do not ingest germs. Instead they produce antibodies that detoxify bacterial poisons (toxins) or may cause bacteria to stick together and clump. The clumped bacteria can then be readily engulfed by the phagocytes. A severe bacterial or viral infection usually causes an increase in the number of white blood cells to about 25,000 cells per microliter. In the blood of people suffering from leukemia, or cancer of the blood, the number of white blood cells may increase to about one million per microliter.

Thrombocytes (platelets). *Thrombocytes* are cell fragments that help the blood to clot. They are formed in the red bone marrow, and are much smaller than either red blood cells or white blood cells. There are about 250,000 platelets in a microliter of blood.

Table 5–2 Summary of Blood Cells

Cell Type	*Function*	*Number Per Microliter*	*Important Facts*
Erythrocytes (red blood cells)	Transport oxygen and carbon dioxide	About 5 million in males About 4.5 million in females	Live for about 120 days Recycled in liver for iron and heme portions Destroyed and removed by phagocytes in liver and spleen
Leukocytes (white blood cells)			
Granulocytes			
1. Neutrophils	Phagocytosis	3000-6000	Neutrophils pass through capillaries to tissues and ingest bacteria
2. Eosinophils	Phagocytosis	About 250	Eosinophils increase in number as a result of allergies
3. Basophils	Liberate anticoagulant heparin	About 50	
Agranulocytes			
1. Lymphocytes	Antibody production; B cells and T cells	About 1800	Lymphocytes produce antibodies to neutralize antigens
2. Monocytes	Phagocytosis	About 400	Monocytes enlarge to become macrophages
Thrombocytes (platelets)	Blood clotting	About 250,000	Pinched off megakaryocytes (giant cells in red bone marrow); source of most blood cells

BLOOD CLOTTING

When tissues or blood vessels are injured, cells and blood platelets release an enzymelike substance called *thromboplastin*. In the presence of calcium ions, thromboplastin changes *prothrombin* (a protein in the blood) to the enzyme *thrombin*. The liver manufactures prothrombin with the help of vitamin K, which is synthesized by bacteria in the large intestine. Thrombin acts on *fibrinogen* (another protein in the blood), changing it to interlacing threads of *fibrin*. The meshwork of fibrin threads traps escaping blood cells and a clot is formed. As the clot forms, a yellow fluid called *serum* oozes out (serum is plasma lacking the clotting substances). The following equations summarize blood clotting:

$$\text{prothrombin} + \text{Ca ions} \xrightarrow[\text{}]{\substack{\text{excess}\\ \text{thromboplastin}}} \text{thrombin}$$

$$\text{fibrinogen} \xrightarrow{\text{thrombin}} \text{fibrin}$$

$$\text{fibrin} + \text{blood cells} + \text{dried serum} \longrightarrow \text{blood clot}$$

Table 5–2 summarizes blood cell functions. Body defenses against disease are discussed in detail in Chapter 6, Disease and Immunity.

CLOSED AND OPEN CIRCULATORY SYSTEMS

Unicellular and simple organisms, such as the ameba and the hydra, obtain needed materials directly from the external environment without special transport systems. Multicellular and complex animals have developed either a closed or an open circulatory system, because their cells are not in close contact with the external environment. In a *closed circulatory system*, the blood remains in blood vessels that are connected to each other in a circuit. In an *open circulatory system*, a heart pumps blood to an aorta that opens into body space. From body spaces, the blood returns to the heart. These systems enable the end products of digestion and oxygen to reach the cells. These systems also transport wastes and other metabolic products away from cells.

Is a closed circulatory system more effective than an open circulatory system? At first glance, it appears that the closed circulatory system of humans and other animals is more effective. A closed circulatory system carries nutrients, wastes, and oxygen. In contrast, the open circulatory system of an insect carries only nutrients and wastes. Because insect species and populations outnumber all other species, we can argue that an open circulatory system is more effective. It is probably more accurate to say that each type of circulatory system is adapted for the way of life of the particular organism.

ABSORPTION AND CIRCULATION IN OTHER ORGANISMS

Table 5–3 summarizes absorption and circulation in protozoans, hydra, earthworms, grasshoppers, and birds.

Table 5–3 Absorption and Circulation in Other Organisms

Organism	Absorption	Circulation
Protozoans (ameba)	Dissolved oxygen and water diffuse into the protist through the plasma membrane; food is engulfed	No system; cyclosis (streaming of cytoplasm), diffusion, and channels of endoplasmic reticulum distribute materials within the cell
Hydra	Same as in protozoans; also diffusion of end products resulting from extracellular digestion	No system; method in each cell is the same as in protozoans
Earthworm	End products of digestion pass from typhlosole of intestine into the blood; oxygen enters through the moist skin, which is supplied with capillaries	Closed circulatory system consisting of hearts, arteries, capillaries, and veins; transports nutrients, oxygen, liquid wastes, carbon dioxide
Grasshopper	End products of digestion pass from the stomach into the blood; oxygen enters tissue cells through moist walls of the tracheae	Open circulatory system consisting of heart, aorta, sinuses; transports nutrients and liquid wastes
Bird	End products of digestion pass from intestine into the blood, and from the blood to all tissue cells	Closed circulatory system consisting of a four-chambered heart, arteries, veins, capillaries, and different circuits, such as systemic and renal

Protozoans. Most free-living protozoans, such as the ameba and the paramecium, live in a watery environment. As a result, osmosis and diffusion readily occurs between these unicellular organisms and their environments. Oxygen, dissolved minerals, and water enter the protozoan through its cell membrane. Metabolic wastes, such as carbon dioxide and ammonia, also leave through the cell membrane. The end products of digestion and other substances are distributed to all parts of the organism by *cyclosis* (streaming of cytoplasm).

Hydra. All of the cells of this simple multicellular animal are in contact with water. Thus, the processes of osmosis and diffusion easily exchange substances between the cells of a hydra and its watery environment. Oxygen and carbon dioxide diffuse between the watery environment and cell layers. The end products of digestion diffuse from cell to cell; metabolic wastes diffuse directly into the surrounding environment (Figure 5–10).

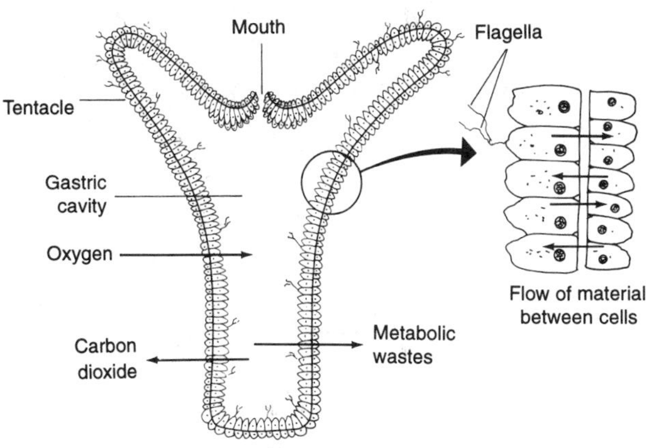

Figure 5–10 Osmosis and diffusion in the hydra

Earthworm. An earthworm is a complex, multicellular animal that has a simple closed circulatory system. A series of five pumps, called aortic arches, function as a heart by pumping blood into dorsal and ventral blood vessels. These muscular blood vessels are connected to each other by many capillaries, which are in close contact with all tissues. As blood is pumped through the capillaries by the aortic arches, oxygen, carbon dioxide, digested nutrients, water, wastes, and other substances are exchanged. The blood eventually returns to the "heart" to be pumped through the system again (Figure 5–11).

Grasshopper. A grasshopper is a complex, multicellular animal that has an open circulatory system. Its blood, called *hemolymph*, circulates in open-ended vessels that empty into the body cavity, or *hemocoel* (Figure 5–12). The heart of a grasshopper is a muscular tube that receives blood from the rear of the insect's body and squeezes it to the front. As the blood moves forward, it seeps through tissues, where exchanges of substances

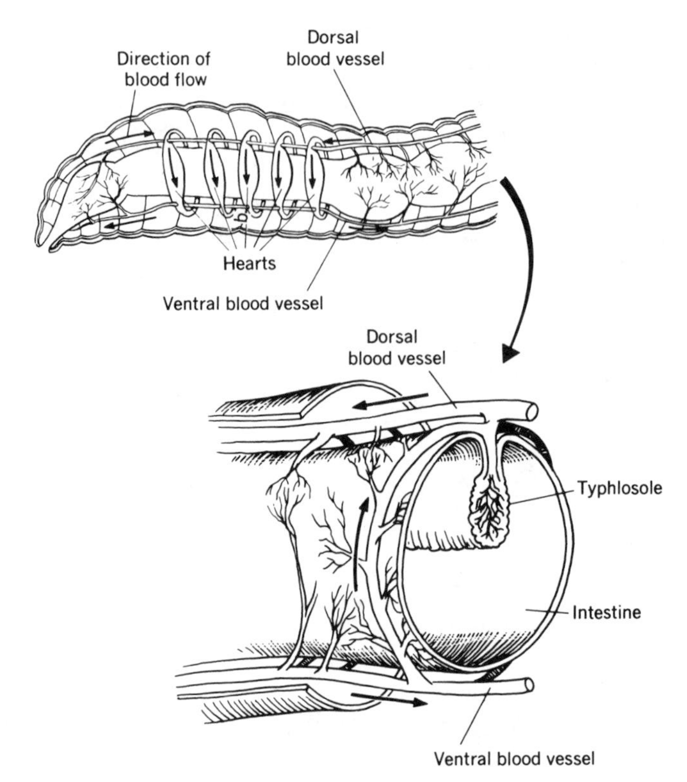

Figure 5–11 Transport in the earthworm

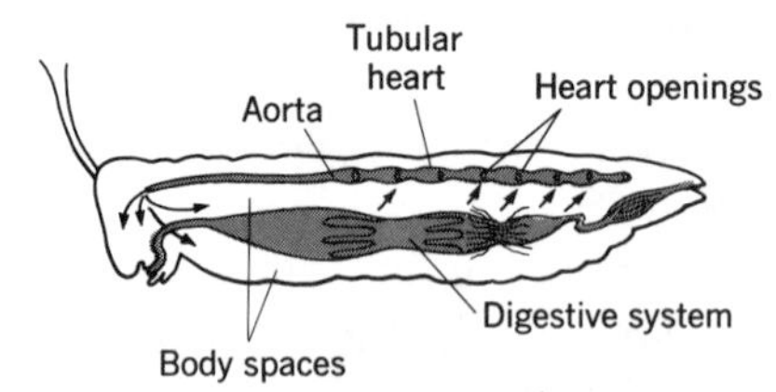

Figure 5–12 Transport in the grasshopper

occur. The blood then returns to the heart through *ostia*, which are small openings in the heart wall. Breathing movements help move blood throughout the insect's body.

Bird. A bird is a warmblooded vertebrate with a closed circulatory system that closely resembles that of mammals, including humans. The closed circulatory system of a bird includes a four-chambered heart and pulmonary, systemic, coronary, hepatic portal, and adrenal circuits. A four-chambered heart is an advantage to warmblooded animals for two reasons:

1. Blood returning to the heart is sent to the lungs where an exchange of oxygen and carbon dioxide occurs.
2. Oxygenated blood from the lungs is distributed by the systemic circulation to all parts of the body. (Refer back to Figure 5–7.)

Section Quiz

1. About 90 percent of the blood's plasma is composed of (*a*) proteins (*b*) hormones (*c*) water (*d*) oxygen.
2. Vessels that carry blood from capillaries to veins are called (*a*) venules (*b*) arterioles (*c*) lacteals (*d*) portals.
3. Two types of white blood cells that help fight infections are (*a*) leukocytes and thrombocytes (*b*) lymphocytes and erythrocytes (*c*) lymphocytes and phagocytes (*d*) phagocytes and thrombocytes.
4. Which organism relies on cyclosis as a way of transporting nutrients, gases, and waste products? (*a*) grasshopper (*b*) earthworm (*c*) paramecium (*d*) bird.
5. Two organisms that have closed circulatory systems are the (*a*) bird and grasshopper (*b*) ameba and earthworm (*c*) hydra and paramecium (*d*) earthworm and bird.

Chapter Review Questions

The following questions will help you check your understanding of the material presented in the chapter.

1. An important function of a transport system is to (*a*) receive stimuli from the external environment (*b*) bring materials from the external environment into contact with all cells (*c*) hydrolyze nutrients so the cells can utilize them (*d*) remove solid waste materials from the digestive system.
2. Transport in mammals involves absorption and (*a*) circulation (*b*) adsorption (*c*) transpiration (*d*) assimilation.
3. Materials enter cells passively by diffusion, or they are absorbed actively as a result of (*a*) a difference in concentration gradient (*b*) differential permeability (*c*) the expenditure of energy by the cell (*d*) plasmolysis of the cell.

4. One of the processes involved in the transport of molecules within cells is (*a*) osmosis (*b*) hydroloysis (*c*) pinocytosis (*d*) cyclosis.
5. Active transport of certain proteins into a cell from its environment is most closely associated with the (*a*) cell membrane (*b*) cell wall (*c*) ribosome (*d*) nucleolus.
6. The human circulatory system is most similar in structure and function to that of the (*a*) paramecium (*b*) hydra (*c*) grasshopper (*d*) bird.
7. The human heart is divided into a right and left side to (*a*) prevent blood clotting (*b*) provide for an open circulatory system (*c*) provide for a separation of oxygenated and deoxygenated blood (*d*) prevent reabsorption in the nephrons.
8. The thick, muscular vessels that transport blood away from the heart are called (*a*) atria (*b*) arteries (*c*) veins (*d*) ventricles.
9. The backward flow of blood in veins is prevented by (*a*) muscles (*b*) heart action (*c*) valves (*d*) lymphatics.
10. In humans, oxygen and nutrients normally leave the blood through thin-walled structures called (*a*) veins (*b*) capillaries (*c*) arteries (*d*) lymph vessels.

For questions *11 through 13,* select the letter of the circulatory term from the list below that is most closely associated with that question. A letter may be used more than once or not at all.

Circulatory Terms

(*a*) platelets	(*c*) red blood cells
(*b*) plasma	(*d*) white blood cells

11. Which contains a pigment that combines with oxygen?
12. Which initiates the clotting process when a blood vessel is ruptured?
13. Which carries dissolved food substances to the cells?
14. In human metabolism, which are found in the blood but not normally in the intercellular fluid? (*a*) end products of digestion (*b*) mineral salts (*c*) metabolic wastes (*d*) red blood cells.
15. In many animals, the oxygen-carrying capacity of the blood is increased by the presence of (*a*) ATP (*b*) hemoglobin (*c*) platelets (*d*) white blood cells.

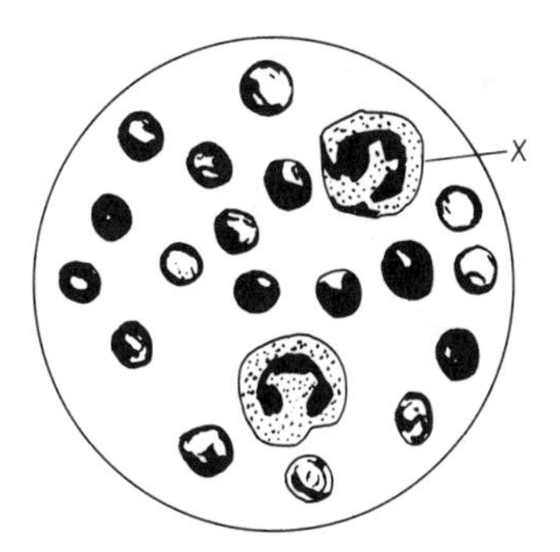

16. The sketch shown above represents a stained sample of human blood as seen under a compound microscope. Which may be a function of cell X? (*a*) to transport glucose to cells (*b*) to transport oxygen in chemical combination with hemoglobin (*c*) to initiate the blood-clotting process (*d*) to protect the body against certain germs by phagocytosis.
17. Which substance makes up most of the lymph? (*a*) water (*b*) hemoglobin (*c*) protein matter (*d*) mineral matter.
18. Which can diffuse directly into the blood without being digested? (*a*) proteins (*b*) starch (*c*) cellulose (*d*) glucose.
19. The human body adapts to living at high altitudes by producing (*a*) more white blood cells (*b*) more red blood cells (*c*) fewer white blood cells (*d*) fewer red blood cells.
20. The amino acid content of the blood increases as it flows through the (*a*) kidney (*b*) gallbladder (*c*) pancreas (*d*) small intestine.
21. The renal circulation of vertebrates flows through the (*a*) liver (*b*) kidneys (*c*) spleen (*d*) ovaries.
22. Which would not be associated with a grasshopper? (*a*) jointed appendages (*b*) closed circulatory system (*c*) chitinous exoskeleton (*d*) respiratory tubules.
23. Which animal does not have a specialized system for the transport of oxygen? (*a*) grasshopper (*b*) hydra (*c*) earthworm (*d*) human.
24. Which two organisms have a closed circulatory system? (*a*) earthworm and human (*b*) earthworm and protozoan (*c*) grasshopper and human (*d*) grasshopper and earthworm.
25. In hydra, the movement of materials into and out of individual cells is regulated by the (*a*) cell walls (*b*) tentacles (*c*) cell membranes (*d*) food vacuoles.

Biology Challenge

The following questions will provide practice in answering SAT II-type questions.

Part I

Based on this text, and your knowledge of biology, select the letter that best completes the statement or answers the question.

1. When blood containing carbon monoxide is exposed to air, the color changes quickly from cherry red to dull red. The best explanation for the color change is that (*a*) oxyhemoglobin formed (*b*) carbon dioxide–hemoglobin formed (*c*) coagulation occurred (*d*) agranulocytes agglutinated (*e*) many red blood cells cytolyzed.
2. As part of a lab investigation, a student observed a stained sample of human blood with a compound microscope; some cells were larger than others and possessed nuclei, and had colored granules within the cytoplasm. The student identified these cells as (*a*) thrombocytes (*b*) erythrocytes (*c*) phagocytes (*d*) agranulocytes (*e*) lymphocytes.
3. Which device would be used to separate the cellular portion of blood from the liquid portion? (*a*) dialysis machine (*b*) pipette (*c*) sphygmomanometer (*d*) centrifuge (*e*) graduated cylinder.
4. Which statement is correct? (*a*) The heart extracts oxygen and nutrients from blood within the atria and ventricles. (*b*) Removal of all nerves leading to the heart slows down the heartbeat. (*c*) The SA node (pacemaker) in the wall of the right atrium generates a rhythmic impulse. (*d*) Frogs, toads, and salamanders possess a two-chambered heart. (*e*) Hemoglobin is the only substance capable of transporting oxygen and carbon dioxide in the blood.
5. An important function of lymph is to (*a*) provide a medium for the reproduction of white blood cells (*b*) regulate blood pressure within arteries, veins, and capillaries (*c*) transport nutrients to all cells (*d*) store gamma globulins (*e*) help maintain homeostasis between the interior and exterior of cells.

Part II

Base your answers to questions *1 through 4* on library research and on the diagram shown below, which shows a small goldfish placed in a petri dish containing a small amount of water. Most of the fish's body is wrapped with wet gauze and cotton. The fish is lying on a glass slide and its tail is covered by another glass slide.

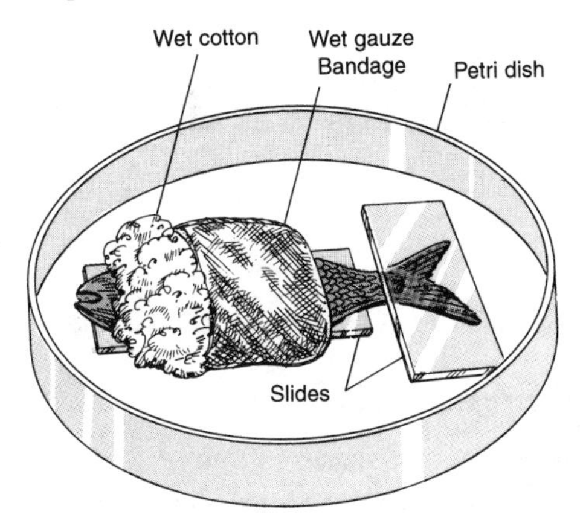

1. The fish's head is wrapped with wet cotton to (*a*) keep its gills moist (*b*) prevent the fish from becoming frightened (*c*) maintain the fish's body temperature (*d*) absorb metabolic wastes (*e*) prevent fungus infection.
2. Which part of the fish do you think the student wants to study? (*a*) the head (*b*) the tail (*c*) the midsection (*d*) the reproductive organ (*e*) the scales.
3. A student observed the petri dish and its contents under low power. The student arranged the petri dish so that the thinnest part of the tail was over the opening in the microscope stage. The primary objective of this procedure is to (*a*) observe tail bones and their arrangement (*b*) estimate how long it will take before the tail flips and upsets the slide covering it (*c*) calculate the total number of scales by extrapolation (*d*) observe capillaries and blood flow (*e*) observe hemopoiesis in the tail bones.
4. The observer should also be able to (*a*) distinguish between small arteries and small veins (*b*) count the fish's heartbeat (*c*) view red blood cells (*d*) determine the direction of blood flow (*e*) make all of the above observations.

6

Disease and Immunity

Learning Objectives

When you have completed this chapter, you should be able to:

- **Identify** the causes of diseases in humans.
- **Describe** some adaptations shown by pathogenic organisms.
- **Discuss** some adaptations shown by humans that enable them to defend themselves from organisms that cause disease.
- **Explain** the importance of the human immune system.
- **Distinguish** between the human blood groups, and recognize the importance of blood types in transfusions.
- **Identify** diseases that occur when the human immune system fails.
- **Discuss** the effect of AIDS on the human immune system.

OVERVIEW

The dictionary defines disease as a departure from health. In the past, diseases had two important effects upon people. First, they made people ill. And second, since the causes of disease were poorly understood, the "idea" of disease filled people with fear.

Today, medical science offers a better understanding of the nature of disease. Medicine also offers cures for many of the diseases that were once a certain sentence for death. And most important, with an understanding of the causes of disease came freedom from the fear of disease. That may be medicine's greatest gift of all.

In this chapter, you will learn about the various causes of disease and about the ways in which your body protects itself from the effects of disease-causing organisms.

OCCURRENCE AND TRANSMISSION OF DISEASE

Some diseases, including rickets and scurvy, are caused by a deficiency of particular vitamins. A balanced diet usually cures vitamin *deficiency diseases* or prevents their occurrence. Other diseases, called *communicable diseases*, are passed from one person to another. A *contagious disease* is a type of communicable disease that spreads readily from one person to another. An *infectious disease* is caused by a *pathogenic* (disease-causing) microorganism. Yet other disorders and diseases are caused by alterations in chromosome number and gene defects. For example, Down syndrome is caused by an abnormal number of chromosomes, and sickle-cell anemia results from a defective gene. These disorders are not communicable or infectious. (See Chapter 15 for more information on genetic disorders.)

When your body is invaded by infectious microorganisms, you become the host for specific *pathogens*, or parasites. These parasites rely on your body for nourishment and energy to satisfy their metabolic needs. After the pathogens invade your body, an infection often follows that produces specific symptoms and immunological signs. A physician can readily distinguish measles from typhoid fever by the symptoms and immunological signs associated with the different pathogens and their effects on you.

There are four general kinds of disease: sporadic, endemic, epidemic, and pandemic. A disease such as typhoid fever, which occurs occasionally in a population, is called a *sporadic disease*. A disease such as the common cold, which is always present in a population, is called an *endemic disease*. A disease such as AIDS (acquired immunodeficiency syndrome), which affects many people in a relatively short time, is called an *epidemic disease*. When diseases such as AIDS and influenza spread worldwide, they are called *pandemic diseases*.

How Diseases are Transmitted

Pathogenic microorganisms frequently are transmitted to a host by contact, vectors, and common vehicle transmission. *Contact* involves person-to-person transmission. For example, kissing can spread the common cold and mononucleosis, commonly referred to as "mono." Sexual contact can spread AIDS, as well as venereal diseases such as gonorrhea and syphilis. *Droplet infection* may spread infectious pathogens through the air when someone sneezes or coughs. A violent sneeze can often discharge a visible mist of mucus particles from the nose and mouth. Each tiny mucus droplet may contain a variety of microorganisms that can be inhaled by others and cause disease.

Vectors include animals such as insects and ticks, which may spread pathogenic microorganisms from one host to another. For example, a housefly comes in contact with many pathogens in its search for food. Feces and decomposing garbage attract the housefly, which subsequently picks up pathogenic microorganisms on its feet while feeding. In this way, a housefly can transmit typhoid fever and dysentery. Fleas are vectors for typhus fever and bubonic plague. Ticks are vectors for Lyme disease and Rocky Mountain spotted fever.

Common vehicle transmission spreads pathogens from a common source such as water, food, blood, and drugs to an individual. About 10 percent of all hospital patients, nursing-home residents, and people in other health-related facilities become infected during their stay. Some even die because of their weakened condition and lack of immunological defenses. These infections are called *nosocomial* (from a Greek word meaning "hospital"). Nosocomial diseases include respiratory, urinary, intestinal, heart, and skin infections.

Most nosocomial diseases are spread by direct contact from patient to patient and from hospital staff to patients. Hospitals and other health-related facilities employ certain control measures designed to prevent the spread of disease. These measures include educating nonmedical staff about basic infection controls; practicing *aseptic* (germ-free) procedures; publicizing the importance of frequent handwashing and other sanitary measures; and being constantly on guard against the danger of a potential epidemic.

Section Quiz

1. Which statement is correct? (*a*) Scurvy is an infectious disease. (*b*) All microbes are pathogens. (*c*) Droplet infection spreads typhoid fever. (*d*) Nosocomial diseases may be spread by contact between hospital patients.

2. An infection usually results from (*a*) an invasion and growth of pathogens in the body (*b*) direct contact with a hospital patient (*c*) a bite by a flea or tick (*d*) using aseptic procedures.

3. Which is an example of an infectious disease that is also contagious? (*a*) typhoid fever (*b*) appendicitis (*c*) typhus (*d*) malaria.

4. A disease such as the common cold is considered (*a*) sporadic (*b*) pandemic (*c*) epidemic (*d*) endemic.

5. Fleas and ticks are called vectors because they (*a*) spread pathogens from common sources, including food and water (*b*) transfer pathogens from one host to another (*c*) always carry pathogens (*d*) remain in the same host until they die.

ADAPTATIONS OF PATHOGENIC MICROORGANISMS

Pathogenic microoganisms such as bacteria and viruses possess physical characteristics that enable them to survive in a particular environment. Frequently, the human body is a host for pathogens. Humans, however, also possess adaptations, or natural defenses to counteract and destroy pathogens. The following are three adaptations of pathogens that resist the natural defenses of a host:

1. The *cell wall* of certain bacteria, such as *Streptococcus pyogenes*, which causes scarlet fever, contains a protein that resists heat and acid. This protein also helps the bacterium attach itself to a cell's plasma membrane and resist *phagocytosis* (engulfment and destruction) by white blood cells.
2. Another adaptation that enables certain bacteria to resist phagocytosis by white blood cells is called a *bacterial capsule*. This is a jellylike material that covers the cell wall. As a result of the bacterial capsule, the *virulence*, or disease-causing ability of bacteria, is magnified. They can grow, multiply, and spread readily throughout the host's body. Bacteria that have bacterial capsules include *Streptococcus pneumoniae* (one cause of pneumonia) and *Neisseria gonorrhoeae* (the cause of gonorrhea).
3. Some bacteria produce enzymes that help them survive in a host. *Bacterial enzymes* include *hemolysins*, which destroy red blood cells; *leucocidins*, which destroy certain white blood cells (macrophages or neutrophils); and *kinases*, which dissolve blood clots, formed by the body to isolate an infection from healthy tissues. These and other adaptations help pathogens overcome body defenses and damage tissues.

A condition characterized by the presence of bacteria in the bloodstream is called a *bacteremia*. Bacteria that grow, multiply, and excrete their waste products in the bloodstream cause a disease known as *septicemia*. When bacteria release toxins into the bloodstream, a condition called *toxemia* occurs. A *toxin* is a poisonous substance produced by a bacterium, such as *Clostridium tetani*, which causes tetanus. A condition caused by the presence of viruses in the bloodstream is called a *viremia*.

Viruses are not classified in any of the five kingdoms of living things (the classification of viruses will be discussed in Chapter 18). Viruses are submicroscopic particles composed of a nucleic acid (RNA or DNA) enveloped by a protein coat called a *capsid.* A virus is the ultimate parasite because it cannot survive outside of a host cell. A virus grows and multiplies by taking over a cell's metabolic structures. The invasion of a cell by a virus usually results in cellular destruction. For example, the human immunodeficiency virus (HIV), which causes AIDS, destroys T cells (lymphocytes) of the immune system. Other viruses cause cancer by disrupting normal cell growth and mitosis.

ADAPTATIONS OF THE HUMAN BODY

To counteract the adaptations of pathogens, the human body features two main lines of defense. These consist of adaptations that combat the actions of disease-causing bacteria, viruses, and other foreign substances.

First Line of Defense

The first line of defense consists of three adaptations:

1. A *physical shield*, such as the skin, which prevents most pathogens from entering the bloodstream and tissues.
2. A *chemical reaction*, such as the destruction of bacteria by secretions containing acids and enzymes (for example, hydrochloric acid in the stomach and lysozyme in tears).
3. A *biological* response, such as the aggressive action of granulocytes and monocytes, which destroy invading pathogens by phagocytosis.

When there is a break in the skin, monocytes usually gather, increase in size, and ingest any foreign bodies. These large white blood cells are called *macrophages.* Other cells, called *mast cells,* secrete *histamine*, a chemical substance that attracts other phagocytes to the wound. However, there are undesirable side effects of histamine production on living tissue. They include inflammation, swelling, mucus production, tearing of eyes, itching, and sneezing.

Second Line of Defense

The second line of defense involves the *immune system,* which is composed of two lymphoid organs—the spleen and thymus— as well as lymphatic vessels, lymph nodes, and white blood cells. This line of defense responds to *antigens*, or foreign substances, such as bacteria, viruses, proteins, and other complex molecules.

The immune system responds in two ways to combat antigens:

1. *Production of specific antibodies*, which are chemical substances produced by the lymphocytes. Antibodies help prevent disease by making antigens harmless. Antibodies produced by the lymphocytes include:

- *Antitoxins* that neutralize bacterial toxins
- *Agglutinins* that clump bacteria in groups
- *Opsonins* that stimulate phagocytosis of clumped bacteria
- *Lysins* that dissolve the cell walls of bacteria
- *Precipitins* that coagulate chemical products of bacteria

Antibodies respond to specific antigens by recognizing them as foreign substances, attacking them, and finally by destroying and disposing of them. Antibodies are organic chemicals called *immunoglobulins.* For example, immunoglobulins IgA antibodies are mainly found in body secretions, such as sweat and saliva; the most abundant group of immunoglobulins, called IgC, is present in blood and lymph; and IgM antibodies are the first to appear at the site of an infection, usually in the skin.

2. *Production of lymphocytes* from red bone marrow. There are two main types of lymphocytes: (1) *B cells* (B for blood) that enter the blood and (2) *T cells* (T for thymus) that are stored in the thymus gland before moving into the blood. B cells produce specific antibodies; T cells regulate phagocytosis by the granulocytes and monocytes (see Figure 6–1).

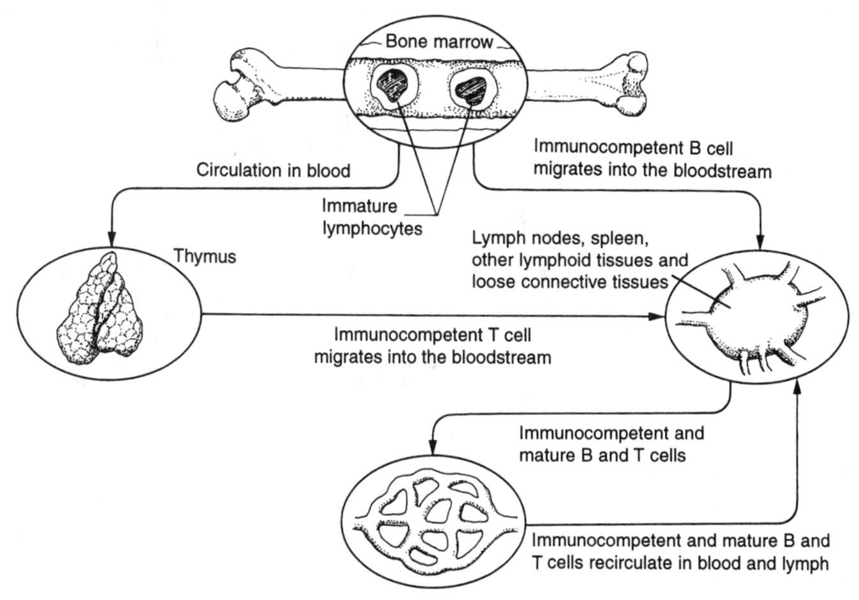

Figure 6–1 B cell and T cell origins

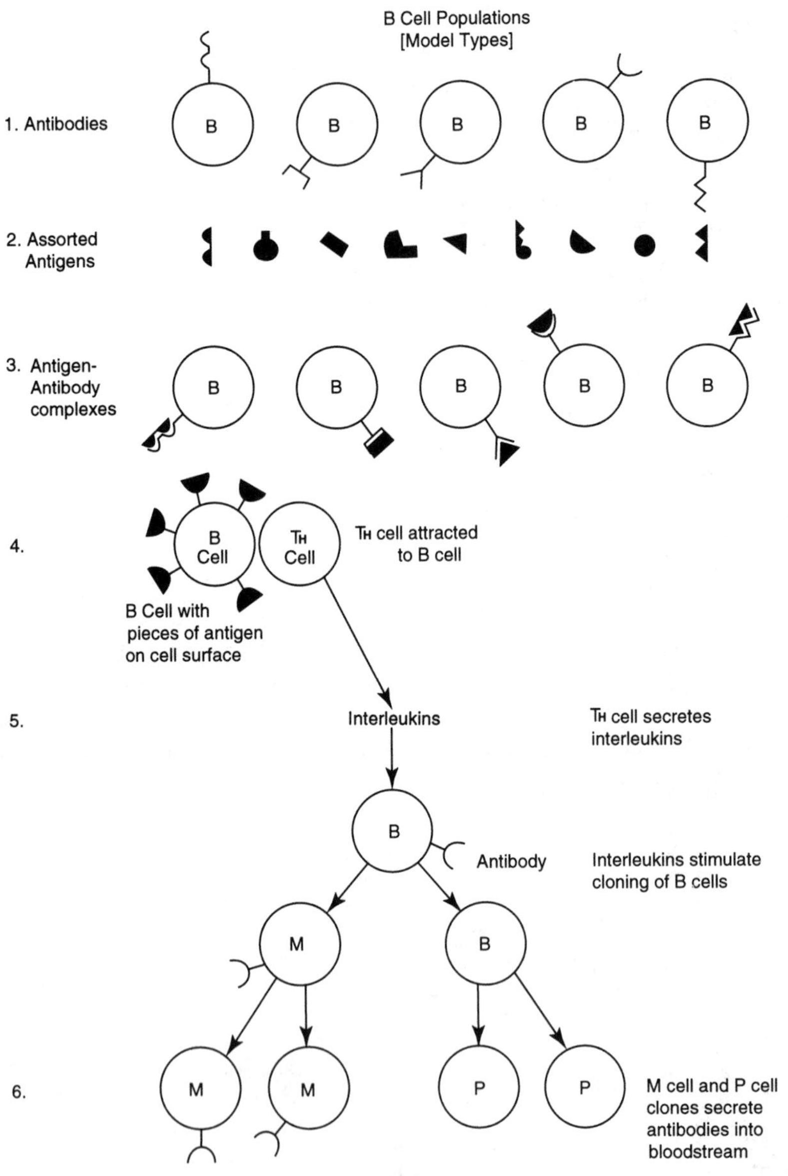

Figure 6–2 B cell populations

B cell functions. Among the trillions of white blood cells in the body are different populations of B cells. Each B cell population carries its own type of *antibody* to counteract a specific *antigen*. One population of B cells may counteract an antigen found on the carbohydrate cell wall of a particular bacterium and another population of B cells may counteract the protein coat (*capsid*) of a virus particle. In effect, many different kinds of B cells, each bearing a specific antibody, defend the body against a formidable array of antigens.

An antigen-antibody mechanism, similar to the lock and key concept of enzyme-substrate action, occurs when a B cell makes contact with a particular antigen. Contact between a B cell and an antigen results in *lysis*, or fragmentation of the antigen. The antigen fragments stick to the B cell's surface and attract specific T cells, called *TH cells* or *helper T cells*. The TH cells cling to the antigen fragments and secrete *interleukins*, chemical substances that cause B cells to reproduce. The B cells produced are *clones* because their genetic makeup is identical. Two types of clones are formed: *memory cells* (M cells) and *plasma cells* (P cells). Memory cells are long-lasting and can produce antibodies to specific antigens when necessary, even years later. Memory cells are the basis of active immunity (page 146). Plasma cells are short-lived and produce additional antibodies to combat the antigen that initiated a B cell's response. Figure 6–2 shows B cell types and their functions.

T cell functions. Unlike B cells, T cells do not produce antibodies. T cells attack and kill antigen-bearing cells, stimulate the reproduction of B cells, and suppress antibody production by slowing down B cell reproduction when an infection has been suppressed by the immune response. T cells that destroy antigen-bearing cells, especially cancer cells, are called *killer T cells*. When surrounded by killer T cells, virus-infected cells secrete *interferons*, which defend healthy body cells from the virus.

Section Quiz

1. Which of these cells provides the body with long-lasting immunity? (*a*) granulocytes (*b*) memory cells (*c*) helper T cells (*d*) plasma cells.
2. The cloning of B cells to produce memory cells and plasma cells is triggered by (*a*) histamines (*b*) prostaglandins (*c*) interleukins (*d*) epinephrine.
3. Substances secreted by virus-infected cells that help other cells resist the same virus are called (*a*) interferons (*b*) interleukins (*c*) macrophages (*d*) antigens.

4. Which defense mechanism is aided by the presence of opsonins? (*a*) antigen-antibody response (*b*) B cell cloning (*c*) TH cell action on active B cells (*d*) phagocytosis.
5. The main function of B cells is to (*a*) produce interferon (*b*) produce antibodies (*c*) change into TH cells (*d*) destroy cancer cells.

Blood Transfusions

In 1900, *Karl Landsteiner* discovered four basic types of blood in humans. He also discovered that the failure of some blood transfusions was due to *agglutination*, or clumping, of the red blood cells in the blood vessels of the recipient. Later, this clumping of red blood cells was discovered to be an antigen-antibody reaction. In a successful transfusion, no clumping of red blood cells occurs because the blood types of the recipient and the donor are *compatible*, or match.

Blood types. The four groups, or types, of blood are A, B, AB, and O. Each *blood type* is named for the kind of antigen, called an *agglutinogen*, found on the plasma membrane of a person's red blood cells. The antibodies for these agglutinogens are called *agglutinins*, and are found in the plasma. The agglutinins are represented by the letters *a* and *b*. Table 6–1 shows the agglutinogen and agglutinin composition of the blood types.

Table 6–1 Blood Types

Type	*Agglutinogens on Red Blood Cells*	*Agglutinins in Plasma*
A	A	*b*
B	B	*a*
AB (universal recipient)	AB	None
O (universal donor)	None	*a* and *b*

Note that type A blood has agglutinin *b* in the plasma. If a person with type B blood is given a transfusion of type A blood, the agglutinin *b* of the type A blood will cause clumping of the type B blood cells. The blood of type B is not compatible with, or does not match, the blood of type A. Similarly, if a person with type A blood receives type B blood, clumping will occur because of the A-*a* combination of antigen and antibody. Type AB blood does not contain agglutinins; therefore a type AB person can receive blood from anyone, and is called a *universal recipient.* Type O blood has red blood cells that lack antigens A and B. Because these red blood cells cannot be clumped by the agglutinins in the plasma of any recipient, a type O person is called a *universal donor*. See Table 6–2.

Table 6–2 Summary of Blood Compatibilities

Group	*Can Receive*	*Can Give to*
AB (universal recipient)	All four types	AB only
A	A or O only	A or AB only
B	B or O only	B or AB only
O (universal donor)	O only	All four types

The typing of blood. A drop of anti-A serum (containing agglutinin *a*) is placed near one end of a slide, and a drop of anti-B serum (containing agglutinin *b*) is placed near the opposite end. Two drops of unknown blood are mixed with each type of serum—one drop of blood with the anti-A serum and the other drop of blood with the anti-B serum. If clumping occurs in the anti-A serum, the unknown blood is type A. If clumping occurs in the anti-B serum, the unknown blood is type B. If clumping occurs in both serums, the unknown blood is type AB. If clumping does not occur in either serum, the unknown blood is type O. Figure 6–3 summarizes these reactions.

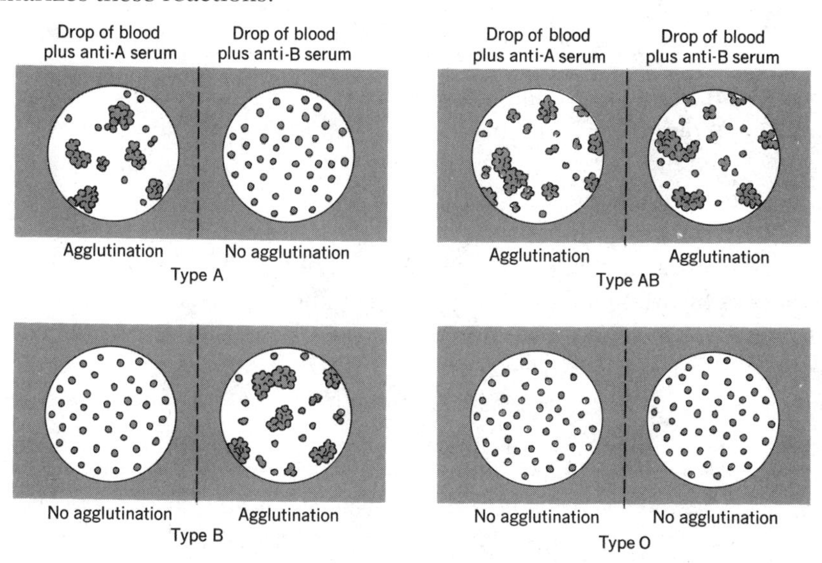

Figure 6–3 How blood type is determined

Rh factor. In 1940, Landsteiner and *Alexander Wiener* discovered that human red blood cells contain another antigen, called the *Rh factor*. People that have the Rh antigen are called *Rh positive* (Rh+); those that lack the antigen are called *Rh negative* (Rh-).

About 85 percent of the population in the United States is Rh positive. People who are Rh positive and Rh negative do not normally possess antibodies for the Rh antigen. However, if the tissues or blood of an Rh negative person comes in contact with Rh positive blood, the Rh negative person will usually manufacture the antibodies that destroy Rh positive blood cells.

If a pregnant woman is Rh negative and her developing child is Rh positive, her child's life may be endangered. As red blood cells from the embryo enter the mother's blood, the mother's body may produce antibodies against them. The antibodies may then enter the embryo's bloodstream and destroy its red blood cells. As a precaution, physicians routinely determine the Rh type of a pregnant woman. If she is Rh negative, the father's Rh type is also determined. If he is Rh positive, either the woman may be injected with substances that neutralize the antibodies or preparations may be made for the transfusion of Rh negative blood from a donor into the newborn baby. If the father is Rh negative, there is no need for these measures.

These precautionary measures are especially important if it is the woman's second or third pregnancy. There may not have been enough antibodies formed during the first pregnancy to do harm. The concentration of antibodies builds up, however, during later pregnancies.

Types of immunity. An inherited immunity to a disease such as tuberculosis and diphtheria is natural, or not acquired during a person's life. *Natural immunity* is passed on from parents to their children. *Acquired immunity* may be either active or passive.

Active immunity is obtained when your body is invaded by pathogenic microorganisms or other foreign proteins. Your body counteracts these pathogens by producing specific antibodies and specialized lymphocytes (B cells and T cells). Active immunity is long-lasting for many diseases, such as smallpox and measles. Active immunity also can be developed by *vaccination*, or injecting a specific antigen from a certain pathogen. The injected material, or *vaccine*, may be a weakened toxin, called a *toxoid*. Vaccines also can contain dead or weakened bacteria and viruses. Thus, diphtheria toxoid can provide immunity against diphtheria; dead typhoid bacteria can provide immunity against typhoid fever; a weakened polio virus can provide immunity against polio. Active immunity was first used almost 200 years ago when *Edward Jenner* (in 1796) developed a vaccine against smallpox using live cowpox virus.

Vaccines are strong enough to cause an active immune response, but not strong enough to cause the disease. Live and weakened virus vaccines, such as the Sabin polio vaccine, provide long-lasting immunity without the need for additional "booster" shots. Booster shots are not required because the live virus multiplies, increasing the original dose. In contrast, bacterial vaccines (such as those for cholera and typhoid fever) are not as effective or long-lasting as viral vaccines and generally require booster shots.

Passive immunity is obtained when you are injected with antibodies taken from an animal or human that has immunity to a particular disease. The injected material consists of blood serum containing *gamma globulin*, a protein that contains antibodies. The serum gives a person some immediate protection against diseases such as measles, rabies, and hepatitis. The protection, however, is not long-lasting because memory cells are not produced as in active immunity.

Section Quiz

1. People with type O blood are called universal donors because their (*a*) red blood cells lack agglutinogens A, B, and AB (*b*) red blood cells contain agglutinogens A, B, and AB (*c*) blood plasma lacks agglutinins a and b (*d*) blood plasma contains only agglutinin a.
2. A person with type AB blood (*a*) lacks agglutinins a and b in the plasma (*b*) has agglutinins a and b in the plasma (*c*) can donate blood to a person with type O blood (*d*) is called a universal donor.
3. An Rh negative mother is pregnant with a second Rh positive fetus. The expectant mother (*a*) may give birth to a child with clumped red blood cells (*b*) should be given a transfusion of Rh negative blood (*c*) should be given a transfusion of Rh positive blood (*d*) may risk the danger of having her red blood cells destroyed.
4. Which of these statements is correct? (*a*) Passive immunity may be acquired by having a disease. (*b*) Active immunity may be acquired by having a disease. (*c*) Passive immunity is associated with memory cells. (*d*) Active immunity involves a person receiving antibodies produced by another organism.
5. Immunoglobulins that appear first in response to an antigen at an infection site are referred to as (*a*) IgC (*b*) IgA (*c*) IgM (*d*) IgE.

Failures of the Immune System

Autoimmune diseases. The body's immune system normally recognizes and distinguishes its antigens (proteins) from the antigens of foreign substances. At times, however, the body's immune system loses the ability to tell the difference between the antigens of foreign substances and its own antigens. As a result, an *autoimmune disease* occurs in which antibodies and T cells attack body tissues.

The following are some examples of autoimmune diseases and their effects on the body:

- *Addison's disease*—destruction of adrenal glands' cells.
- *Graves' disease*—oversecretion of thyroid hormone.
- *Rheumatoid arthritis*—inflammation of joints.
- *Multiple sclerosis (MS)*—destruction of the myelin sheath covering axons of nerve cells.
- *Lupus erythematosus (SLE)*—antibodies form against DNA, RNA, red blood cells, and other components of the body. Death may occur due to large antibody deposits in organs such as the kidneys.

Recent evidence suggests that autoimmune diseases are triggered by viral infections in which specific viruses cause the body's immune system to attack healthy tissues and cells. Other studies indicate that Grave's disease and Addison's disease may be inherited.

Organ and Tissue Transplants

The first kidney transplant was performed in 1954. Since then, the transplanting of kidneys has become a routine procedure. A donated kidney that comes from an identical twin does not trigger an immune response because donor and recipient have the same genetic makeup. However, a kidney transplant between nonidentical twins usually triggers an immune response because the tissues of donor and recipient are not compatible. Their noncompatibility is due to genetic differences in MHC protein. Initially, T cells appear at the transplant site. Then, many antibodies follow and eventually destroy the foreign tissue. Rejection of foreign organs and tissues can be controlled by *immunosuppressive drugs* such as *cyclosporine.* An immunosuppressive drug works best when the donor organ is composed of cells that contain the same type or a similar type of MHC protein as the cells of a prospective recipient. MHC protein enables T cells to recognize a foreign protein. Cyclosporine (derived from a mold) suppresses an immune response by controlling the action of T cells. After an organ or tissue is accepted and is functioning properly, cyclosporine is withdrawn and normal immunity returns.

Allergies

The immune system may overreact to a variety of proteins and foreign substances, including pollen, animal dander, penicillin, chocolate, and milk. This overreaction is a type of hypersensitivity called an *allergy*, and the offending substances are called *allergens*. The reason why an individual is allergic to a particular allergen has still not been discovered.

As discussed earlier, mast cells secrete histamines in the presence of foreign substances, which produce familiar allergic symptoms such as sneezing, tearing of eyes, and inflammation of tissues.

Histamines may also cause *asthma*. A person suffering from asthma cannot breathe normally because of narrowed air passageways in which mucus accumulates. Drugs containing the hormone *epinephrine* are taken or *antihistamines* are used to open air passageways and restore normal breathing. Antihistamines counteract the effects of histamines; epinephrine widens the air passageways by relaxing smooth muscle in the bronchioles.

If a person who has been previously exposed (sensitized) to an injected antigen (for example, penicillin) receives another injection of the same antigen, *anaphylactic shock* can occur. The production of histamine and prostaglandins, which accompanies anaphylactic shock, causes surface blood vessels to *dilate*, or increase in diameter. Blood pressure drops rapidly and shock may cause death if epinephrine is not administered immediately. To prevent dangerous allergic reactions, a person should avoid making contact with sensitizing antigens. To diagnose sensitivity, a physician injects a small amount of a suspected allergen under the skin; sensitivity is shown by a *wheal*, or patch of skin that reddens, swells, and itches.

The immune system defends the body when the skin and other organs fail to prevent invasion by pathogens. Phagocytes engulf and ingest bacteria, viruses, and other foreign substances. B cells and T cells either release antibodies or destroy foreign cells. Immune responses, however, can cause allergies as a result of hypersensitivity and trigger autoimmune diseases.

AIDS, or Acquired Immune Deficiency Syndrome

AIDS is caused by a *retrovirus* (contains double-stranded RNA) called *HIV*, or the *human immunodeficiency virus*. HIV attacks and destroys helper T cells (TH cells). AIDS is classified as an immunodeficiency disease because an infected person cannot produce antibodies to counteract other pathogens or their toxic products. Thus, people who have AIDS often suffer from a variety of opportunistic infectious diseases, such as tuberculosis and pneumonia, as well as from a rare form of cancer called *Kaposi's sarcoma*. Medical researchers still do not know how HIV originated. Current speculation is that HIV originated in African monkeys, mutated, and was then spread throughout the world. Figure 6–4 shows the structure of HIV.

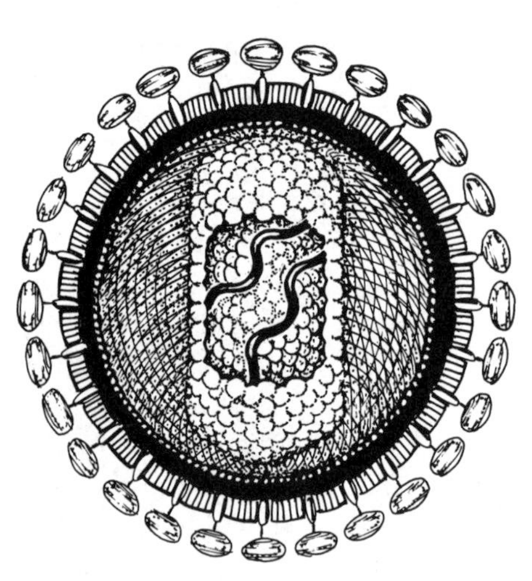

Figure 6–4 The Human Immunodeficiency Virus (HIV)

An infected person produces antibodies to HIV. Thus, when a person is given a blood test for the antibody, the result is "HIV positive." Surprisingly, the antibodies to HIV do not attack the virus. They act as "markers," revealing the presence of the virus that causes AIDS. The early symptoms of AIDS are similar to those of infectious mononucleosis. They include fever, body aches, and a persistent feeling of fatigue. A person who is HIV positive can spread the virus during an extended incubation period, lasting more than ten years. For that reason, blood, organ, and sperm donations should not be made by people who have been diagnosed as HIV positive. In time, HIV positive people develop swollen lymph nodes, body rashes, and loss of appetite and weight. Most people with fully developed AIDS die within 10 months to a year of pneumonia and other infectious diseases.

Transmission of AIDS. AIDS is usually transmitted by an exchange of body fluids, such as blood, vaginal secretions, and semen, that contain HIV. Thus, the two main ways that an HIV positive person can transmit the virus that causes AIDS is through sexual contact and the sharing of hypodermic needles by intravenous drug users. Before 1986, people sometimes contracted AIDS after receiving blood transfusions tainted with HIV. Thus, blood screening for HIV was instituted by hospitals and blood banks. This measure has greatly reduced the chance of contracting AIDS by a blood transfusion. There is no evidence that AIDS is transmitted by social contact with an infected person, such as touching or using the same towel.

Treating AIDS. At present, the drug AZT (azidothymidine) is used to treat people with AIDS. AZT interferes with the virus's ability to reproduce in TH cells. Other drugs to treat AIDS have been developed, too. One class of drugs, the *protease inhibitors*, prevents the AIDS virus from entering TH cells. At present, complicated regimens of various drugs are used to deal with the virus's ability to mutate. These drugs are not effective in treating all AIDS patients, but they do help many people.

Preventing AIDS. The only sure ways to prevent HIV infection are to abstain from having sexual relations and to avoid the exchange of body fluids, especially the blood-to-blood contact that may result from the sharing of needles during intravenous drug use. The use of a condom may also help to prevent the spread of HIV during sexual relations.

Section Quiz

1. An autoimmune disease, such as rheumatoid arthritis, is caused by (*a*) a hypersensitivity to an allergen (*b*) an excess of B cells (*c*) an attack on body tissues by T cells and antibodies (*d*) an excess of histamine.
2. Anaphylactic shock is caused by (*a*) a type of hypersensitivity (*b*) a lack of antibodies (*c*) a type of virus (*d*) an excess of T cells.
3. Which of the following is an immunosuppressive drug? (*a*) epinephrine (*b*) cyclosporine (*c*) histamine (*d*) prostaglandin.
4. Kaposi's sarcoma is a rare form of skin cancer associated with (*a*) the AIDS virus (*b*) an overexposure to ultraviolet radiation (*c*) the ingestion of large amounts of processed foods (*d*) an excess of dietary fat.
5. To prevent contracting AIDS, a person should (*a*) avoid touching a person with AIDS (*b*) ingest AZT before and after having sexual intercourse (*c*) always use a condom when having sexual relations (*d*) receive a series of vaccinations with a modified retrovirus.

Chapter Review Questions

The following questions will help you check your understanding of the material presented in the chapter.

1. Which statement is correct? (*a*) The immune system has a "memory." (*b*) Diseases such as measles never recur. (*c*) Vaccination usually results in active immunity. (*d*) All of the above.
2. Blood typing is a procedure that relies on (*a*) an antigen-antibody reaction (*b*) a precipitins reaction (*c*) platelet agglutination (*d*) DNA differences in granulocytes.
3. Suppressor T cells (*a*) moderate production of B cells (*b*) limit antibody production (*c*) gain immunocompetency in the thymus gland (*d*) all of the above.
4. Which statement is correct? (*a*) Cyclosporin is an effective antibiotic. (*b*) Transplant rejection rarely occurs among members of the same family. (*c*) Rheumatoid arthritis affects only elderly people. (*d*) All of the above.
5. The immune system is seriously impaired by (*a*) the AIDS virus (*b*) multicellular parasites (*c*) protozoan parasites (*d*) all of the above.
6. A substance that causes an immune response when introduced into the human body is (*a*) glucose (*b*) insulin (*c*) an antibody (*d*) an antigen.
7. The protein coat of a virus is called a (*a*) capsule (*b*) cyst (*c*) cell wall (*d*) capsid.
8. The AIDS virus affects the body by (*a*) destroying lymphocytes called T cells (*b*) destroying lymphocytes called B cells (*c*) increasing the number of granulocytes in the blood (*d*) decreasing the concentration of immunoglobulins in the blood.
9. Histamine production by mast cells helps the body by (*a*) reducing swelling at an infection site (*b*) controlling itching and sneezing (*c*) deactivating antigens (*d*) attracting phagocytes to an infection site.
10. The main function of interleukins is to (*a*) activate antigen-antibody complexes (*b*) depress the production of plasma cells (*c*) stimulate the cloning of B cells (*d*) stimulate the cloning of TH cells.

11. Which statement about plasma cells is correct? (*a*) They are short-lived. (*b*) They are also known as helper T cells. (*c*) They initiate lysis of antigens. (*d*) They are not part of the immune system.
12. After the body's immune system successfully fights off invading pathogens, (*a*) plasma cells change into memory cells (*b*) T cells suppress antibody formation by B cells (*c*) T cells secrete interferons (*d*) the population of B cells decreases.
13. The Sabin polio vaccine provides long-lasting immunity because (*a*) it requires booster shots at appropriate intervals (*b*) it is made from a live virus that multiplies (*c*) it is made from a dead virus that stimulates continuous antibody production (*d*) it is composed of synthetic antigens that activate phagocytosis.
14. The early symptoms of AIDS resemble those of (*a*) gonorrhea (*b*) syphilis (*c*) infectious mononucleosis (*d*) genital herpes.

Biology Challenge

The following questions will provide practice in answering SAT-II type questions.

Part I

Select the letter of the statement that best completes the sentence.

1. Allergic symptoms are accompanied by an increase in the number of (*a*) monocytes (*b*) lymphocytes (*c*) neutrophils (*d*) eosinophils (*e*) macrophages.
2. An example of an autoimmune disease is (*a*) AIDS (*b*) multiple sclerosis (*c*) Tay-Sachs disease (*d*) Lyme disease (*e*) polio.
3. Common vehicle transmission spreads pathogens to individuals by means of (*a*) blood-to-blood contact (*b*) faulty air conditioning ducts (*c*) improperly cooked food (*d*) vectors, such as houseflies (*e*) droplet infections, such as sneezing.
4. Immunological enzymes, such as hemolysins and kinases, (*a*) are produced by agranulocytes to counteract a pathogen's defenses (*b*) destroy white blood cells (*c*) destroy bacteria and viruses (*d*) are bacterial adaptations to overcome a host's defenses (*e*) only attack viruses.

5. Choose the correct statement. (*a*) At the site of an infection, monocytes enlarge and change into macrophages. (*b*) Mast cells are leukocytes that secrete histamine. (*c*) Anaphylaxis is the immune response of a sensitized person. (*d*) None of the above. (*e*) All of the above.

Part II

Base your answers to the questions below on the following graph, which shows the relationship of *antigens, antibodies,* and *memory cells.*

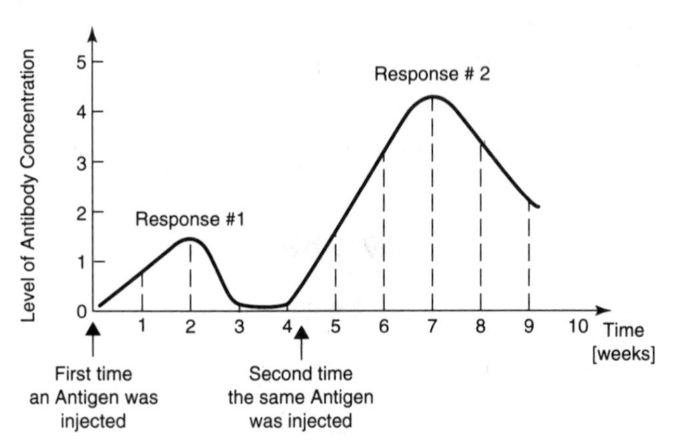

1. Which part of the immune system accounts for response #2? (*a*) helper T cell activity (*b*) increased B cell activity (*c*) memory cell activity (*d*) plasma cell activity (*e*) interleukin activity.
2. How long does it take for significant amounts of antibody to appear after the first antigen injection? (*a*) 2–3 days (*b*) 10–14 days (*c*) 18–20 days (*d*) 21 days (*e*) 22–28 days.
3. During which time period were memory cells formed? (*a*) weeks 4–9 (*b*) weeks 4–7 (*c*) weeks 3–4 (*d*) weeks 1–3 (*e*) week 1 only.
4. What do you predict will happen to antibody concentration during weeks 9–13? (*a*) Probably decrease to level 1, form a plateau, and remain there. (*b*) Probably decrease to the base line and remain there. (*c*) Probably decrease to a point midway between level 1 and level 2 and remain there. (*d*) Remain at level 2. (*e*) Data given is insufficient to make a prediction.

7

Respiration and Excretion

Learning Objectives

When you have completed this chapter, you should be able to:

- **Describe** the functions of the organs that make up the human respiratory system.
- **Contrast** internal and external respiration.
- **Distinguish** between aerobic and anaerobic respiration.
- **Contrast** human respiration with respiration that occurs in other representative organisms.
- **Define** excretion as it relates to the removal of waste products from cells.
- **Discuss** the functions of the organs that make up the human excretory system.
- **Discuss** the adaptations of other organisms that enable them to eliminate waste products.

OVERVIEW

All organisms require energy to carry out their life processes. They get this energy from the end products of digestion by performing a process called respiration. Excess materials and toxic substances are removed from the bodies of living things as wastes by a process called excretion. Normal cell metabolism produces wastes, including excess water, carbon dioxide, mineral salts, ammonia, urea, and uric acid. In this chapter, you will learn that if either respiration or excretion does not function properly, homeostasis is disturbed, and the organism begins to die—cell by cell.

RESPIRATION IN HUMANS

There are two kinds of respiration that all living things use. *Aerobic respiration* is used by humans, and most other organisms, which require a continuous supply of oxygen. *Anaerobic respiration* is used by organisms that can live without oxygen (some bacteria only). In either kind of respiration, the exchange of respiratory gases must take place through a moist *respiratory surface*, which enables osmosis and diffusion to occur.

Humans have developed specific adaptations that carry out respiration and excretion. Breathing is a body function that performs both respiration and excretion. Each time you inhale, your body gets a new supply of oxygen; when you exhale, your body removes excess water and carbon dioxide. Most of the body's excess water, nitrogenous wastes, and mineral salts are removed by the urinary system and the skin.

The cardiovascular system and the respiratory system supply oxygen and remove carbon dioxide from a human. The respiratory system consists of organs that facilitate gas exchange, or the intake of oxygen and the elimination of carbon dioxide. The cardiovascular system transports oxygen and carbon dioxide in the blood in a circuit between the lungs and body cells. The exchange of gases between the atmosphere, blood, and body cells is called *respiration.* There are three stages that characterize the process of respiration:

1. *Breathing*—exchange of oxygen and carbon dioxide between the atmosphere and the lungs.
2. *External respiration*—exchange of oxygen and carbon dioxide between the lungs and the blood.
3. *Internal respiration*—exchange of oxygen and carbon dioxide between the blood and body cells.

In humans, the respiratory surface is the moist inner lining of the *lungs*, which are enclosed within the ribs, muscles of the chest cavity, and the diaphragm. In this part of the body, the respiratory surface always remains moist. Oxygen in the air reaches the respiratory surface by passing through the organs of the respiratory system (Figure 7–1).

The Respiratory System

The respiratory system consists of the nose, pharynx, larynx, trachea, bronchi, and lungs.

Nose. The two openings in the nose, or *nostrils*, lead to narrow spaces called the *nasal passages.* These are lined with a mucous membrane and

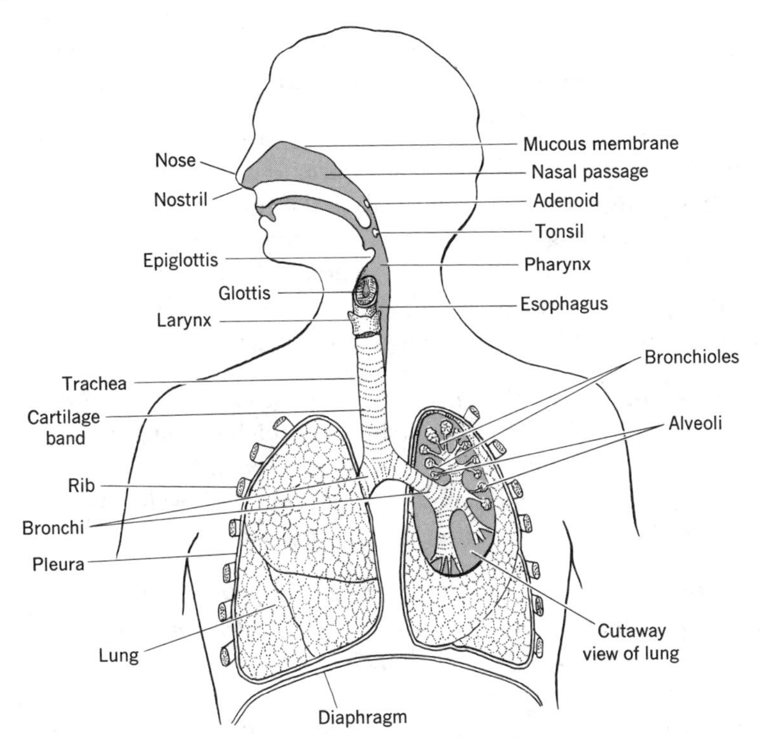

Figure 7–1 Human respiratory system

cells that are covered with cilia. The *mucous membrane* moistens the air and traps dust and bacteria. The cilia move dust-laden mucus and bacteria toward the nostrils. When you sneeze, mucus is expelled from the nose. Hairs that line the nostrils filter dust from inhaled air. Beneath the mucous membrane, blood in the capillaries gives off heat that warms cold air to body temperature (37°C) before air reaches the lungs.

The pharynx. Air passes through the nasal passages into the *pharynx*, or throat, which is the cavity at the back of the mouth. In the wall between the mouth and the pharynx are large lymph glands called *tonsils*. Other lymph glands, called *adenoids*, are located in the rear wall of the pharynx. Both of these lymph glands help protect the body against infection.

The larynx. The *larynx*, commonly called the voice box or Adam's apple, is located at the upper part of the windpipe. The *glottis* is a slitlike opening in the larynx where air enters the windpipe. Above the glottis is a flaplike structure called the *epiglottis*. When you swallow, the epiglottis covers the glottis, usually preventing food from entering the windpipe. Below the glottis are the *vocal cords*. Sounds are produced as air from the windpipe passes over the vocal cords making them vibrate.

The trachea (windpipe). Air passes through the larynx into a tube called the *trachea,* or windpipe. The trachea is about 10 centimeters long and about 2.5 centimeters wide. The walls of the trachea are strengthened by horseshoe-shaped bands of cartilage. The inner wall of the trachea also is lined with a mucous membrane and ciliated (cilia-bearing) cells. Dust and bacteria-laden mucus are carried by the cilia to the throat, where the mucus that accumulates is either coughed up or swallowed.

The bronchi. The trachea branches into two *bronchi* (singular, bronchus). Each bronchus leads into a lung in which the bronchus branches many times in a treelike fashion to form smaller air tubes called *bronchioles.* Like the trachea and nasal passages, the inner walls of the bronchi and bronchioles are lined with a mucous membrane and ciliated cells.

The lungs. The *lungs* are two spongy organs that occupy most of the chest cavity. Each lung is covered with a two-layered protective membrane, called the *pleura.* Inside each lung, the bronchioles branch into still smaller tubes that eventually open into *air sacs*, which consist of a group of tiny saclike *alveoli* (singular, alveolus)(Figure 7–2). Each alveolus is microscopic, consisting of a single layer of epithelial cells surrounded by a network of capillaries. The epithelial cells of the alveoli form the surface of the lungs, through which oxygen (present in the alveoli) and carbon dioxide and water (present in the blood of the capillaries) are exchanged.

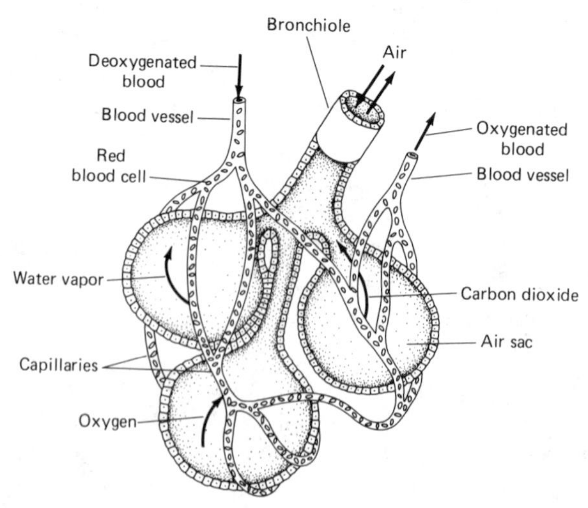

Figure 7–2 Alveoli (magnified)

If the membranes of the alveoli dry out, the exchange of respiratory gases stops and suffocation occurs. Thus, a person trapped in the hot, dry air of a fire may die from a lack of oxygen because the diffusion of oxygen cannot occur across a dry respiratory surface.

Breathing

Breathing consists of drawing air into the lungs *(inhalation)* and then expelling it from the lungs *(exhalation)*. Figure 7–3 and Table 7–1 show how inhalation and exhalation move air into and out of the lungs.

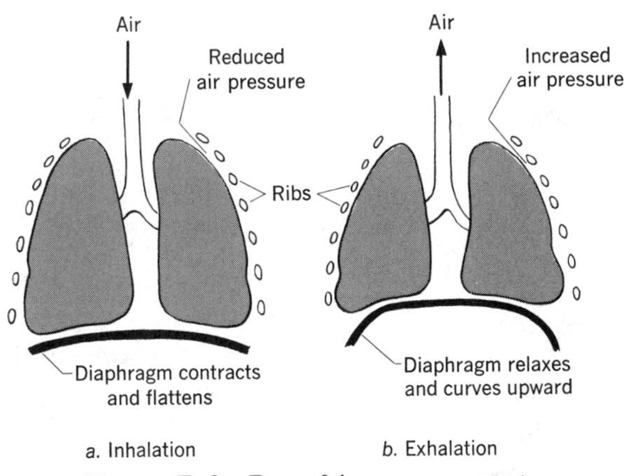

Figure 7–3 Breathing movements

The normal rate of breathing varies from 12 to 18 times a minute. The breathing rate is controlled by the amount of carbon dioxide in the blood. Excess carbon dioxide in the blood causes it to become more acidic. Thus, as the volume of carbon dioxide increases in the blood, the pH decreases. This acidic condition of the blood stimulates the *respiratory center* in the brain, which then stimulates the nerves that carry electrochemical impulses to the diaphragm and muscles of the chest wall. These organs respond to the impulses from the respiratory center by increasing the rhythm of breathing movements. As a result, there is an increase in breathing rate, which causes the release of more carbon dioxide from the blood into the lungs. The removal of carbon dioxide from the blood raises the pH and the rate of breathing returns to normal. This automatic control of breathing is another example of homeostasis.

Table 7–1 Breathing Movements

Breathing Stage	*Rib Movements*	*Movement of Diaphragm*	*Size of Chest Cavity*	*Pressure in Chest Compared to Outside Pressure of Air*	*Movement of Air*
Inhalation	Upward and outward	Contracts and flattens	Increases	Less	Flows into lungs
Exhalation	Downward and inward	Relaxes and curves upward	Decreases	Greater	Squeezed out of lungs

Stages of Respiration

In addition to breathing, respiration involves external respiration and internal respiration.

External respiration involves the breathing movements and the exchange of oxygen and carbon dioxide between the alveoli of the lungs and the blood. Oxygen passes from the alveoli into the capillaries, and carbon dioxide and water enter the alveoli from the capillaries by diffusion. The exchange of gases and water is caused by different concentrations of oxygen, carbon dioxide, and water; molecules of oxygen, carbon dioxide, and water move from a region of greater concentration to a region of less concentration. Thus, when you inhale, oxygen enters the blood; and when you exhale, carbon dioxide and excess water are removed.

Internal respiration involves the exchange of oxygen and carbon dioxide between the blood and the body cells. Red blood cells carry oxygen to the body cells as oxyhemoglobin. When blood containing oxyhemoglobin reaches oxygen-poor tissues, the oxygen separates from the hemoglobin and diffuses into the oxygen-poor cells. As this happens, carbon dioxide diffuses out of the tissue's cells and enters the blood. Carbon dioxide combines with water in the blood to form bicarbonate ions that are carried by both the plasma and the red blood cells. In red blood cells, some carbon dioxide combines with hemoglobin, forming carbon dioxide–hemoglobin (carboxyglobin). In the lungs, carbon dioxide separates from the hemoglobin and from bicarbonate ions in the plasma and then diffuses into the carbon dioxide-poor alveoli.

Cellular respiration involves anaerobic and aerobic oxidation reactions that occur within cells. Anaerobic respiration occurs within the cytoplasm; aerobic respiration occurs within the mitochondria.

Anaerobic respiration (glycolysis). In *anaerobic respiration,* certain enzymes break apart the chemical bonds in glucose between carbon atoms and between molecules made of carbon and hydrogen atoms. Figure 7–4 summarizes the reactions of glycolysis.

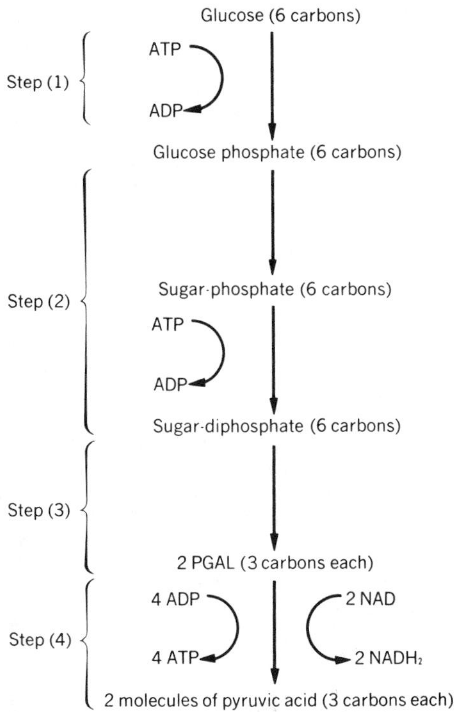

Figure 7–4 Summary of major steps in glycolysis

In steps 1 and 2, as glucose is decomposed, two molecules of adenosine triphosphate (ATP) release energy as they are changed to adenosine diphosphate (ADP). In step 3, the energy is used to form phosphoglyceraldehyde (PGAL). In step 4, four molecules of ATP are produced. At the same time, a hydrogen atom is removed from each molecule of PGAL and is held temporarily by another molecule called a *hydrogen acceptor,* or *hydrogen carrier. Nicotinamide-adenine dinucleotide* (NAD) is a hydrogen acceptor that accepts hydrogen atoms and becomes NADH (reduced NAD). As a result of step 4, pyruvate, a salt of pyruvic acid, is formed. (In living things, acids are usually changed to salts of the acid, such as pyruvate.) At this point, more ATP is produced than is used in these reactions. Thus, two molecules of ATP are used and four molecules are produced. This gives a net gain of two molecules of ATP.

The anaerobic stage of respiration ends with the formation of pyruvic acid. In some anaerobic organisms, however, the breakdown of glucose may go beyond pyruvic acid, producing additional energy. For example, when yeast grows in a culture that contains glucose in the absence of oxygen, pyruvic acid is broken down into ethyl alcohol (C_2H_5OH) and carbon dioxide (CO_2). This process is called *alcoholic fermentation.*

Aerobic respiration. In *aerobic respiration,* pyruvic acid molecules formed by glycolysis pass into the mitochondria. Here, the aerobic reactions occur in two phases. One phase is called the *citric acid cycle,* or *Krebs cycle;* the other phase is called the *hydrogen transport system.* Both phases require specific enzymes found in the folds of mitochondria. Figure 7–5 summarizes the major steps of the Krebs cycle and Figure 7–6 provides a model for the hydrogen transport system.

Phase 1: the Krebs cycle. During the Krebs cycle, pyruvic acid loses a molecule of carbon dioxide and changes into the compound *acetyl coenzyme* A. This newly formed compound then combines with another compound to form citric acid (citrate). This begins the cycle, in which citric acid, after several enzyme-controlled steps, changes to the compound that combined with acetyl coenzyme A to form citric acid. During the Krebs cycle, some energy, carbon dioxide molecules, and hydrogen atoms are released. The carbon dioxide is carried away and excreted; the released energy is stored in the high-energy bonds of ATP molecules; and the hydrogen atoms are stored temporarily in $NADH_2$, later released to enter the hydrogen transport system.

The same series of reactions is repeated as more pyruvic acid enters the cycle. Think of turning a bicycle wheel by pushing with your hand. As long as you push the wheel with your hand, the wheel turns. In the Krebs cycle, the continuous supply of pyruvic acid into the cycle is similar to the pushing action of your hand.

Phase 2: the hydrogen transport system. In the hydrogen-transport phase, hydrogen and electrons (released by other reactions) are picked up and transferred among enzymes called *flavins* and *cytochromes.* Each transfer of hydrogen and electrons releases energy and the energy is stored as chemical energy when ADP molecules change into ATP molecules. The hydrogen atoms and electrons are finally accepted by oxygen to form water. The water is carried away by the blood and either excreted by the lungs and kidneys or used by the cells.

The transfer of hydrogen along the hydrogen transport system is similar in effect to a ball at the top of a stairway that receives a slight push.

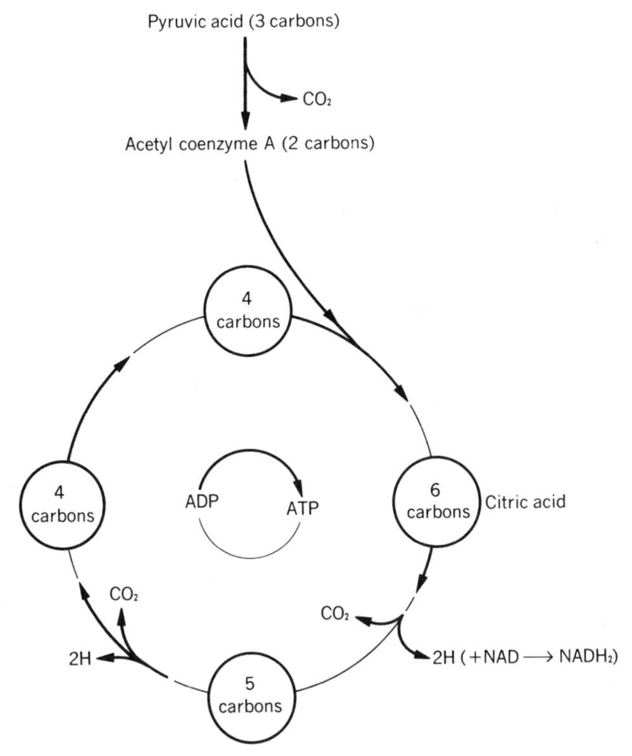

Figure 7–5 Summary outline of the Krebs cycle

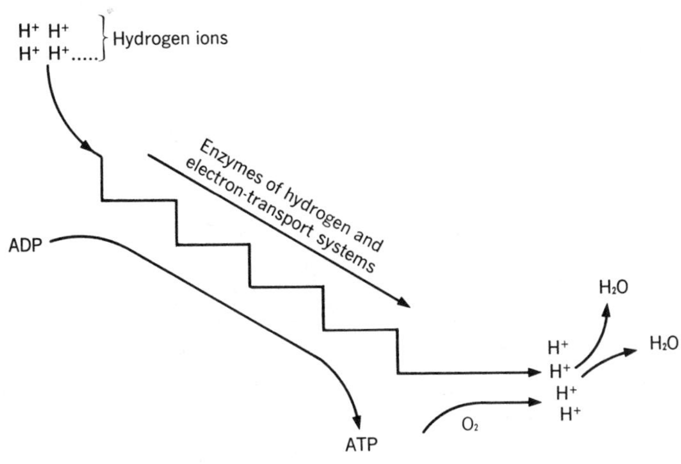

Figure 7–6 Scheme of the hydrogen transport system

A ball rolls down one step at a time until it is at the bottom and comes to rest. At the top of the stairs, the ball has 100 percent potential energy and zero percent kinetic energy. As the ball falls lower and lower, its potential energy is converted to kinetic energy. Likewise, when a hydrogen ion and an electron travel from one acceptor to another, the potential energy decreases. When hydrogen finally unites with oxygen, the potential energy is exhausted, or zero percent.

Energy relations in cellular respiration. Studies indicate that 36 molecules of ATP are formed by the Krebs cycle and the hydrogen transport system. Adding the two molecules of ATP formed by glycolysis, there is a net gain of 38 ATP molecules from one molecule of glucose. These energy relations are shown in Figure 7–7.

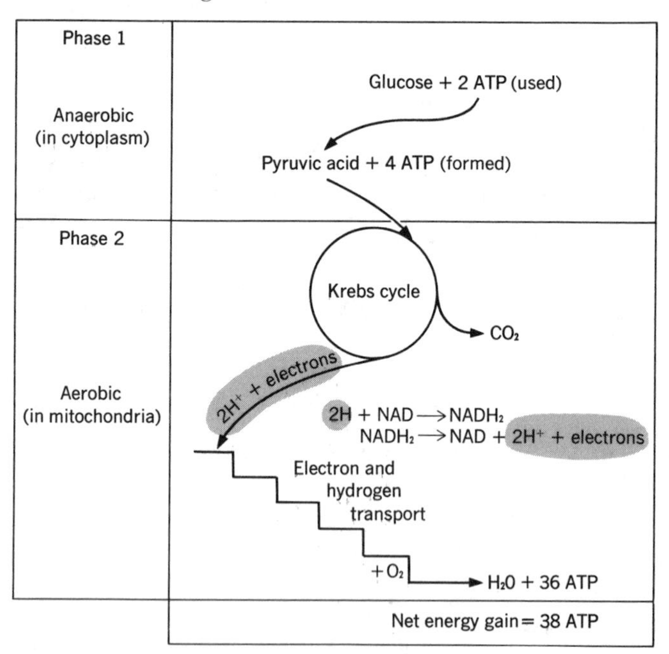

Figure 7–7 Schematic summary of energy release in cells

In the laboratory, when 180 grams of glucose are oxidized, or combine with oxygen, about 673,000 calories of heat are rapidly released. In cells, when the same amount of glucose is oxidized, about 380,000 calories are stored as chemical energy in the form of ATP molecules. The remaining 293,000 calories are released as heat. This means that the cells gain about 56 percent of the energy resulting from the oxidation of glucose. In fact, a cell is more efficient than an electric power plant, which can extract only

30 percent of the total energy from its fuel. Some of the heat released by warmblooded animals provides the optimum temperature for enzyme-controlled reactions. Thus the efficiency of cellular respiration in humans is greater than 56 percent.

Section Quiz

1. Aerobic respiration (*a*) produces a net gain of 2 ATP molecules (*b*) does not require oxygen (*c*) produces a net gain of 36 ATP molecules (*d*) ends with the formation of pyruvic acid.
2. The lymph glands in the rear wall of the pharynx that help the body fight infections are the (*a*) glottis and the epiglottis (*b*) tonsils and adenoids (*c*) alveoli and pleura (*d*) trachea and glottis.
3. Excess carbon dioxide in the blood (*a*) causes the pH of the blood to increase (*b*) has no effect on the pH of blood (*c*) causes the glottis to swell (*d*) causes the pH of the blood to decrease.
4. Which organic compound is an end product of glycolysis? (*a*) pyruvic acid (*b*) citric acid (*c*) acetyl coenzyme A (*d*) phosphoglyceraldehyde.
5. The number of ATP molecules formed from a molecule of glucose by aerobic respiration (Krebs cycle) and anaerobic respiration (glycolysis) is (*a*) 2 (*b*) 4 (*c*) 36 (*d*) 38.

RESPIRATION IN OTHER ORGANISMS

Throughout the history of life on Earth, adaptations for respiration have developed in different kinds of organisms. Whatever the adaptation, however, differences in concentration gradients have facilitated the exchange of oxygen and carbon dioxide.

Protozoans

The respiratory surface of a protozoan, such as ameba, is the plasma membrane. Dissolved oxygen enters directly by diffusion through the plasma membrane, and carbon dioxide leaves the same way. Within the body of a protozoan, oxygen is transported to all parts by diffusion and cyclosis (cytoplasmic movement) and is absorbed by the mitochondria. Waste products of respiration, including carbon dioxide and excess water, diffuse out of the cytoplasm through the plasma membrane into its watery environment.

Hydra

All of a hydra's cells are in direct contact with water. As in protozoans, dissolved oxygen and carbon dioxide diffuse in and out of the plasma membrane of each cell.

Earthworm

The respiratory surface of an earthworm is its skin. A layer of mucus on an earthworm's skin helps keep it moist. Oxygen diffuses from the air through the moist skin and enters capillaries just beneath the skin. Inside of these capillaries, oxygen combines with the hemoglobin dissolved in the plasma to form oxyhemoglobin. Hemoglobin is not present in an earthworm's blood cells, as it is in humans. The blood transports oxyhemoglobin to the tissue cells, where oxygen is released and carbon dioxide enters the blood. From capillaries beneath the skin, carbon dioxide diffuses through the skin and into the air. After a heavy rain, earthworms rise to the soil's surface because air pockets in the soil fill with water. An earthworm will drown if it is trapped in a flooded pocket. If it remains too long on the surface, an earthworm will die of suffocation because its skin (respiratory surface) will dry out.

Grasshopper

The blood of a grasshopper does not transport oxygen and carbon dioxide. These respiratory gases are transported to and from cells by means of air tubes called *tracheae* (Figure 7–8). Ten pairs of openings, or *spiracles*, lead into the tracheae. The innermost ends of the tracheae are thin and moist and are in close contact with the tissue cells. The innermost ends of a grasshopper's tracheae comprise its respiratory surface.

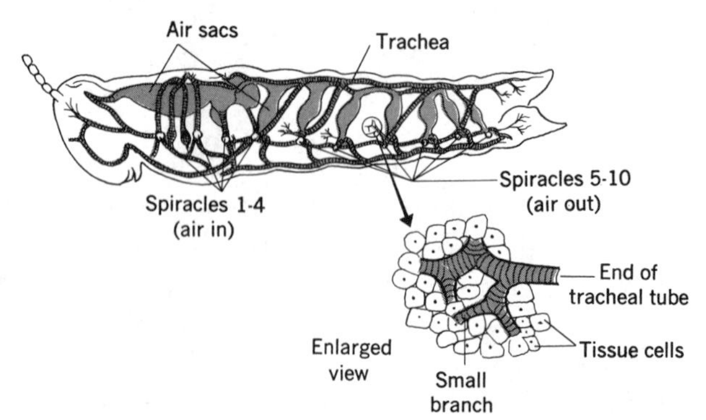

Figure 7–8 Respiration in the grasshopper

As air enters the first four pairs of spiracles, valves shut the last six pairs. When valves close the first four pairs, the valves of the last six pairs open and air is forced out. Pulsating movements of the abdomen of the grasshopper, as well as walking and flying, help move air through the tracheae.

Bird

The flight muscles of flying birds use a lot of energy and therefore require a large volume of oxygen. For this reason, the respiratory system of birds has developed large air sacs, which supply additional oxygen (Figure 7–9). Breathing in birds occurs in two stages:

1. In the first stage, inhaled air passes through the trachea and enters the *posterior* (rear) *air sacs*: exhaled air flows from the posterior air sacs into the lungs and then out of the body.
2. In the second stage, inhalation draws air into the *anterior* (front) *air sacs*; then, an exhalation pushes air out of the anterior air sacs, into the trachea, and finally out of the body.

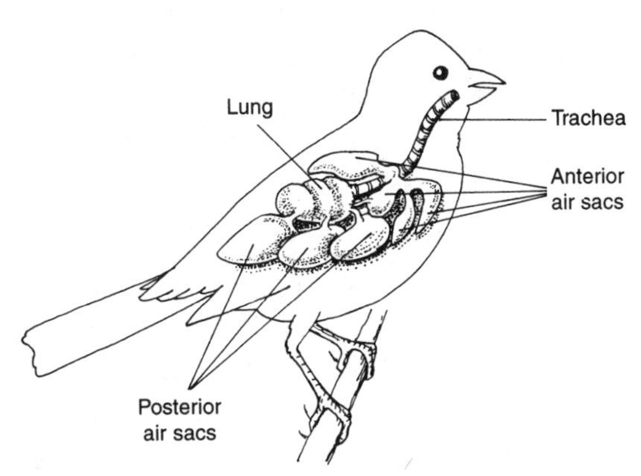

Figure 7–9 Respiratory organs of the bird

Most birds have little difficulty breathing at an elevation of 3,000 meters. A rabbit of equal weight, similar metabolic rate, and blood chemistry cannot breathe at that same elevation. In fact, birds have developed the most efficient breathing adaptations of all aerobic organisms.

Table 7–2 compares the respiratory surfaces of the protozoan, hydra, earthworm, grasshopper, bird, and human.

Table 7–2 Respiratory Surfaces

Organism	*Type of Respiratory Surface*
Protozoan	Plasma membrane
Hydra	Plasma membrane of each cell
Earthworm	Moist skin
Grasshopper	Network of tracheae
Bird	Lungs and air sacs
Human	Air sacs in lungs

EXCRETION IN HUMANS

The process that removes metabolic wastes from the cells of an organism is called *excretion.* If metabolic wastes accumulate, homeostasis is upset and the cells die. Organisms have developed different adaptations to excrete waste products. The following sections describe excretion in humans and several other organisms.

In humans, the organs of excretion include the lungs, liver, kidneys, skin, and large intestine.

The lungs. The tiny droplets of water that form when you exhale on a cold surface, such as a windowpane, show that the lungs excrete water. You can show that the lungs excrete carbon dioxide by blowing gently through a straw into a solution of bromthymol blue or into a solution of limewater. In the presence of carbon dioxide, bromthymol blue turns yellow and clear limewater turns milky.

The liver. One function of the *liver* is to remove old red blood cells from the blood. The hemoglobin of the red blood cells is recycled to make bile. In addition, iron is removed from hemoglobin and reused to make hemoglobin for new red blood cells. The liver also removes excess amino acids from the blood. Unlike excess glucose, which is stored in the liver as glycogen, excess amino acids are not stored. One of the processes used by the liver to change excess amino acids to urea is called deamination.

Deamination removes the nitrogen-containing group (NH_2) and an atom of hydrogen from an amino acid to form ammonia (NH_3), which is toxic to cells. The cells of aquatic animals are safe from the poisonous effect of ammonia because it quickly dissolves in water and immediately diffuses out of cells into the surrounding water. Fish, for example, excrete ammonia through their gills. Humans and other terrestrial (land-dwelling) animals must change the toxic ammonia into the relatively harmless waste urea and water (urine).

The kidneys. The main organs of the urinary system include the kidneys, ureters, urinary bladder, and the urethra (see Figure 7–10).

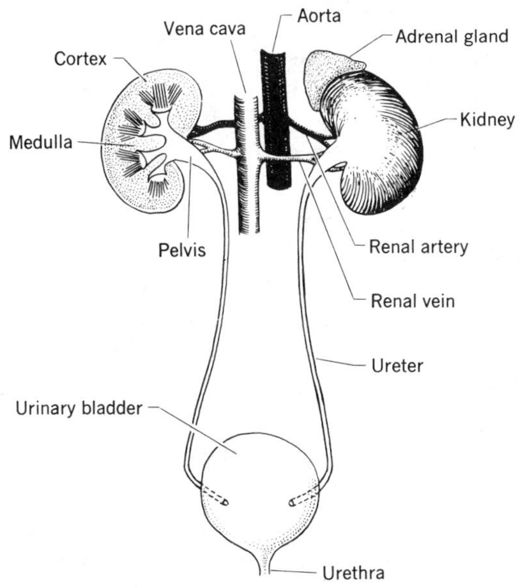

Figure 7–10 Human urinary system

The *kidneys* are bean-shaped, reddish-brown organs at the rear of the abdominal cavity just below the diaphragm. Each kidney consists of an outer layer called the *cortex*, an inner layer called the *medulla*, and a sac-like chamber called the *pelvis*.

The cortex contains about a million *nephrons*. Each nephron has a system of coiled tubes associated with a rich supply of capillaries (Figure 7–11). Branches of the renal artery carry blood to a tuft of capillaries, or *glomerulus* (plural, glomeruli), which is enclosed by a double-layered cup of epithelial cells called *Bowman's capsule*. Bowman's capsule leads into a U-shaped tubule that joins a larger tubule called the *collecting tubule*. The medulla is composed mainly of collecting tubules. Urine in the collecting tubules drains into the pelvis of the kidney, and then down into the *ureter*, which

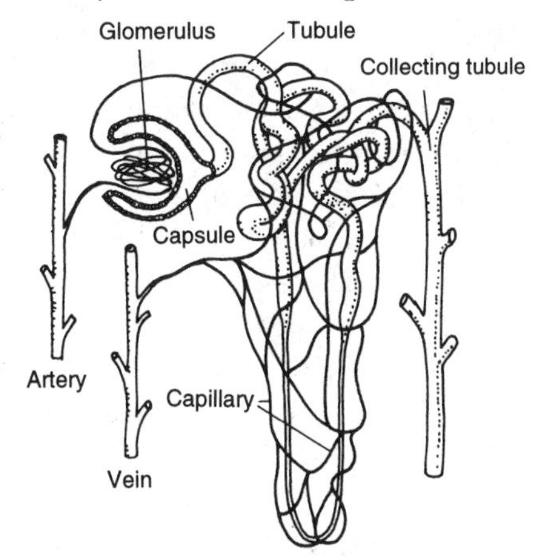

Figure 7–11 Nephron (magnified)

empties into the *urinary bladder.* From the urinary bladder, which temporarily stores urine, the urine leaves the body by the *urethra.*

From the capillaries of a glomerulus, the blood joins the blood from capillaries that surround the tubule of a nephron. These capillaries finally join a branch of the renal vein, which eventually takes blood back to the heart through the inferior vena cava. Blood in the renal vein contains less urea, water, and mineral salts than blood in the renal artery. The removal of these wastes and the formation of urine involves the processes of *filtration* and *reabsorption.*

Filtration. Filtration occurs in the glomerulus. Here, the pressure of the blood within the capillaries forces a plasmalike fluid, called the *nephric filtrate,* to pass from the capillaries into Bowman's capsule. The nephric filtrate consists mainly of water, glucose, amino acids, mineral salts, and urea. The filtrate does not contain molecules of plasma proteins. These molecules are too large to pass through the plasma membrane of the cells.

In 24 hours, about 180 liters of filtrate flow from the glomeruli into the U-shaped tubules. Since the human body contains about 5.7 liters of blood and normally expels about 1.4 liters of urine daily, water must be reabsorbed somewhere in the nephrons and pass through the tubules over and over.

Reabsorption. As the nephric filtrate leaves a glomerulus and passes into the U-shaped tubule, useful substances including water, glucose, amino acids, and some sodium and potassium salts are reabsorbed. Reabsorption features ordinary diffusion (passive transport) and diffusion against a concentration gradient (active transport). Reabsorbed substances enter the capillaries surrounding the tubule and become part of the blood; the remaining fluid, called urine, enters a collecting tubule.

The amount of water absorbed by a nephron depends upon the interaction between the concentrations of water and salts in the blood and a hormone, called the *antidiuretic hormone* (*ADH*). ADH regulates the water content of the body by controlling the reabsorption of water in nephrons. This adaptation has enabled land animals to survive in a dry external environment, yet still have an adequate internal supply of water.

The skin. The skin consists of two main layers, the epidermis and the dermis (Figure 7–12). The *epidermis* has outer layers of dead epithelial cells and inner layers of live epithelial cells. The major function of the epidermis is to protect the body, especially by preventing bacteria from

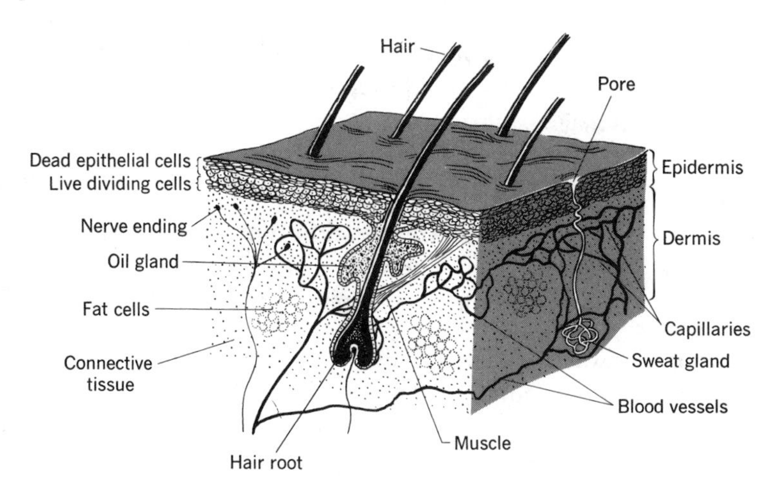

Figure 7–12 Structure of the skin

entering the body and preventing the evaporation of too much water from the body.

The *dermis* is a thicker layer, composed mainly of connective tissue, numerous blood vessels, lymph vessels, fat cells, nerves, and different sense organs. Oil glands in the dermis discharge secretions to the surface of the skin by means of a tiny tube. The oil secretions soften the epidermis and prevent the skin from cracking. Sweat glands in the dermis remove water, mineral salts, and urea from the blood. This mixture, called *sweat*, is excreted by each sweat gland. Sweat flows through a narrow tube from a sweat gland to an opening at the surface of the skin. These openings on the skin, which excrete sweat, are called *pores*.

The amount of sweat produced usually depends on the temperature and humidity in a particular environment. When the body is too warm, small arteries beneath the skin dilate, which increases the volume of blood flow. This in turn, stimulates the sweat glands to excrete more sweat. The blood loses heat when sweat evaporates from the skin and by radiation.

When the body is too cold, arteries beneath the skin contract, which reduces the volume of blood flow. This adaptation reduces sweat production and heat loss by radiation. The sweat glands maintain homeostasis by excreting metabolic wastes and regulating body temperature. The sense organs in the dermis that can detect temperature changes and pressure changes are discussed in Chapter 8.

The large intestine. The main function of the *large intestine* is to temporarily store the solid wastes of digestion and later expel them from the body. A minor function is performed by the cells of the mucous membrane lining the large intestine. They excrete excess salts of metals such as iron, calcium, and magnesium.

EXCRETION IN OTHER ORGANISMS

Over time, organisms have developed different adaptations for excretion. The type of adaptation developed by an animal is closely related to the amount of available water in its environment.

Protozoans. The wastes of protozoans include excess water, carbon dioxide, mineral salts, and ammonia. These are excreted through the plasma membrane by simple diffusion. Freshwater protozoans excrete excess water by active transport and by the pumping of the contractile vacuoles. In this way, the water balance of a freshwater protozoan is maintained. Protozoans that live in salt water lack contractile vacuoles.

Hydra. Like protozoans, hydra excrete metabolic wastes by simple diffusion and active transport. The metabolic wastes excreted by hydra are the same as those excreted by protozoans.

Earthworm. In earthworms, carbon dioxide and water are excreted by diffusion from the capillaries in the skin into the soil. First a mixture of water, ammonia, urea, and mineral salts enters the blood from the tissue cells by diffusion. Then, this mixture diffuses into the fluid of the body cavity. Figure 7–13 shows how these wastes are removed from the fluid within a segment of the body cavity. The segment shown in the figure contains a kidneylike tubule called a *nephridium* (plural, nephridia). Most

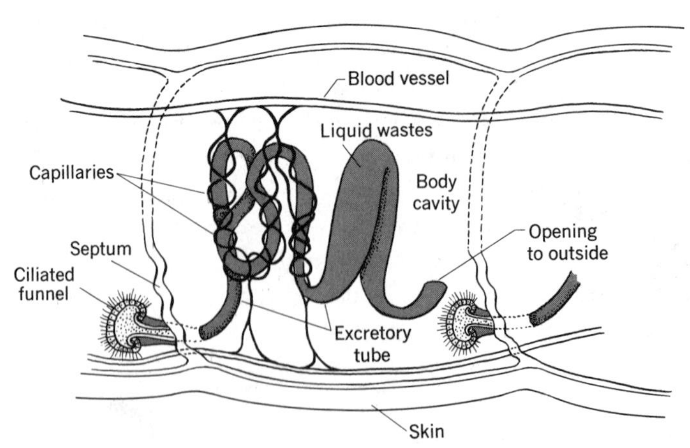

Figure 7–13 Nephridium in a segment of an earthworm

of an earthworm's segments contain two nephridia. The ciliated, funnel-shaped entrance of the nephridium draws in fluid containing metabolic wastes from the body cavity. As these wastes pass through the tubule, essential substances are reabsorbed into the capillaries surrounding the nephridium. The remaining material is expelled through a pore into the soil. Note that the opening of the nephridium is located in one segment while most of the nephridium and the excretory pore are located in the adjacent segment. Because the volume of water available to an earthworm is relatively small, ammonia is changed to urea inside its excretory system.

Grasshopper. In grasshoppers, metabolic wastes such as water, mineral salts, and urea diffuse from tissue cells directly into the blood. *Malpighian tubules* (Figure 7–14) are the excretory organs of the grasshopper. They absorb wastes and change urea to *uric acid*—an insoluble substance. The formation of uric acid is an adaptation for conserving water in organisms

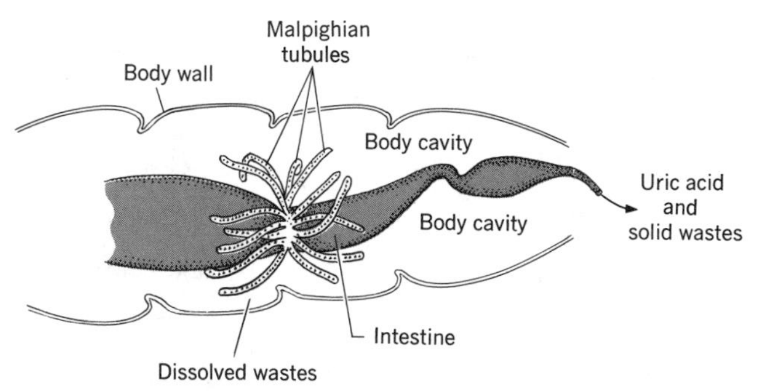

Figure 7–14 Excretory organs of the grasshopper

that either live in a dry environment or do not have a ready supply of water. Uric acid passes from the Malpighian tubules into the intestine and is eliminated with the solid wastes of digestion.

Bird. Unlike most aquatic organisms, terrestrial animals, such as birds, cannot diffuse ammonia directly into water. Like terrestrial animals, birds must conserve water and cannot excrete toxic ammonia in urine. Birds (as well as reptiles and insects) remove ammonia from the circulation before it accumulates by the action of kidneys. Then, ammonia is changed to *uric acid*, which is not as toxic as ammonia and much less soluble in water. Uric acid and fecal materials exit a bird through a common tube, called the *cloaca*.

Guano is an accumulation of bird droppings often deposited on islands. The high volume of uric acid in guano makes it rich in nitrogen. For

Table 7–3 Adaptations for Excretion

Organism	*Adaptations for Excretion*	*Metabolic Wastes Excreted*
Protozoan	Plasma membrane	Water, carbon dioxide, salts, ammonia
	Contractile vacuoles	Water
Hydra	Plasma membranes	Water, carbon dioxide, salts, ammonia
Earthworm	Nephridia	Water, salts, ammonia, urea
	Skin	Water, carbon dioxide
Grasshopper	Malpighian tubules and intestine	Salts, uric acid, very little water
	Tracheal tubes	Carbon dioxide and a little water
Bird	Lungs and air sacs	Carbon dioxide
	Kidneys	Uric acid, minor amount of water, salts
	Liver	Uric acid
	Cloaca	Uric acid, fecal material
Human	Lungs	Carbon dioxide, water
	Skin and kidneys	Water, salts, urea
	Liver	Urea
	Large intestine	Some salts

that reason, guano is collected and used as fertilizer. Table 7–3 summarizes the adaptations for excretion in organisms discussed in this chapter.

The survival of an organism depends in part upon its adaptations for detecting changes in its environment and reacting to the changes in ways that maintain homeostasis. Such adaptations are discussed in the three chapters that follow.

Section Quiz

1. In humans, urine is composed of water, salts, and (*a*) ADH (*b*) urea (*c*) bile pigments (*d*) ammonia.

2. Kidneys help maintain homeostasis by (*a*) excreting urea and uric acid (*b*) regulating blood volume (*c*) regulating blood pH (*d*) all of these.
3. Which animal excretes ammonia directly into its environment? (*a*) bird (*b*) insect (*c*) fish (*d*) whale.
4. In birds, the need for a large volume of oxygen is supplied by (*a*) large air sacs (*b*) a long trachea (*c*) large flight muscles (*d*) a four-chambered heart.
5. An increase in the secretion of antidiuretic hormone (ADH) by the body (*a*) increases urine volume (*b*) decreases urine volume (*c*) increases reabsorption of sodium ions (*d*) decreases secretion of potassium ions.

Chapter Review Questions

The following questions will help you check your understanding of the material presented in the chapter.

1. In humans, gas exchange between the blood and the external environment occurs primarily across the moist membranes of the (*a*) alveoli (*b*) oral cavity (*c*) trachea (*d*) nasal passages.
2. If the respiratory enzymes of an animal are destroyed, which process will probably not occur in the cell membrane of that animal? (*a*) osmosis (*b*) diffusion (*c*) molecular motion (*d*) active transport.
3. Atmospheric air contains approximately 0.04 percent carbon dioxide, and exhaled air contains 4.5 to 5.0 percent carbon dioxide. The process responsible for the difference is (*a*) photosynthesis (*b*) secretion (*c*) hydrolysis (*d*) respiration.
4. In humans, an increase in the rate of respiration and heartbeat occurs when the blood contains (*a*) insufficient glucose (*b*) insufficient protein (*c*) excess nitrogenous waste (*d*) excess carbon dioxide.
5. Which process is represented by the equation below? (*a*) aerobic respiration (*b*) anaerobic respiration (*c*) photosynthesis (*d*) chemosynthesis.

$$C_6H_{12}O_6 + O_2 \xrightarrow{\text{enzymes}} H_2O + CO_2 + ATP$$

6. Collectively, all of the chemical reactions in a cell that break down food molecules in order to release potential energy are called (*a*) cellular respiration (*b*) carbon fixation (*c*) absorption (*d*) excretion.
7. During aerobic respiration, the principal hydrogen acceptor is (*a*) pyruvic acid (pyruvate) (*b*) lactic acid (lactate) (*c*) carbon (*d*) oxygen.

Base your answers to questions 8 *through* 12 on the equation below.

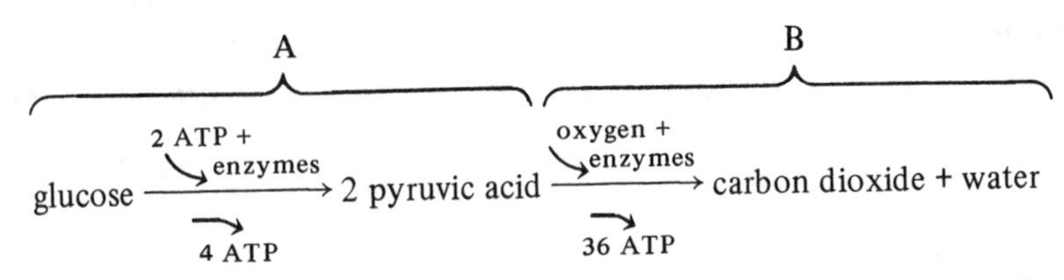

8. The group or groups of reactions that may result in an accumulation of lactic acid are represented by (*a*) A only (*b*) B only (*c*) A and B.
9. Which statement concerning the energy yield of reactions is true? (*a*) It is greater in A than in B. (*b*) It is greater in B than in A. (*c*) It is the same in A and B.
10. Respiration involves the group or groups of reactions represented by (*a*) A only (*b*) B only (*c*) A and B.
11. During fermentation, yeast carries on the group or groups of reactions represented by (*a*) A only (*b*) B only (*c*) A and B.
12. Which group or groups of reactions generally occur in humans? (*a*) A only (*b*) B only (*c*) A and B.
13. During respiration, chemical bond energy is transferred from C-C and C–H bonds of carbohydrates to chemical bonds in the formation of molecules of (*a*) DNA (*b*) protein (*c*) ATP (*d*) oxygen.
14. Two phases that must occur in the complete oxidation of glucose are (*a*) the anaerobic phase and the aerobic phase, both involving molecular oxygen (*b*) the anaerobic phase without molecular oxygen and the aerobic phase with molecular oxygen (*c*) the anaerobic phase with molecular oxygen and the aerobic phase without molecular oxygen (*d*) the anaerobic phase and the aerobic phase, neither involving molecular oxygen.
15. Which is used by cells to activate the anaerobic phase of respiration? (*a*) light (*b*) heat (*c*) ATP (*d*) DNA.

16. The final stages of cellular respiration occur (*a*) along the endoplasmic reticulum (*b*) throughout the cytoplasm (*c*) on the surface of ribosomes (*d*) inside the mitochondria.
17. Which process yields lactic acid and a small amount of usable energy? (*a*) photosynthesis (*b*) dehydration synthesis (*c*) aerobic respiration (*d*) anaerobic respiration.
18. An ameba obtains oxygen for respiration directly from the oxygen that is (*a*) combined with hydrogen (*b*) dissolved in water (*c*) in carbon dioxide molecules (*d*) in glucose molecules.
19. Yeast cells undergoing anaerobic respiration produce carbon dioxide and (*a*) water (*b*) ethyl alcohol (*c*) nitrogen (*d*) glucose.
20. Lactic acid fermentation by certain bacteria occurs as a result of (*a*) dehydration synthesis (*b*) anaerobic respiration (*c*) plasmolysis (*d*) photosynthesis.
21. Which two organisms possess a system of respiratory tubes used for gas exchange? (*a*) earthworm and hydra (*b*) earthworm and grasshopper (*c*) human and grasshopper (*d*) human and earthworm.
22. Nitrogenous wastes are produced by the metabolism of (*a*) lipids (*b*) sugars (*c*) proteins (*d*) starches.
23. The principal nitrogenous waste excreted by humans is (*a*) urea (*b*) mineral salt (*c*) ammonia (*d*) carbon dioxide.
24. Many mitochondria are observed in the tubule cells of nephrons. This observation suggests that the nephron is involved in (*a*) active transport (*b*) passive transport (*c*) osmosis (*d*) diffusion.
25. The Malpighian tubules are associated with excretion in a(n) (*a*) paramecium (*b*) hydra (*c*) grasshopper (*d*) earthworm.
26. Which is the major nitrogenous waste produced by water-dwelling organisms such as parameciums and hydra? (*a*) uric acid (*b*) ammonia (*c*) urea (*d*) nitrates.
27. Water is conserved by the conversion of nitrogenous wastes to uric acid in (*a*) ameba (*b*) humans (*c*) earthworms (*d*) grasshoppers.
28. Wastes are removed from an earthworm by (*a*) nephrons (*b*) nephridia (*c*) Malpighian tubules (*d*) contractile vacuoles.
29. Which invertebrate structures are closest in function to the nephrons of humans? (*a*) tracheae (*b*) pseudopods (*c*) gastrovascular cavities (*d*) Malpighian tubules.

30. The function of the contractile vacuole is to (*a*) exchange gases between the earthworm and its environment (*b*) store most of the grasshopper's hydrolytic enzymes (*c*) transmit impulses in a hydra's nerve net (*d*) maintain water balance between a protozoan and its environment.

Biology Challenge

The following questions will provide practice in answering SAT II-type questions.

Part I

Base your answers to questions *1 through* 5 on the following paragraph, library research, and on your knowledge of biology.

In humans, about 5.7 liters of blood are filtered through the kidneys every 45 minutes. Therefore, in 24 hours, the kidneys filter about 180 liters of blood. In the same period (24 hours), however, a normal adult excretes only about 1.4 liters of urine. This shows that the kidneys reabsorb more than 99 percent of the water that passes through them, and eliminate less than 1 percent as urine. More urine is excreted when the fluid intake is large, and less when it is small.

1. The main function of a kidney is to (*a*) regulate the volume of blood in the body (*b*) keep the percent of urea in the blood below that of uric acid (*c*) reabsorb water (*d*) help circulate blood (*e*) standardize the urinary output.
2. A part of the blood usually not found in the urinary filtrate is (*a*) red blood cells (*b*) urea (*c*) uric acid (*d*) sodium chloride (*e*) glucose.
3. An adult who excretes 3 liters of urine in a 24-hour period most likely (*a*) excretes about the same amount daily (*b*) has about twice the normal volume of blood (*c*) has an increased output of antidiuretic hormone (*d*) has a decreased output of antidiuretic hormone (*e*) has an inflammation of the urinary bladder.
4. According to the paragraph, (*a*) the heart pumps about 6 liters of blood every 45 minutes (*b*) the heart pumps about 20,000 liters of blood in one day (*c*) nearly 6 liters of blood pass through the kidneys in 45 minutes (*d*) urine and blood are similar in chemical composition (*e*) a decrease in urinary output signifies poor blood circulation.

5. In humans and birds, the organ mainly responsible for elimination of excess proteins is the (*a*) spleen (*b*) kidney (*c*) caecum (*d*) gall bladder (*e*) liver.

Part II

Each of the following questions consists of a *statement* in the left column and a *reason* in the right column. Select the correct letter based on the following conditions:

A—The statement is true, but the reason is false.
B—The statement is false, but the reason is true (as a statement of fact).
C—Both statement and reason are false.

Statement	*Reason*
1. The respiratory surface of an earthworm consists of internal gills.	1. Only a moist respiratory suface can function properly.
2. The main wastes excreted by protozoans and hydra include water, carbon dioxide, mineral salts, and ammonia.	2. Both kinds of organisms excrete their wastes mainly through contractile vacuoles.
3. Grasshoppers are to spiracles as snakes are to lungs.	3. Spiracles and lungs are both respiratory sufaces.
4. An earthworm immersed in water for several hours will die.	4. Starvation is the probable cause of death.
5. Most terrestrial animals change ammonia into uric acid.	5. Ammonia is highly toxic to living tissues.

8

Nervous System

Learning Objectives

When you have completed this chapter, you should be able to:

- **Identify** the structure of the human nervous system and relate it to its functions.
- **Describe** the ways in which the nervous system contributes to homeostasis.
- **Explain** the structure of the human eye and the part it plays in gathering environmental stimuli.
- **Relate** the structure of the human ear to its role in gathering sounds.
- **Identify** receptors in other organisms.
- **Explain** how impulses travel through the parts of the human nervous system.
- **Distinguish** between the central nervous system and the peripheral nervous system.
- **Compare** the nervous systems of various representative organisms.
- **Describe** patterns of behavior in humans.
- **Discuss** the effects of drugs on the human body and mind.

OVERVIEW

Organisms possess many adaptive characteristics that enable them to maintain homeostasis. One of these characteristics is a nervous system that helps an organism adjust to changes in both its internal and external environments. In this chapter, you will learn about the human nervous system and compare it with nervous systems found in some other representative organisms.

RECEPTORS

A nervous system functions by detecting *stimuli*, or changes; sending stimuli, as electrochemical signals, to a control center; and relaying responses that reverse, or cancel, original stimuli. A mechanism that reverses an original response triggered by a stimulus is called a *feedback system.*

Planaria, a flatworm usually found in ponds, has photoreceptors, or light-sensitive structures. These photoreceptors are stimulated by light, which causes a planaria to respond by moving to a darker location. Like the planaria's response to light, some human responses to stimuli are automatic. When you enter a brightly lit room, the pupils in your eyes automatically contract, preventing too much light from entering your eyes; when you enter a dimly lit room, your pupils dilate (enlarge), allowing more light to enter your eyes.

A nervous system usually consists of receptors, control centers, and effectors. *Receptors* are either sensitive cells or a collection of tissues, called *sense organs*, which detect specific changes in the environment. The skin, nose, tongue, eyes, and ears have receptors that detect stimuli outside the body (external stimuli). Types of receptors include (1) *chemoreceptors,* which detect molecules and ions dissolved in body fluids (taste and smell); (2) *mechanoreceptors,* which detect changes in pressure, position, and acceleration (hearing and equilibrium); (3) *photoreceptors,* which detect visible light and light intensities (vision); and (4) *thermoreceptors,* which detect variations in radiant energy, or heat.

The types and numbers of receptors vary among animals. In general, animals without backbones, or invertebrates, lack the complex, highly specialized receptors that are found in vertebrates, or animals with backbones. A receptor changes a stimulus into an electrochemical signal, or impulse. An electrochemical signal can be detected and measured by a galvanometer, a device that can measure small amounts of voltage.

Most animals distinguish nutritious substances from toxic substances and register different odors by using *chemoreceptors.* Taste receptors, called *taste buds,* are chemoreceptors found in the mouths of many vertebrates. Taste buds usually are located on the tongue. The structure of a taste bud is similar to an onion bulb—layers of specialized epithelial cells surround receptor cells, which are attached to sensory nerves that reach

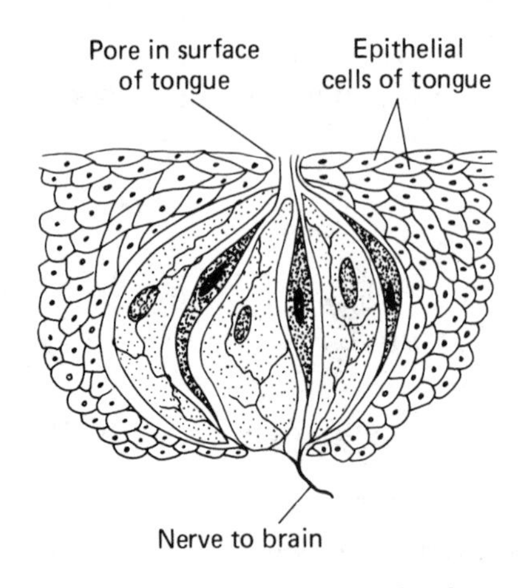

Figure 8–1 A taste bud

the brain (Figure 8–1). In humans, taste buds detect four basic sensations: salty, sweet, bitter, and sour. These basic sensations, in combination with the aroma of food, produce flavor. *Olfactory receptors* detect gas molecules that have distinctive odors. Detecting and responding to odors play a significant role in the survival of many species. For example, when a fox is hunting, it can locate a rabbit by its odor. On the other hand, a male rabbit can use odor to detect female rabbits and select one for mating. Insects and other animals emit and respond to *pheromones*, or odor molecules. Pheromones are used to communicate with other members of the same species, attract a mate, and to mark territorial boundaries.

Taste and smell supplement each other, producing a combination of sensations when interpreted by the brain. In humans, olfactory receptors are located in the mucous membrane of the nasal cavity (Figure 8–2). You

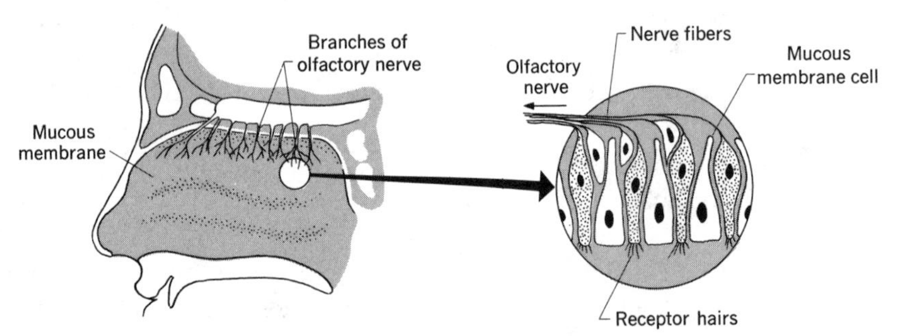

Figure 8–2 Receptors in the nose

can see in the figure that certain cells end in a tuft of receptor cilia. A *stereochemical* theory of smell states that active sites on receptor cilia only respond to odor molecules of a particular size and shape. This is similar to how an enzyme joins with its substrate (lock-and-key theory). Each active site generates a different signal, or impulse, which passes into the olfactory nerve, and then to the brain. The olfactory center in the brain interprets the signal as a specific odor or combination of odors.

Mechanoreceptors detect mechanical pressure or stretching. A cat that falls from a roof twists its body in flight so that it lands feet first. Like most animals, a cat senses the natural position of its body and makes adjustments to gravity, acceleration, and position. In many cases, a cat's acute sense of balance and hearing helps it survive falls if not from too great a height. Most vertebrates, including humans, can sense balance and equilibrium; not all vertebrates hear sounds. Invertebrates such as jellyfish possess an organ of equilibrium called a *statocyst,* which consists mainly of ciliated cells and tiny stones, or *statoliths.* When a jellyfish's body tilts to one side, statoliths also move. As the statoliths move, they bend the cilia, which causes the jellyfish to return to its normal position.

Most vertebrates possess a *vestibular apparatus,* made up of a closed system of channels and fluid-filled sacs inside the ears. The vestibular apparatus, commonly called *semicircular canals,* provides information on the position of the head, posture, and balance. Semicircular canals are discussed further in the section on the human ear.

Photoreceptors detect visible light and ultraviolet light. The reception of light, or photoreception, involves receptor cells and pigment molecules that absorb light energy. Impulses are relayed, interpreted, and monitored into responses. Some invertebrates can respond to light without photoreceptors or pigmented cells. Orientation to light, however, is not vision. For example, protozoans usually stop moving when subjected to light of high intensity; they respond to light, but do not have vision. Vision requires an organ that is highly sensitive to different light intensities. Sight occurs when light is assembled into a pattern of electrochemical signals and then interpreted by a *control center,* or brain.

Insects and crabs have *compound eyes* made up of thousands of closely packed photosensitive cells called *ommatidia* (singular, ommatidium). The image received is a *mosaic*—an unclear composite of the light detected by each ommatidium. Squids, octopuses, and cuttlefish are mollusks that have large eyes similar to those of vertebrates, consisting of a cornea, iris, pupil, lens, vitreous body (jellylike material), and a retina composed of photoreceptors. The well-developed eyes of these mollusks form clear images. Vertebrate eyes form very clear images as a result of *accommodation,* which involves focusing light rays on the retina. This is accomplished by adjusting the position of the lens, as in some fishes and birds of prey, or by adjusting the curvature of the lens, as in humans.

Thermoreceptors detect changes in temperature. Most birds and mammals possess two kinds of thermoreceptors in their skin: one sensitive to low temperatures and the other to high temperatures. In each case, impulses reach a brain part called the *hypothalamus,* which responds by sending messages that either conserve or dissipate heat. Humans conserve heat by shivering and dissipate heat by perspiring. Most thermoreceptors are naked nerve endings located in the skin (Figure 8–3).

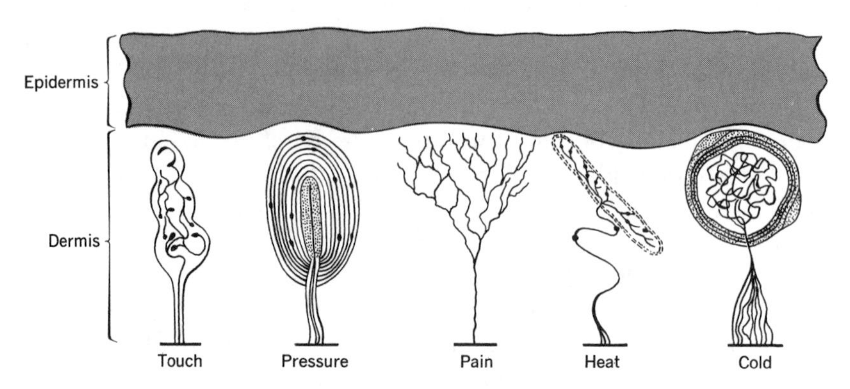

Figure 8–3 Receptors in the skin

THE HUMAN EYE

Each eye fits into a protective bony socket in the skull. Other protective adaptations include a cushion of shock-absorbing fat behind each eye; tear glands that keep the eyeball moist and dust-free; and eyelids and eyelashes, which keep dust out of the eye. Three pairs of muscles that attach the eyeball to the walls of its socket enable the eye to move in several directions.

Structure and Functions

The eyeball can be divided into three layers: (1) an outer layer, composed of the cornea and sclera; (2) a middle layer, composed of the choroid, ciliary body, and iris, and (3) the innermost layer, or retina (see Figure 8–4). Just behind the pupil and iris, within the eye cavity, is the lens, which divides the interior into two chambers. One chamber is the *aqueous humor,* a space between the iris and cornea filled with watery fluid. The other chamber is the *vitreous humor,* a space filled with a clear, jellylike substance.

Cornea. The *cornea* is a transparent tissue that covers the pigmented iris. The curvature of the cornea helps focus light, and thus functions as a lens. A protective membrane, called the *conjunctiva,* covers the cornea. An inflammation of the conjunctiva (conjunctivitis) usually results from an infection or irritation.

Sclera. The "white" of the eye, or *sclera,* is a strong layer of connective tissue that covers the entire eye excluding the cornea. The sclera provides shape and firmness, and protects the delicate, inner parts of the eye.

Choroid. Most of the inner surface of the sclera is covered by the *choroid.* It contains an abundance of blood vessels and cells bearing the pigment *melanin,* which makes the choroid appear reddish-brown. Melanin prevents hazy images by absorbing diffuse light rays. In this way, the choroid functions like the black interior of a camera, which prevents the fogging of film by absorbing stray light rays. The choroid also provides the retina with nutrients and removes wastes.

Ciliary body. Near the front of the eye, the choroid forms the *ciliary body,* which is composed mainly of ciliary muscle. *Ciliary muscle* changes the shape of the lens for near or far vision. The ciliary body also secretes the fluid of the aqueous humor.

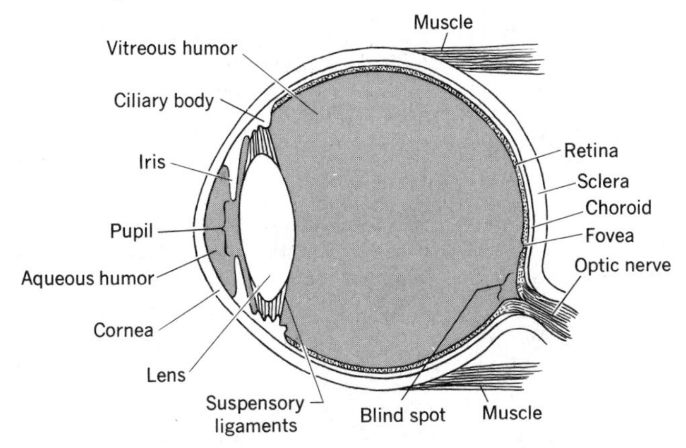

Figure 8–4 Structure of an eye

Iris. The *iris* is the colored part of the eye. The *pupil* is the opening in the center of the iris; the diameter of the pupil is regulated by the amount of light that enters the eye. Bright light stimulates certain muscle fibers in the iris to contract, which decreases the pupil's diameter. Dim light causes other muscle fibers in the iris to contract, increasing the pupil's diameter. For this reason, finding your seat in a darkened theater is often difficult until your pupils adjust to the dim light.

Retina. The inner layer of the eyeball is the *retina,* which consists of a pigmented epithelium and a neural, or visual, section. Like the choroid, the *pigmented epithelium* section absorbs diffuse light rays. The *neural section* of the retina transmits impulses to the thalamus and then to the visual cortex of the brain for interpretation. Two types of photoreceptors, rods and cones, convert light rays into electrochemical signals (see Figure 8–5). *Rods* provide black-and-white vision in dim light, enable us to distinguish between different shades of dark and light, and detect movement. *Cones* provide color vision and visual clearness. A *blind spot*, or optic disk, is where nerve fibers exit the retina and enter the optic nerve. The blind spot does not contain rods or cones. Thus, an image is not produced when light rays strike the blind spot. The *macula lutea* ("yellow spot") is located in the center of the retina, at the visual axis of the eye. A central depression in the macula lutea, called the *fovea,* contains a high density of cones. As a result, this area provides the sharpest vision. (Rods are not found in the fovea.)

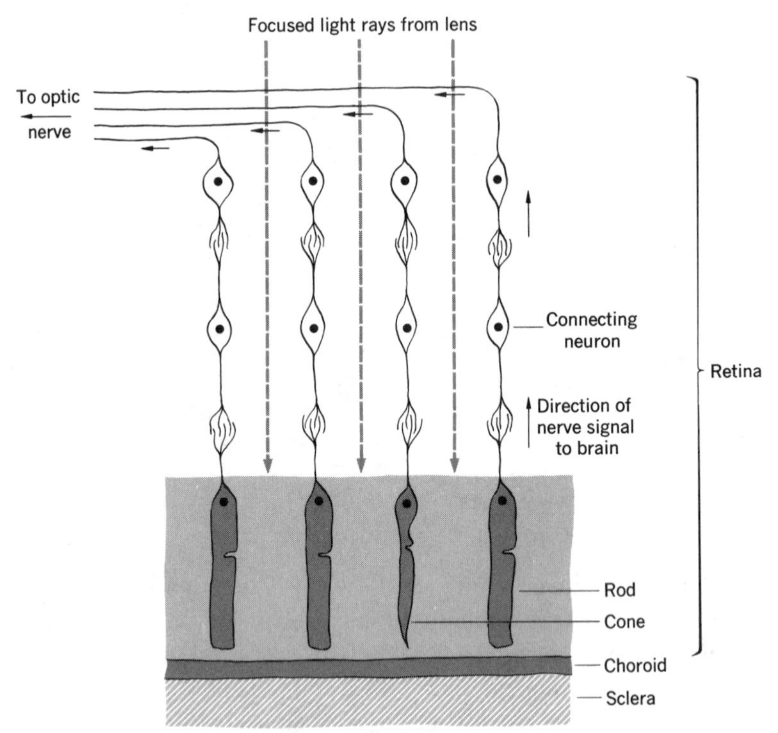

Figure 8–5 Receptor cells in the retina (schematic)

Lens. The *lens* is located directly behind the pupil and iris. The lens is transparent; does not contain blood vessels; and is enclosed by a clear capsule held in position by *suspensory ligaments.* These elastic ligaments adjust the curvature of the lens, focusing light rays on the retina for either near or far vision.

How an image is formed. In many ways, an eye functions like a camera: the cornea and lens focus light rays (or images) on the light-sensitive retina, which is similar in function to film in a camera. Changes in the shape of the lens focus objects more finely, while changes in the pupil's diameter provide proper light "exposure." Light rays that pass through the cornea are *refracted*, or bent. Both surfaces of the lens further refract light rays as they emerge from the cornea and focus them on the retina. The image formed on the retina is inverted (upside down); the brain interprets the image as right side up (see Figure 8–6).

With aging, the lens hardens and loses some of its ability to change shape. If the lens become too hard, a condition called *presbyopia,* or farsightedness, may occur. In a farsighted person, light rays converge to form

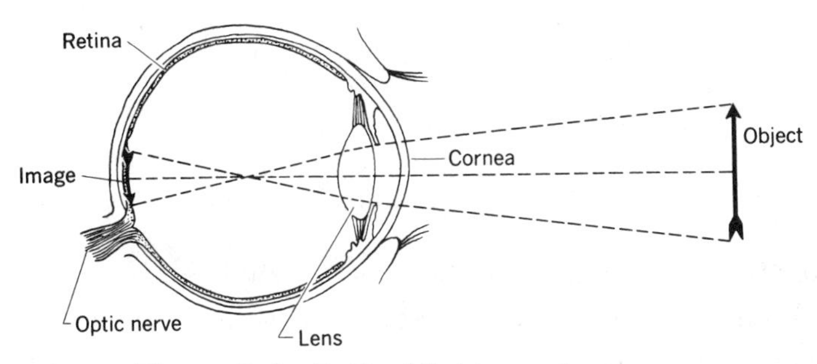

Figure 8–6 Path of light rays in an eye

an image behind the retina. Convex lenses correct farsightedness by focusing light rays directly on the retina. If the eye becomes elongated, light rays form an image in front of the retina. This condition is called *myopia*, or nearsightedness. Concave lenses correct nearsightedness by focusing light rays directly on the retina.

Color blindness is the result of a genetic defect caused by a lack of photopigment in cone cells of the retina. A person with color blindness cannot distinguish between certain colors. For example, a person with red-green color blindness, the most common type, cannot distinguish red from green. Inheritance of color blindness is discussed in Chapter 15.

THE HUMAN EAR

The ability to hear sounds at different frequencies varies among animals. Bats, dogs, and some invertebrates respond to high-frequency sounds and elephants respond to low-frequency sounds that are all beyond the range of human hearing. The human ear has receptors that can

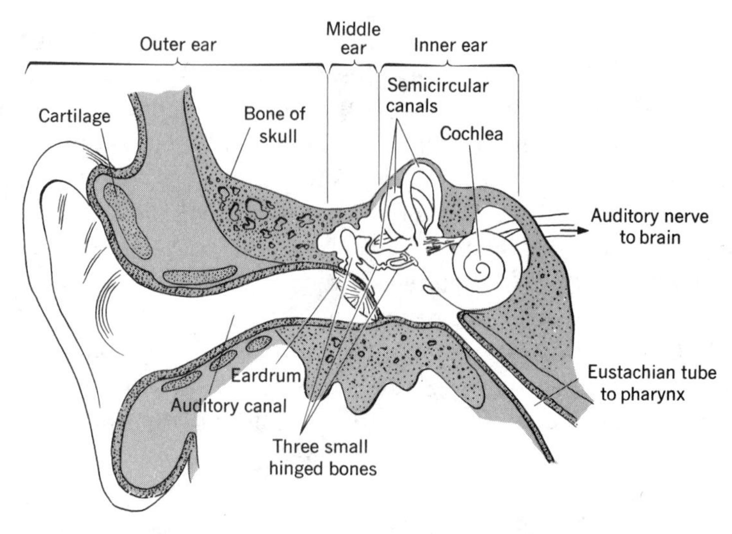

Figure 8–7 Structure of an ear

detect sound frequencies ranging from 16 vibrations per second to 20,000 vibrations per second. Figure 8–7 shows the structure of a human ear.

Structure and Functions

The human ear consists of an outer ear, a middle ear, and an inner ear. The *outer ear* features a funnel-shaped flap of cartilage, an auditory canal, and a *tympanum*, or eardrum. The *middle ear* includes the Eustachian tube and three small bones. Each bone is hinged to the next one. The outer bone is connected to the eardrum; the inner bone touches a membranous part of the inner ear. The *Eustachian tube* connects the middle ear to the pharynx and equalizes air pressure between the middle ear and outer ear.

The *inner ear* houses a coiled *cochlea*, which contains the sound receptors. Two small sacs and three *semicircular canals* are attached to the cochlea at right angles to each other. The semicircular canals and sacs are discussed later in this chapter.

Hearing sounds. Initially, sound of a certain frequency enters the ear canal, causing the eardrum to vibrate. The small bones of the middle ear transfer the vibrations to the membranous part of the cochlea, causing it to vibrate. Vibrations of the membranous part of the cochlea then cause a fluid in the cochlea to vibrate. The vibrating fluid causes hairs on receptor cells to vibrate, which produce signals. The signals pass into the *auditory nerve*, which transfers them to the brain. The brain interprets the signals as different sounds.

Other Receptors

The body also has internal receptors that are sensitive to changes in and around the internal organs and help maintain homeostasis. These internal receptors include the semicircular canals and the two sacs associated with them, pH receptors, rate-of-heartbeat receptors, and blood pressure receptors.

The semicircular canals. The semicircular canals and the sacs associated with them have receptors sensitive to body movements, and to changes in the position of the head. In the sacs, tiny mineral crystals coat the hairs that grow out of the receptor cells. When you move your head, the crystals change position, causing the hairs to move. Movement of the hairs produces signals that enter the brain, which sends signals to muscles that adjust your head to its original position. Like the sacs, the semicircular canals contain receptor cells with hairs, but lack mineral crystals. Movement in the fluid that fills the canals causes the hairs to move. The movement of the hairs produces signals that go to the brain. The brain interprets the signals and activates muscles that help maintain balance.

pH receptors. A group of specialized cells, called the *breathing center*, is located in the medulla of the brain. The breathing center is sensitive to the amount of carbon dioxide in the blood. An excess of carbon dioxide causes the pH of the blood to drop below normal. When this occurs, the breathing center sends signals to the diaphragm and rib muscles. The signals cause the diaphragm and rib muscles to contract more rapidly, which increases the breathing rate. This action continues until the excess carbon dioxide is eliminated. When the pH of the blood returns to normal, the rate of breathing also returns to normal.

Rate-of-heartbeat receptors. These receptors are located in the walls of the heart and are connected to the *heartbeat center* in the medulla of the brain. When the heart beats too quickly, the heartbeat center sends signals through the *vagus nerve* to the heart, which make the heartbeat rate decrease. When the heartbeat is too slow, the heartbeat center sends signals through the *accelerator nerves* to the heart, which make the heartbeat rate increase.

Blood pressure receptors. The walls of some arteries have pressure-sensitive receptors. When the pressure within one of these arteries exceeds the normal pressure, signals are sent to the heartbeat center. The heartbeat center then sends signals that slow down the heartbeat, reducing the pressure within the artery.

RECEPTORS IN OTHER ORGANISMS

Protists

In most protists, the plasma membrane is the receptor for external stimuli such as light, chemicals, and water. A paramecium usually swims away from bright light and ejects trichocysts when subjected to an irritating chemical, such as dilute vinegar (acetic acid). If its watery environment begins to dry up, an ameba surrounds its body with a waterproof wall (cyst). When water becomes plentiful again, the ameba breaks out of the cyst. Euglena (a protist with chloroplasts) has an orange-colored particle in its cytoplasm called an *eyespot*, which enables the euglena to detect light.

Hydra

Both of the cell layers of hydra contain receptor cells sensitive to touch, light, and chemicals. These receptor cells are part of a network of nerve cells that helps to coordinate body movements. When a hydra's tentacles are stimulated by touch or by chemicals, such as meat juice, the tentacles discharge their stinging cells. If placed in a dark environment adjacent to a light source, the hydra will always move toward the light.

Earthworm

An earthworm's skin contains receptors for touch, light, and chemicals. When placed on the surface of soil, an earthworm responds to light by burrowing into the soil. Earthworms prefer to live in soil with a low pH (5.0–6.0). When placed in a box that contains alkaline soil and acid soil, an earthworm responds by moving to the acid soil.

Grasshopper

Several of a grasshopper's complex receptors are located in its head. Two large *compound eyes* form many images in bright light, and three small *simple eyes* function in dim light. The antennae and the fine hairs on the grasshopper's body are touch receptors. Odor receptors are located in small pits at the base of the antennae and around the mouthparts. A pair of sound receptors called *tympanic membranes* are located in the first segment of a grasshopper's abdomen.

Table 8–1 compares receptors of the organisms discussed in this chapter.

Table 8–1 Comparison of Receptors

Organism	*Receptors*		*Sensitive to*
Paramecium	Plasma membrane		Many external stimuli, including light
Euglena	Plasma membrane		Many external stimuli
	Eyespot		Light
Hydra	Nerve endings in ectoderm and endoderm		Touch, temperature, chemicals, light
Earthworm	Nerve endings in skin		Touch, temperature, chemicals, light, moisture
Grasshopper	Compound and single eyes		Light
	Antennae		Touch, odor
	Body hairs		Touch
	Tympanic membranes		Sound
Human	External	Skin	Touch, pressure, heat, cold, pain
		Nose	Odor
		Tongue	Taste
		Eyes	Light
		Ears	Sound
	Internal	Semicircular canals	Body movement and position
		Nerve endings in blood vessels and internal organs	Pressure, pH, oxygen concentration, carbon dioxide concentration

Section Quiz

1. Which of the following statements is correct? (*a*) The eyes of an octopus resemble the eyes of a dog. (*b*) Bats can fly in the dark because they consume a great deal of vitamin A. (*c*) The receptor cells of a hydra are not sensitive to light or touch. (*d*) Frogs cannot taste foods.

2. Pheromones are (*a*) organs of balance in most invertebrates (*b*) tiny stones located in the semicircular canals of reptiles (*c*) odor molecules. (*d*) mechanoreceptors.
3. The choroid layer of the eye (*a*) refracts or bends light (*b*) provides nutrients to eye tissues (*c*) detects color wavelengths of light (*d*) provides the aqueous humor with fluid.
4. The tympanum of a grasshopper has a function similar to the (*a*) ommatidia of a fly (*b*) antennae of a butterfly (*c*) semicircular canals of a bird (*d*) eardrum of a human.
5. Defective cones may be a primary cause of (*a*) myopia (*b*) presbyopia (*c*) conjunctivitis (*d*) color blindness.

NERVOUS SYSTEMS

Sense organs detect changes in the environment and translate them into nerve signals. In many animals, such as humans, these nerve signals are transmitted to different parts of a nervous system that adjusts the signals, interprets the signals, and then activates muscles or glands (effectors).

The Neuron

Figure 8–8 shows a nerve cell, or *neuron,* which is the unit of structure and function of the nervous system. Numerous cytoplasmic branches, or *dendrites*, extend from one end of the *cell body*, or *cyton*. A long fiber of cytoplasm called the *axon* extends from the opposite end and terminates in a tuft of small branches, the *end brush.* The axons of some neurons are uncovered, or bare; others are covered by an inner, fatty covering, the

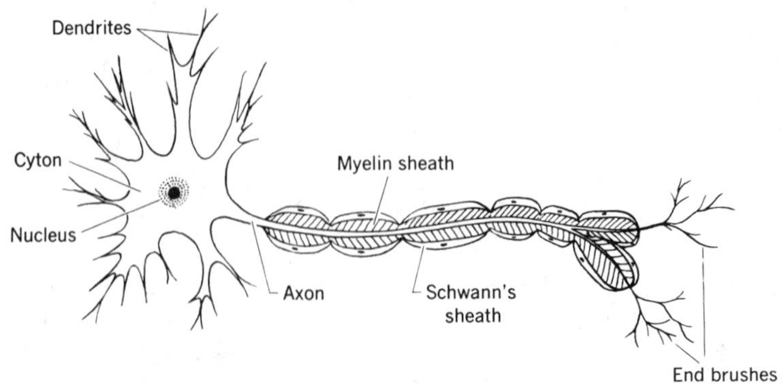

Figure 8–8 Structure of a neuron

myelin sheath, and an outer covering of flattened cells called *Schwann's sheath*. The myelin sheath protects and insulates the axon; Schwann's sheath helps repair injured axons. At birth, your body contains a fixed number of neurons. These remain with you throughout life.

The Nerve Impulse

The electrochemical signal that travels along a neuron is called an *impulse*. A neuron at rest has a negative internal charge and a positive external charge (Figure 8–9). Active transport of positively charged sodium ions (Na^+) from within the neuron to the outside is mainly responsible for the difference in charge between the inside and outside of a neuron. During this activity, the plasma membrane of a neuron is called a *polarized membrane*.

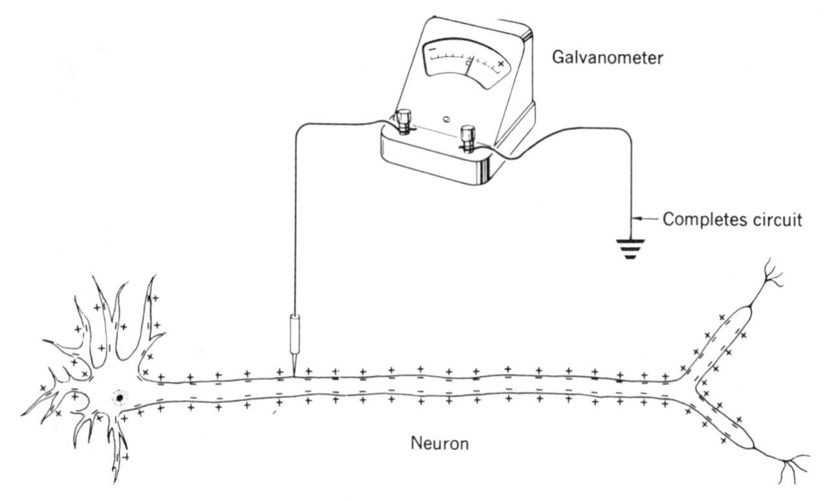

Figure 8–9 Electrical charges in an unstimulated neuron

When a neuron is stimulated, the permeability of the plasma membrane changes, permitting sodium ions to enter the neuron. This produces an internal positive charge at the point of entry (Figure 8–10a) and *depolarizes*, or upsets, the normal condition of the membrane. The upsetting of the normal electrical condition of the neuron at one point upsets the electrical condition of an adjacent point. This is the beginning of the electrochemical signal, or impulse, which continues along the axon (Figure 8–10 b and c). A nerve impulse does not travel as fast as a wave of electricity. Electrons in an ordinary electric current travel at the speed of light, about 300,000 kilometers (186,000 miles) per second; in humans, a nerve impulse travels at about 100 meters (330 feet) per second.

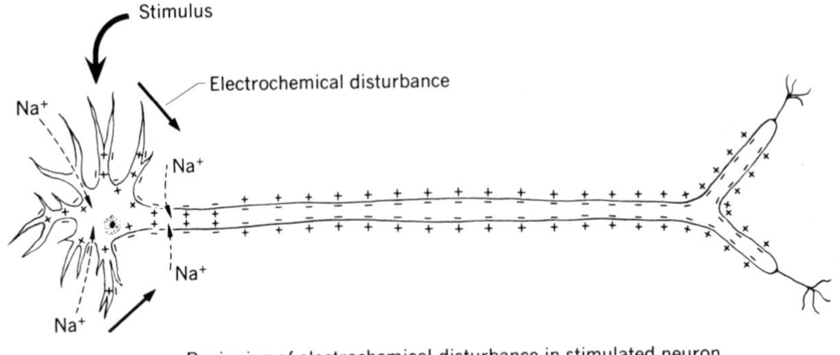

a. Beginning of electrochemical disturbance in stimulated neuron

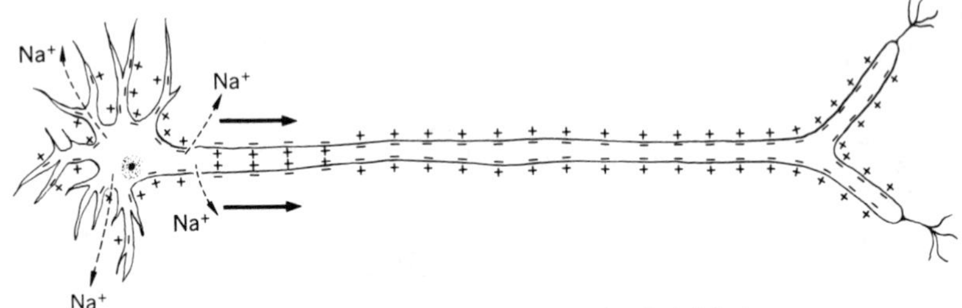

b. Later stage of movement of electrochemical disturbance

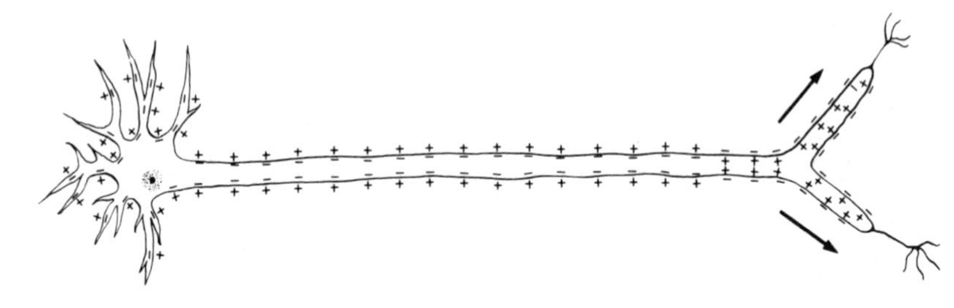

c. Final stage of movement of electrochemical disturbance

Figure 8–10 Movement of a nerve impulse along a neuron

As the nerve impulse moves along the neuron, the sodium ions are pumped out by active transport, thus restoring the normal electrical condition of the neuron and repolarizing the plasma membrane. As a result, the neuron can conduct another nerve impulse. A stimulus must be just strong enough to start a nerve impulse; a stronger stimulus does not produce a stronger impulse, and a weaker stimulus does not produce any impulse. This electrochemical property of neurons is called the *"all or none"* response.

The Synapse

The axon of a neuron is seldom long enough to connect directly with a muscle or a gland. For this reason, the nerve impulse in one neuron usually is passed on to another neuron or neurons. The "connection" between two neurons, called a *synapse*, includes the end brush of one neuron, a gap, and the dendrites of the other neuron (Figure 8–11).

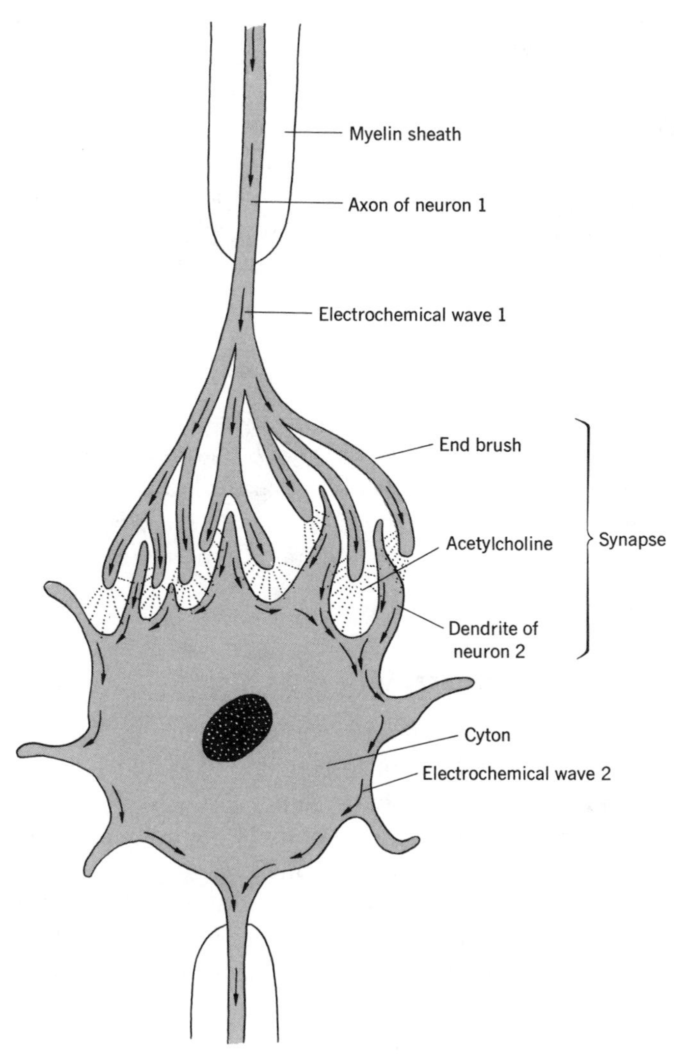

Figure 8–11 Nerve impulse crossing a synapse

The German scientist *Otto Loewi* determined how a nerve impulse crosses the gap of a synapse. Studying the hearts of frogs, he showed that the ends of the vagus nerves in the heart secrete a *neurohumor* (a hormonelike compound called *acetylcholine)* that slows down the heartbeat. Loewi also showed that the ends of the accelerator nerves in the heart secrete a neurohumor called *epinephrine* (adrenalin), which speeds up the heartbeat. Additional research has revealed that when an impulse reaches the tips of an end brush, the brush secretes acetylcholine into the synapse gap. Acetylcholine changes the permeability of the membrane of the dendrites of successive neurons. This starts the sequence of events that depolarizes the plasma membrane and starts a new nerve impulse in the neuron. Nerve impulses always travel from axon to dendrites because the dendrites do not secrete acetylcholine. Following the passage of nerve impulse from one neuron to another through a synapse, an enzyme decomposes the acetylcholine in the synapse gap. As a result, the membranes of the end brush and the dendrites become polarized and can transmit another nerve impulse.

Types of Nerves

A *nerve* consists of parallel axons of many neurons held together by a covering of connective tissue. *Sensory nerves* such as the optic nerves and auditory nerves carry impulses to the spinal cord or brain. *Motor nerves* carry impulses from the spinal cord or brain to muscles or glands. The nerves that extend from the brain and spinal cord into the arms and legs are examples of motor nerves.

THE HUMAN NERVOUS SYSTEM

The main divisions of the human nervous system are the central nervous system and the peripheral nervous system. Figure 8–12 outlines the general organization of the nervous system.

The *central nervous system,* consisting of the brain and spinal cord, integrates and correlates sensory information; generates thoughts and emotions; and forms and stores memories. Nerve impulses that stimulate muscle contractions and glandular secretions begin in the central nervous system. The *peripheral nervous system* connects the central nervous system to muscles, glands, and sensory receptors in other parts of the body by means of *cranial nerves* from the brain and *spinal nerves* from the spinal cord. The peripheral nervous system features the *sensory*, or *afferent,*

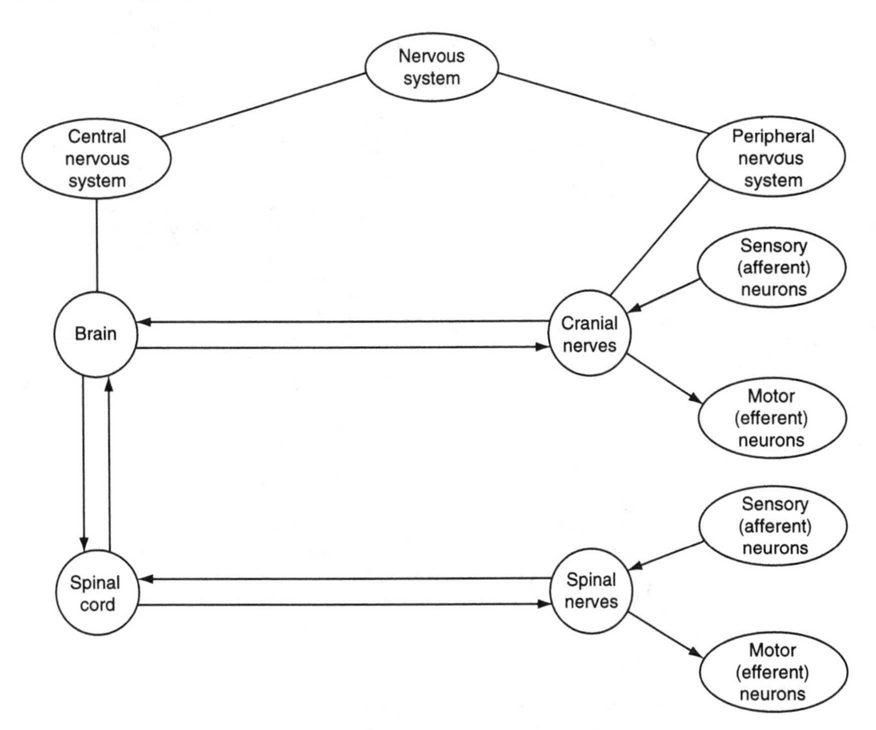

Figure 8–12 Organization of the nervous system

neurons, which carry messages from receptors to the spinal cord and brain; *motor,* or *efferent, neurons*, which carry messages from the spinal cord and the brain to effectors; and *associative neurons,* which connect sensory neurons to motor neurons (Figure 8–13).

Central Nervous System

The brain. The *brain* is the control center, or integrator, for most animals. In humans, the brain controls most of the subconscious and conscious activities. Subconscious activities include respiratory rate, regulation of blood pressure, posture, and balance; conscious activities include voluntary movement, learning, memory, and abstract reasoning.

An adult brain weighs about 1.5 kilograms (3 pounds) and consists of about 100 billion neurons. The brain fits into the cranium of the skull and is divided into four main parts: the medulla, or brain stem; the thalamus and hypothalamus, located above the brain stem; the cerebrum, which is divided into right and left *cerebral hemispheres;* and the cerebellum,

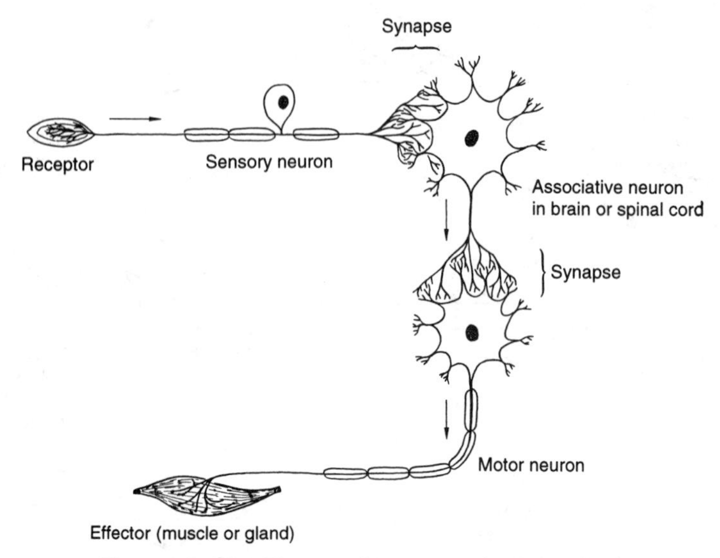

Figure 8–13 Types of neurons (schematic)

which is behind the cerebrum (Figure 8–14). The brain is protected by cranial bones and tough membranes called *meninges*.

The medulla. The *medulla,* or brain stem, connects cerebrum and cerebellum to the spinal cord. The outer region of the medulla consists of *white matter* (mainly axons) and the inner region consists of *gray matter* (mainly cytons). Centers for certain *reflex actions* of the head are located in the gray matter. These reflex actions are inborn automatic responses

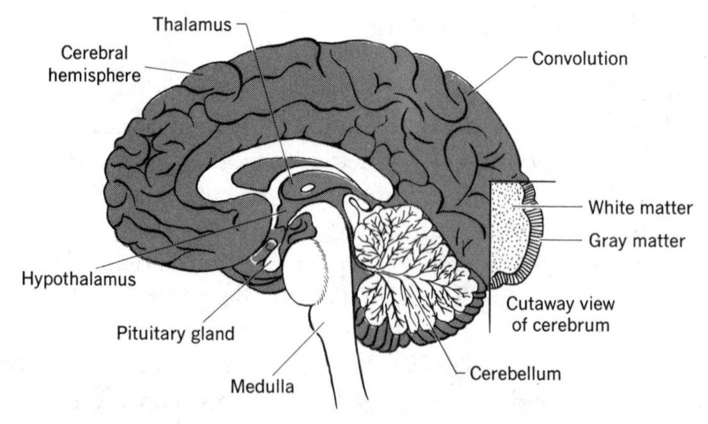

Figure 8–14 Brain (cut lengthwise)

such as sneezing and blinking. The respiratory center, the heartbeat center, and the center that controls the size of arteries are also located in the gray matter.

The thalamus and hypothalamus. The relationship of the thalamus and hypothalamus to other parts of the brain is shown in Figure 8–14. The *thalamus* is an oval structure about 3.0 cm (½ inch) long and consists mainly of gray matter. The thalamus relays sensory impulses from the spinal cord, medulla, cerebellum, and certain parts of the cerebrum to the cerebral cortex. In effect, the thalamus functions as a relay station for the transmission of many sensory impulses. The *hypothalamus*, located beneath the thalamus, is composed mainly of gray matter. The hypothalamus:

- Controls and integrates activities of the autonomic nervous system.
- Develops feelings of rage and aggression.
- Generates feelings of hunger (before eating) and well-being (after eating).
- Generates a feeling of thirst.
- Stimulates production of major hormones of the pituitary gland, which are discussed in Chapter 9.

The cerebrum. Many folds, or *convolutions,* increase the surface area of the *cerebrum.* The presence of convolutions is an adaptation that enables billions of neurons to fit in the relatively small space inside the cranium.

The *cortex*, or gray matter, is the outer layer of the cerebrum. Most of the cortex contains cytons of neurons, associative neurons, and synapses between these neurons. The axons of the neurons extend below the cortex, and comprise most of the white matter of the cerebrum. Axons from the right hemisphere of the brain cross over to the left side of the body; axons from the left hemisphere of the brain cross over to the right side of the body. An injury to neurons on one side of the brain causes paralysis of muscles on the opposite side of the body.

Figure 8–15 shows the main areas of the cerebrum. The *sensory area* receives nerve impulses from sense organs and interprets them as sight, sound, smell, taste, pain, pressure, heat, cold, and touch. The *motor area* controls voluntary body movements. The *speech area* coordinates speaking and writing with seeing and hearing words. The *association area* is involved with memory, intelligence, and emotions, including love, hate, and fear. All of these areas of the cerebrum are connected to each other by an intricate network of associative neurons.

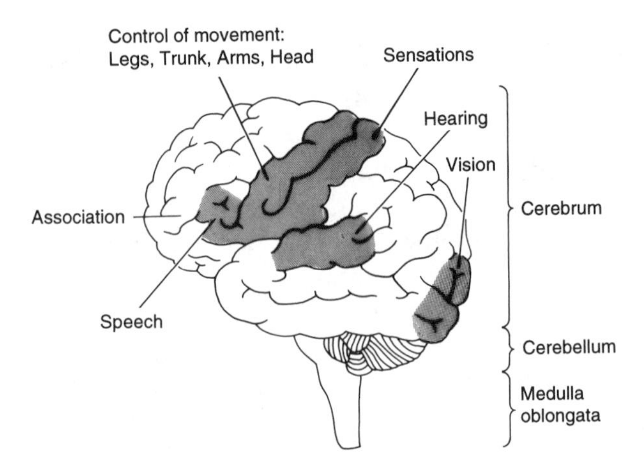

Figure 8–15 Major areas of the brain

The cerebellum. Like the cerebrum, the *cerebellum* is divided into two parts called *cerebellar hemispheres.* The surface of the cerebellum consists of gray matter; the interior consists of a mixture of gray and white matter. The cerebellum coordinates skilled movements and regulates posture and balance.

The spinal cord. The *spinal cord* and spinal nerves help the body make rapid responses to environmental changes. A rapid response to an environmental stimulus is called a *reflex.* The spinal cord also is an important pathway for sensory impulses going to the brain and motor impulses going to effectors or to spinal nerves. The spinal cord is a thick, tubelike structure about 2.0 cm in diameter, located in a canal within the spinal column. Two distinct regions of gray and white matter surround the small canal in the center of the spinal cord (Figure 8–16). Like the brain, the spinal cord is covered by protective meninges.

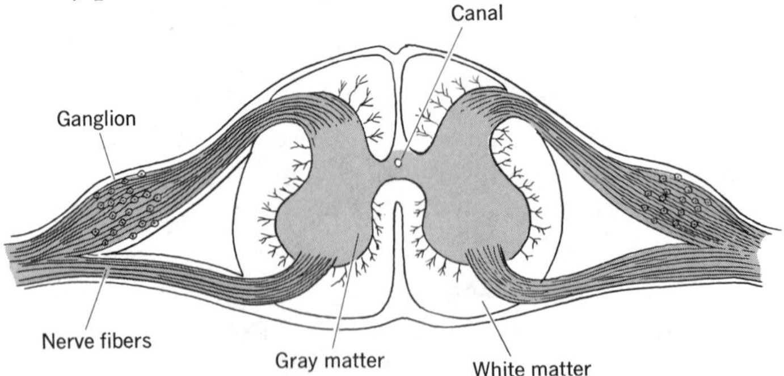

Figure 8–16 Spinal cord (cross section)

Gray matter in the spinal cord, as in the cerebrum and cerebellum, is mainly composed of cell bodies and dendrites and is involved with reflex actions. White matter in the spinal cord is composed mainly of axons bundled into sensory nerve tracts going to the brain and motor nerve tracts going toward the rest of the body. Table 8–2 summarizes the major structures of the central nervous system and their functions.

Table 8–2 Summary of Structures and Functions of the Human Central Nervous System

Structure	*Functions*
Brain	
Cerebrum	Receives and interprets sensory impulses; controls voluntary muscular movement; center of consciousness, intelligence, association, memory, speech, and some emotions
Cerebellum	Coordinates muscular activity; controls balance
Medulla	Control center of automatic nervous system; contains centers for control of breathing, diameter of blood vessels, heartbeat, reflex acts above the neck
Thalamus	Relay station for impulses going to cerebral cortex
Hypothalamus	Contains centers for control of body temperature, blood pressure, water balance, sleep; center for some emotions
Spinal cord	Contains centers for control of reflex acts below the neck; major pathway for impulses between brain and peripheral nerves; connecting center between sensory and motor neurons

Peripheral Nervous System

The peripheral nervous system (PNS) consists of *cranial nerves* in the brain and *spinal nerves* that emerge from the spinal cord. The input part of the peripheral nervous system consists of sensory (afferent) neurons, which transmit nerve impulses from sensory receptors to the central nervous system. Motor (efferent) neurons originating within the central nervous system conduct impulses to muscles and glands. The central nervous system is connected to sensory receptors, glands, and muscles located throughout the body. The peripheral nervous system is made up of the somatic nervous system and the autonomic nervous system. (Interestingly, it has been discovered that some types of neurons in the peripheral nervous system can repair themselves if damaged.)

The *somatic nervous system* involves the action of sensory neurons in the skin and skeletal muscles, as well as motor neurons, which carry responses to skeletal muscles. The somatic nervous system controls reflex actions and voluntary movement. The *autonomic nervous system* involves the action of sensory neurons originating in internal organs and ending in the central nervous system, and motor neurons from the central nervous system that stimulate cardiac muscle, smooth muscle, and glands. The responses made by these tissues and organs are *involuntary*, or not consciously controlled (see Figure 8–17). Note that a *ganglion* contains a mass of cytons, and a *plexus* is a sheetlike group of ganglions connected by nerves.

The motor part of the autonomic nervous system is divided into the *sympathetic system* and *parasympathetic system*. Although both systems are connected to all internal organs, the effects of the sympathetic and

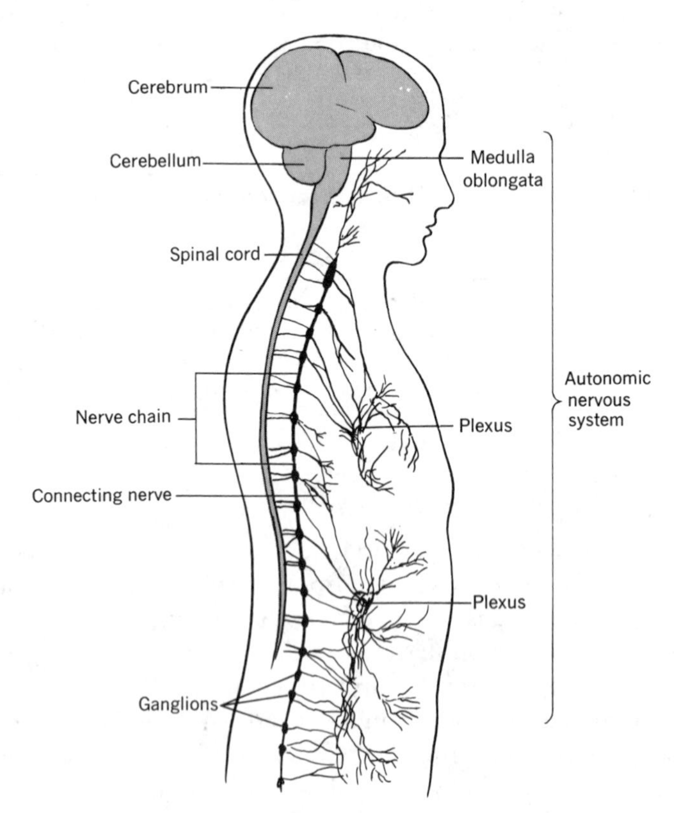

Figure 8–17 Autonomic nervous system

parasympathetic system are antagonistic. Thus, when one system *stimulates,* the other represses, or *inhibits.* Table 8–3 summarizes the effects of the sympathetic and parasympathetic systems on some internal organs.

Table 8–3 Effects of the Autonomic Nervous System on Some Organs

Division	*Pupil of Eye*	*Salivary Glands*	*Heart*	*Bronchioles*	*Adrenal Glands*
Sympathetic	Widens	Inhibits	Speeds up	Widens	Stimulates
Para-sympathetic	Narrows	Stimulates	Slows down	Narrows	Inhibits

The sympathetic and parasympathetic systems maintain homeostasis because of their antagonistic effects. For example, when your body detects danger, signals from the sympathetic system speed up the heartbeat, widen the bronchioles, and stimulate the adrenal glands to secrete a hormone that releases glucose from the liver. As a result, your body is ready for an emergency. After the danger has passed, signals from the parasympathetic system cause the body to return to normal.

Section Quiz

1. *Subconscious* activities include control of (*a*) respiratory rate, voluntary movements, memory, and balance (*b*) respiratory rate, blood pressure, posture, and balance (*c*) blood pressure, voluntary movements, learning, and memory (*d*) voluntary movements, learning, memory, and abstract thought.
2. Reflex actions such as sneezing, blinking, and breathing are controlled by centers in the (*a*) cerebrum (*b*) cerebellum (*c*) medulla (*d*) spinal cord.
3. In the brain, feelings of hunger, thirst, and rage are generated by the (*a*) thalamus (*b*) hypothalamus (*c*) left hemisphere (*d*) right hemisphere.
4. *Convolutions* of the cerebrum enable (*a*) billions of neurons to fit inside a limited space (*b*) fast reflex responses to environmental stimuli (*c*) skillful movements and maintenance of good balance (*d*) endocrine glands to function properly.

5. The *peripheral nervous system* consists of the autonomic nervous system and the (*a*) sympathetic system (*b*) somatic nervous system (*c*) parasympathetic system (*d*) central nervous system.

NERVOUS SYSTEMS OF OTHER ORGANISMS

Protozoans

Protozoa do not have a nervous system. In an ameba, cytoplasm carries information from receptive sites on the plasma membrane throughout the organism. For example, when one end of an ameba touches a sharp object, the opposite end forms pseudopods and the ameba moves away. A paramecium possesses a network of fine channels under the plasma membrane that connect the cilia. This network allows a paramecium to use cilia for ingestion and for locomotion.

Hydra

A hydra has a simple nervous system, or *nerve net,* composed of a network of receptor cells and nerve cells that lacks central control (Figure 8–18). For example, a stimulus applied to any part of a hydra starts an impulse that spreads to all parts of its body at the same time. As a result, a hydra usually responds by contracting its entire body.

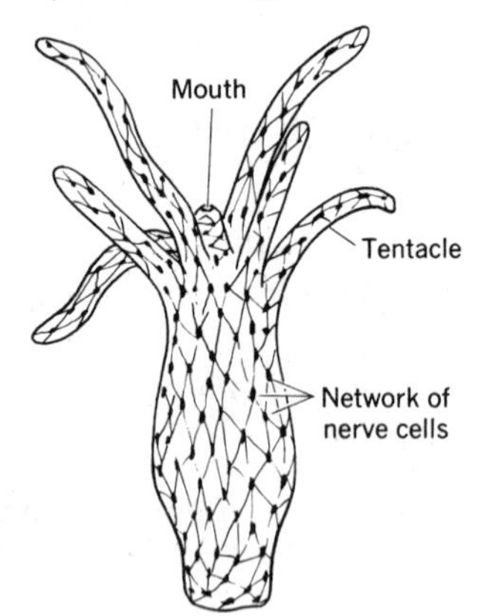

Figure 8–18 Nerve net of hydra

Earthworm

In an earthworm, the nervous system is centralized into a brain and a nerve cord with definite pathways (Figure 8–19). An earthworm's brain consists of two fused ganglions located above the mouth, and is connected to the ventral nerve cord by a ring of nerve fibers. The ventral nerve cord

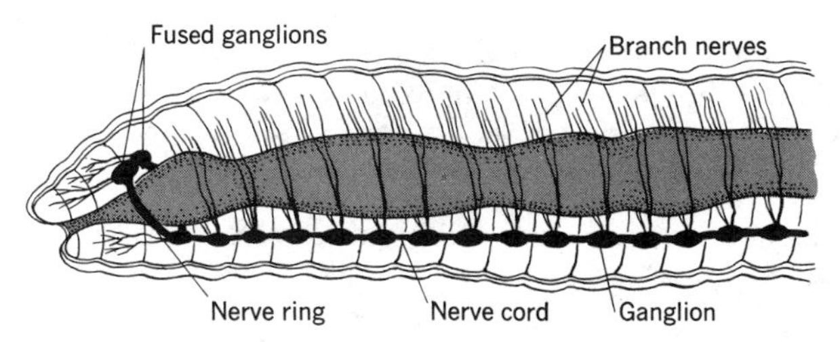

Figure 8–19 Nervous system of the earthworm

consists of a chain of ganglions. Nerves that emerge from the ganglions connect the central nervous system to sense organs, muscles, and glands.

Grasshopper

The structure of a grasshopper's nervous system is similar to that of an earthworm (Figure 8–20). However, the brain of a grasshopper is larger than that of the earthworm, and receives signals directly from different receptors on the grasshopper's head. There are fewer, but larger, ganglions in the ventral nerve cord of a grasshopper than in an earthworm. Table 8–4 compares the regulatory adaptations of the organisms described.

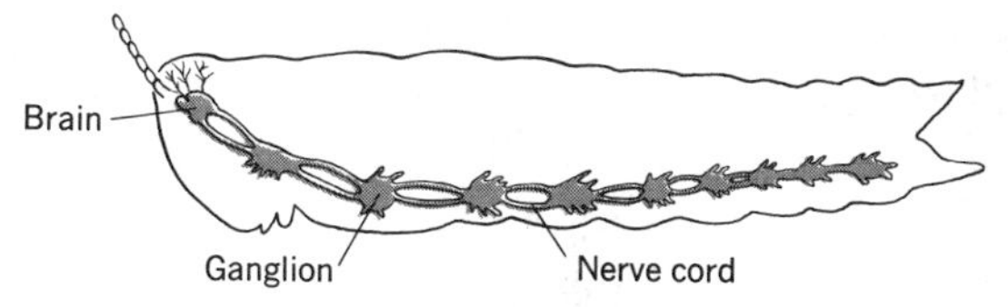

Figure 8–20 Nervous system of the grasshopper

Table 8–4 Regulatory Adaptations of Selected Organisms

Organism	*Regulatory Adaptation*
Protozoan (ameba)	General ability of protoplasm to receive and respond to stimuli; no nervous system
Hydra	Nerve network that has no central control
Earthworm	Central nervous system that includes a brain, nerve cord, ganglions, nerve fibers, and sense organs
Grasshopper	Central nervous system similar to that of earthworm; larger brain and better developed ganglions than in earthworm

The central nervous system of *invertebrates* features a ventral nerve cord; *vertebrates*, including humans, possess a dorsal, tubelike nerve cord. In invertebrates, the brain develops from fused ganglions. In vertebrates, the brain develops as an extension of the nerve cord. In humans, the brain, especially the cerebrum, is the most highly developed part of the nervous system.

PATTERNS OF BEHAVIOR

The *behavior* of an organism involves its responses to different stimuli originating in the external and internal environments. These responses enable the organism to adjust to its environment and maintain homeostasis. Behavior may be either *inborn* or *learned* (acquired during the lifetime of the organism).

Inborn Behavior

Three types of inborn behavior found in animals are taxes (singular, taxis), simple reflexes, and instincts.

Taxes. A *taxis* is an inborn, automatic response in which an entire organism moves either toward or away from a stimulus. For example, a paramecium will move toward an acidic region (*positive chemotaxis*), and a cockroach will move away from light (*negative phototaxis*).

Simple reflexes. A *simple reflex* act is an inborn, automatic response to a stimulus, in which only part of an organism responds to a stimulus. In humans, simple reflex acts include blinking and withdrawing a hand after contact with a hot object. In a simple reflex act, nerve impulses pass from receptors to effectors through a pathway, called a *reflex arc* (Figure 8–21). A reflex arc includes a sensory neuron that carries the impulse from the receptors; an associative neuron in the spinal cord that carries the impulse to other neurons; and a motor neuron that carries the impulse from the spinal cord to the appropriate effector. Associative neurons form synapses inside the gray matter of the spinal cord with the sensory neurons, motor neurons, and with the neurons leading to the brain.

Reflex behavior supports the survival of an organism because the responses usually are beneficial, rapid, and automatic. Reflex behavior does not require thought. For example, the flow of tears in an irritated eye keeps the eye moist and clean, and vomiting ejects toxic substances from the stomach.

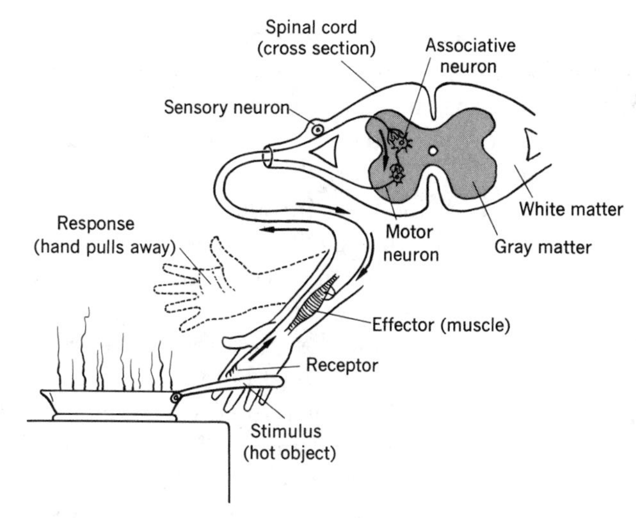

Figure 8–21 Scheme of a reflex arc

Instincts. An *instinct* is a complex pattern of many reflex acts. The completion of the first reflex act is the stimulus for the second, and so on until an entire pattern of behavior is completed. Animals display many different types of instincts, including nestmaking and migration in birds, spinning of webs by spiders, and the suckling response of newborn mammals shortly after birth. Humans appear to rely less on instinct than do other animals because much of our behavior is modified by our reasoning abilities.

Learned (Acquired) Behavior

Three types of learned behavior found in animals are imprinting, conditioning, and habit formation.

Imprinting. A type of learned behavior in which a fixed pathway is established in the nervous system by the stimulus of the *first* object or condition to which an animal was exposed is called *imprinting.* Examples of imprinting include newborn geese following their mother, and adult salmon returning to breed in the river in which they were hatched. In 1935, the Austrian zoologist *Konrad Lorenz* discovered that imprinting is a type of learned response. Lorenz observed that when newly hatched geese (goslings) were exposed immediately to a slow-moving object instead of their mother, the goslings would instinctively follow the object.

Conditioning. Behavorial *conditioning* occurs when an organism responds to a stimulus that is a substitute for the one that originally triggered the response. In 1900, the Russian scientist *Ivan Pavlov* discovered that conditioning is a type of learned response. He observed that when one of his assistants entered the laboratory, the dog he was studying began to secrete saliva, although no food was present. Pavlov hypothesized that, since this assistant always fed the dog, it associated the appearance of the assistant with the offering of food. In this case, the assistant who fed the dog became a substitute stimulus for food.

To test his hypothesis, Pavlov designed a series of experiments in which different stimuli were present just before a dog was fed. In one experiment, Pavlov rang a bell to which the dog responded by barking. Later, he rang the bell and then fed the dog. After constant repetition of these two stimuli, the dog stopped barking; instead, it secreted saliva when the bell was rung. Later, when the bell was rung, the dog secreted saliva even when food was not offered. Thus, the ringing of the bell became a substitute stimulus for food.

To understand conditioning, it is important to know what happens in the nervous system. The reflex arcs for the barking reflex and the flow-of-saliva reflex are connected by associative neurons. Many repetitions of these reflexes lower resistance at their synapses which permit a nerve impulse that ordinarily triggers the barking reflex to bypass its own reflex arc and enter the reflex arc that triggers the flow-of-saliva reflex.

According to Pavlov, the following rules are necessary for successful conditioning to occur:

1. The substitute stimulus must be given before, or at the same time as, the original stimulus so that the animal will associate the two stimuli.
2. The time interval between the original and the substitute stimulus must be brief.
3. The response to the substitute stimulus can be established by rewarding the animal after each trial.

Learning in organisms that possess a complex nervous system relies on successful conditioning. For example, each word in your vocabulary was acquired, or learned, as the result of conditioning. You probably learned the word "book" when an object was shown to you and the word "book" was spoken. After many repetitions, the sight of a book and your association of the object with the word "book" helped you learn the concept. In your nervous system, impulses from the vision area of the brain travel to the sound area and then to the speech area. The coordination of these areas enables you to say "book" when you see a book.

Many of our fears are the result of conditioning. For example, some adults are afraid of snakes although they have never touched a snake and were never bitten by a snake. Fear of snakes occurs when a person continually associates unpleasant situations or fear with snakes. In time, the repetition of this feeling results in an actual fear of snakes.

Habit formation. A *habit* is a learned, automatic response that initially requires thinking and voluntary action. But, as a result of repetition, a habit becomes automatic and involuntary. Habit formation occurs as the result of a series of conditioned reflexes that are continually repeated. Habits include writing your name, biting your fingernails, and stopping for a red light.

To illustrate the difference between a habit and a voluntary act, write your name. Then, write your name with the hand you do not ordinarily use. In the first trial, writing you name is rapid and easy; in the second trial, writing is slow and difficult. Like writing your name, more than 85 percent of your daily behavior is habitual, making it easier for you to perform everyday tasks.

Harmful habits. Some people acquire harmful habits such as smoking cigarettes and abusing drugs.

Smoking. The lungs receive a mixture of nicotine, carbon monoxide, ash particles, and many *carcinogenic* (cancer-causing) chemicals when a person smokes cigarettes. Heavy smokers, who acquire a one-pack, two-pack, or three-pack a day habit, often suffer from respiratory diseases such as *bronchitis*, a chronic inflammation of the bronchi; *emphysema,* damage to the elastic tissue of the lungs; and lung *cancer.* Good health and a more productive life are two incentives that may help a person stop smoking.

Abusing drugs. A *drug* is any ingested substance lacking nutritive value, but with the ability to cause physiological reactions. Aspirin, antibiotics, cough medicine, alcohol, caffeine, and other familiar "medications" are drugs. A *psychoactive drug* affects the brain and parts of the central nervous system that control consciousness and behavior. The four main kinds of psychoactive drugs are depressants, stimulants, analgesics, and psychedelics and hallucinogens.

Depressants. *Depressants* are psychoactive drugs that slow down, or depress, the central nervous system. They can reduce anxiety, induce sedation (sleep), and reduce pain. Depressant drugs include alcohol, valium, and a variety of barbiturates. When combined, depressants can be extremely harmful. For example, drinking alcohol and taking barbiturates together magnifies behavioral depression. Thus, a person may become suicidal under the influence of several depressants.

Although a person who drinks alcohol gets an initial "*rush*" or "*high*," this effect is misleading. In fact, alcohol deadens nerves, slows down reaction time, disturbs coordination of body movements, dulls reflexes, and interferes with proper reasoning. Consequently, many communities in the United States have established maximum blood alcohol levels for driving a motor vehicle. Driving while intoxicated (DWI) is one of the leading causes of traffic fatalities in the United States. Long-term use of alcohol may damage the liver *(cirrhosis)*, permanently impair brain function, and destroy neurons. During pregnancy, the excessive use of alcohol by a mother can result in *fetal alcohol syndrome* (FAS). FAS is the third most common cause of mental retardation in the United States.

Stimulants. Cocaine, amphetamines, nicotine, and caffeine are stimulant drugs that increase alertness and sometimes make a person feel more powerful. In time, regular use of stimulants leads to loss of appetite, depression, other psychological disorders, and abnormal physiological functions. For example, cocaine, and its stronger form *crack*, can cause irritability, paranoia, and aggressive behavior. Caffeine (an ingredient of chocolate, cola drinks, coffee, and tea) may affect proper reasoning and contribute to improper functioning of the medulla oblongata, which causes the "jitters," or uncoordinated body movements. Nicotine affects the central nervous system and peripheral nervous system by increasing heartbeat rate, blood pressure, and water retention in body tissues. Studies have shown that smoking during pregnancy can result in newborns with heart problems, as well as poor growth and mental abilities.

Analgesics. Drugs that are used to reduce pain and provide relief are called analgesics. *Endorphins* and *enkephalins* are natural analgesics produced in your body. In 1889, *aspirin* (acetyl salicylic acid) was introduced and soon became the world's largest selling analgesic. Recent studies have shown that aspirin prescribed in controlled doses prevents blood clotting in small arteries, which may reduce the risk of suffering a heart attack. Aspirin is a nonaddictive analgesic. However, narcotic analgesics, such as *heroin, opium,* and *codeine,* are extremely addictive and require increasingly larger doses to produce the desired results.

Psychedelics. Psychedelics, such as *lysergic acid diethylamide (LSD,* commonly called "acid") and *marijuana* (commonly called "grass") are hallucinogenic drugs. They induce hallucinations, trigger psychoses, and produce colorful psychedelic visions. Small doses of LSD can dramatically distort normal perceptions.

Marijuana and hashish are derived from the plant *Cannabis sativa*, a species of hemp. Small doses of marijuana depress body functions, slow down metabolism, and create a state of *euphoria,* or general feeling of well-being. Regular use of marijuana can produce disorientation, anxiety, panic, delusions, and hallucinations.

Table 8–5 summarizes dangerous drugs and their effects.

Table 8–5 Dangerous Drugs and Their Effects

Drug Type	*Examples*	*Effects*
Depressants	Alcohol, valium, barbiturates (i.e., nembutal, seconal)	Slow central nervous system; modify behavior (loss of anxiety); induce sleep; slurred speech; anesthesia; delirium; death from overdose
Stimulants	Cocaine, crack, amphetamines, caffeine, nicotine	Give "highs"—increased alertness to sound and sight; reduce fatigue; elevate mood. Lead to depression; increased heart rate; loss of appetite; death from overdose
Analgesics	Heroin, opium, codeine	Sedate body; reduce pain; some are extremely addictive; death from overdose
Psychedelics	LSD, marijuana, hashish	Induce hallucinations and psychoses; alter perceptions; poor motor control; confusion

Table 8–6 Summary of Behavior Patterns

Behavior Pattern	*Characteristics*	*Example*
Inborn responses		
Taxes	Automatic; inborn; whole animal responds to stimulus	Hydra moves to light
Simple reflexes	Automatic; inborn; beneficial; part of animal responds to stimulus	Sneezing
Instincts	Automatic; series of reflexes that lead to a beneficial response	Nest building in birds
Learned responses		
Imprinting	Response to first object or condition newborn is exposed to	Baby geese following a human
Conditioning	Substitute stimulus triggers a response originally triggered by another stimulus	Dog runs to refrigerator when he hears its door close
Habits	At first, require thinking; after repetition, become automatic; series of conditioned responses	Brushing teeth
Voluntary responses	Conscious activities under control of the will	Solving a problem

Section Quiz

1. Which of these drugs is *not* a stimulant? (*a*) dexedrine (*b*) nicotine (*c*) heroin (*d*) caffeine.
2. Aspirin is classified as a (an) (*a*) analgesic (*b*) hypnotic (*c*) stimulant (*d*) depressant.

3. Regular use of psychoactive drugs usually produces (*a*) total brain damage (*b*) psychological and physiological dependency (*c*) anesthesia and coma (*d*) constant euphoria and "highs."
4. Drugs that alter visual and auditory perceptions are called (*a*) analgesics (*b*) hypnotics (*c*) hallucinogens (*d*) stimulants.
5. Mental retardation in newborns can be caused by the mother's (*a*) excessive ingestion of coffee, tea, and cola drinks (*b*) smoking marijuana (*c*) frequent use of aspirin (*d*) addiction to alcohol.

Voluntary Behavior

Voluntary behavior includes consciously controlled actions such as reasoning, remembering, making decisions, and evaluating. The cerebrum is the control center for all voluntary actions, as well as the part of the brain that is responsible for intelligence, or the ability to learn and adjust to different conditions. The well-developed cerebrum of humans enables us to associate symbols with meanings, memorize facts, solve problems, gain insights, and make conclusions.

Table 8–6 summarizes the behavior patterns of animals. The next chapter will describe how the endocrine system regulates the internal activities of the body.

Chapter Review Questions

The following questions will help you check your understanding of the material presented in the chapter.

1. Which structure is involved primarily in the transmission of nerve impulses? (*a*) platelet (*b*) neuron (*c*) vacuole (*d*) nephron.
2. In humans, the simplest type of response is the (*a*) inborn reflex (*b*) conditioned reflex (*c*) habit (*d*) tropism.
3. The nerve impulse consists of a (*a*) series of sensations (*b*) series of electrochemical changes (*c*) flow of neurons (*d*) flow of protons.

Use the diagram below to answer statements *4 through 8*. Write the number of the part of the human nervous system that is most closely associated with each statement. A number may be used more than once or not at all.

4. This structure is known as the cerebellum.

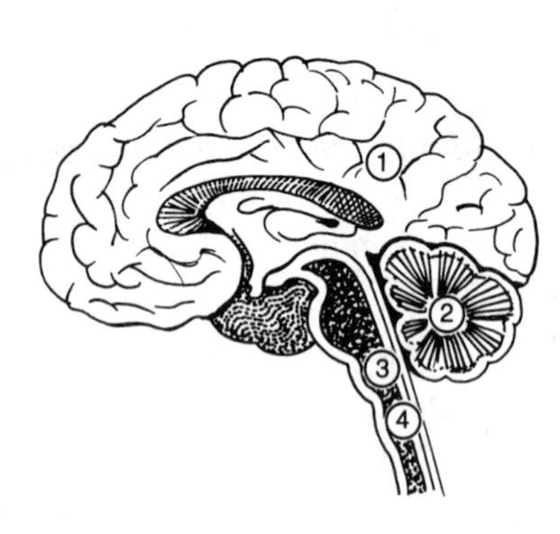

5. This part of the nervous system serves as a memory bank and the center for thought.

6. This structure regulates heartbeat and breathing.

7. This structure is the main site for the coordination of motor activities.

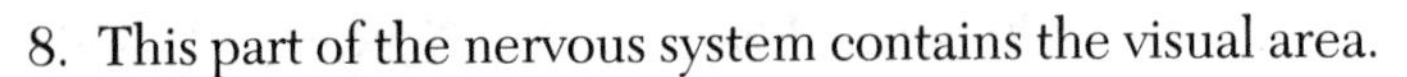

8. This part of the nervous system contains the visual area.

9. The nerve cells of humans and earthworms (*a*) are organized into a nerve-net type structure (*b*) secrete chemical messengers called neurohumors (*c*) cause slower responses than endocrine responses (*d*) transmit electrochemical impulses called synapses.

10. In humans, responses controlled by the nervous system (*a*) are under the control of excretions from ribosomes (*b*) are associated with homeostasis (*c*) are slower than endocrine reactions (*d*) are not dependent upon the presence of ATP molecules.

11. Compounds that are secreted at the ends of nerve cells and influence nerve impulse transmission are (*a*) neurohumors (*b*) antibodies (*c*) narcotics (*d*) myelins.

12. The spaces between adjacent neurons are called (*a*) axons (*b*) neurohumors (*c*) synapses (*d*) dendrites.

13. Which of the following sequences best describes the path of a reflex arc? (*a*) effector, stimulus, synapse, receptor, sensory neuron, motor neuron (*b*) stimulus, receptor, sensory neuron, synapse, motor neuron, effector (*c*) receptor, sensory neuron, stimulus, effector, motor neuron, synapse (*d*) stimulus, receptor, sensory neuron, synapse, effector, motor neuron.

14. An environmental change that causes an organism to respond is called (*a*) a reflex (*b*) a synapse (*c*) a stimulus (*d*) an electrochemical impulse.

15. The conduction of a nerve impulse across a synapse involves (*a*) electrons jumping a gap (*b*) secretion of a neurohumor (*c*) physical contact between neurons (*d*) stimulation of a neuron by a catalyst.

16. The involuntary responses of the heart are controlled by the (*a*) nervous system (*b*) lymphatic system (*c*) digestive system (*d*) excretory system.

17. The main difference between the nerve net of a hydra and the nervous system of an earthworm is that the earthworm (*a*) has a definite pathway for impulses (*b*) is protected by a backbone (*c*) does not require neurons to effect a response (*d*) responds to external stimuli only.

18. Which of the following has a nervous system consisting of a relatively simple anterior brain, a ventral nerve cord, and connecting nerves? (*a*) mouse (*b*) human (*c*) frog (*d*) grasshopper.

19. The nervous system of a grasshopper is most similar to the nervous system of a (an) (*a*) earthworm (*b*) hydra (*c*) bird (*d*) human.

20. Nerve impulses may travel in either direction over a pathway in the nervous system of (*a*) grasshoppers (*b*) hydras (*c*) earthworms (*d*) humans.

21. Which group of drugs contains only depressants? (*a*) alcohol, valium, and cocaine (*b*) alcohol, valium, and nembutal (*c*) heroin, alcohol, and cocaine (*d*) heroin, opium, and cocaine.

22. Long-term addiction to alcohol destroys neurons, impairs mental functioning, and causes cirrhosis of the (*a*) heart (*b*) skin (*c*) liver (*d*) kidneys.

23. A stimulant usually found in chocolate, coffee, and tea is (*a*) cocaine (*b*) dexedrine (*c*) nicotine (*d*) caffeine.

24. Which group of drugs contains only addictive analgesics (painkillers)? (*a*) heroin, opium, and cocaine (*b*) heroin, opium, and codeine (*c*) heroin, cocaine, and aspirin (*d*) heroin, cocaine, and hashish.

25. The scientific name for the psychedelic drug commonly called "acid" is (*a*) lysergic acid diethylamide (*b*) enkephalin (*c*) *Cannabis sativa* (*d*) acetyl salicylic acid.

Biology Challenge

The following questions will provide practice in answering SAT II-type questions.

Part I

Circle four items in column II that are related to one item in column I. Then, underline the related item in column I.

Example:

I	II
X (A) enzyme	X (1) ptyalin
(B) hormone	X (2) steapsin
(C) vitamin	(3) lymph
	X (4) hexokinase
	X (5) amylase

I	II
1. (A) central nervous system	(1) hypothalamus
(B) peripheral nervous system	(2) parasympathetic neurons
(C) limbic system	(3) sympathetic neurons
	(4) spinal nerves
	(5) cranial nerves

I	II
2. (A) brain	(1) cerebrum
(B) spinal cord	(2) pons
(C) receptors	(3) thalamus
	(4) pineal gland
	(5) statocyst

I	II
3. (A) thermoreceptor	(1) pigmented cell
(B) mechanoreceptor	(2) retina
(C) photoreceptor	(3) maculabutea
	(4) ommatidium
	(5) motor end plate

4. (A) analgesic
   (B) depressant
   (C) psychedelic

   (1) alcohol
   (2) valium
   (3) nembutal
   (4) caffeine
   (5) barbiturate

5. (A) hypothalamus
   (B) thalamus
   (C) medulla

   (1) autonomic nervous system (ANS)
   (2) anti-diuretic hormone (ADH)
   (3) taxis
   (4) oxytocin
   (5) hunger

Part II

Use your knowledge of biology and the diagram below to answer questions *1 to 7*. Each numbered box represents a different part of the human nervous system. Several boxes already are labeled to help you answer the following questions.

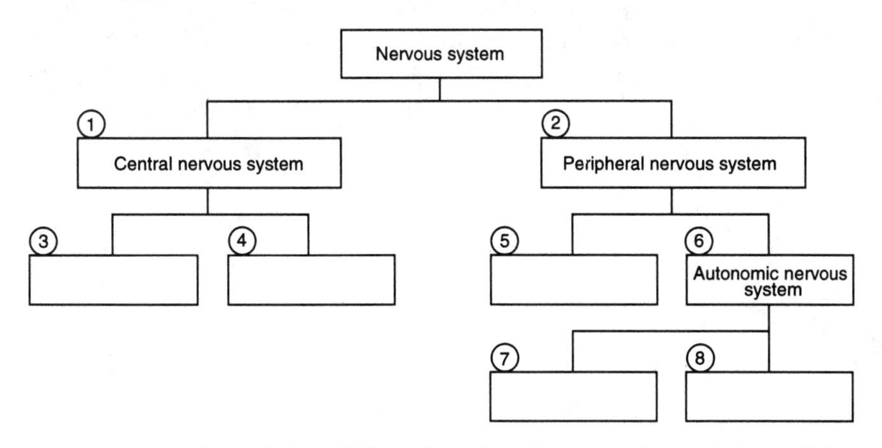

1. The *spinal cord* should be placed in box number (*a*) 7 (*b*) 6 (*c*) 5 (*d*) 4 (*e*) not in the schematic.
2. The *sensory neurons* should be placed in box number (*a*) 3 (*b*) 4 (*c*) 5 (*d*) 7 (*e*) 8.

3. The *parasympathetic system* should be placed in box number (*a*) 3 (*b*) 4 (*c*) 5 (*d*) 8 (*e*) not in the schematic.
4. Which numbered box should have the name of a part of the nervous system that prepares the body for intense physical exertion? (*a*) 3 (*b*) 4 (*c*) 5 (*d*) 6 (*e*) 7.
5. The *brain* should be placed in box number (*a*) 3 (*b*) 4 (*c*) 5 (*d*) 6 (*e*) not in the schematic.
6. Which numbered box contains a part of the nervous system that coordinates motor activities and helps maintain balance? (*a*) 2 (*b*) 3 (*c*) 4 (*d*) 5 (*e*) 6.
7. The interpretation of nerve impulses from light and sound is performed by a part of the nervous system numbered (*a*) 2 (*b*) 3 (*c*) 4 (*d*) 5 (*e*) 6.
8. The nervous system of most complex animals is derived from embryonic tissue called (*a*) ectoderm (*b*) epiderm (*c*) endoderm (*d*) mesoderm (*e*) periderm.
9. Certain snakes, called pit vipers, possess heat-sensitive pits located on both sides of their head. This adaptation enables the pit vipers to hunt and prey on (*a*) grasshoppers (*b*) lizards (*c*) frogs (*d*) mice (*e*) salamanders.
10. An evolutionary trend illustrated by the nervous system of an insect and a more advanced form found in vertebrates is (*a*) development of a ventral nerve cord (*b*) formation of a nerve net (*c*) concentration of ganglia in the head region (*d*) increasing control of voluntary actions by secondary ganglia (*e*) large olfactory lobes.

9

Endocrine System

Learning Objectives

When you have completed this chapter, you should be able to:

- **Compare** duct glands and ductless glands.
- **List** the functions and effects of human endocrine glands.
- **Discuss** the importance of other endocrine secretions.
- **Describe** the effects of hormones in other organisms.

OVERVIEW

The nervous system sends messages rapidly, producing very fast responses to stimuli in the body and from the external environment. In contrast, the endocrine system sends messages that travel slower, but have longer-lasting effects on the body. Humans and other mammals have an endocrine system that consists of ductless glands, hormones, a circulatory pathway, and target organs. In this chapter, you will learn how the endocrine system regulates body functions.

ACTION OF HORMONES

An *endocrine gland* is ductless; its secretions, called *hormones*, pass directly into the blood (Figure 9–1). Once in the blood, a hormone is carried all over the body, but it affects only specific organs called *target organs*. Thus, a hormone acts as a messenger that may either stimulate or

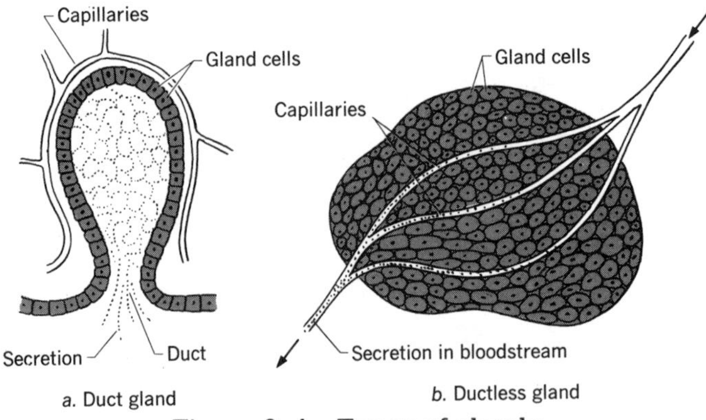

Figure 9–1 Types of glands

inhibit (repress) the activity of its target organs. For example, partly digested food that passes out of the stomach as chyme stimulates specialized cells at the entrance to the small intestine to produce the hormone secretin. This hormone stimulates the pancreas to secrete through a duct pancreatic juice, which contains enzymes that digest food in the small intestine. The differences between duct glands and ductless glands are summarized in Table 9–1.

Table 9–1 Comparison of Duct Glands and Ductless Glands

Gland	*Secretion*	*Method of Transport*	*Where Secretion Functions*
Exocrine (duct gland)	Juice, often containing enzymes	Duct	Organ near gland
Endocrine (ductless gland)	Hormone	Bloodstream	Organ distant from gland

An endocrine gland can be stimulated or inhibited by a nerve impulse, by the presence of glucose or other substances in the blood, and by a hormone from another gland. For example, when you are frightened, impulses from the brain travel to the autonomic nervous system, which relays impulses to the adrenal glands. The impulses stimulate the adrenal glands to secrete the hormone *epinephrine,* which affects the liver, one of its target organs. The liver responds to epinephrine by changing glycogen to glucose. The stomach, another target organ of epinephrine, responds by slowing peristalsis.

Earl W. Sutherland, 1971 Nobel Prize winner, explained how hormones affect target organs. According to his hypothesis, a hormone acts on particular receptor sites located on the secreting cells of the target organ. This results in a receptor-hormone reaction that stimulates the formation of an enzyme within the cell, which helps form *cyclic adenosine monophosphate,* or AMP, from ATP. Cyclic AMP either triggers or restrains the secretion of a specific hormone by the secreting cells. For example, when the thyroid-stimulating hormone (TSH) from the pituitary gland reaches the secreting cells of the thyroid gland, cyclic AMP is formed and stimulates the cells of the thyroid gland to secrete thyroxin (Figure 9–2).

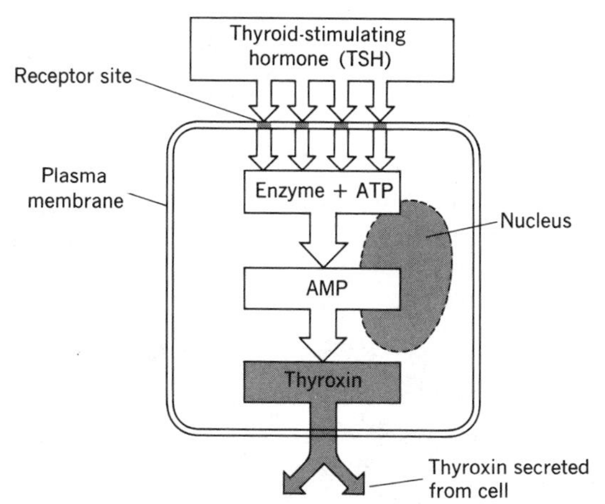

Figure 9–2 TSH acting on a cell of the thyroid gland

HUMAN ENDOCRINE GLANDS

Figure 9–3 shows the major glands of the human endocrine system.

Mucous Membranes of the Digestive Tract

Stomach. The mucous membrane lining the *stomach* contains certain cells that secrete the hormone *gastrin,* which stimulates production of hydrochloric acid.

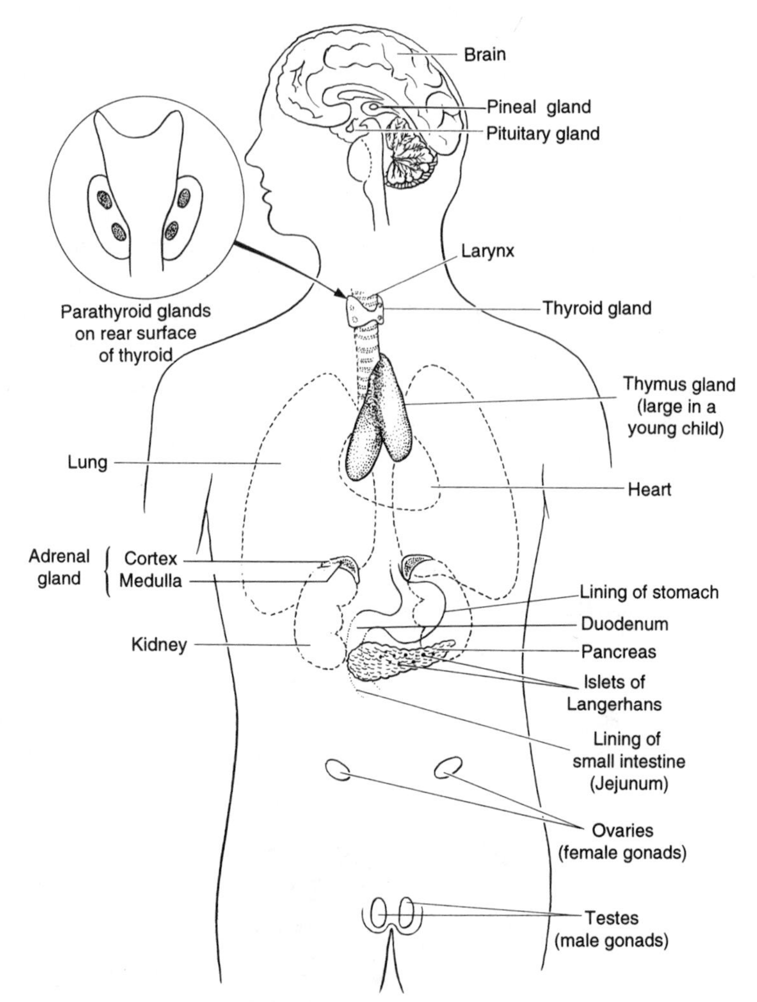

Figure 9–3 Human endocrine system

Duodenum. The initial 5 to 10 cm of the *duodenum* includes cells that are stimulated by acidic chyme leaving the stomach. These cells secrete the hormone *secretin*, which stimulates pancreatic cells to secrete digestive juices and liver cells to secrete bile.

Jejunum. The mucous membrane of the *jejunum* is stimulated by partially digested proteins to secrete the hormone *cholecystokinin.* As a result, bile is released from the gall bladder into the duodenum, and pancreatic juice (with its full complement of digestive enzymes) mixes with bile.

The Pituitary Gland

The pea-sized *pituitary gland* (about 1.3 cm in diameter) lies in a bony cavity of the skull (Figures 9–3, 9–4). The pituitary gland consists of the *anterior lobe,* which accounts for about 75 percent of its weight, and the *posterior lobe.* The anterior lobe secretes several hormones; the posterior lobe stores two hormones that are released by nerve impulses from the brain. A third region, called the *intermediate lobe,* is small in adults and has no known function. The pituitary gland was formerly considered the "master" endocrine gland because of its apparent "governing" effect on the actions of other endocrine glands. Actually, the master organ is the *hypothalamus,* a small region in the brain that maintains homeostasis in the nervous system and in the endocrine system.

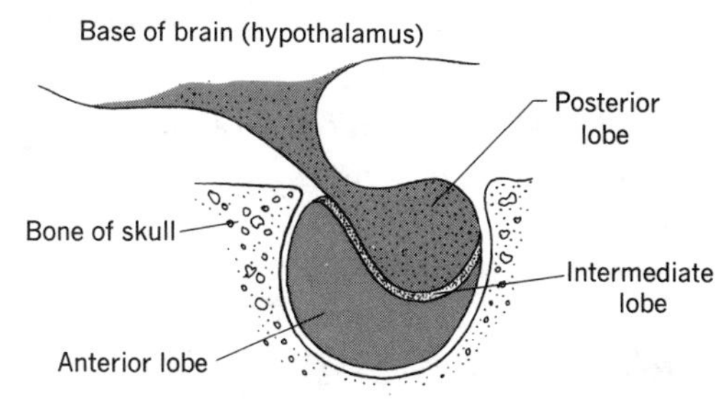

Figure 9–4 Pituitary gland (enlarged)

The anterior lobe. Six major hormones are secreted by different types of cells in the anterior lobe.

1. *Growth hormone* (*GH*). The secretion of GH is controlled by releasing and inhibiting hormones of the hypothalamus. GH has target cells throughout the body, especially those that make up muscles and bones. Muscle growth results from protein synthesis and cell growth; bone growth results mainly from controlled growth and multiplication of bone cells. *Dwarfism* is caused by an undersecretion of GH during the growth years. Today, young children who suffer from dwarfism are given GH produced by bacteria which were altered using recombinant DNA techniques. *Giantism* is caused by an oversecretion of GH during the growth years. *Acromegaly* is a growth disorder caused by an oversecretion of GH during adulthood. People suffering from acromegaly have enlarged bones of the hands, feet, cheek, and jaws.

2. *Thyroid-stimulating hormone* (*TSH*). The secretion of TSH is regulated by thyrotropic-releasing hormone (TRH) released by the hypothalamus. The target tissue of TSH is the secretory tissue of the thyroid

gland, which then produces the hormone *thyroxin*. When thyroxin reaches a certain level in the blood, the hypothalamus stops releasing TRH, which checks the secretion of TSH by the anterior lobe.

3. *Adrenocorticotropic hormone* (*ACTH*). The secretion of ACTH is controlled by corticotropic-releasing hormone (CRH) released by the hypothalamus. ACTH controls secretions of the adrenal cortex, or outer layer of the adrenal gland. Hormones of the adrenal cortex reduce CRH output when levels of ACTH are higher than normal.

4. *Follicle-stimulating hormone* (*FSH*). The secretion of FSH is controlled by the hormone gonadotropin-releasing hormone (GRH) released by the hypothalamus. FSH stimulates gonads (sex glands). In females, FSH stimulates development of egg follicles (pockets) in the *ovaries*; in males, FSH stimulates sperm cell production in the *testes*. GRH and FSH release is checked by female sex hormones, or *estrogens*, and by the male sex hormone called *testosterone*. Again, a feedback system ensures normal hormone levels.

5. *Luteinizing hormone* (*LH*). The secretion of LH also is controlled by gonadotropin-releasing hormone (GRH). LH stimulates estrogen production by ovarian cells; regulates *ovulation*, the release of an egg cell by an ovary; and stimulates the formation of a *corpus luteum* (yellow body) after an egg follicle discharges an egg cell. A corpus luteum produces the hormone *progesterone*, which together with estrogens prepare the lining of a uterus for implantation of a fertilized egg cell. The secretion of LH is regulated by the same hormones that regulate FSH secretion.

6. *Prolactin* (*lactogenic hormone*). The secretion of prolactin is controlled by releasing and inhibitory hormones that are affected by levels of estrogen and progesterone. Prolactin (in combination with estrogens, progesterone, thyroxin, and human growth hormone) prepares the mammary glands of a pregnant woman for milk production and stimulates milk secretion after birth.

The posterior lobe. The posterior lobe of the pituitary gland does not synthesize hormones; it stores and releases the hormones oxytocin and antidiuretic hormone (ADH). These hormones are produced by specialized cells in the hypothalamus and then transported by nerve fibers to the posterior lobe (Figure 9–4).

Oxytocin has two target organs, the uterus and the mammary glands of a pregnant woman. During childbirth, oxytocin stimulates contraction of the uterus. After childbirth, oxytocin stimulates the flow of milk from the mammary glands.

Antidiuretic hormone (*ADH*), or vasopressin, acts on the kidneys. When the body is dehydrated (lacks water), ADH causes the kidneys to decrease urine output. However, when the concentration of water in the blood is above normal, the release of ADH is checked and the kidneys respond by increasing urine output.

The intermediate lobe. The intermediate pituitary lobe secretes the hormone *intermedin*. In fishes, frogs, and reptiles, intermedin regulates the size of certain color-bearing cells in the skin. In humans, the hormone has no known effect.

The Thyroid Gland

The *thyroid gland*, located just below the larynx, consists of two lobes connected by a narrow bridge of tissue. The thyroid gland secretes the hormones *thyroxin* and *calcitonin*.

Thyroxin, which contains the element iodine, stimulates the rate of cellular oxidation and influences growth and development. In a young child, an undersecretion of thyroxin results in *cretinism*—a condition characterized by dwarflike features and below-normal intelligence. If started early enough, continued treatment with either thyroid extract or thyroxin usually corrects cretinism. In an adult, an undersecretion of thyroxin causes puffiness of the body and mental sluggishness. This condition, called *myxedema*, also is treated with thyroxin.

The thyroid gland enlarges if a person's diet lacks iodine. This condition, called *simple*, or *nontoxic, goiter*, may be treated or prevented by eating food that contains iodine or by using iodized salt. Another condition in which the thyroid gland enlarges is called *toxic*, or *exophthalmic, goiter*, which results in an oversecretion of thyroxin. In this condition, the rate of metabolism increases dramatically and the person consumes a great deal of food, loses weight, and is nervous and restless. A characteristic symptom of toxic goiter is bulging eyeballs. Toxic goiter may be corrected by surgery, treatment with X rays or radioactive iodine, or drugs that interfere with the synthesis of thyroxin.

Calcitonin, a more recently discovered hormone, interacts with the parathyroid hormone to maintain a constant level of calcium in the blood and in bones.

The Parathyroid Glands

The *parathyroid glands* are four small, oval glands embedded in the back of the thyroid gland. They secrete the hormone *parathormone*, which regulates the level of calcium and phosphate salts in the blood. An undersecretion of parathormone results in a deficiency of calcium salts in the blood and an increase in the excretion of calcium salts in the urine. A lack of calcium salts may result in a disorder called *tetany*, which makes motor nerves leading to skeletal muscles overly sensitive and causes muscular spasms. Tetany is usually treated by giving injections of calcium salts and by increasing dietary intake of vitamin D. Vitamin D aids in the absorption of calcium salts from the small intestine into the blood. An oversecretion of parathormone removes calcium salts from bones, causing them to soften.

The Thymus Gland

The *thymus gland* consists of two lobes located in the upper part of the chest, near the heart. The thymus gland, which is large at birth and during childhood, gradually shrinks after adolescence. For this reason, some biologists think that a thymus hormone called *thymosin* inhibits sexual development during childhood.

Experiments with young mice indicate that the thymus gland plays an important part in the body's ability to form antibodies. Scientists think that a young thymus secretes a hormone that enables some of its cells to leave the thymus gland, enter the circulation, and travel to different lymph glands where they become established. When an infection occurs, these cells, called *lymphocytes,* leave the lymph glands and combat the infection by producing antibodies.

In one experiment, the thymus gland was removed from a mouse embryo. As a result, the mouse did not produce antibodies against the bacteria that cause pneumonia and quickly died. When a thymus gland, or a thymus gland wrapped in a permeable membrane, was grafted into another mouse embryo whose own thymus gland had been removed, the mouse produced antibodies and survived.

The Pancreas

The *pancreas* is both an exocrine, or duct, gland and an endocrine gland. As an *exocrine gland,* it secretes pancreatic juice used in digesting food. As an endocrine gland, it secretes two hormones, *insulin* and *glucagon,* from clusters of cells known as the *islets of Langerhans.* The cells of the islets of Langerhans are of two types, *alpha cells* and *beta cells.*

Insulin is secreted by the beta cells of the islets of Langerhans. The main functions of insulin are (1) regulating the passage of glucose into and out of cells, and (2) converting glucose to glycogen in the liver and in muscles.

An undersecretion of insulin results in a disorder called *diabetes mellitus,* in which the liver and muscles cannot absorb glucose and change it to glycogen. As a result, glucose accumulates in the blood, and some is changed to poisonous compounds. These compounds (such as acetone) can cause an untreated diabetic person to become unconscious. A person suffering from diabetes mellitus may be treated by regulating carbohydrate intake and by injections of insulin. An overdose of insulin may cause insulin shock, in which the blood glucose level drops dangerously, and convulsions, unconsciousness, and death may occur if untreated. For this reason, diabetic persons who give themselves injections of insulin usually carry sugar or candy to counteract the effects of an overdose of insulin.

Glucagon is secreted by the alpha cells of the islets of Langerhans. Glucagon increases the level of glucose in the blood by helping to change glycogen into glucose. The normal level of glucose in the blood is maintained mainly by the antagonistic actions of glucagon and insulin.

The Adrenal Glands

There are two *adrenal glands,* one on top of each kidney. Each adrenal gland is composed of two different regions. The outer region, or *adrenal cortex,* makes up most of the gland. It surrounds an inner region called the *adrenal medulla.*

The adrenal cortex. The adrenal cortex is divided into three different zones that produce hormones called *corticosteroids.* All of the corticosteroids are synthesized from cholesterol. The outer zone secretes a group of hormones that maintain the normal mineral balance in the blood. The main hormone of the group, called *aldosterone,* increases both the reabsorption of sodium by the kidney tubules and excretion of potassium by the kidney tubules. The middle zone secretes a group of hormones that regulate the level of glucose in the blood. The main hormone of this group is *cortisol.* The hormones of the middle zone promote normal metabolism, provide resistance to stress, and help reduce inflammations from allergies, arthritis, and rheumatism. The innermost zone also secretes trace amounts of *estrogens* and *androgens.* Estrogens are chemically similar to the female sex hormones produced by the ovaries. Androgens are responsible for the same masculine characteristics as testosterone, secreted by the testes.

The adrenal medulla. The adrenal medulla secretes two hormones, *epinephrine* (*adrenalin*) and *norepinephrine* (*noradrenalin*). About 80 percent of the secretion is epinephrine and 20 percent is norepinephrine. The two hormones have similar but not identical functions. The functions of epinephrine, which are best noted when a person is under stress, are:

- Stimulating liver cells to change glycogen to glucose. This provides tissue cells with enough fuel for cellular oxidation.
- Widening the small arteries in skeletal muscles, in the heart, and in the brain, thus increasing the blood supply to these organs.
- Constricting the small arteries leading to the skin and the digestive organs, which increases blood flow to skeletal muscles.
- Increasing the rate and strength of the heartbeat, which increases the blood flow to active muscles.
- Causing muscles of the air passages to relax, which allows more air to enter the lungs.
- Increasing the rate at which blood clots, which prevents excessive blood loss.

The main function of norepinephrine, a weaker form of epinephrine, is to constrict small arteries.

The Sex Glands (Gonads)

The male sex glands are the *testes* and the female sex glands are the *ovaries*. Some cells of the sex glands produce sex cells (either egg cells or sperm cells); other cells secrete sex hormones.

The testes. Under the influence of LH, certain cells of the testes secrete the male sex hormone, *testosterone*. This hormone causes the development of male secondary sex characteristics such as facial hair, a deep voice, and a muscular body. FSH (follicle-stimulating hormone) stimulates the formation of sperm cells.

The ovaries. Under the influence of FSH, certain cells of the ovaries form a follicle in which an egg cell matures. Cells that line the follicle secrete the female sex hormone estradiol, an *estrogen*. This hormone causes the development of female secondary sex characteristics, mammary glands, a high-pitched voice, and wider hips than males. After an egg has been discharged from the follicle, LH causes the follicle to change into the *corpus luteum,* which secretes *progesterone.* This hormone prepares the wall of the uterus to receive a fertilized egg cell. When found in sufficient amounts, estrogen and progesterone inhibit target cells in the anterior lobe of the pituitary gland from secreting FSH and LH.

The Pineal Gland

In humans, the *pineal gland* is located in the midbrain (in the cleft between the cerebrum and cerebellum); it secretes a hormone called *melatonin* (refer back to Figure 9–3). Animal studies show that melatonin governs maturation of the sex glands and that more melatonin is secreted in darkness than in light. By varying the amount of light, researchers discovered that the pineal gland varies melatonin concentration in an animal's body. As a result, biorhythms, including reproductive cycles, were affected. For example, certain rodents increase their sexual activity during long summer days; the rodents decrease their sexual activity during long winter nights. The migratory instincts of geese, which depend on the length of day and night, are upset when the pineal gland is removed from their brain. Some fishes and salamanders respond to variations in light and color by secreting melatonin in their skin cells, causing a change in pigmentation. In humans, melatonin concentrations are highest in young children and decrease by up to 75 percent at adolescence. Human mating habits are not significantly affected by seasonal changes.

Other Endocrine Secretions

Enkephalins and *Endorphins.* Researchers have recently learned that the brain produces hormonelike secretions called *neurohumors,* or *neuropeptides.* These secretions, called *enkephalins* and *endorphins,* consist of up to 40 amino acids linked together. Both are extremely strong pain relievers, often called the body's "natural" painkillers. Enkephalins and endorphins also are linked to improved memory and learning, feelings of pleasure, sex drive, and mental illnesses, including depression.

Erythropoietin. The kidneys produce the hormone *erythropoietin,* which stimulates bone marrow to form erythrocytes (red blood cells).

Prostaglandins. These hormones found in semen are synthesized by the seminal vesicles. Prostaglandins regulate blood pressure, influence the body's immune response, and regulate feelings of pain.

Table 9–2 summarizes the functions of human endocrine glands.

Table 9–2 Human Endocrine Glands and Their Functions

Gland	*Hormone*	*Function*
Mucous membrane of stomach	Gastrin	Stimulates production of hydrochloric acid (HCl)
Mucous membrane of small intestine	Secretin	Stimulates secretion of pancreatic juice and bile
	Cholecystokinin	Stimulates flow of bile from gall bladder
Pituitary:		
Anterior lobe	Growth hormone	Stimulates growth of skeleton and muscles
	Follicle stimulating hormone (FSH)	Stimulates follicle formation in ovaries and sperm cell formation in testes
	Luteinizing hormone (LH)	Stimulates production of estrogen by ovary, release of ovum from ovary (ovulation), and production of progesterone from corpus luteum of ovary
	Thyroid stimulating hormone (TSH)	Stimulates secretion of thyroxin from thyroid gland
	Adrenocorticotropic hormone (ACTH)	Stimulates secretion of a group of cortical hormones (see adrenal cortex)
	Prolactin	Helps prepare mammary glands for milk production; stimulates secretion of milk in mammary glands after birth
Posterior lobe	Antidiuretic hormone (ADH)	Regulates water balance by decreasing urine output
	Oxytocin	Stimulates uterine muscle during childbirth, and milk secretion in mammary glands after birth

Table 9–2 *(Continued)*

Gland	*Hormone*	*Function*
Thyroid	Thyroxin	Controls rate of metabolism and physical and mental development
	Calcitonin	Controls calcium metabolism
Parathyroids	Parathormone	Increases blood calcium and magnesium levels. Stimulates production of *calcitriol*, which increases rate of calcium and magnesium absorption from the intestines
Thymus	Thymosin	Stimulates formation of antibody system
Islets of Langerhans:		
Beta cells	Insulin	Promotes storage and oxidation of glucose
Alpha cells	Glucagon	Releases glucose into blood
Adrenals (cortex)		
1. Outer zone	Aldosterone	Regulates mineral homeostasis
2. Middle zone	Cortisol	Regulates glucose metabolism; promotes normal metabolism; provides resistance to stress; reduces rheumatoid inflammations
3. Inner zone	Sex steroids: estrogens and androgens	Estrogens are chemically similar to ovarian estrogen and function in the same way (see gonads); androgens produce the same effects as testosterone from testes (see gonads).

(*Continued on next page*)

Table 9–2 *(Continued)*

Gland	*Hormone*	*Function*
Medulla	Epinephrine (adrenalin) and norepinephrine (noradrenalin)	Releases glucose into bloodstream, increases rate of heartbeat, increases rate of respiration, reduces clotting time, relaxes smooth muscle in air passages
Gonads		
Ovaries, follicle cells	Estrogen	Controls female secondary sex characteristics
Corpus luteum cells	Progesterone	Helps maintain attachment of embryo to uterine wall
Testes	Testosterone	Controls male secondary sex characteristics
Pineal	Melatonin	Regulates sexual maturity and biorhythms in some vertebrates (rodents, geese, fish, and possibly humans)
Brain	Enkephalins Endorphins	Natural painkillers, also improve memory and learning; affect sex drive
Kidneys	Erythropoietin	Stimulates red blood cell formation in bone marrow
Seminal vesicles (and possibly other parts of body)	Prostaglandins	Involved in responses to pathogens such as bacteria, viruses, and foreign proteins; can raise and lower blood pressure

Section Quiz

1. Secretory cells of the hypothalamus produce (*a*) neuropeptides (*b*) ecdysone (*c*) releasing hormones (*d*) aldosterone.
2. The hormone erythropoietin (*a*) is produced by the kidneys (*b*) directs red blood cells to the liver (*c*) stimulates production of leukocytes (*d*) regulates water balance.
3. A hormone whose production depends on light and dark cycles is (*a*) secretin (*b*) melatonin (*c*) androgen (*d*) estrogen.
4. Partly digested proteins in the small intestine trigger the release of cholecystokinin, which (*a*) stimulates beta cells to secrete insulin (*b*) regulates intestinal peristalsis (*c*) helps maintain an optimum pH in the ileum (*d*) causes the gall bladder to eject bile.
5. Which is a natural painkiller produced in the brain? (*a*) prostaglandins (*b*) progesterone (*c*) endorphins (*d*) oxytocin.

ENDOCRINE GLANDS IN OTHER ORGANISMS

In protozoans and hydra, specific hormones have not been isolated. Some hormones have been isolated in mollusks, crustaceans, and insects.

The life cycle of moths and butterflies illustrates how hormones affect the development of insects. Moths and butterflies develop in four stages: egg, larva (caterpillar), pupa, and adult. *Metamorphosis* (change in form) of one stage to the next is controlled by the interaction of two hormones: *ecdysone* and *juvenile hormone*. Ecdysone stimulates the periodic shedding, or *molting*, of the larva's outer covering as it grows. The same hormones in the pupa stimulate cells to specialize and change into adult tissues and organs. Juvenile hormone promotes growth of the larva and prevents it from molting. If the glands that secrete the juvenile hormone are removed, a larva rapidly metamorphoses into a pupa. If an extract of these glands is injected into a larva about to metamorphose into a pupa, the larva continues to grow in size rather than change into a pupa.

Apparently, both the nervous system and the endocrine system help maintain homeostasis. Both systems involve chemical messengers: neurohumors in the nervous system and hormones in the endocrine system. The systems are unalike in two significant ways. In the nervous system, nerve fibers carry messages; and in the endocrine system, hormones carry messages. Also, responses of the body to nervous impulses are short and specific, whereas responses of the body to hormone messages last longer and are more generalized.

Chapter Review Questions

The following questions will help you check your understanding of the material presented in the chapter.

1. Which glands secrete substances directly into the circulatory fluid? (*a*) endocrine glands (*b*) salivary glands (*c*) sweat glands (*d*) tear glands.
2. Which substance is a secretion? (*a*) insulin (*b*) carbon monoxide (*c*) urea (*d*) lactic acid.
3. The substance secreted by a ductless gland is called a (an) (*a*) exocrine juice (*b*) hormone (*c*) enzyme (*d*) extract.
4. A gland that has both duct and ductless portions is the (*a*) pituitary (*b*) thymus (*c*) pineal (*d*) pancreas.
5. Hormones and enzymes are similar in that both (*a*) are always secreted directly into the digestive system (*b*) affect the rate of physiological processes (*c*) are secreted only by endocrine glands (*d*) convert carbohydrates to amino acids.
6. The pituitary gland is located (*a*) in a bony cavity of the skull (*b*) on the thyroid (*c*) in the ovaries (*d*) under the breastbone.
7. An abnormally high level of calcium in the blood suggests a disorder of the (*a*) liver (*b*) pancreas (*c*) parathyroids (*d*) large intestine.
8. Insulin is secreted by the (*a*) pancreas (*b*) kidney (*c*) liver (*d*) small intestine.
9. A basal metabolism test determines functioning of the (*a*) islets of Langerhans (*b*) thymus (*c*) thyroid (*d*) corpus luteum.
10. To study thyroid function in a cow, a scientist may feed the cow a compound containing (*a*) iodine-131 (*b*) strontium-90 (*c*) carbon-14 (*d*) oxygen-18.
11. Although the concentration of iodine in the blood is low, cells in the thyroid gland continue to accumulate iodine by the process of (*a*) mechanical digestion (*b*) hydrolysis (*c*) active transport (*d*) phagocytosis.
12. Glucose in the urine may be an indication of (*a*) diabetes (*b*) cretinism (*c*) giantism (*d*) goiter.
13. In a person suffering from acromegaly there is an excess of (*a*) gastrin (*b*) growth hormone (*c*) juvenile hormone (*d*) prolactin.

14. The normal reproductive cycle of the human female involves the interaction of the (*a*) oviduct, thyroid gland, and ovary (*b*) pituitary gland, ovary, and uterus (*c*) adrenal gland, ovary, and placenta (*d*) placenta, pituitary gland, and uterus.
15. In experiments, scientists have produced roosters that build nests, cluck like hens, and try to hatch artificial eggs. This is probably the result of (*a*) dietary changes (*b*) parthenogenesis (*c*) light changes (*d*) hormone injections.
16. A doctor can control a patient's menstrual cycle by administering regulated doses of (*a*) enzymes (*b*) hormones (*c*) antibiotics (*d*) vitamins.

Questions *17 through 21* pertain to specific human hormones. For each statement, select the number of the hormone from the list below that is most closely associated with the statement.

Hormones

(1) adrenalin	(3) insulin
(2) ACTH	(4) estrogen

17. Influences the development of female secondary sex characteristics.
18. Stimulates activities in the adrenal glands.
19. Promotes the outflow of sugar from the blood into the muscles.
20. Strongly influences regulation of heartbeat, blood sugar levels, and blood clotting rates.
21. Secreted by the pituitary gland.

Biology Challenge

The following questions will provide practice in answering SAT II-type questions.

Part I

Select the letter of the answer that best completes each sentence.

1. Anabolic steroids are derived mainly from (*a*) parahormone (*b*) cholecystokinin (*c*) testosterone (*d*) growth hormone (*e*) adrenocorticotropic hormone.

2. An organ of the central nervous system that influences hormone secretion is the (*a*) hypothalamus (*b*) thalamus (*c*) limbic system (*d*) cerebellum (*e*) cerebrum.
3. Many hormones are composed of (*a*) lipoproteins (*b*) triglycerides (*c*) enzymes (*d*) vitamins (*e*) amino acids.
4. A person with diabetes mellitus takes medication to replace the hormone (*a*) insulin (*b*) glucagon (*c*) ACTH (*d*) epinephrine (*e*) thyroxine.
5. A hormone that stimulates growth and molting in insects is (*a*) pheromone (*b*) ecdysone (*c*) indoleacetic acid (*d*) oxytocin (*e*) melatonin.

Part II

Each of the following questions consists of a *statement* in the left column and a *reason* in the right column. Select the correct letter based on the following conditions:

A — The statement is true, but the reason is false.
B — The statement is false, but the reason is true (as a statement of fact).
C — Both statement and reason are false.

Statement	*Reason*
1. Hormones function independently of one another.	1. Different hormones have different target organs.
2. The pineal hormone, melatonin, decreases in concentration as a child enters adolescence.	2. A reciprocal relationship exists between an increase in sex hormones and a decrease in pineal hormone.
3. The posterior lobe of the pituitary gland secretes and stores the hormones oxytocin and ACTH.	3. The posterior lobe of the pituitary gland possesses neurosecretory cells.
4. Adrenal cortical sex steroids (estrogen and androgens) are chemically similar to ovarian estrogens and testicular testosterone.	4. The adrenal cortex is derived from endoderm, the same embryonic tissue that forms male and female sex organs.
5. Enkephalins and endorphins are neurohumors that function as impulse transmitters in nerve cells.	5. Both neurohumors are strong pain killers that are transported by nerve cells to painful areas.

10

Muscular-Skeletal System

Learning Objectives

When you have completed this chapter, you should be able to:

- **List** the adaptive advantages of movement.
- **Relate** the structure of parts of the human skeletal and muscular systems to movement.
- **Describe** the functions of the human skeletal system.
- **Identify** parts of the axial and appendicular skeletons.
- **Describe** ways in which muscles move bones.
- **Discuss** movement in other representative organisms.

OVERVIEW

A paramecium uses cilia to capture its prey. A grasshopper spots a predatory frog and escapes by hopping away. A person runs toward a bus stop as he or she spots an approaching bus. Organisms receive information about changes in the environment. The information is usually translated into a response that maintains homeostasis and promotes survival.

Movement involves part or all of an organism's body. Although plants do not move from place to place, plants move as they grow. For example, stems and leaves grow up toward the light, and roots grow down into the soil. However, most organisms do move from place to place, or carry out locomotion. In this chapter, you will learn how the muscular and skeletal systems provide body support and enables organisms to move.

MOVEMENT IN LIFE

What do swimming, walking, flying, hopping, crawling, leaping, jumping, and wriggling have in common? These movements require a shift in position of an appendage (arm, leg, wing, fin) or an entire body against a particular resistance such as water, air, and soil. In addition, movement usually involves contracting muscles. In arthropods, such as crabs and insects, muscles are attached to an outer skeleton, or *exoskeleton*. Vertebrates, including humans, have an inner skeleton, or *endoskeleton*, and their muscles are attached to bones, which articulate with each other. The possession of a flexible skin also facilitates movement. For example, contrast the graceful swimming movements of a fish, which has skin and an endoskeleton, with the stiffer movements of a beetle.

Locomotion as a life process enables organisms to:

- Find food by moving from one region to another.
- Find shelter for nesting, resting, or during unfavorable environmental conditions.
- Escape from enemies and predators.
- Find mates.
- Move from unsafe regions to areas more favorable to survival.

MOVEMENT IN HUMANS

In humans, locomotion depends upon the combined actions of the *skeletal system* and the *muscular system*. As muscles contract, they pull on

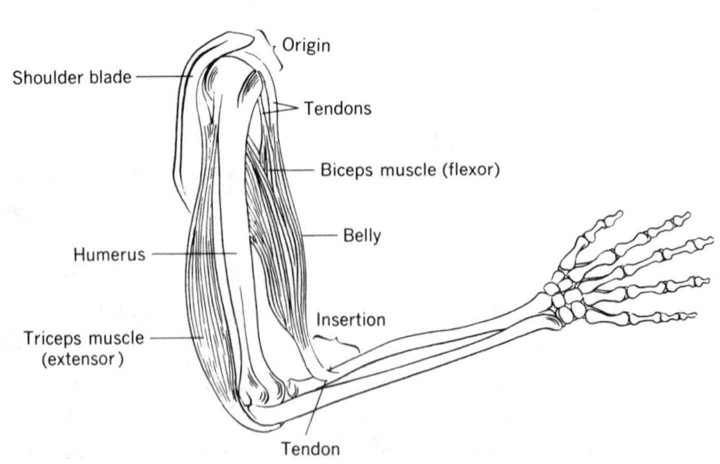

Figure 10–1 Antagonistic muscles of the upper arm

bones that are connected to each other at joints, such as the knee joint. When a muscle attached to a bone contracts, the bone works as a lever. Most muscles are arranged in opposing pairs called *antagonistic muscles.* Each pair consists of a *flexor* muscle and an *extensor* muscle. The flexor muscle bends a part of the skeleton; the extensor muscle straightens a part of the skeleton and restores the part to its original position (Figure 10–1). For example, in the upper arm, the biceps muscle is a flexor muscle and the triceps muscle is an extensor muscle.

The Skeletal System

In humans, the skeletal system consists of 206 bones, and cartilage and ligaments—different types of connective tissue. *Cartilage*, which is found at the tip of the nose, in the earlobes, and at the ends of many bones, particularly at joints, is softer and more flexible than bone. *Ligaments* are tough, fibrous bands of elastic tissue that bind bones to each other.

Functions of the skeletal system and bone tissue. Muscles are attached to bones; thus, when muscles contract, they pull on bones, causing movement. The following are other functions of bone tissue:

1. *Support.* A bony skeleton is the structural framework for the body, providing support for soft tissues and anchor points for the skeletal muscles.
2. *Protection.* Bones protect internal organs. For example, the cranial bones protect the brain, the rib cage protects the heart and lungs, and the vertebrae of the spinal column protect the spinal cord.
3. *Blood cell production. Red marrow* is usually located at the end of a bone (epiphysis). Red marrow found in many bones produces red blood cells, white blood cells, and platelets. The process of blood cell formation is called *hemopoiesis.*
4. *Homeostasis.* Minerals, such as calcium and phosphorus salts, are stored in bone tissue. These minerals are necessary to normal muscle and nerve activity. When needed, minerals are released and distributed to parts of the body to maintain a constant internal environment.
5. *Energy storage. Yellow marrow* is located within the main shaft of a bone (diaphysis). Yellow marrow is composed mainly of fat tissue—a primary source of stored energy in the form of lipids.

Bone. Many bones of the human skeleton are long, such as the *femur*, or thigh bone (Figure 10–2). The main parts of a long bone include the *diaphysis*, two *epiphyses*, *periosteum*, and the *cartilage*. The *periosteum* is a protective membrane made of dense connective tissue that covers the surface of a bone. The periosteum, which contains many blood vessels, promotes bone growth, and repair, and distributes essential nutrients. *Cartilage* covers the ends of a bone where it forms a joint with another bone. The smooth, flexible cartilage reduces friction and serves as a shock absorber.

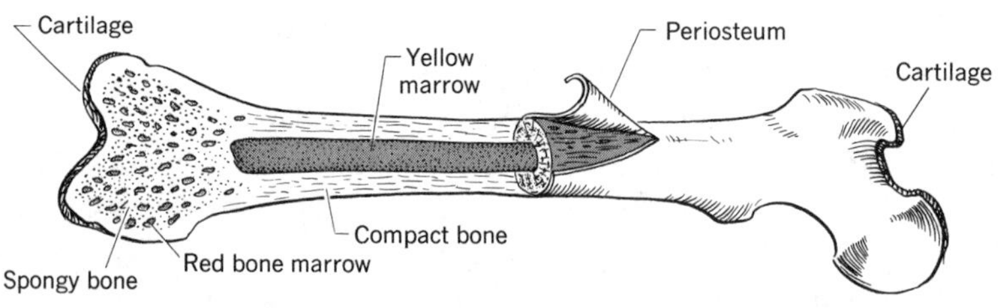

Figure 10–2 Femur (dissected)

The following are useful facts about bone:

1. Bone is connective tissue with a large intercellular matrix that surrounds widely scattered bone cells called *osteocytes*. About half of the matrix is composed of mineral salts (mainly hydroxyapatite and trace amounts of magnesium fluoride). The remainder consists of protein fiber and water.
2. Bone tissue is either spongy or compact. *Spongy bone*, which consists of thin plates called *trabeculae*, is located at the ends of bones. It is here that blood cells are produced and circulated. Compact bone, the cylinder of bone around the central canal, is composed of *osteons*, or *Haversian systems;* each osteon features concentric rings of hard, mineralized matrix called *lamellae*.
3. Bone cells are found in small spaces in the matrix called *lacunae*. These small spaces are interconnected by a system of canals called *canaliculi*, which bring nourishment to cells and remove wastes. Figure 10–3 shows the organization of compact bone and spongy bone.

Bone growth. Bone is a living tissue that has the ability to grow and repair itself. *Osteoclasts*, or bone-destroying cells, dissolve the hydroxyapatite of the hard matrix, releasing calcium salts and phosphorus salts into the

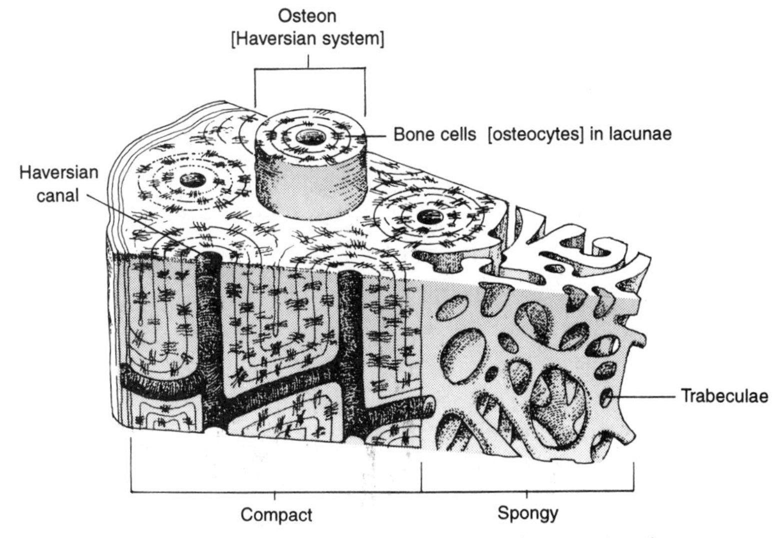

Figure 10–3 A section of bone (magnified)

blood. Gravity and normal body activities regulate the interaction between osteoblasts (which rebuild bone), osteocytes, and osteoclasts. In a healthy person, there is equilibrium between bone formation and destruction, which maintains homeostasis. Bone loss is common in elderly people, which causes their bones to become brittle and more easily broken.

Recent studies show that the elements magnesium, manganese, and boron are essential for normal bone growth and maintenance. Vitamins A, C, and D also promote the growth and repair of bone tissue. Vitamin A regulates osteoblast and osteoclast activities during bone development; vitamin C helps maintain a normal matrix and other connective tissues; and vitamin D increases absorption of calcium salts by the intestines.

The skeleton. Figure 10–4 shows the major bones of the human skeleton. The human skeleton is usually divided into an *axial skeleton* and an *appendicular skeleton.*

Axial Skeleton

The axial skeleton consists of bones that surround the axis of the body. The *axis* is an imaginary line that bisects the body—from the head down through the space between the feet. Bones of the axial skeleton include the skull, ribs, breastbone, and backbone.

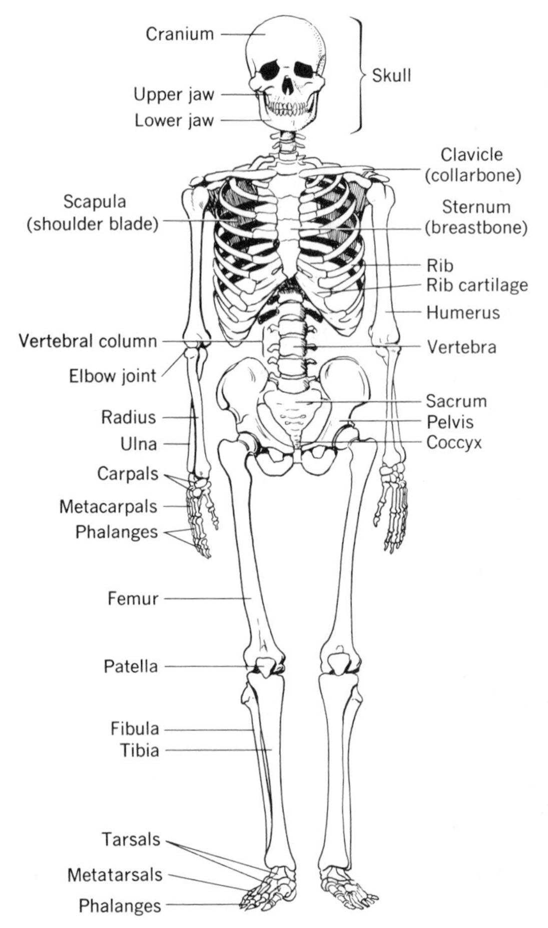

Figure 10–4 Human skeleton

Skull. The *skull,* which consists of 22 bones, rests on the end of the vertebral, or spinal column. The *cranium*, which protects the brain, is formed by the *cranial bones.*

Facial bones include the ear bones, jawbones, cheekbones, and nasal bones. At birth and during infancy, there are *fontanels,* or spaces, between an infant's cranial bones. The fontanels, commonly called "soft spots," enable the skull to grow and change in shape during gestation and infancy. The joints between cranial bones are called *sutures.*

Vertebral column. The *vertebral,* or *spinal column* (backbone), *sternum* (breastbone), and the *ribs* form the *trunk* of a human skeleton. In an adult, the vertebral column contains 26 *vertebrae.* Five vertebrae are fused into a bone called the *sacrum,* and four vertebrae are fused into one or two bones called the *coccyx.* The sacrum and coccyx are the only immovable vertebrae. The other vertebrae are linked by ligaments and interlocking bone projections, or *processes.* The *intervertebral disks* are cartilage pads found between the vertebrae that function as shock absorbers. The spinal cord is enclosed within the spinal column. The convex (outward) and concave (inward) structures of the spinal column increase its strength, absorb shocks, and facilitate balance.

Thorax. The entire chest, including the sternum (breastbone) and ribs, is called the *thoracic cage.* It encloses and protects the heart, lungs, and the upper abdominal cavity. The thorax also supports the bones of the pectoral girdle. The sternum consists of three parts; the largest part connects the *collarbones.* The other parts serve as the front attachment for ten pairs of ribs. The last two pairs of ribs, called *floating ribs,* are not connected to the sternum.

Appendicular Skeleton

The *appendicular skeleton* consists of the pectoral girdle, pelvic girdle, and *appendages,* or arms and legs. At the *pectoral girdle,* a *scapula* (shoulder blade) connects to a *clavicle* (collarbone). The upper end of the humerus is also connected to a pectoral girdle by a *ball-and-socket joint.* In this case, the pectoral girdle provides the socket for the spherical end of the humerus. A pitcher winding up and throwing a baseball illustrates how a ball and socket joint allows freedom of arm motion.

Each *pelvic girdle* features a *hip bone,* which forms a ball-and-socket joint with the femur (thigh bone). You can see that the arrangement of bones in an arm is similar to that of a leg (Figure 10–4). A hip bone consists of an *ilium, ischium,* and *pubis.* The right and left hip bones support the legs which, in turn, bear the weight of the upper body. The *pelvis* is a combination of the hip bones, sacrum, and coccyx. The female pelvis, adapted for pregnancy and childbirth, is larger and more oval in shape than the male pelvis.

Section Quiz

1. The soft, flexible connective tissue found at the ends of many bones is called (*a*) marrow (*b*) cartilage (*c*) lacunae (*d*) tendon.
2. Bands of fibrous elastic tissue that bind bones to each other are called (*a*) epiphysis (*b*) diaphysis (*c*) ligaments (*d*) cartilage.

3. Cells that release calcium and phosphorous salts into the blood are (*a*) osteoclasts (*b*) osteocytes (*c*) osteoblasts (*d*) thrombocytes.
4. Fontanels and sutures are features of (*a*) broken bones (*b*) cranial bones (*c*) pelvic bones (*d*) vertebral bones.
5. Which bones are present in the axial skeleton? (*a*) ribs, skull, and hip bone (*b*) ribs, backbone, and appendages (*c*) backbone, hip bone, and appendages (*d*) skull, ribs, and backbone.

The Muscular System

Muscles are attached to the skeleton by strong bands of connective tissue called *tendons.* These *skeletal muscles* also are called *voluntary muscles,* because their movement is consciously controlled. Observed under a microscope, skeletal muscle reveals stripes, or striations, in its cells. Consequently, another name for muscle is *striated muscle.*

Muscle tissue has the ability to contract (shorten) and relax (lengthen). Motion occurs when chemical energy, in the form of ATP, is changed to mechanical energy. An outstanding characteristic of muscle tissue is its *excitability*, or its ability to respond instantly to electrochemical signals delivered by nerve cells or by hormones carried in the blood.

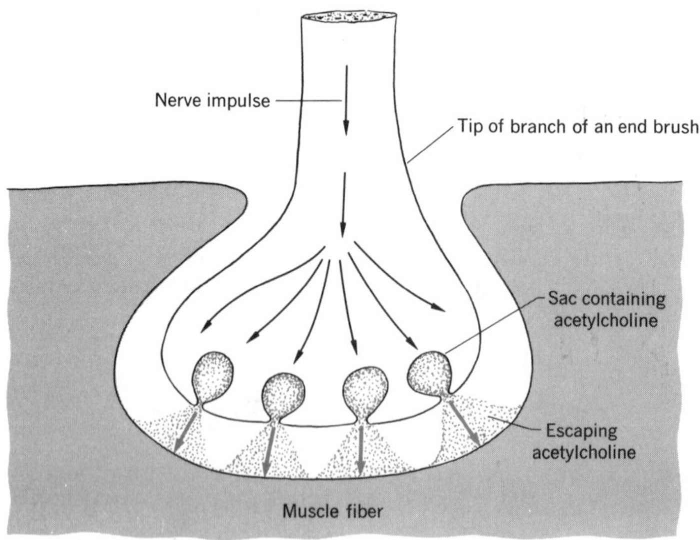

Figure 10–5 Motor end plate

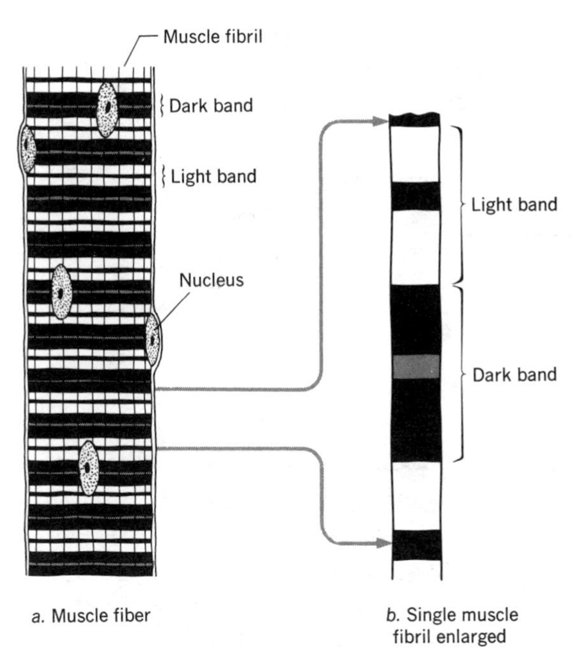

Figure 10–6 Striated muscle

Muscle contraction. When an impulse in a motor nerve reaches a muscle, the end brush of the axon secretes acetylcholine into the *motor end plates* (Figure 10–5). This starts a series of complex chemical reactions that cause the muscle to contract. The strength of the contraction depends upon the number of muscle fibers stimulated.

A muscle fiber contains many banded, or striated, fibrils, called *myofibrils*, arranged parallel to each other in the cytoplasm (Figure 10–6a). There are alternate dark and light bands in each myofibril that are related to differences in the chemical composition of the bands (Figure 10–6b). Figure 10–7 shows that a dark band contains a central core of rods of a protein called *myosin* which alternates with thinner rods of another protein called *actin*.

When an impulse stimulates a muscle fiber, ATP (in the presence of calcium salts) releases energy. This enables the actin rod to shift and combine temporarily with the myosin, forming an *actomyosin complex*. As a result, the myofibrils shorten and the muscle contracts. When muscles relax, the actomyosin complex separates into actin and myosin in their original positions. Glycolysis (anaerobic respiration) provides the initial source of energy for muscle contraction by pro-

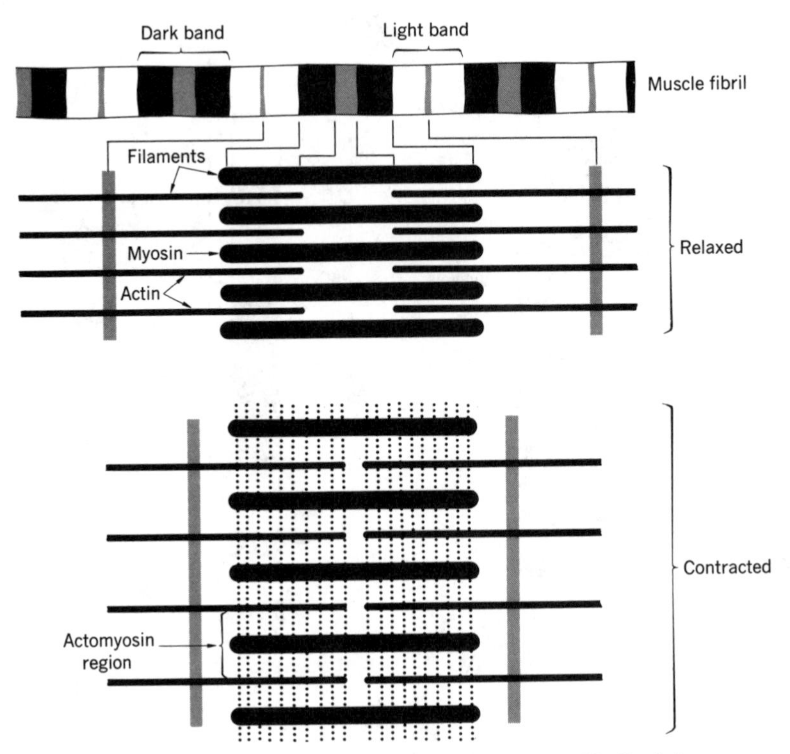

Figure 10–7 Relaxed and contracted muscle fibrils (electron-microscope view)

ducing ATP. During moderate exercise, however, more ATP is required than can be supplied by glycolysis. Additional ATP is supplied by the Krebs cycle and the hydrogen-transport system (aerobic respiration).

Lactic acid formation. Strenuous exercise produces more pyruvic acid than can normally enter the Krebs cycle. Most of this pyruvic acid is changed to lactic acid. The temporary accumulation of lactic acid in the muscles during strenuous exercise causes muscle fatigue. Later, the lactic acid is changed back into pyruvic acid, which eventually enters the aerobic reactions of the Krebs cycle and the hydrogen-transport system. The discomfort associated with muscle fatigue is relieved as lactic acid changes back to pyruvic acid.

Section Quiz

1. The cytoplasm of a skeletal muscle fiber contains many banded, or striated, (*a*) sarcomeres (*b*) lamellae (*c*) myofibrils (*d*) canaliculi.
2. The skeleton is covered by striated muscle tissue that is (*a*) voluntary (*b*) involuntary (*c*) smooth (*d*) spongy.
3. When actin combines with myosin, a muscle usually (*a*) expands (*b*) contracts (*c*) relaxes (*d*) twitches.
4. After exercise, muscle fatigue occurs due to the temporary accumulation of (*a*) calcium salts (*b*) pyruvic acid (*c*) lactic acid (*d*) acetylcholine.
5. The strength of a muscle's contraction depends upon the (*a*) number of muscle fibers that have been stimulated (*b*) number of dark and light bands in a muscle fiber (*c*) amount of calcium salts available for the ATP (*d*) size of the motor end plates affected.

LOCOMOTION IN OTHER ORGANISMS

Protozoans

An ameba moves by forming pseudopods out of its body. A paramecium moves by the coordinated action of its cilia. Euglena moves by lashing its whiplike flagellum. In all protozoans, the energy for locomotion is provided by either adenosine monophosphate (AMP) or ATP.

Hydra

One way a hydra moves is by gliding on its base (foot). This is done by amebalike cells that form pseudopods. A hydra also can "somersault" into a new environment. The hydra bends its body and grasps an object with its tentacles. Then the hydra loosens its base, flips its body over, and attaches its base to the new location.

Earthworm

Each segment of an earthworm has two pairs of *setae*, or bristles, on each side. The setae protrude from the body and can be extended or withdrawn by special muscles attached to their inner ends. An earthworm also possesses two layers of antagonistic muscles: a *circular* layer of muscle fibers that cause the animal to become longer when the fibers contract, and a *longitudinal* layer of muscle fibers that cause the animal to shorten when the fibers contract. These muscular movements are coordinated by impulses from the ventral nerve cord.

When an earthworm moves, setae in the rear segments anchor this part of the worm in the soil. Then, the circular layer of muscle fibers contracts, causing the front end of the worm to move forward. Setae in the front segments anchor this part of the worm in the soil. At the same time, the setae in the rear segments are released and the longitudinal layer of muscle fibers contracts, causing the rear of the earthworm to be pulled forward.

Grasshopper

A grasshopper uses its wings to fly and its legs to walk, crawl, or hop. These movements result from the coordinated action of sets of antagonistic muscles attached to its exoskeleton, or outer skeleton. Figure 10–8a shows how the wings are used in flying and Figure 10–8b shows how the antagonistic muscles are used in leg movements.

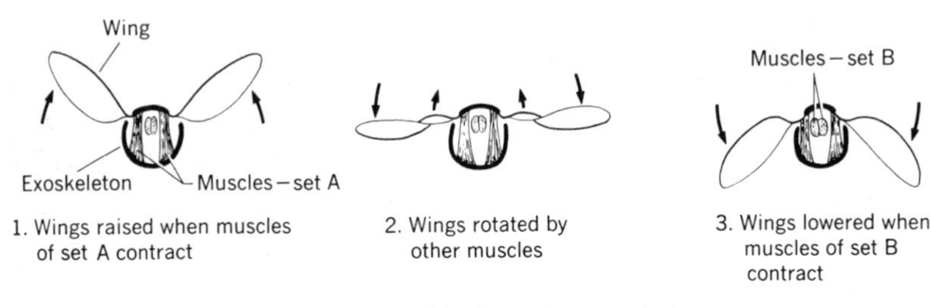

a. Thorax of a flying insect (cross section)

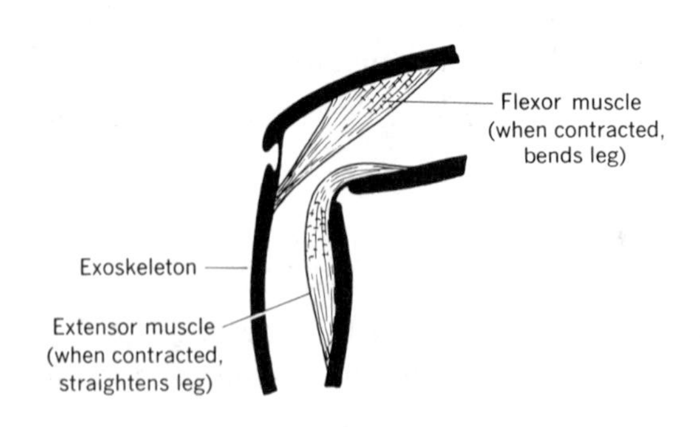

b. Joint of insect leg (internal view)

Figure 10–8 Organs of locomotion of insects

Table 10–1 summarizes the adaptations for locomotion of organisms discussed in this chapter.

Table 10–1 Adaptations for Locomotion

Organism	*Adaptation*
Ameba	Pseudopods
Paramecium	Cilia
Euglena	Flagellum
Hydra	Pseudopods; contractile cells
Earthworm	Setae; two sets of antagonistic muscles (circular and longitudinal)
Grasshopper	Wings; legs; antagonistic sets of muscles attached to exoskeleton
Humans and most other vertebrates	Antagonistic sets of muscles attached to internal skeleton (endoskeleton); movable bones that function as levers; arms and legs

Section Quiz

1. Structures used for locomotion by some protozoans are (*a*) vacuoles (*b*) flagella (*c*) tentacles (*d*) nematocysts.
2. An earthworm is adapted to move by using (*a*) circular muscles and longitudinal muscles (*b*) circular muscles and paired setae (*c*) setae, circular muscles, and longitudinal muscles (*d*) paired setae, involuntary muscles, and mucus.
3. In insects such as the grasshopper, muscles (*a*) interact with an exoskeleton (*b*) interact with an endoskeleton (*c*) are neither antagonistic nor voluntary (*d*) are activated by hormones.
4. The coordinated action of *cilia* enable movement in the (*a*) ameba (*b*) paramecium (*c*) hydra (*d*) euglena.
5. Movement by *pseudopods* is a characteristic of (*a*) ameba and euglena (*b*) ameba and paramecium (*c*) ameba and hydra (*d*) euglena and hydra.

Until now, we have discussed the life processes that enable organisms to maintain homeostasis and survive. But animals are heterotrophs and therefore depend upon other species for inorganic and organic materials. Directly or indirectly, the materials needed by heterotrophs come mainly from plants. The ability of these chlorophyll-bearing autotrophs to synthesize organic materials from inorganic materials is the primary characteristic that distinguishes plants from animals. The following chapter discusses specific adaptations that help plants carry out life processes.

Chapter Review Questions

The following questions will help you check your understanding of the material presented in the chapter.

1. Red blood cells and white blood cells are formed in long bones, specifically in the (*a*) lymph (*b*) periosteum (*c*) marrow (*d*) plasma.
2. To provide the skeleton with calcium and phosphate salts, the diet should contain adequate amounts of vitamin (*a*) A (*b*) B_1 (*c*) C (*d*) D.
3. Your hand is part of your (*a*) sternum (*b*) pelvic girdle (*c*) appendages (*d*) axial skeleton.

Use the following diagram and your knowledge of biology to answer questions *4 to 7*.

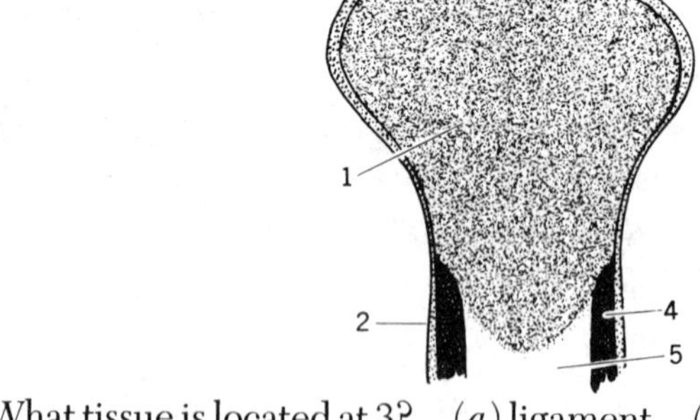

4. What tissue is located at 3? (*a*) ligament (*b*) periosteum (*c*) cartilage (*d*) marrow.
5. What type of bone is located at 4? (*a*) compact bone (*b*) spongy bone (*c*) yellow marrow (*d*) red marrow.
6. Where are blood cells formed? (*a*) 1 (*b*) 3 (*c*) 4 (*d*) 5.

7. The membrane shown by 2 is (*a*) cartilage (*b*) ligament (*c*) spongy (*d*) periosteum.
8. Which is the initial source of the energy used by muscle cells when they contract? (*a*) ADP (*b*) ATP (*c*) DNA (*d*) RNA.
9. Muscle fatigue is accompanied by an accumulation of (*a*) ATP (*b*) pyruvic acid (*c*) lactic acid (*d*) glycogen.
10. Extensors and flexors are related to (*a*) locomotion (*b*) chemical regulation (*c*) synthesis (*d*) respiration.
11. In humans, which tissues join muscles to bones? (*a*) tendons (*b*) setae (*c*) flexors (*d*) ligaments.

Use the following diagram and your knowledge of biology to answer questions *12 to 16.*

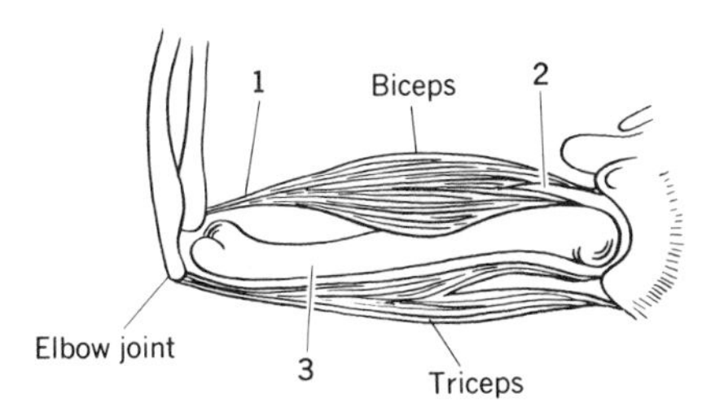

12. The bone shown by 3 is the (*a*) radius (*b*) ulna (*c*) humerus (*d*) scapula.
13. The origin of the biceps muscle is near number (*a*) 1 (*b*) 2 (*c*) 3 (*d*) none of these.
14. To straighten the arm, which muscle(s) must contract? (*a*) biceps (*b*) triceps (*c*) both biceps and triceps (*d*) neither biceps nor triceps.
15. When the biceps muscle contracts, the triceps muscle (*a*) contracts (*b*) relaxes (*c*) neither contracts nor relaxes (*d*) develops more myosin.
16. These muscles are connected to the bones by (*a*) tendons (*b*) cartilage (*c*) motor end plates (*d*) joints.
17. Which is not used for locomotion by protozoans? (*a*) cilia (*b*) contractile vacuoles (*c*) flagella (*d*) pseudopods.

18. Hydra glides along on its base by using cells that possess (*a*) flagella (*b*) pseudopods (*c*) cilia (*d*) nematocysts.
19. Locomotion in earthworms is accomplished by the interaction of muscles and (*a*) setae (*b*) glands (*c*) nephridia (*d*) cartilaginous appendages.
20. Movement in grasshoppers is accomplished by the interaction of muscles and (*a*) setae (*b*) bones (*c*) appendages (*d*) tentacles.

Biology Challenge

The following questions will provide practice in answering SAT II-type questions.

Part I

Based on this text and your knowledge of biology, select the letter of the statement that best completes the sentence or answers the question.

1. Insect flight differs from bird flight in that (*a*) insect wings lack muscles (*b*) insect wings are attached to an exoskeleton that functions as a fulcrum (*c*) insect wings are rotated by sets of muscles (*d*) *a* and *c*, only (*e*) *a*, *b*, and *c*.
2. In a muscle fiber, the region between two Z bands is (*a*) composed of repeating bands of actin and myosin (*b*) a motor end plate (*c*) lacking in mitochondria (*d*) a storehouse of acetylcholine (*e*) responsible for muscle growth.
3. Which statement *correctly* describes skeletal muscle? (*a*) A result of aging is a loss of muscle tissue. (*b*) Muscle fibers regenerate quickly after injury. (*c*) Responsiveness to a stimulus is delayed after an initial contraction. (*d*) A muscle fiber contains no more than five nuclei. (*e*) All fibers of a muscle contract as one, regardless of strength of the stimulus.
4. Haversian systems are also called (*a*) lacunae (*b*) trabeculae (*c*) lamellae (*d*) osteons (*e*) phalanges.

Part II

Base your answers to questions *1 through 5* on the graph and the following paragraph dealing with muscle responses and electrical pulses.

The graph below summarizes data gathered in an investigation involving the gastrocnemius (calf) muscle of a frog and a series of electrical pulses. The graph shows the relationship between intensity of electrical stimuli and muscle response. The stimuli varied in intensity from just enough to fire a few muscle fibers to large enough to fire all of the fibers in the muscle.

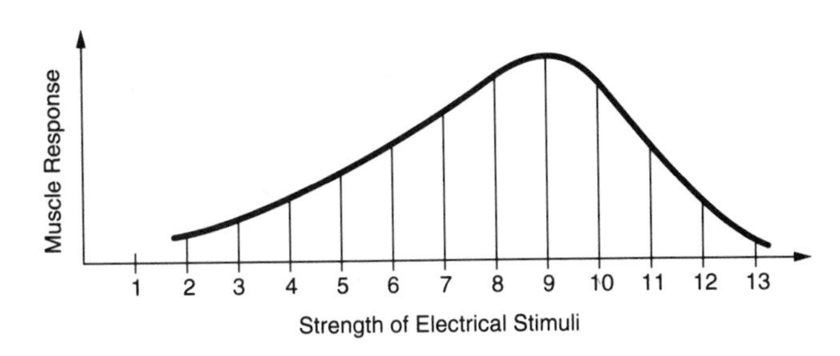

1. Sustained contraction of muscle fibers occurred at electrical intensities (*a*) 6, 7, 8 (*b*) 8, 9, 10 (*c*) 10, 11, 12 (*d*) 11, 12, 13 (*e*) not shown in the graph.
2. Muscle responses indicated by electrical intensities numbered 11, 12, and 13 represent (*a*) relaxation time (*b*) inadequacy of electrical pulses (*c*) muscle fatigue (*d*) myofibril breakdown (*e*) need for lactic acid.
3. At which electrical intensity does the muscle contain the most ATP? (*a*) 1 (*b*) 7 (*c*) 9 (*d*) 11 (*e*) 13.
4. Which electrical intensity best illustrates the "all or none" law of muscle contraction? (*a*) 1 (*b*) 9 (*c*) 10 (*d*) 12 (*e*) 13.
5. Lactic acid accumulation is the reason why the curve of the graph (*a*) begins at electrical intensity 1 (*b*) rises from electrical intensity 2 through 8 (*c*) reaches a plateau at electrical intensities 8, 9, and 10 (*d*) drops steeply after 10 (*e*) assumes a flattened bell shape.

11

Plant Structure and Function

Learning Objectives

When you have completed this chapter, you should be able to:

- **Describe** characteristics of several important groups of plants.
- **Relate** the structures of plant tissues to their functions.
- **Discuss** various experiments that relate to the process of photosynthesis.
- **Explain** the process of respiration in plants.
- **Identify** differences in monocots and dicots.
- **Describe** methods by which water and other materials are moved throughout plant tissues.
- **Compare** the life activities of plants and animals.
- **Describe** the process of photosynthesis and other important metabolic activities in plants.

OVERVIEW

If you could travel back in time to 450 million years ago, you would immediately notice a lack of familiar plants such as oak trees, grapevines, and grasses. Instead you would observe a swampy landscape inhabited mainly by multicellular green algae. Moving forward in time to about 300 million years ago, the same landscape would contain a variety of plants, including mosses, ferns, as well as cone-bearing and flower-bearing plants. Today, there are more than 250,000 different plant species—all of which have evolved from green algae. Mosses and ferns still display their ancestral relationship to water by preferring damp environments, such as swamps and bogs.

Over hundreds of millions of years, plants have evolved specialized physical and chemical adaptations that enable them to thrive in various environments. This chapter will discuss how the special adaptations of flowering plants have helped them overcome their lack of locomotion and succeed as autotrophs. The distinguishing characteristics of mosses, ferns, and cone-bearing plants are discussed in Chapter 18.

PLANTS: UNITY AND DIVERSITY

The great variety of plant species alive today inhabit marine, freshwater, and terrestrial environments. Plants can be found in all of Earth's climates—from warm, moist tropical regions to regions of extreme cold. Plants are classified as nonvascular (nonconductors) and vascular (conductors).

Bryophytes

Mosses, liverworts, and *hornworts* are nonvascular land plants because they lack specialized conducting tissue. These *nonvascular plants,* classified as *bryophytes,* do not possess true leaves, roots, and stems. They have leaflike and stemlike structures that are usually covered with a waxy cuticle to prevent water loss. Rootlike structures, called *rhizoids,* anchor bryophytes to the soil; water is absorbed directly through epidermal cells. Their lack of a vascular system prevents bryophytes from moving water upward against the force of gravity. Consequently, mosses and other bryophytes are small and usually live near water.

Tracheophytes

The most familiar and common plants are vascular. *Vascular plants* possess true stems, leaves, and roots, as well as well-developed conducting vessels that transport water and nutrients. Although all plants need water, this conducting tissue greatly reduces the plant's need to live in very moist habitats. Vascular plants are classified as *tracheophytes,* a term derived from the *tracheids,* which are thick-walled, tubelike cells that make up the water-conducting tissue called *xylem.* Tracheophytes include *ferns* and all *seed-bearing plants.* Seed-bearing plants are classified as either gymnosperms or angiosperms. *Gymnosperms* are the cone-bearing plants (conifers), such as pines, firs, and spruces. *Angiosperms* are the flowering plants, such as apple trees and rose shrubs. All of the angiosperms (about 230,000 species) have true roots, stems, leaves, and well-developed vascular systems.

Flowering plants are divided into two groups: monocots and dicots. A *monocot* is a plant whose embryo has one cotyledon, or seed leaf; *dicot* plants have embryos with two cotyledons. Figure 11–1 illustrates some major differences between monocot and dicot plants. Figure 11–2 shows the general body plan of a typical dicot plant.

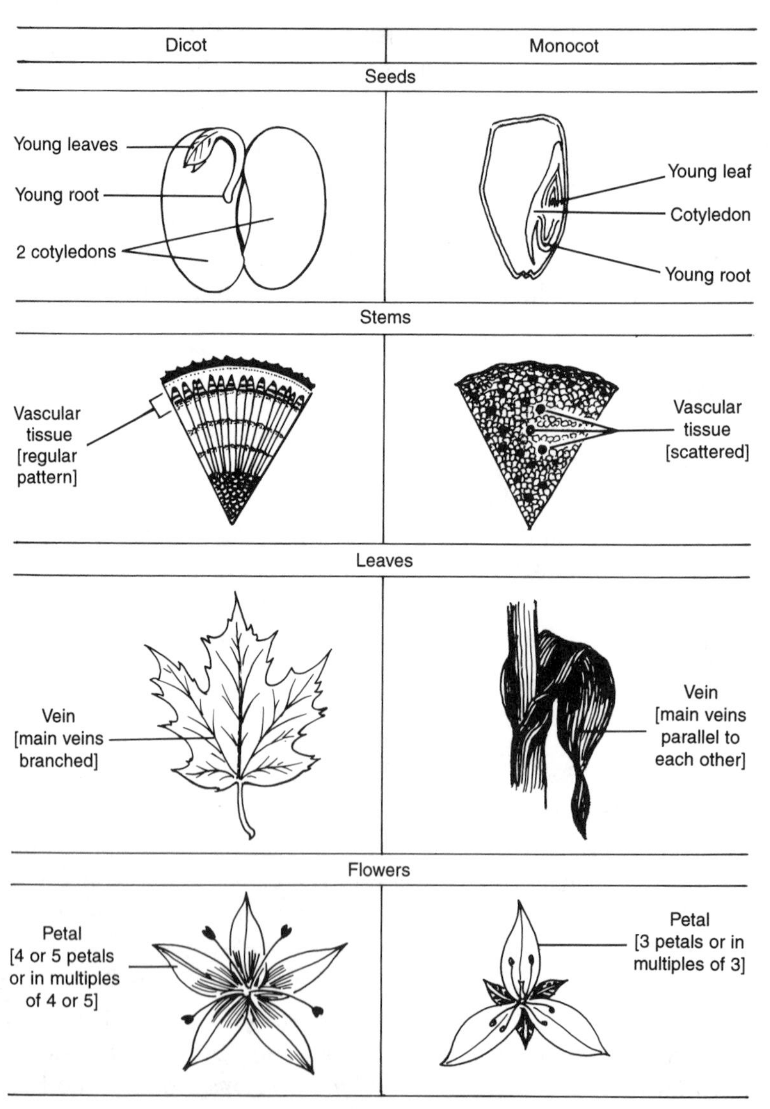

Figure 11–1 A comparison of dicots and monocots

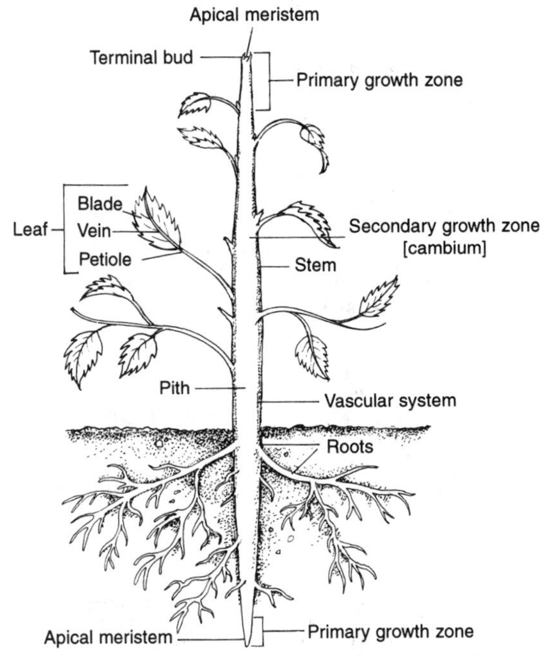

Figure 11–2 Typical dicot plant structure

PLANT STRUCTURES AND FUNCTIONS

In Chapter 2 you learned about different plant cells and tissues without considering their organization. *Leaves, stems,* and *roots* are types of plant tissue that enable plants to carry out photosynthesis and other life processes. See Table 11–1 for a review of plant tissues and their special functions.

Table 11–1 Plant Tissues and Functions

Plant Tissue	*Function*
Chlorenchyma (mesophyll)	Photosynthetic (food-making)
Collenchyma	Supports the stem and some leaves
Dermal (epidermis)	Protects exterior of plant's body
Meristem (cambium)	Embryonic, undifferentiated; growth; forms other tissues and cambium
Parenchyma	Most common tissue; bulk; storage; secretion; some regeneration
Sclerenchyma	Supports and adds strength
Vascular (xylem and phloem)	Conducts water, dissolved substances and nutrient materials

Leaves

A leaf is adapted to capture light energy, carry out photosynthesis (foodmaking), exchange gases with the atmosphere, and regulate a plant's water content.

Structure. Most leaves possess a flat, broad *blade* and a narrow stalk, or *petiole,* which connects the blade to the stem. Conducting tissue within the petiole extends into the blade as the veins of a leaf. The conducting tissue of the petiole also extends into the stem of the plant, where it joins the conducting tissue of the stem.

Figure 11–3 shows the internal structure of a leaf as observed under the low power lens of a microscope. The structure of a leaf resembles a sandwich—two *epidermal* layers of flat cells make up the "bread"; a layer of chloroplast-bearing cells, called the *mesophyll,* or *chlorenchyma,* makes up the "filler."

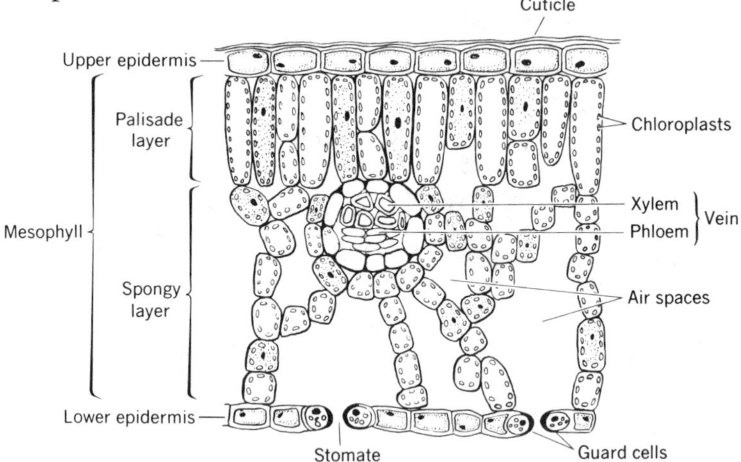

Figure 11–3 Internal structure of a leaf

The upper and lower epidermis. The cells of both upper and lower epidermal layers are flat and fit closely together. Both epidermal layers are alive but usually lack chloroplasts. The upper epidermis of most leaves is covered with a thin coat of wax called a *cuticle.* The cuticle prevents the evaporation of water and the entrance of disease-causing organisms. Scattered among the cells of the lower epidermis are tiny openings called *stomates.* Each stomate lies between two *guard cells* that contain chloroplasts (Figure 11–4). Guard cells open and close stomates, thus regulating the exchange of gases between the leaf and the atmosphere, and the evaporation of water from inside the leaf.

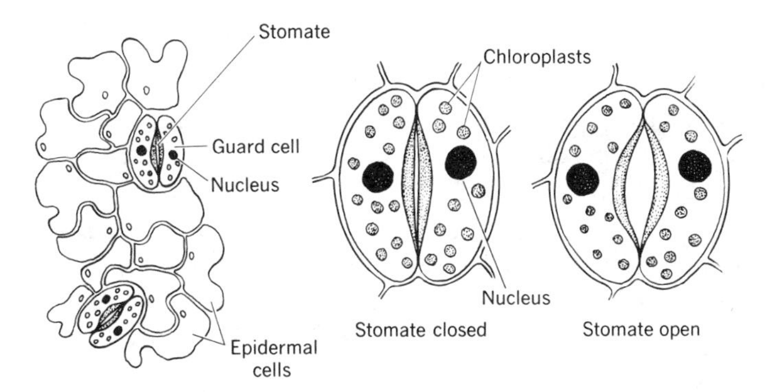

Figure 11–4 Stomates in the lower epidermis of a leaf

Guard cells swell because of an increase in *turgor pressure*, or internal pressure. As a result, the opening between the guard cells widens; when turgor pressure decreases, there is no opening between guard cells. Turgor pressure increases when water fills the guard cells and decreases when guard cells lose water. Recent studies show that during photosynthesis the carbon dioxide level within a leaf is low. The low level of carbon dioxide apparently triggers a potassium "pump" in guard cells to actively transport potassium ions (K^+) into the cells. As a result, a concentration gradient is established that causes water to enter, increasing the turgor pressure and opening the stomates (see Figure 11–5). High carbon dioxide levels, which occur when photosynthesis does not take place, cause stomates to close. In any case, stomates will not open unless there is an adequate supply of water.

The mesophyll. This part of the leaf is composed of the palisade layer, spongy layer, and veins.

1. The *palisade layer* consists of closely packed boxlike cells located directly beneath the upper epidermis. The location and arrangement of cells in the palisade layer enable them to receive maximum light. Because these cells contain many chloroplasts and absorb much light, most photosynthesis occurs in the palisade layer.
2. The *spongy layer,* located underneath the palisade layer, consists of loosely

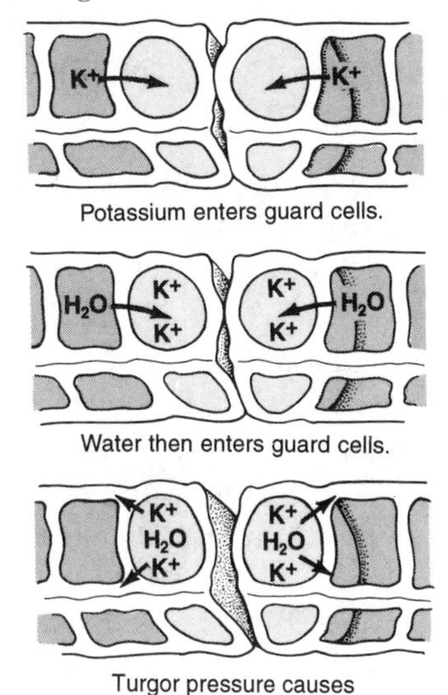

Figure 11–5 The potassium pump in stomates

arranged, somewhat rounded cells. Air spaces in the spongy layer temporarily store gases that are released by the leaf cells and that enter the leaf from the atmosphere. Cells of the spongy layer have fewer chloroplasts than cells of the palisade layer. Thus, less photosynthesis occurs in the spongy layer.

3. The *veins* (*conducting tissue*) consist of xylem and phloem tubes, which transport fluids in the leaf. Xylem transports water and dissolved minerals to leaf cells from the stem and roots; phloem transports materials manufactured in leaf cells to the stem and roots.

Functions. A leaf exchanges oxygen and carbon dioxide between the plant and the atmosphere. Leaves regulate the water content of a plant by eliminating excess water and by conserving water during dry periods. A leaf is also a chemical factory that changes inorganic substances to organic substances by photosynthesis.

Photosynthesis. In this process, chloroplasts absorb light energy. This energy is used by a plant to combine carbon dioxide and water to synthesize carbohydrates. Oxygen, a by-product of photosynthesis, either diffuses into the air through the stomates, remains in the air spaces, or is used by a plant's cells during respiration. The following experiments help demonstrate the conditions that are necessary for photosynthesis as well as the presence of substances formed during photosynthesis.

1. *Presence of carbohydrates in a leaf.* Obtain a green geranium leaf from a plant that has been grown in sunlight. Boil it in alcohol in a hot-water bath heated by a hot plate, until the chlorophyll dissolves and the leaf turns white. *Caution: Do not boil over a direct flame.* Rinse the leaf in water and then soak it in iodine solution. A blue or blue-black color indicates the presence of starch, which does not dissolve in either alcohol or water.
2. *The need for chlorophyll.* Use a variegated (green and nongreen) leaf of a geranium plant that has been exposed to light. Detach the leaf, and boil it in alcohol until all the chlorophyll dissolves and the leaf turns white. Rinse the leaf in water and then soak it in iodine solution. Where the leaf was green (possessed chlorophyll), it turns blue-black. Where it was white or another color, the leaf takes on the color of the iodine solution. Thus, this experiment indicates that starch is found in a leaf only in areas that contain chlorophyll (Figure 11–6).

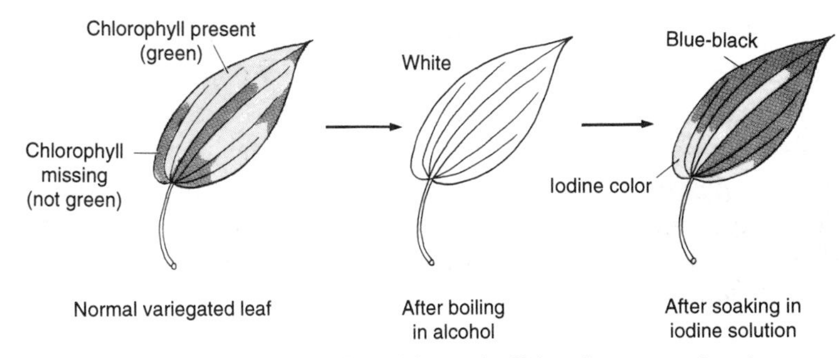

Figure 11–6 Need for chlorophyll in photosynthesis

3. *The need for light.* Cover one-half of a green geranium leaf with black paper and expose the entire plant to sunlight or artificial light for a few days. Detach the leaf and remove the paper. Boil the leaf in alcohol, rinse in water, and soak the leaf in iodine solution. Where the leaf was exposed to light, it turns blue-black. Where the leaf was covered, it takes on the color of the iodine solution. This indicates that starch is found in the part of a leaf exposed to light (Figure 11–7).
4. *The need for water.* Like all living things, plants require water to survive. In the classroom, it is difficult to show that water is needed for photosynthesis. However, the need for water in photosynthesis will become apparent when the details of photosynthesis are discussed.
5. *Use of carbon dioxide.* Expose an elodea plant to light. Prepare a pale blue solution of the indicator bromthymol blue. Place some of the bromthymol blue solution in a test tube and make the solution turn yellow by blowing into it with a soda straw. Carbon dioxide in the exhaled air acidifies the solution and causes it to turn yellow. When the carbon dioxide is removed, the solution will change back to blue.

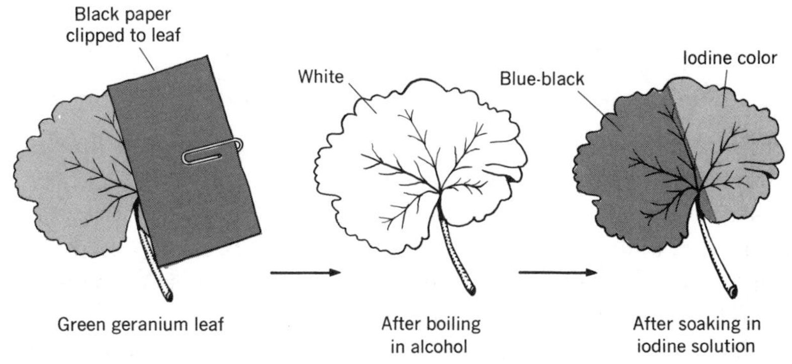

Figure 11–7 Need for light in photosynthesis

Place a small sprig of elodea into the yellow solution and seal the test tube with a rubber stopper. Prepare a second test tube of elodea in the same manner. Wrap one of the test tubes in black paper. Set both sealed test tubes in a well-lighted area for several hours. Observe that in the exposed test tube, the yellow solution turns blue. In the unexposed test tube, the yellow solution remains yellow. This experiment shows that in the presence of light, carbon dioxide is absorbed by a plant (Figure 11–8).

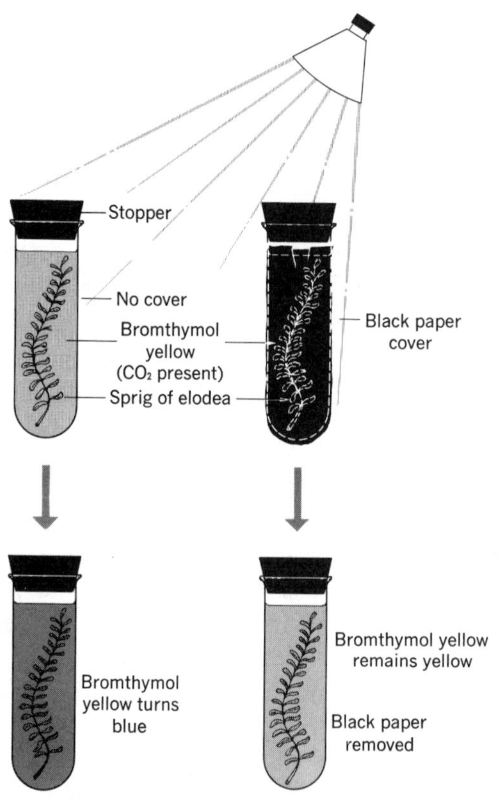

Figure 11–8 Absorption of carbon dioxide by a plant

6. *Release of oxygen.* Place a few sprigs of elodea under a funnel immersed in water. Over the funnel, a test tube of water has been inverted as shown in Figure 11–9. Repeat this procedure with another group of elodea sprigs. Place one setup in a dark closet and the other under a bright light. Soon bubbles of gas accumulate in each test tube. When a glowing splint is inserted into the test tube exposed to light, the splint bursts into flame, showing the presence of oxygen. When a glowing splint is inserted into the other test tube, the glowing splint does

not burst into flame; instead, it goes out. (The bubbles of gas are probably oxygen molecules coming out of solution.) This experiment shows that a plant releases oxygen in the presence of light.

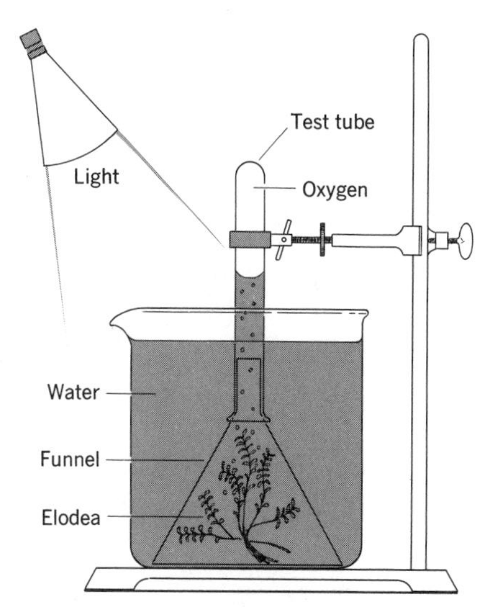

Figure 11–9 Release of oxygen by a plant

Respiration. Like animal cells, plant cells carry on anaerobic and aerobic respiration continuously. In daylight, however, when a plant carries on photosynthesis, the plant produces more oxygen than its cells can use, and takes in more carbon dioxide than its cells produce in respiration. At dawn, a plant first uses the carbon dioxide formed by aerobic respiration during the night to carry on photosynthesis. This reduces the amount of carbon dioxide in the air spaces of the leaves. It also produces a concentration gradient between the carbon dioxide in the air spaces in the leaf and in the air around the leaf. Consequently, carbon dioxide diffuses from the air into the air spaces, and then into the mesophyll cells. As photosynthesis continues, the amount of oxygen inside the leaf increases. This produces a concentration gradient between oxygen in the leaf and the oxygen in the air outside the leaf. This results in the diffusion of oxygen out of the leaf into the air.

At night, photosynthesis stops but respiration continues. Then, the concentration gradient for oxygen is from the air outside into the leaf; the gradient for carbon dioxide is from the leaf into the air outside.

Transpiration. The escape of water vapor from the stomates is called *transpiration.* As water molecules evaporate from the surface of a leaf, water molecules in the tracheids move up to take the place of the escaped water molecules. The evaporating water molecules appear to exert a pull, or upward force, called *transpirational pull,* on the molecules below them. Transpirational pull may be demonstrated by using the apparatus shown in Figure 11–10. As water molecules evaporate from the leaves, water rises in the glass tube. This allows the mercury from the reservoir to rise in the tube.

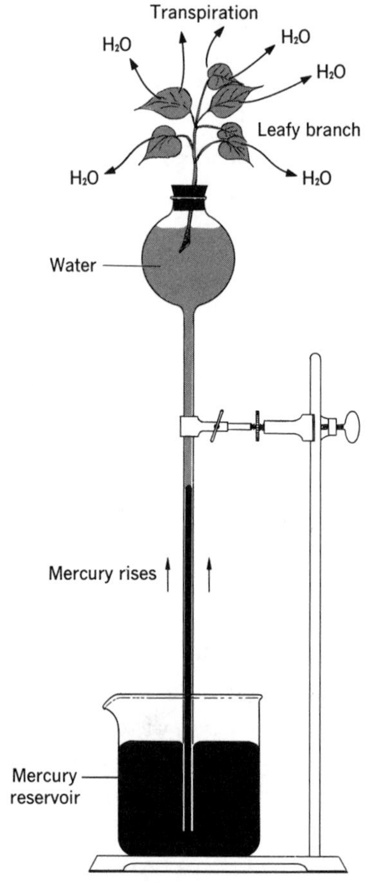

Figure 11–10 Demonstration of transpirational pull

Section Quiz

1. Which plant is nonvascular? (*a*) oak tree (*b*) palm tree (*c*) grass (*d*) moss.
2. All angiosperm plants have (*a*) cones (*b*) flowers (*c*) spores (*d*) rhizoids.
3. The pattern of veins in all dicot plants is (*a*) netlike (*b*) parallel (*c*) triangular (*d*) boxlike.
4. Embryonic plant tissue that remains undifferentiated is called (*a*) parenchyma (*b*) collenchyma (*c*) meristem (*d*) epiderm.
5. The layer of chloroplast-bearing cells in leaf tissue is called the (*a*) xylem (*b*) mesophyll (*c*) epidermis (*d*) collenchyma.

Stems

The stems of monocots and dicots contain growing tissue called *meristem.* Both have *apical meristem,* at the tips of their stems. Dicots have *cambium,* or *lateral meristem,* which monocots lack. Otherwise, the tissues found in the two plant groups are the same; only the arrangement of the tissues differs.

Structures. Figure 11–11 shows two wedges, one from a cross section of a young corn stem (monocot) and the other from a cross section of a young ash tree stem (dicot).

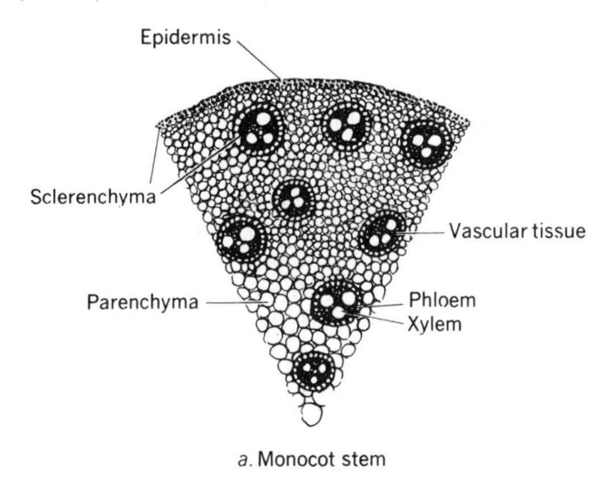

a. Monocot stem

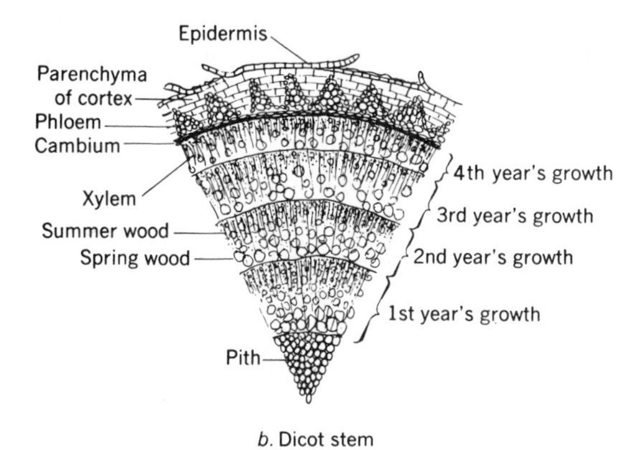

b. Dicot stem

Figure 11–11 Plant stem structure

Arrangement of tissues. Tissue arrangement in monocot stems differs from that in dicot stems. Recall that *epidermal tissue* protects underlying plant tissues and prevents loss of water; *sclerenchyma tissue* aids in supporting the plant; *parenchyma tissue* stores water and manufactured food;

conducting tissue transports water upward (*xylem*) and transports manufactured products downward (*phloem*); and *pith* stores materials.

1. *Monocot stem*. The stem of a monocot consists mainly of epithelial tissue, sclerenchyma, and parenchyma. Many bundles, consisting of xylem and phloem, are scattered throughout the parenchyma. Each bundle is enclosed and strengthened by the surrounding sclerenchyma cells. New tissues are not formed year after year. Thus, many monocots live for one or two growing seasons and then die.
2. *Dicot stem*. A dicot stem consists of the same tissues as a monocot with the addition of pith. However, the tissues of a dicot stem form a pattern that easily distinguishes it from a monocot stem. Underneath the epidermis of a dicot stem is the cortex, which consists mainly of parenchyma. Masses of phloem cells form an incomplete layer within the cortex. Inside the phloem layer is the cambium, which is composed of undifferentiated (not specialized) cells. Some cambium cells remain undifferentiated and continue to divide. Others differentiate (become specialized) and produce both phloem cells and xylem cells. The cambium enables dicot plants to live for many seasons. The continual growth and division of cambium cells account for the increase in thickness of dicot stems. The xylem is located in a circle next to the inner surface of the cambium layer. By counting and studying the xylem rings of a dicot tree, you can determine the age of the tree and the weather conditions under which it grew. Small xylem tubes indicate dry seasons and large xylem tubes indicate wet seasons. Pith stores food, water, and dissolved mineral salts in parenchyma cells.

Functions. The main functions of a stem are support, transport, storage, and photosynthesis.

Support. The stem supports the leaves and the reproductive organs, or flowers, of the plant. The stem and its branches expose the leaves to light and air, which enables a plant to make food and exchange gases with the air. The stem and its branches also expose the flowers to insects, other pollinating animals, and wind. This enables a plant to receive pollen from other plants and to disperse (spread) its own pollen.

Transport (or *translocation*). The xylem and phloem of the stem form continuous passageways between the roots and leaves.

Xylem develops from the cambium as living, elongated cells that are stacked one on top of another. As the cells mature, they develop thick walls, die, lose their cytoplasm, and become hollow. Openings appear in the ends of the adjoining cells, forming long, hollow tubes that extend from the roots to the leaves.

Phloem develops from the cambium in the same manner as xylem. However, the walls of phloem cells are thinner. In phloem, the cytoplasm of the tubular cells remains alive, but the cell nuclei disappear. Openings in the ends of adjoining phloem cells permit the cytoplasm to be continuous from cell to cell. *Companion cells,* with abundant cytoplasm and a nucleus, are closely associated with phloem tubes. These cells help the phloem tubes conduct food materials.

Water containing dissolved minerals first enters the root xylem, then passes through the stem xylem, and finally reaches the leaf xylem. In the leaves, phloem transports manufactured substances to the stem, which, in turn, transports them to other parts of the plant.

How water is translocated. In the xylem, water and dissolved minerals rise by the combined action of capillarity, root pressure, and transpirational pull.

1. *Capillarity* is the upward movement of liquid in a very narrow tube such as a blood capillary. Water movement is caused by the combined effects of cohesion and adhesion. *Cohesion* is the attractive force that similar molecules exert on each other. *Adhesion* is the attractive force that unlike molecules exert on each other. Because of adhesion, water molecules creep up the sides of the xylem tube; at the same time, because of cohesion, these water molecules pull other water molecules behind them.
2. *Root pressure* develops in the root xylem as water flows into the roots by osmosis.
3. *Transpirational pull* begins when the evaporation of water through the stomates decreases the amount of water in the leaf cells. As a result, water from the leaf xylem moves into these cells by osmosis. Then the cohesion between water molecules pulls water upward from the xylem of the petiole into the leaf blade. Thus, a series of water transfers draws water upward from the xylem of the petiole, stem, and roots.

How synthesized materials are translocated. The products of photosynthesis usually are transported downward by the sieve cells of the phloem. The translocation of materials within sieve cells is the result of diffusion, active transport, and cytoplasmic streaming.

Storage. The parenchyma cells of a stem store manufactured products such as sugars and starches. In the United States, most table sugar, for example, is derived from the sugarcane plant. The plant stem stores most of the sugar manufactured by the leaves. The sugar manufactured in the leaves of a potato plant is translocated to underground stems called *tubers.* In a tuber, the sugar is changed to starch and stored in parenchyma cells. Other plants, such as the cactus, store water in their stems.

Photosynthesis. The stem of nonwoody plants, including cactus, geranium, and corn, is green because the epidermis and cortex of the stem contain many cells with chloroplasts. Like the mesophyll cells in leaves, these cells carry on photosynthesis.

Roots

The roots of most plants are similar in structure and function. Roots take up water and dissolved substance from the soil.

Structure. Figure 11–12 shows a longitudinal section and a cross section of a young root under a microscope. The types and arrangement of tissues in a root and dicot stem are similar except for the lack of pith in the root. Other ways in which a root differs from a stem include the following: the xylem is star-shaped and located in the center; clusters of phloem cells are located between the arms of the xylem; the growing tissue (apical meristem of the root) is located in the tip of a root, and it is covered with a

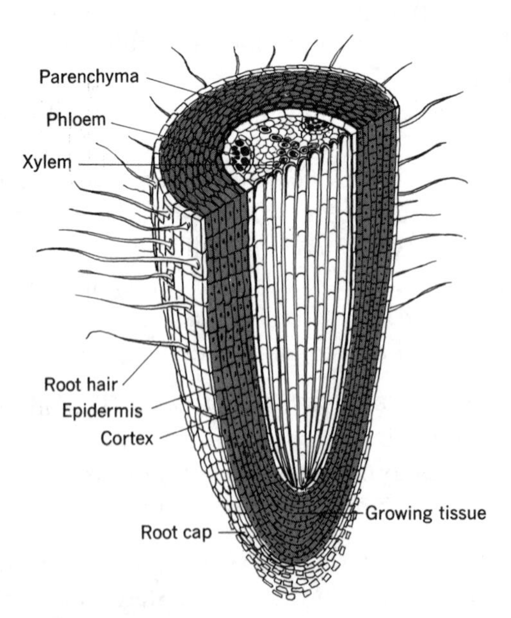

Figure 11–12 Root section

protective layer of cells called the *root cap;* extensions called *root hairs* grow out of many epidermal cells of a root.

Functions. The main functions of a root are anchorage, absorption, transport, and storage.

Anchorage. Figure 11–13 shows two types of roots. In carrots, the *taproot* (main root) and its branches penetrate deeply into the soil. In grasses, the *fibrous roots* (many thin branching roots) spread through the soil. Each root type securely fastens a plant in the soil.

Absorption. Roots are adapted to absorb water from the soil by root hairs (refer back to Figure 11–12). A *root hair* is a long, fingerlike projection of a root epidermal cell. The abundant root hairs covering a root increase the root's surface area. Water and some dissolved mineral salts pass into a root hair by passive transport. Some minerals may enter by active transport.

Transport. Water travels upward from the root xylem into the stem xylem and then to the leaves. Manufactured materials from the leaves pass into the leaf phloem and move downward through the stem to the root phloem.

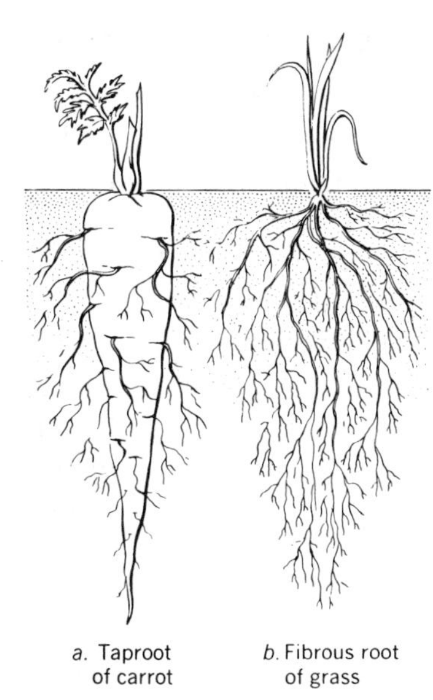

Figure 11–13 Types of root systems

Storage. Carrots, beets, and turnips are actually roots that store food materials in parenchyma cells. In beet and carrot roots, glucose is changed to more complex sugars. In the turnip root, glucose is changed to complex sugars and starch.

Section Quiz

1. Which tissue enables some dicot plants to live for many years? (*a*) collenchyma (*b*) cambium (*c*) sclerenchyma (*d*) pith.
2. The tissue in a stem that stores food and water is the (*a*) chlorenchyma (*b*) parenchyma (*c*) collenchyma (*d*) meristem.

3. The type of thick, main root that extends deeply into the soil is called a (*a*) root cap (*b*) root hair (*c*) taproot (*d*) fibrous root.
4. Companion cells assist the part of a plant's vascular system called the (*a*) phloem (*b*) xylem (*c*) meristem (*d*) tracheids.
5. The opening and closing of stomates is controlled mainly by (*a*) oxygen concentration within a leaf (*b*) increase and decrease in turgor pressure (*c*) osmosis resulting from potassium pumps in the spongy layer of a leaf (*d*) emission of wavelengths of red light from irradiated chlorophyll molecules.

PLANT METABOLISM AND MAINTENANCE

The life activities of plants and animals are similar, and include digestion, respiration, transport, and synthesis. Like animals, plants have tissues and organs specialized for different functions. Also, like animals, plants respond to stimuli and are internally controlled by hormones. Unlike animals, which are heterotrophs, green plants are autotrophs: they obtain organic materials essential for survival by synthesizing them from inorganic compounds present in their environment. Scientists believe that the similarities between plants and animals have their origin in the first forms of life, which were simple protists. Two other basic similarities in plants and animals—cell division and formation of reproductive cells—are discussed in the next chapter.

Chemical Needs of Plants

Plants also are similar to animals in their need for basic nutrients. As plants grow, their need for chemicals to synthesize important organic compounds, such as proteins and nucleotides, increases. For example, protein synthesis requires nitrogen; chlorophyll synthesis requires magnesium; nucleotide synthesis requires phosphorus; active transport requires potassium; and calcium is needed for cell metabolism.

Laboratory studies show that plants need nine basic elements, called *macronutrients*. The macronutrients are *sulfur, phosphorus, carbon, oxygen, hydrogen, nitrogen, potassium, calcium,* and *magnesium*. Plants also need other elements, including iron, copper, manganese, and chlorine. These are called *micronutrients*, or *trace elements*. (Table 11–2 summarizes macronutrients and micronutrients and their functions in plants.) However, plants differ from animals in their ability to manufacture food. The foodmaking process, *photosynthesis*, occurs in plants, algae, and in some species of bacteria.

Table 11–2 Plant Macronutrients and Micronutrients

Element	Chemical Form	Function
Macronutrients		
Carbon Oxygen Hydrogen	Molecules of: CO_2 H_2O; O_2 H_2O	Basic elements needed to synthesize organic compounds (for example, $C_6H_{12}O_6$)
Nitrogen	Nitrate ions (NO_3^-) Ammonium ions (NH_4^+)	Synthesis of chlorophyll proteins, nucleotides, nucleic acids, enzymes
Potassium	Potassium ions (K^+)	Opening and closing of stomates; synthesis of enzymes and some proteins; active transport
Calcium	Calcium ions (Ca^{2+})	Cell wall construction; cell division; some enzyme actions; cell membrane maintenance
Magnesium	Magnesium ions (Mg^{2+})	Metallic part of chlorophyll molecules; activates enzymes, protein synthesis
Phosphorus	Phosphate ions (PO_4^{3-})	Part of ADP, ATP, nucleic acids, phospholipids, and enzymes
Sulfur	Sulfate ions (SO_4^{2-})	Protein synthesis; part of some amino acids and enzymes
Micronutrients		
Iron	Iron Ions (Fe^{2+}, Fe^{3+})	Synthesis of chlorophyll and cytochromes; part of enzymes
Chlorine	Chloride ions (Cl^-)	Passive transport; osmosis
Boron	Borate ions (BO_3^{3-})	Transport in phloem; synthesis of nucleic acids
Manganese	Manganese ions (Mn^{2+})	Activates enzymes used in Krebs cycle
Zinc	Zinc ions (Zn^{2+})	Activates some enzymes; auxin synthesis
Copper	Copper ions (Cu^+), (Cu^{2+})	Activates some enzymes
Molybdenum	Molybdate ions (MO_4^{2-})	Nitrogen fixation

Photosynthesis

The experiments described previously indicate the conditions for photosynthesis. However, they do not provide much detailed information about this process. Although investigations into the nature of photosynthesis began about 200 years ago, in-depth understanding of this complex process is relatively recent.

History of photosynthesis. The following are important contributions of scientists to the understanding of photosynthesis.

Priestley. In 1774, Joseph Priestley, an English chemist, demonstrated that green plants give off a gas that "restored the air that had been destroyed" by a burning candle or by a mouse in a closed jar. Priestley concluded that a candle burned longer and a mouse lived longer in the presence of a plant. Today, it is known that oxygen released by the plant "restores" the air.

Ingen-Housz. In 1779, Jan Ingen-Housz, a Dutch physician, repeated Priestley's candle and mouse experiments. He discovered that the ability of plant parts that contain chlorophyll to give off oxygen depends upon exposure to sunlight. Just as important, Ingen-Housz discovered that as green plants add oxygen to the air, they also remove carbon dioxide.

An early chemical equation. In the 19th century, scientists discovered that starch is a product of photosynthesis. By this time it was known that a green plant exposed to light combines carbon dioxide and water, forming carbohydrates. As a result of this chemical reaction, oxygen is released as a by-product. The entire process was represented by the following equation, later to be corrected:

$$6\ H_2O + 6\ CO_2 \xrightarrow[\text{(sunlight)}]{\text{(chlorophyll)}} C_6H_{12}O_6 + 6\ O_2$$

Engelmann. At the end of the 19th century, Theodore W. Engelmann, a German scientist, investigated the effect of different colors of light on photosynthesis. He reasoned that the rate of photosynthesis could be measured by the amount of oxygen given off by a plant. To determine this amount, Engelmann used a green alga and an aerobic bacteria that could move. First he placed a strand of green alga and the bacteria in water. Then he illuminated the alga and the bacteria with the colors of the spectrum (or rainbow) by transmitting a beam of light through a glass prism. The spectrum consists of a band of the colors red, orange, yellow, green, blue, and violet. After exposing the alga and the bacteria to the spectrum, Engelmann found the bacteria clustering in the region illuminated by the red and blue-violet light (Figure 11–14). The bacteria moved to these regions because the concentration of oxygen was greater there than elsewhere along the spectrum. Engelmann concluded that the red part of the spectrum was most effective in photosynthesis, the blue-violet parts next most effective, and the yellow-green parts least effective.

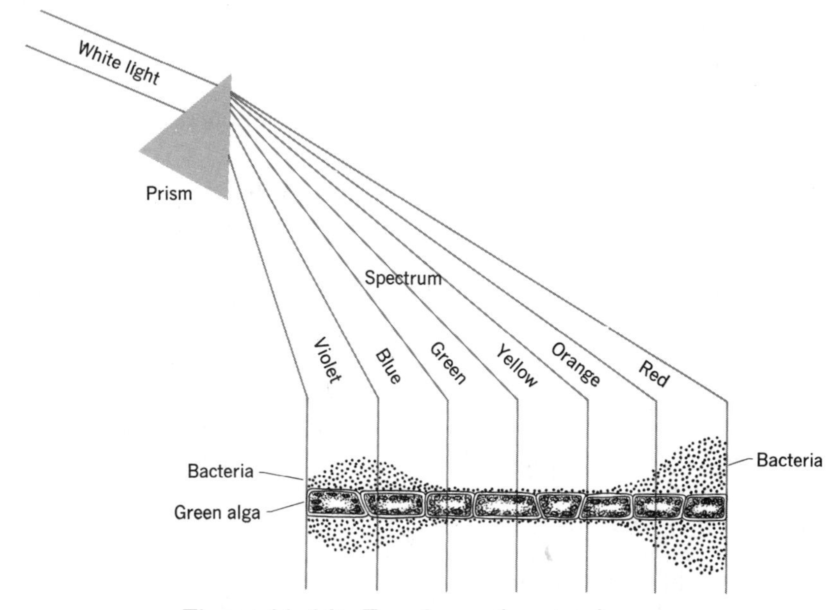

Figure 11–14 Engelmann's experiment

Blackman and Hill. At the beginning of the 20th century, F. F. Blackman, a British plant physiologist, investigated the effects of continuous and flashing light on photosynthesis. He discovered that the amount of oxygen given off by a plant exposed to a flashing light was the same as the amount given off by a similar plant exposed to continuous illumination. Blackman concluded that the rate of photosynthesis of a green plant was unaffected by brief periods of darkness. These results led to the idea that photosynthesis probably consists of two phases—one that does not require light.

In 1937, Robin Hill, a British biochemist, removed the chloroplasts from some plant cells. When the chloroplasts were placed in a solution that contained iron salts and exposed to light, the mixture produced substantial amounts of oxygen. This showed that the release of oxygen by chloroplasts can occur without carbon dioxide. Hill's experiment provided additional evidence that photosynthesis is a two-phase process in which only one phase requires light. In the light-dependent phase of photosynthesis, carbon dioxide is not used and carbohydrates are not formed.

Kamen and Ruben. In 1941, Martin Kamen, Samuel Ruben, and other American scientists investigated the release of oxygen during photosynthesis. They supplied a plant with carbon dioxide that contained the ordinary isotope of oxygen (oxygen-16) and with water that contained another isotope of oxygen (oxygen-18) as a tracer element. They discovered that only the oxygen-18 was released during photosynthesis. This experiment showed that water, not carbon dioxide, was the source of oxygen in photosynthesis.

Calvin. The major steps in the formation of carbohydrates in photosynthesis were identified after World War II by Melvin Calvin and a team of scientists at the University of California. During the late 1940s and early 1950s, radioactive carbon-14 became available as a tracer element for experimentation and research. Calvin and his associates described in detail the chemical pathway of carbon from carbon dioxide to carbohydrates. The steps in photosynthesis that result in carbohydrate formation are known as the *Calvin cycle.*

Factors in photosynthesis. The experiments discussed earlier in this chapter and the contributions of these scientists have revealed that the four essential factors in photosynthesis are *light, carbon dioxide, water,* and *chlorophyll.*

Light. Light is a type of energy that travels in the form of wavelengths. Ordinary white light is a mixture of different colors, or wavelengths, of the spectrum. Each wavelength of light has a different amount of energy. For example, red light has less energy than violet light. Photosynthesis is most rapid when energized by red-orange and blue-violet light (Figure 11–15). These wavelengths of light are absorbed by chlorophyll when white light strikes a leaf. Green and yellow wavelengths of light are reflected and not absorbed, thus making plants appear green.

Carbon dioxide. Photosynthesis does not occur in the absence of carbon dioxide. As the amount of carbon dioxide available to a plant increases, the rate of photosynthesis increases. Too much carbon dioxide, however, poisons a plant and photosynthesis stops.

Water. During periods of drought, photosynthesis stops. Under normal weather conditions, water and dissolved minerals rise to the leaves, where—in most plants—most of the chloroplasts are located.

Chlorophyll. Chlorophyll is a complex molecule that contains magnesium. The arrangement of the atomic bonds in a chlorophyll molecule

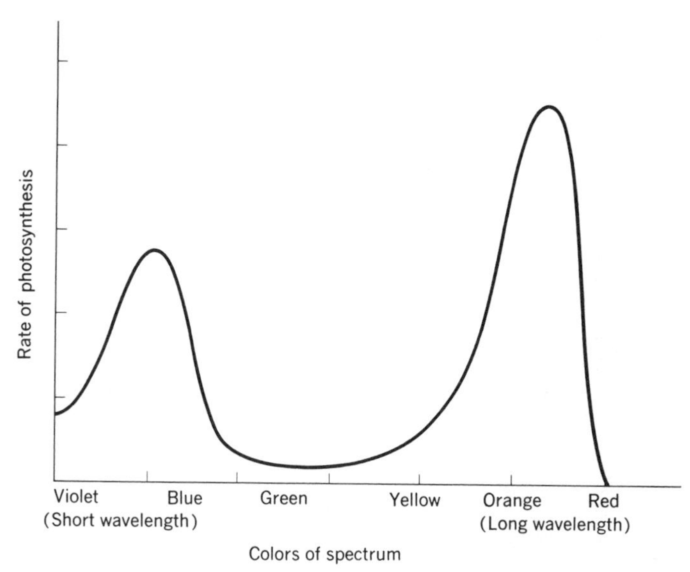

Figure 11–15 Wavelengths of light and rate of photosynthesis

permits electrons to move to higher energy levels when light of the proper wavelengths and intensity strike the chlorophyll.

Energy is released when electrons fall from higher energy levels to lower energy levels. An amount of energy is emitted equal to the amount needed to jump electrons to higher levels. This is the energy used by a plant in the first step of photosynthesis. To demonstrate the energy emitted from irradiated chlorophyll, prepare an alcoholic solution of chlorophyll. (*Follow safety precautions.*) Pour off about 30 milliliters of the solution into a test tube and then seal it with a rubber stopper. After the test tube has cooled to room temperature (22°C), expose the test tube and its contents to the sun or a strong white light for about five minutes. Examine the test tube in subdued light and you will observe that the green chlorophyll has become bloodred. The bloodred color results from emitted energy in the form of electromagnetic radiation (light rays) in the red spectrum. The red color of the solution soon fades, leaving the original green color.

Steps in photosynthesis. Photosynthesis is a two-step process. One step requires light and is called the *light phase.* In this phase, light energy splits water molecules into hydrogen and oxygen. This process is called *photolysis.* The other step does not require light and is called the *dark phase.* In this phase, carbohydrates are formed from carbon dioxide and hydrogen. This process is called *carbon dioxide fixation.*

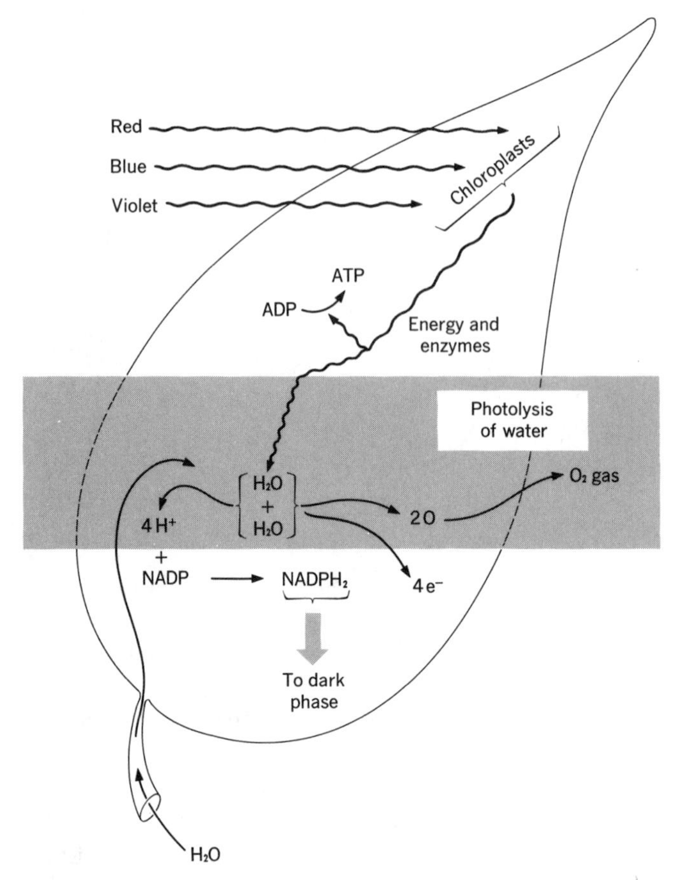

Figure 11–16 Outline of the light phase of photosynthesis

Photolysis of water: the light phase (Figure 11–16).

1. In the chloroplasts, red, blue, and violet light rays are absorbed by chlorophyll. As a result, light energy is transformed into chemical energy.
2. The chemical energy splits, or decomposes, water into hydrogen ions and oxygen atoms, and electrons are released.
3. These electrons help provide stored energy in the form of ATP and NADP (nicotinamide adenine dinucleotide phosphate).
4. NADP accepts hydrogen ions (from decomposed water) and becomes $NADPH_2$.
5. Oxygen is liberated as a gas.

Photolysis also is called *photophosphorylation* because (1) electrons travel in a one-way direction from water to form $NADPH_2$, and (2) radiant energy helps form ATP molecules. Photophosphorylation occurs in the *thylakoid* membranes of grana. Thylakoids are flattened membranous sacs that make up a single granum (see Chapter 2).

Carbon dioxide fixation (formation of carbohydrates): the dark phase (Figure 11–17). ATP supplies energy for the enzyme-controlled reactions in this phase.

1. In chloroplasts, carbon dioxide combines with a five-carbon sugar phosphate, RDP (ribulose diphosphate), forming a compound that splits into two molecules of the three-carbon compound PGA (phosphoglyceric acid).
2. The PGA combines with the hydrogen $NADPH_2$, forming PGAL (phosphoglyceraldehyde). NADP is released and used again in the light phase.

   Water is a product of reaction 1 and 2 in this phase.

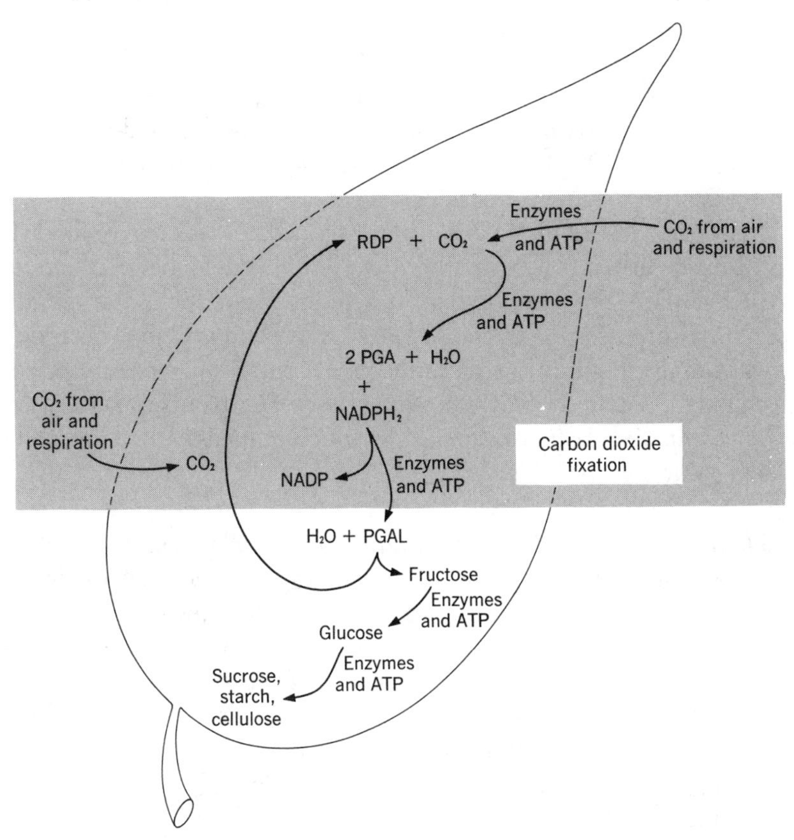

Figure 11–17 Outline of the dark phase of photosynthesis

3. Fructose, a six-carbon sugar, is formed from PGAL. At the same time, the five-carbon sugar is formed again and is available, together with additional carbon dioxide and hydrogen, to form more PGAL. This process continues as long as the necessary factors are present.
4. Fructose is changed to glucose. From glucose, complex carbohydrates are synthesized.

Modern equation of photosynthesis. The modern equation of photosynthesis below shows the use and release of water:

$$6\,CO_2 + 12\,H_2O \xrightarrow[\text{chlorophyll, ATP, enzymes}]{\text{red, blue, violet light}} C_6H_{12}O_6 + 6\,H_2O + 6\,O_2$$

From this equation, you can see that more water is used than is released. Recall that a major result of photolysis is the release of hydrogen atoms from water, and that the hydrogen atoms are used in the dark phase. Thus, the number of reactant atoms of hydrogen equals the number of hydrogen atoms in the products.

Plants that use the CO_2 fixation cycle are called C_3 plants, and usually are adapted to live in temperate climates. The plants are named C_3 because PGA, a *three-carbon compound,* is the first molecule detected. The CO_2 cycle is not 100 percent efficient. It loses about one-half of its fixed carbon in other reactions involving oxygen.

In effect, a loss of fixed carbon can be translated as a loss of food. The loss of fixed carbon is temperature dependent: the higher the temperature, the greater the reaction rate. In regions where the air temperature frequently surpasses 30°C, the loss of fixed carbon is high. This reduces crop yields in agricultural areas. Some plants, such as sugar cane and corn, are called C_4 plants. The CO_2 fixation phase of C_4 plants involves a *four-carbon compound*, called *oxaloacetate,* which reduces carbon loss and results in greater crop yields.

Plants use carbohydrates as building blocks. Plants and animals display unity, especially in their use of similar organic chemicals. In regard to nutrition, animals obtain all the organic molecules they need from other living things. In contrast, plants manufacture sugars and use them as a base for synthesizing the different types of organic molecules they need to carry out their life processes. Thus, proteins, lipids, enzymes, cellulose, nucleotides, pigments, vitamins, and a host of other organic compounds have their origin in fructose—a simple sugar initially manufactured by photosynthesis.

Other Life Activities of Plants

Excluding photosynthesis, plants carry on life processes that are similar to those of animals.

Digestion. Complex nutrient molecules are manufactured by photosynthesis. However, before they can be used by plant cells, the complex molecules must be digested. In plant cells, digestion is intracellular. Amylases, proteases, and lipases change complex nutrient molecules into sugars, amino acids, and fatty acids.

Respiration. Unity in living things also is reflected by the processes of anaerobic respiration (glycolysis) and aerobic respiration (Krebs cycle and hydrogen transport system). Both processes occur in plants and animals. Table 11–3 compares photosynthesis and aerobic respiration in plants.

Both cellular respiration and photosynthesis occur in plant cells. Respiration occurs continuously; photosynthesis occurs only in light. Energy stored in ATP molecules is used to mobilize carbohydrates as storage products. At night, these storage products are used for cellular metabolism.

Table 11–3 Comparison of Photosynthesis and Aerobic Cellular Respiration

Photosynthesis (*Chloroplast*)	*Aerobic Cellular Respiration* (*Mitochondrion*)
ADP → ATP (energy storage)	ADP → ATP (energy storage)
O_2 released	O_2 used (combined)
$CO_2 \rightarrow C_6H_{12}O_6$ (CO_2 fixed)	$C_6H_{12}O_6 \rightarrow CO_2$ (CO_2 released)
$NADPH_2$ → NADP (coenzyme reduced)	NAD → $NADH_2$ (coenzyme oxidized)

Excretion. Plants, unlike animals, reuse most of their wastes and lack special organs of excretion. Much of the carbon dioxide generated by aerobic respiration in plant cells is used in the dark phase (carbon dioxide fixation) of photosynthesis. Part of the oxygen released in the light phase (photolysis) is used in aerobic respiration. At night, some carbon dioxide is stored in the air spaces of a leaf and some is excreted through the stomates. During the day, oxygen is released through the stomates and some is stored in the air spaces of a leaf. Roots excrete some carbon dioxide by diffusion through root hair cells and other epidermal cells.

Although plants require water, they also produce water in respiration, in dehydration synthesis reactions, and in photosynthesis. This water, along with excess water that has been absorbed from soil is excreted in transpiration.

Excess amino acids are changed to ammonia and other nitrogenous substances and are reused.

Regulation. Although plants lack a nervous system, they do respond, or orient, to external stimuli such as light, gravity, and touch. In each case, a type of plant hormone called *auxin* promotes cellular growth, which in turn activates a response. A response by a plant to an external stimulus is called a *tropism* (see Figure 11–18).

Response to light (phototropism). A response by a plant that moves the entire plant or a part of the plant toward an external stimulus is called

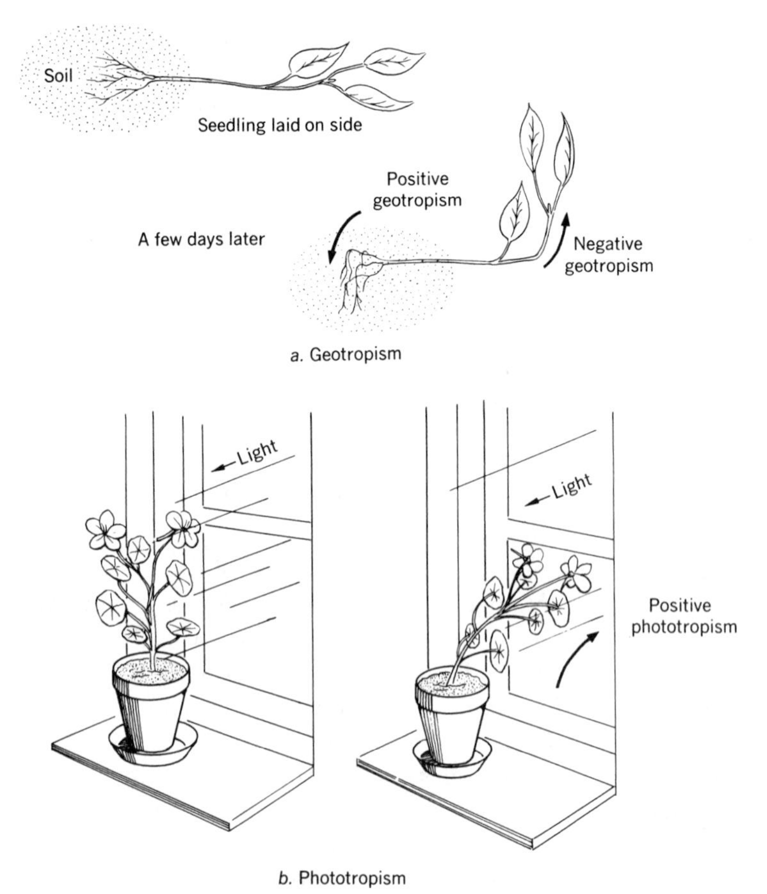

Figure 11–18 Plant tropisms

a positive tropism. For example, a plant that bends toward light is *positively phototrophic*, benefiting from the response by obtaining more light. Auxin is produced in cells facing the light. The hormone then diffuses to cells of the dark side and causes them to elongate. As a result of the difference in cell length, a stem or leaf bends toward the light. The chemical name for an important auxin is *indoleacetic acid (IAA)*, which is synthesized from the amino acid tryptophane. A response that moves a plant away from an external stimulus is called a negative tropism. For example, roots are *negatively phototropic,* which enables them to probe deep in the soil and reach water and dissolved minerals. Both the positive and the negative phototropic responses of a plant increase the potential output of photosynthesis.

Response to gravity: gravitropism (geotropism). When the stem and leaves of a plant grow upward and its roots grow downward, it responds to gravity by being *negatively gravitropic* (growing away from gravity) and *positively gravitropic* (growing toward gravity) at the same time. Stems and leaves growing upward receive more light; roots growing downward are more likely to find water with dissolved minerals. The positive and negative tropisms of stems, leaves, and roots result from auxin action. When a young plant is placed horizontally on soil, differences in auxin concentration develop between the upper and lower sides of the plant. Auxin tends to stunt root growth and stimulate stem growth. As a result, a small amount of auxin causes the root to curve away from the side where the auxin concentration is greatest, and grow downward.

Response to touch (thigmotropism). Thigmotropism, which is still not completely understood, involves the hormones auxin and *ethylene.* Ethylene is a normal metabolic product of plants. Vines usually respond to physical contact by clinging or winding around solid objects. Tendrils of plants also respond in a similar manner. Venus flytrap, a carnivorous plant, obtains a rich supply of proteins and minerals by eating insects. The plant possesses a trigger mechanism that, when activated by an insect, quickly encloses its prey within two movable leaves. Later, the insect is digested by juices secreted by special cells in the leaf.

Other Plant Hormones

Cytokinins are plant hormones that combine with auxin to stimulate cell division (growth) and differentiation of embryonic cells into specific tissues. Cytokinins are produced in roots and are transported to all parts of a plant.

Gibberellins are plant hormones that stimulate the elongation of stems, especially of mature shrubs and trees. *Gibberellins* are produced in root tip cells and stem tip cells. Recent studies show that gibberellins aid in seed germination and flower formation.

Ethylene, a gaseous product of plant metabolism, combines with auxin to suppress stem and root elongation. Ethylene also helps form an *abscission layer* (separation layer of dead cells between a stem and the petiole of a leaf) in the autumn. Ethylene also influences the ripening of fruit. Today, ethylene gas is used commercially to ripen tomatoes, bananas, oranges, and lemons.

Abscissic acid is a hormone produced in green leaves, fruit, and in root caps. Abscissic acid promotes the aging of leaves and the retardation of bud growth. This hormone and the potassium pumps in guard cells work together to open and close stomates. Recent research indicates that abscissic acid interferes with protein synthesis in cells and alters the chemistry of a cell's plasma membrane. Table 11–4 summarizes the functions of plant hormones.

Table 11–4 Plant Hormones and Their Functions

Hormone	*Function*	*Part of Plant That Produces Hormones*
Auxins (IAA)	Stimulate growth of vascular tissue; elongate stem cells; tend to stunt root growth	Tips of stems rich in meristem tissue
Cytokinins	Stimulate cell division and tissue differentiation	Roots; then is transported to all parts of plant
Gibberellins	Stimulate elongation of stem cells (growth); aid in seed germination and flower formation	Root tips and stem tips rich in meristem tissue
Ethylene	Aids in forming abscission layer of leaves, flowers, fruits; suppresses stem and root elongation; aids fruit ripening	Metabolic product of leaves, stems, fruits
Abscissic acid	Promotes aging of leaves; retards bud growth; interferes with protein synthesis	Green leaves, root caps, fruits

Other Plant Responses

Turgor Pressure Responses. Plants generally respond to stimuli such as light (phototropism), gravity (gravitropism), and touch (thigmotropism) by positive or negative tropisms. In each case, one or more hormones is involved. *Turgor pressure responses* do not require hormones. These responses result from changes in turgor pressure of specific cells. For example, the prayer plant, a common decorative houseplant, spreads its leaves during the day and folds them at night. At the base of each leaf of a prayer plant is a *pulvinus* and *motor cells*. When the motor cells either take in or release water, the pulvinus, which operates as a hinge, raises a leaf or lets it droop. Sunflowers and other related plants follow the sun by stem and leaf movements, which are coordinated by turgor mechanisms. The orientation of flowers, stems, and leaves resulting from turgor mechanisms is an adaptation that promotes photosynthesis, and protects a plant against extreme heat, cold, and water loss.

Flowering Responses. Plant growth and development are related to changes in the length of daylight in a 24-hour cycle. A plant response to varying lengths of daylight caused by seasonal change is called *photoperiodism*. *Long-day plants* flower only when daylight is longer than a critical day length (12 to 14 hours); these plants flower in late spring and early summer. Wheat, hollyhocks, irises, and clover are examples of long-day plants. *Short-day plants* flower when days are shorter than the critical length; these plants flower in late summer and autumn. Goldenrod, ragweed, and chrysanthemums are examples of short-day plants. *Day-neutral plants* flower anytime, regardless of day length. Day-neutral plants include tomatoes, roses, and cucumbers. In all cases, it is actually the length of darkness that determines flowering, not the length of day. The chemical basis for flowering responses is a blue pigment called *phytochrome*, which functions as a light receptor. Phytochrome either suppresses or stimulates flowering in plants when triggered by varying wavelengths of red light.

Section Quiz

1. A day-neutral plant, such as a rose (*a*) is unaffected by photoperiodism (*b*) lacks phytochrome (*c*) flowers only when day length exceeds the critical length (*d*) flowers only when day length is less than the critical length.

2. The plant hormone abscissic acid is involved mainly with (*a*) growth of stem tips and root tips (*b*) aging of leaves (*c*) production of ethylene gas (*d*) regulating turgor pressure responses.
3. Plants that contain chlorophyll appear green because (*a*) green light is absorbed (*b*) green light is reflected (*c*) red light and blue light are absorbed (*d*) electrons jump from higher to lower energy levels.
4. When an alcoholic solution of chlorophyll is subjected to strong white light and later removed, a red color will appear for several seconds, due to (*a*) temporary conversion of a chlorophyll molecule to a hemoglobin molecule (*b*) temporary fading of the green pigment, thus letting hidden red pigments appear (*c*) emission of electromagnetic rays in the red spectrum of visible light (*d*) appearance of the pigment phytochrome.
5. Photophosphorylation is a chemical process that (*a*) takes place during the carbon dioxide-fixation phase of photosynthesis (*b*) reduces NAD to $NADH_2$ (*c*) is part of a series of reactions that produces auxins (*d*) changes ADP to ATP during the photolysis phase of photosynthesis.

Chapter Review Questions

The following questions will help you check your understanding of the material presented in the chapter.

1. An organism is classified as an autotroph rather than a heterotroph if it (*a*) gives off carbon dioxide as a waste (*b*) grows only in the daytime (*c*) synthesizes nutrients from inorganic materials (*d*) forms a spindle during cell division.
2. Which gas is excreted by green plants in darkness? (*a*) carbon dioxide (*b*) molecular oxygen (*c*) nitrogen (*d*) hydrogen.
3. Xylem and phloem cells in woody stems are formed by (*a*) epidermal cells (*b*) palisade cells (*c*) guard cells (*d*) cambium cells.
4. Which is characteristic of all vascular plants? (*a*) Conducting tissue is present. (*b*) Flowers are produced. (*c*) Seeds develop within fruits. (*d*) Seeds develop within cones.
5. Transpirational pull in land plants is a physical factor associated with upward transport in the (*a*) phloem (*b*) cambium (*c*) xylem (*d*) cotyledon.

6. The conversion of light energy into chemical energy can best be carried on by cells that contain (*a*) centrioles (*b*) contractile vacuoles (*c*) chlorophyll (*d*) glycogen.
7. In plants, the energy required to convert glucose molecules into starch is derived from (*a*) water (*b*) ATP (*c*) auxins (*d*) minerals.
8. In vascular plants, it is believed that the combined effects of transpirational pull, capillarity, and root pressure are probably responsible for the (*a*) positive hydrotropism of roots (*b*) movement of fluids in the xylem (*c*) discharge of carbon dioxide from root hairs (*d*) active transport of starch by the phloem.
9. Plants that normally grow in dry environments would probably have (*a*) small leaves with few stomates (*b*) stomates that are permanently open (*c*) large leaves with many stomates (*d*) stomates in their stems and roots.
10. Which waste products of respiration are used by most autotrophs? (*a*) carbon dioxide and water (*b*) carbon dioxide and oxygen (*c*) glucose and water (*d*) glucose and oxygen.
11. In which parts would the greatest active cell reproduction most likely be observed? (*a*) root tip, xylem, phloem (*b*) root hair, root tip, cambium (*c*) cambium, root tip, stem tip (*d*) stem tip, xylem, phloem.
12. Which plant part is composed only of undifferentiated cells? (*a*) oak cambium (*b*) maple tree root system (*c*) geranium leaves (*d*) raspberry sepals.
13. Which part of a typical land plant is most directly involved with transpiration? (*a*) chloroplasts (*b*) phloem (*c*) stomates (*d*) cambium.
14. Conversion of light energy into chemical bond energy is directly dependent upon (*a*) structural adaptations of roots (*b*) carbon fixation in green plants (*c*) negative auxin reaction in green plants (*d*) activation of chlorophyll molecules.
15. If a euglena is placed in darkness for long periods of time, its chloroplasts break down. As a result, that euglena will no longer (*a*) reproduce asexually (*b*) utilize oxygen (*c*) ingest organic molecules (*d*) function as an autotroph.
16. Phloem tissue is most directly associated with (*a*) absorption by root hair cells (*b*) transpiration pull (*c*) conduction of organic materials (*d*) capillary action and root pressure.

17. Which statement concerning a geranium plant is true if the plant is first exposed to sunlight and then to pure green light? (*a*) Its rate of photosynthesis will decrease. (*b*) Its rate of photosynthesis will increase. (*c*) It will release more oxygen. (*d*) It will require more carbon dioxide.
18. The passage of oxygen from the air spaces to the cells of a leaf occurs by (*a*) hydrolysis (*b*) diffusion (*c*) osmosis (*d*) capillary action.
19. The closing of stomates on a leaf would have the least effect on the (*a*) diffusion of carbon dioxide into the atmosphere (*b*) release of water vapor from the leaf (*c*) size of the glucose molecules within the cell (*d*) diffusion of oxygen molecules into the leaf.
20. Leaves appear to be green because (*a*) green represents the combining of many other colors (*b*) most of the green wavelength of light is transmitted (*c*) they absorb other wavelengths of light and reflect green (*d*) they absorb only the green wavelength of light.
21. Starch in the root tissues of a bean plant was probably (*a*) synthesized from glucose in the cells (*b*) absorbed from the soil through root hairs (*c*) produced as a by-product of respiration (*d*) secreted from the cell walls.
22. As a result of the function of root hairs, the fluid within the xylem tubes shows an increase in (*a*) chlorophyll (*b*) glucose (*c*) nitrates (*d*) proteins.
23. The structure responsible for the growth in diameter of woody roots and stems is the (*a*) apical meristem (*b*) cotyledon (*c*) hypocotyl (*d*) lateral meristem.
24. Carbon-14 was used in the study of photosynthesis to indicate (*a*) that free oxygen is produced from water (*b*) that green plants give off carbon dioxide in the dark (*c*) the path of carbon in the carbon-fixation reactions (*d*) the role of root hairs in absorbing water.
25. The first step in photosynthesis is the (*a*) formation of ATP (*b*) synthesis of water (*c*) excitation of chlorophyll by light (*d*) fixation of carbon dioxide.
26. During the photochemical reaction (light reaction) of photosynthesis, which action occurs? (*a*) Carbon dioxide is combined with hydrogen. (*b*) Gaseous oxygen is liberated. (*c*) Chlorophyll absorbs carbon dioxide. (*d*) Glucose is converted into PGAL.
27. The oxygen released by plants during photosynthesis comes from (*a*) glucose (*b*) carbon dioxide (*c*) water (*d*) starch.

28. During photosynthesis in a bean plant, which wavelength of light is least effective as an energy source? (*a*) red (*b*) blue (*c*) green (*d*) violet.

29. In photosynthesis, which event normally occurs before the other three? (*a*) oxygen release (*b*) water absorption (*c*) PGAL synthesis (*d*) glucose formation.

30. The basic inorganic materials used during photosynthesis are (*a*) H_2O and O_2 (*b*) O_2 and CO_2 (*c*) H_2O and CO_2 (*d*) $C_6H_{12}O_6$ and CO_2.

31. In plants, respiration differs from photosynthesis because respiration occurs (*a*) all the time, while photosynthesis occurs only in the presence of light energy (*b*) only in the presence of light energy, while photosynthesis occurs all the time (*c*) only in the presence of light, while photosynthesis occurs only in darkness (*d*) only in darkness, while photosynthesis occurs only in the presence of light.

32. Phototropism and geotropism are plant growth responses. These tropisms occur because light and gravity (*a*) cause equal auxin distribution in roots and stems (*b*) are toxic to roots and stems (*c*) slow down the lengthwise growth of individual cells (*d*) result in unequal auxin distribution in roots and stems.

33. Under microscopic examination, cells treated with auxins would probably appear different from untreated control cells mainly in their (*a*) length (*b*) color (*c*) chromosome number (*d*) cytoplasmic organelles.

Biology Challenge

The following questions will provide practice in answering SAT II-type questions.

Part I

Questions *1 through 5* describe certain procedures associated with five different investigations. For each question, select the letter of the statement that best describes the purpose of the procedure.

1. A transparent grease is smeared on the underside of a geranium leaf. The grease also is smeared on the upper surface of another leaf on the same plant. Twenty-four hours later, the leaves are removed, soaked in

warm alcohol to remove the chlorophyll, and placed in iodine solution. (*a*) Light is necessary for photosynthesis. (*b*) Lipids are necessary for photosynthesis. (*c*) Chlorophyll is involved in photosynthesis. (*d*) Carbon dioxide is necessary for photosynthesis. (*e*) Epidermal cells are involved in photosynthesis.

2. The upper tips of ten oat seedlings were cut off and discarded. The cut seedlings were then smeared with a lanolin-auxin mixture. Another group of ten oat seedlings was left untreated. Both groups of seedlings were placed in a cup containing moist soil and exposed to light from one direction. (*a*) Plant hormones, such as auxins, regulate photosynthesis. (*b*) Plant hormones, such as auxins, help a plant respond to environmental stimuli. (*c*) Plant hormones, such as auxins, help absorb and harness light energy. (*d*) Plant hormones, such as auxins, absorb red and blue wavelengths of light and reflect others. (*e*) Plant hormones, such as auxins, promote growth of meristem tissue.
3. The figure shows a student's project concerning photosynthesis. (*a*) A light is necessary for photosynthesis. (*b*) A control is not necessary in certain investigations. (*c*) Determine the rate of photosynthesis. (*d*) Elodea requires unidirectional light for maximum protein synthesis. (*e*) Oxygen gas bubbles expand as pressure decreases.

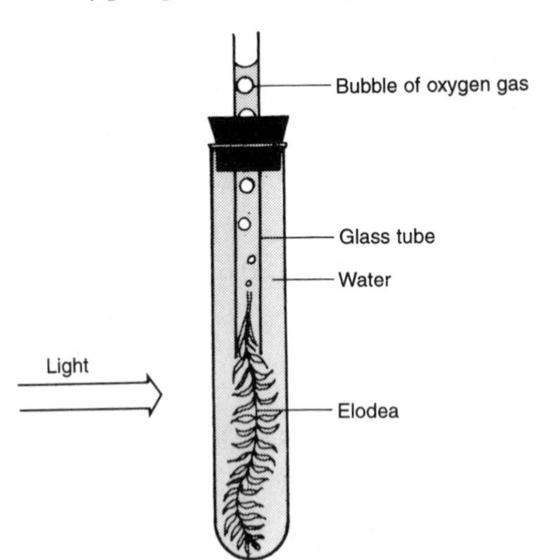

4. The figure on the next page shows two different leaves being considered as specimens in certain investigations involving plant metabolism. (*a*) The geranium leaf will be used to show light is necessary for photosynthesis. (*b*) The geranium leaf will be used to show gas exchange takes place through stomates. (*c*) The dark reaction of pho-

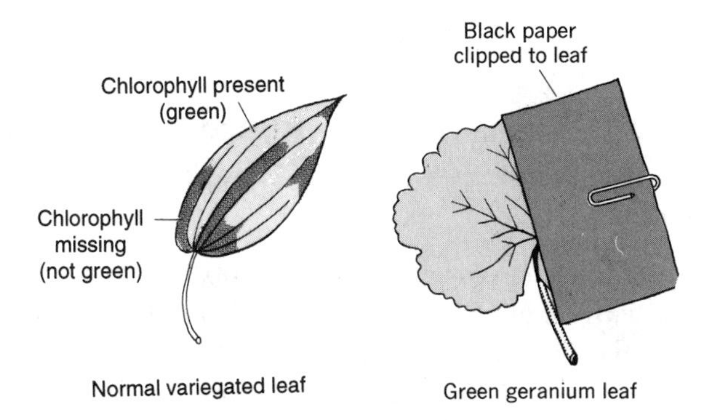

Normal variegated leaf

Green geranium leaf

tosynthesis is speeded up as a result of the black paper. (*d*) The variegated leaf will be used to show starch is present in all parts of the leaf as a result of photosynthesis. (*e*) A variegated leaf cannot be used in any investigation dealing with photosynthesis.

5. The stem of a young corn plant (about 15 cm tall) was cut close to the point where the stem entered the soil. The plant was removed, then left standing in water colored with a red vegetable dye overnight. The next day, the corn plant was removed from the colored water and the stem was rinsed in running water. Then, a sliver was shaved from the bottom of the stem and mounted on a microscope slide. The slide was observed under the low-power objective of a light microscope. The figure below represents a magnified portion of the fields of vision. (*a*) Dissolved minerals are transported upward by tissue number 5. (*b*) Dissolved minerals are transported upward by tissues numbered 1 and 2. (*c*) Dissolved minerals are transported upward by tissue number 3. (*d*) Dissolved nutrients are transported downward by tissues numbered 2 and 3. (*e*) Dissolved nutrients are transported downward by tissue number 4.

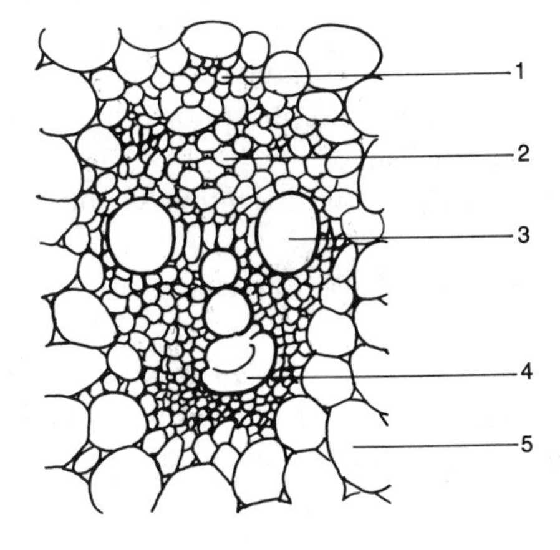

Part II

In 1647, Jean-Baptiste van Helmont, a Flemish physician, described the results of an experiment involving plant growth. The following is an excerpt from his notebook:

> I took an earthenware vessel, placed in it 200 lb. of soil dried in an oven, soaked this with rainwater, and planted in it a willow branch weighing 5 lb. At the end of five years, the tree grown from it weighed 169 lb. and about three ounces. Now, the earthenware vessel was always moistened (when necessary) only with rainwater or distilled water, and it was large enough and embedded in the ground, and, lest dust flying about be mixed with the soil, an iron plate coated with tin and pierced with many holes covered the rim of the vessel. I did not compute the weight of the fallen leaves of the four autumns. Finally, I dried the soil in the vessel again, and the same 200 pounds were formed, minus about two ounces.

Regarding the above paragraph, some of the following statements are correct; others are incorrect. Circle the letter of each correct statement.

A. In van Helmont's time, empirical data usually was gathered by investigative procedures.
B. Van Helmont clearly demonstrated that a plant's body mainly uses soil minerals.
C. The main objective of the "experiment" was to show that mathematics could be used to "prove" a natural phenomenon.
D. Van Helmont clearly demonstrated that most living matter of a plant is not supplied by soil, but by water.
E. Van Helmont considered the possibility that a weightless factor, such as air, could contribute to the plant's growth.
F. A weakness in the "experiment" was the lack of control.
G. Nitrogen, potassium, and phosphorus compounds are required for normal plant metabolism.
H. Current knowledge of photosynthesis supports van Helmont's conclusions.
I. Van Helmont should have used isotopic tracer techniques to confirm his conclusion.
J. When wet, a cover made of "*an iron plate coated with tin*" could constitute an electrochemical cell. In all probability, tiny electrical discharges affected the growth of the willow tree.

12

Cell Division

Learning Objectives

When you have completed this chapter, you should be able to:

- **Compare** the nucleotides found in DNA and RNA.
- **Describe** the structure of DNA and RNA.
- **Explain** how the nucleus of a cell directs the functioning of the cell.
- **Define** replication, transcription, and translation as they relate to protein synthesis.
- **Relate** the importance of chromosomes to cell division.
- **Compare** mitosis in plant and animal cells.
- **Relate** the importance of meiosis to sexual reproduction and the maintenance of a species' chromosome number.
- **Describe** the formation of sperm and egg cells.

OVERVIEW

Growth, repair, and reproduction are the life processes made possible by the division of cells. Cells grow to a certain size and then these mature parent cells divide into new cells called daughter cells. In this chapter, you will learn how a complex set of reactions between the cytoplasm and the genetic information contained in a nucleus controls cell division.

CHEMISTRY OF DNA AND RNA

Eukaryotic organisms possess a chromosome number characteristic of their species. For example, humans have 46 chromosomes, horses have

64, earthworms have 36 chromosomes, and a yellow pine has 24. The number of chromosomes in each cell of an organism does not indicate its complexity or adaptability. In all species, an organism's *genotype* is encoded in DNA. This complex organic molecule carries a program of instructions that is passed on from parents to offspring, from cell to cell, and from generation to generation. When *cell division* occurs, the cytoplasm of the parent cell shares all of its components, including organelles, nutrients, and enzymes, with the daughter cells. As a result, daughter cells possess the structures, compounds, and energy needed to survive.

DNA and RNA are composed of subunits called nucleotides. A nucleotide consists of three parts: (1) a specific nitrogenous base, (2) a pentose, or five-carbon, sugar group (S), and (3) a phosphate group (P). A nucleotide may have any one of the following nitrogenous bases: *adenine* (*A*), *guanine* (*G*), *thymine* (*T*), and *cytosine* (*C*). *Uracil* (*U*) is a fifth nitrogenous base that replaces thymine in an RNA strand. Adenine and guanine are called *purines*; thymine, cytosine, and uracil are called *pyrimidines*.

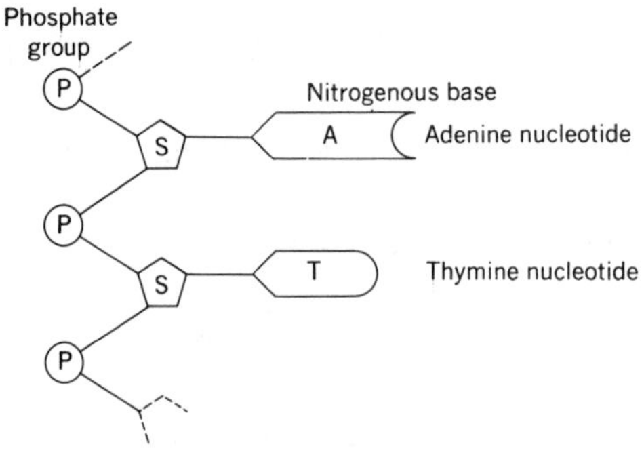

Phosphate group

Nitrogenous base

Adenine nucleotide

Thymine nucleotide

Figure 12–1 Part of a polynucleotide chain

A long strand of either DNA or RNA actually is a union of many nucleotides. When nucleotides join, the phosphate group (P) of one links to the sugar group (S) of another, forming the backbone of a *polynucleotide* strand. The purines and pyrimidines are attached to the sugars and project outward from them (Figure 12–1). The four basic nucleotides of DNA are: P-S-*A*, P-S-*C*, P-S-*G*, and P-S-*T*. The four basic nucleotides of RNA are: P-S-A, P-S-*C*, P-S-*G*, and P-S-*U*.

Structure of DNA

In 1953, geneticists *James Watson* and *Francis Crick* proposed a model that illustrated the structure of DNA (Figure 12–2). This model shows that DNA is a union of two polynucleotide chains that resemble a twisted ladder, or a spiral staircase. You may find it useful to refer to the drawing of a molecule of untwisted DNA (Figure 12–2). In these polynucleotide chains, the five-carbon sugar is deoxyribose (D). The bases between chains are paired in a specific way: adenine (A), a purine, always pairs with thymine (T), a pyrimidine; guanine (G), a purine, always pairs with cytosine, a pyrimidine. Thus, purines always pair with pyrimidines.

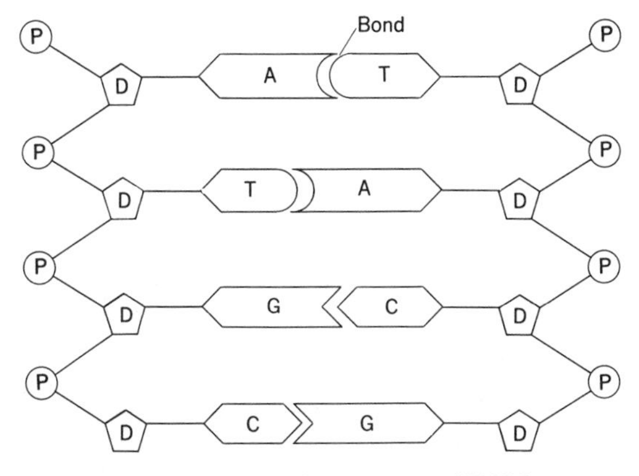

Figure 12–2 Structure of DNA

Weak bonds, called *hydrogen bonds*, join the purine of one chain to the pyrimidine of the complementary (partner) chain.

Structure of RNA

Unlike a DNA molecule, which usually is a double chain, a molecule of RNA is a single chain. Although the backbone of an RNA molecule is similar to that of a DNA molecule (Figure 12–3), the sugar in RNA is

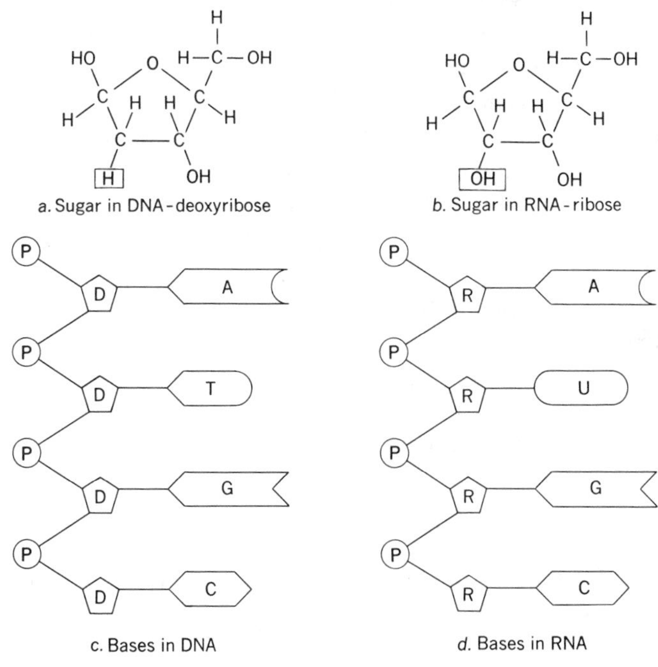

Figure 12–3 Comparison of DNA and RNA

Table 12–1 Comparison of DNA and RNA Molecules

	Number of Chains	*Sugar*	*Bases*	*Phosphate*	*Location*
DNA	2	Deoxyribose	A,T,G,C	Same as RNA	Chromosomes, mitochondria, plasma membrane, chloroplasts
RNA	1	Ribose	A,U,G,C	Same as DNA	Nucleoli, cytoplasm, ribosomes

ribose (R), which has one more oxygen atom per molecule than deoxyribose (D) (Figure 12–3a and b). Also, in RNA, the base thymine (T) is replaced by the base uracil (U) (Figure 12–3c and d). Table 12–1 summarizes the similarities and differences between DNA and RNA.

Sequences of Bases

The nitrogenous bases along a polynucleotide chain occur in specific sequences. The chromosomes of one species will have a characteristic sequence of bases that is not found in any other species. However, most organisms possess some sequences in common, enabling them to perform similar life activities. These genetic similarities provide strong evidence of common ancestry among all living things. The nitrogenous bases in a polynucleotide chain (Figure 12–4a) of a particular species may be

$$\overline{\text{TGC}}\ \overline{\text{AAT}}\ \overline{\text{TCG}}\ \overline{\text{GTC}}$$

The sequence of bases in the complementary chain of the same DNA molecule is

$$\overline{\text{ACG}}\ \overline{\text{TTA}}\ \overline{\text{AGC}}\ \overline{\text{CAG}}$$

In a chain of RNA of the same species (Figure 12–4b), the base sequence is

$$\overline{\text{UGC}}\ \overline{\text{AAU}}\ \overline{\text{UCG}}\ \overline{\text{GUC}}$$

The complementary RNA chain is derived from the DNA chain

$$\overline{\text{ACG}}\ \overline{\text{TTA}}\ \overline{\text{AGC}}\ \overline{\text{CAG}}$$

(Note that in the RNA chain, a T is replaced by a U.)

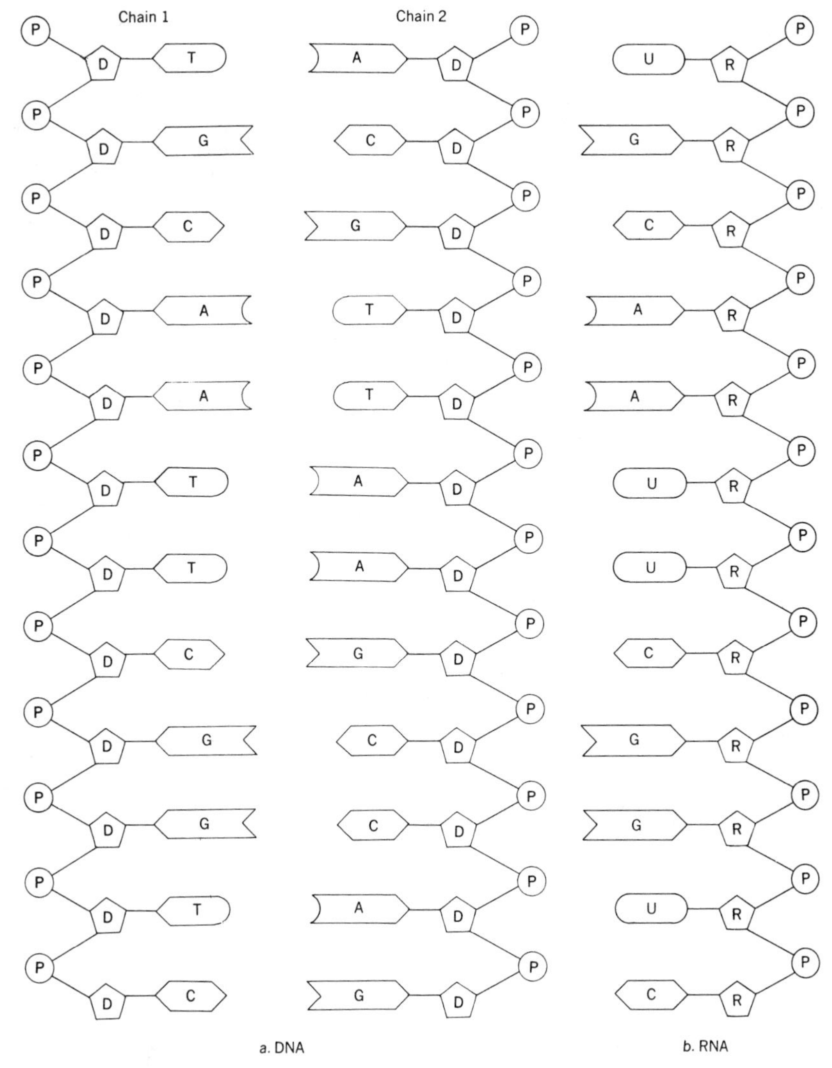

Figure 12–4 Codes of bases in DNA and RNA

The Language of a Cell

A cell "speaks" from its nucleus. Coded messages, or "words" emerging from a nucleus, control cell activities. Thousands of messages stored in the nucleus and its nucleic acids provide vital information, which tells a cell how to perform its life activities. For example, DNA directs a cell to: (1) synthesize enzymes and hormones; (2) extract useful organic chemicals from nutrients; (3) repair cell parts and tissues; (4) divide for growth and reproduction; (5) detect changes in both the internal and external environments and respond to them. When any one or a combination of these activities malfunctions, a cell usually weakens and dies. For example, unlike a white blood cell, which may live for years, a red blood cell only lives for about 120 to 130 days because its nucleus disintegrates shortly after the cell is formed.

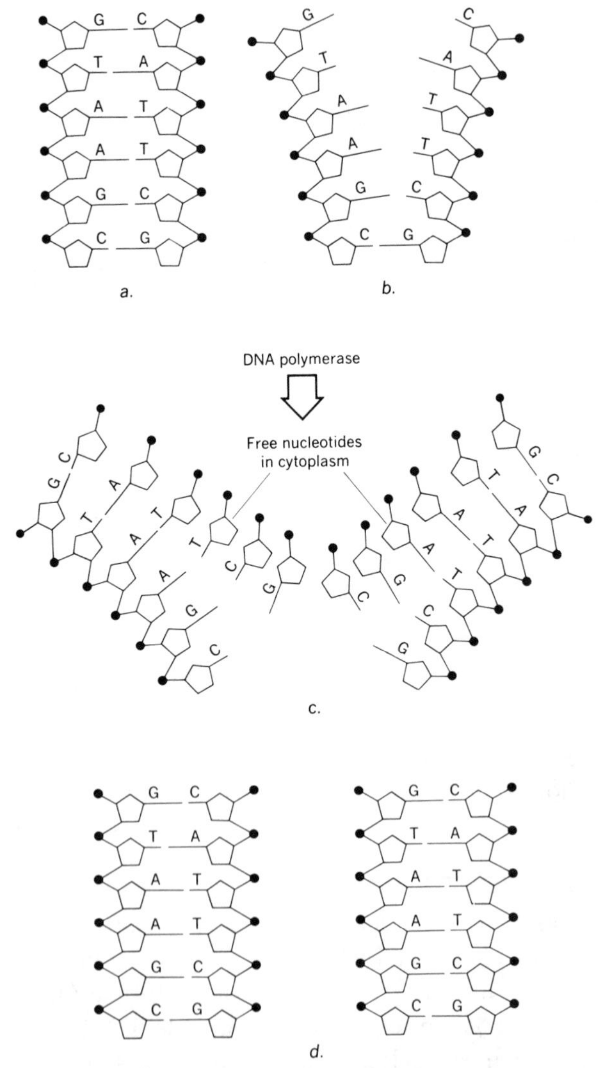

Figure 12–5 DNA replication

DNA Replication

When a cell divides, DNA replication prepares duplicated genetic material for distribution to its two daughter cells. Each daughter cell must receive an identical set of instructions in order to retain the genetic identity of the parent cell. DNA replication begins during the *interphase* stage of a cell's life cycle, or when the cell is maturing but not dividing. This stage often is referred to as the "resting" stage. During interphase, double-stranded DNA separates, producing two single strands of DNA. Then, each single strand replicates, forming two sets of double-stranded DNA. Each new set is identical to the parent strand, as it contains the same number and sequence of bases. The sequence of the replication process is described below:

1. As a DNA molecule unwinds, an enzyme breaks the hydrogen bonds between the nitrogenous bases of the two strands of DNA. As a result, the complementary pair of bases separate. The free ends of each base (Figure 12–5a and b) are then exposed and project from the separated chains of DNA.

2. An enzyme, called *DNA polymerase*, gathers newly formed, unattached nucleotides in the cytoplasm and attaches them to their complementary bases on each single chain of DNA. For example, the nucleotide P-S-*A* of a DNA chain is attracted to an unattached P-S-*T* nucleotide. And a DNA chain's nucleotide P-S-*C* is attracted to an unattached P-S-*G* nucleotide. Then, a bond forms between the sugars (S) and phosphates (P) of the newly attached nucleotides. DNA polymerase joins these sugars and phosphates to form the new chain of nucleotides.

Each separated chain of DNA functions as a template for the formation of its complementary chain. When the original DNA chain unwinds, each chain forms a new complementary chain alongside (Figure 12–5d). Thus, when replication is completed, two identical ladder-like DNA chains have been formed.

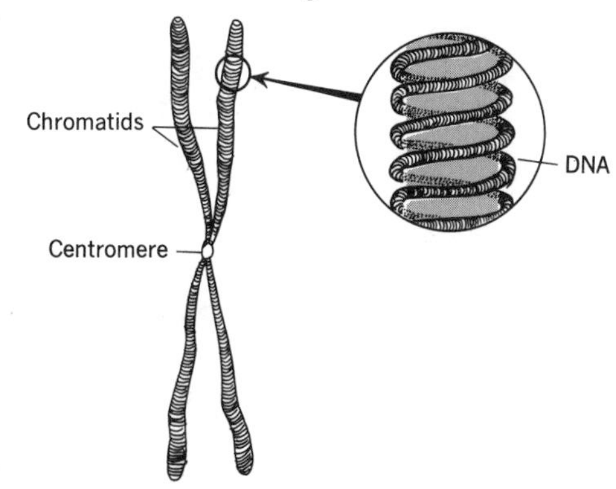

Figure 12–6 Structure of chromosomes

Although DNA replication cannot be observed under a microscope, the process has been verified by chemical studies. It is possible to observe the resulting double strands of chromosome material. Each strand is called a *chromatid*, and is connected to its "sister" chromatid by a *centromere* (Figure 12–6).

DNA replication provides a cell with a double set of genetic information. The cell is then prepared to begin *mitosis*, the process by which equal amounts of DNA are distributed to each daughter cell. The stages of mitosis are discussed later in this chapter.

Transcription Transmits Information

Messages and directions within a cell are sent from its genetic material. For a cell to "understand" a direction or message from its DNA, the message must be converted into "words" that the cell's metabolism

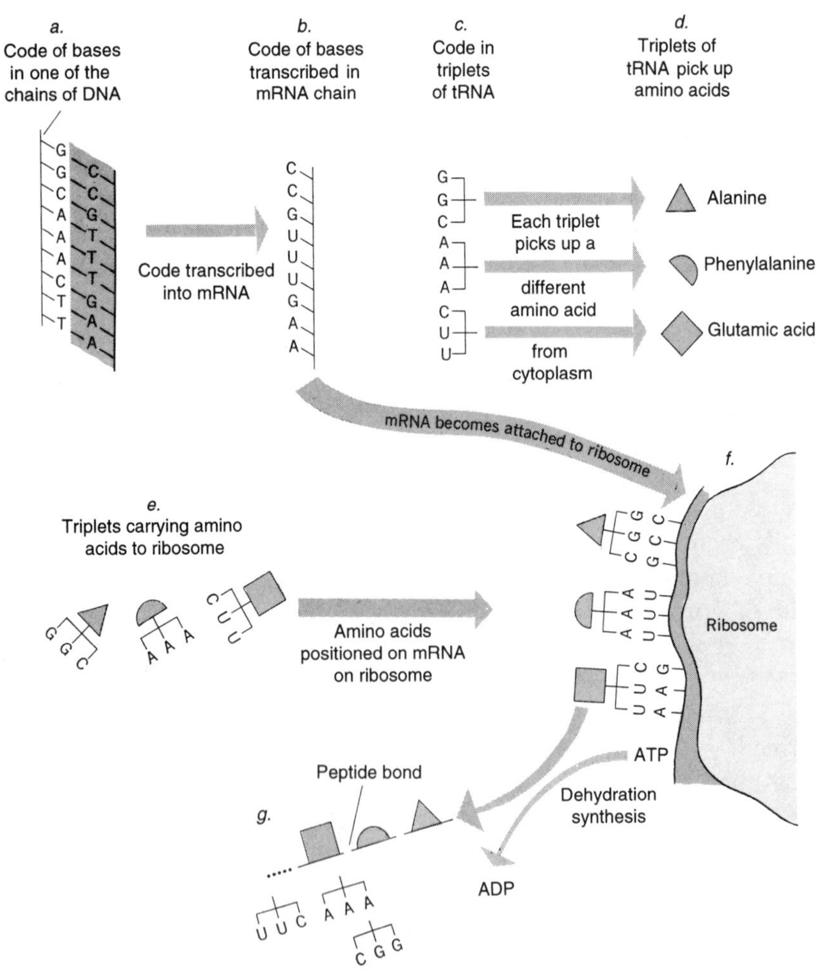

Figure 12–7 Scheme of events in protein synthesis

understands. This dialogue between the nucleus and the rest of a cell is called transcription.

Transcription transfers information (directions) from DNA to RNA. Base-pairing occurs, which is similar to DNA replication. However, a uracil nucleotide (P-S-*U*) replaces a thymine nucleotide (P-S-*T*). As a result, the genetic information in a DNA code is transcribed to an RNA copy. Transcription uses three different polymerases, which work separately to form three different types of RNA: *messenger RNA* (*mRNA*), *transfer RNA* (*tRNA*), and *ribosomal RNA* (*rRNA*). Figure 12–7 illustrates the processes of transcription and *protein synthesis.*

The codes in DNA and RNA are based on groups of three of the four nitrogenous bases in these molecules. Each group of bases in RNA is called a *triplet*, which represents the "words" spoken by DNA. Each triplet codes for one of the 20 amino acids available in the cytoplasm (see Table 12–2). For this reason, a triplet is referred to as a *codon*. Each RNA sequence consists of a series of codons. A DNA sequence of GGC AAA CTT is complemented by the RNA codon sequence CCG UUU GAA. Because they are arranged in a prescribed order, codons can be understood by the cell's synthesizers (ribosomes) for protein synthesis. The codons UAA, UAG, and UGA are stop signals that instruct enzymes to stop synthesis. The codon AUG functions as a *start* signal for *translation*, or the action of genes in protein synthesis. The current definition of *gene* is a series of codons that contains instructions for building a specific protein at a ribosome. Messenger RNA (mRNA) is a single-stranded copy of a gene that "tells" a ribosome the order in which amino acids should be joined to form a particular protein. The information is given in sequences of codons, or nucleotide triplets. Other codons start and stop the process of protein synthesis.

Table 12–2 Some tRNA Triplets and Their Amino Acids

Note: Only one triplet is listed for each amino acid. Other sequences of triplets also code for the same amino acid.

tRNA Triplet Code	*Amino Acid*	*tRNA Triplet Code*	*Amino Acid*
GGC	Alanine	UGG	Histidine
GCG	Arginine	UUU	Lysine
AAC	Cysteine	AAA	Phenylalanine
CUU	Glutamic acid	GGG	Proline
ACC	Glycine		

Transfer RNA (tRNA) is a single strand of nucleotides with bases that are complementary to an mRNA strand. For example, the mRNA sequence CCG UUU GAA appears on transfer RNA (tRNA) as GGC AAA CUU. Transfer RNA molecules pick up designated amino acids and transfer them to a ribosome for protein synthesis. A triplet nucleotide sequence of tRNA that complements a codon on mRNA is called an *anticodon*.

Translation Synthesizes Proteins

Translation is the process of gene action (protein synthesis) at a ribosome. [The function of ribosomal RNA (rRNA), which is involved in the process, is still unknown.] As the tRNA brings amino acids to a ribosome, the messenger RNA (mRNA) slips between the two parts of the ribosome to deliver its protein synthesis instructions. The ribosome "reads" mRNA's order and "tools up" its machinery for synthesizing amino acids and making polypeptides and proteins. At the ribosome, mRNA meets tRNA carrying certain amino acids, and the anticodons of tRNA join with the codons of mRNA. Peptide bonds join amino acids to form polypeptides and, finally, proteins. The proteins are then released into the cytoplasm (see Figure 12–8).

An mRNA strand can be read, or used, many times. Ribosomes function rapidly to join amino acids. Studies have shown that a ribosome in the bacterium *Escherichia coli* can join about 18 amino acids per second.

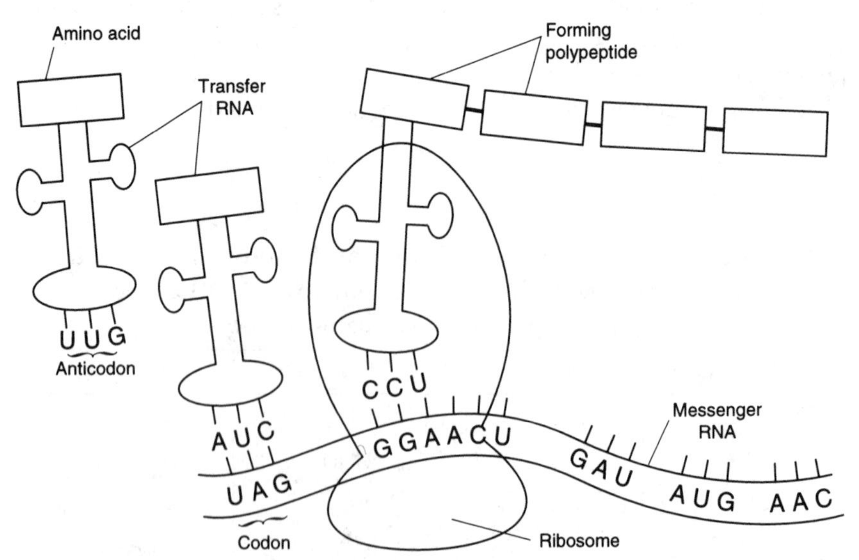

Figure 12–8 Model of translation at a ribosome

Clearly, ribosomes have enormous potential for manufacturing proteins. The following is a summary of protein synthesis:

1. Transcription of DNA to mRNA, tRNA, and rRNA.
2. The mRNA moves to ribosomes and attaches there.
3. The tRNA selects particular amino acids from the cytoplasm and moves to ribosomes.
4. At ribosomes, mRNA and tRNA join, codon to anticodon, and translation takes place.

The genetic information in the nucleus of a parent cell doubles before the cell divides into daughter cells. The following section will discuss how this DNA is distributed during cell division.

Section Quiz

1. Which nucleotide is *not* present in a DNA molecule? (*a*) P-S-A (*b*) P-S-T (*c*) P-S-U (*d*) P-S-G.
2. A nucleotide sequence UGC AAU UCG GUC is complementary to (*a*) ACG TTA AGC CAG (*b*) TCG UUA UGC CAG (*c*) TCG UUA AGC CAG (*d*) ACG TTA TGC CAG
3. Ribonucleic acid (RNA) receives messages regarding protein synthesis from (*a*) ribosomes (*b*) polymerases (*c*) nitrogenous bases (*d*) deoxyribonucleic acid (DNA).
4. A codon that functions as a stop signal is (*a*) AUG (*b*) GAU (*c*) UUU (*d*) UAG.
5. Which statement is correct? (*a*) mRNA is derived from tRNA. (*b*) tRNA is derived from rRNA. (*c*) rRNA is derived from tRNA. (*d*) mRNA, tRNA, and rRNA are derived from DNA.

Cell Division in Prokaryotic Organisms

Although bacteria do not have a true nucleus, they possess DNA. A special site, called the *replication origin,* is rich in certain enzymes that trigger DNA replication. Following replication, two copies of the original DNA are close together and attached to the plasma membrane. A new membrane then grows between each DNA copy. The accompanying growth of the cell wall causes the bacterium to divide. Cell division in bacteria, or *binary fission*, forms two similar daughter cells (Figure 12–9).

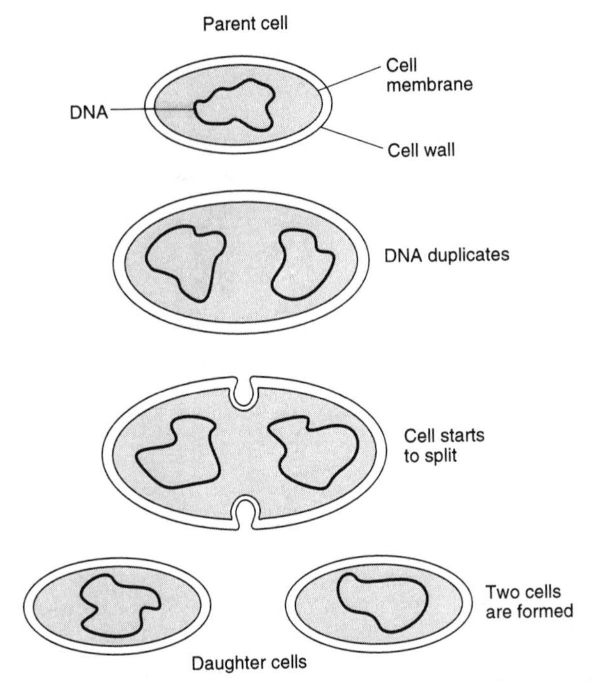

Figure 12–9 Cell division in (a prokaryotic) bacteria

Cell Division in Eukaryotic Organisms

The DNA of eukaryotic organisms is more abundant and complex than bacterial DNA. The DNA usually is contained in sets of chromosomes, one set inherited from the male parent and the other from the female parent. In general, chromosomes are composed of about 60 percent DNA and 40 percent protein. Cell division in eukaryotic cells is a complex process, called *mitosis*. That process replicates chromosomes, providing each daughter cell with the parental number of chromosomes. In humans, the parental number is 46. After mitosis, each daughter cell possesses two sets of 23 chromosomes. Cytoplasmic division, called *cytokinesis,* occurs after mitotic division of the nucleus.

Sexual reproduction involves *meiosis,* a process that halves the parental number of chromosome sets in each daughter cell. Thus, in humans, each sperm cell or egg cell contains 23 chromosomes. Meiosis is discussed more fully on pages 309–312.

Chromosomes. Although one species may have the same number of chromosomes as another, the DNA in the chromosomes usually is quite different. For example, a corn plant, water flea (daphnia), and mealworm

Table 12–3 Chromosome Number in Some Eukaryotes

Organism	*Chromosomes*	*Organism*	*Chromosomes*
Fungi (haploid)		Vertebrates (diploid)	
Penicillium	1–4	Opossum	22
Neurospora	7	Salamander	24
Saccharomyces	18	Frog	26
Insects (diploid)		Vampire bat	28
Mosquito	6	Lungfish	38
Housefly	12	Mouse	40
Honeybee	32	Rat	42
Silkworm	56	Human	46
Plants (ploidy varies)		Chimpanzee	48
Garden pea	14	Cow	60
Corn	20	Horse	64
Bread wheat	42	Black bear	76
Sugarcane	80	Chicken	78
Horsetail	216	Duck	80

(beetle larva) all have 20 chromosomes in each body cell. However, the sizes, shapes, and DNA instructions differ in each organism.

At some point before mitosis begins, each chromosome makes a copy of itself, or replicates, and becomes a doubled chromosome. The original chromosome and its copy are attached to each other by a *centromere* (refer back to Figure 12–6) and are called *chromatids* until they are separated into the two daughter cells. Chromosomes always appear in two sets: humans have two sets of 23; dogs have two sets of 39; fruit flies have two sets of 4; and tobacco plants have two sets of 24 (see Table 12–3). *Homologous chromosomes* are the paired chromosomes of each set. Homologous chromosomes are alike in size, shape, and DNA composition.

A scientist can observe and count chromosomes by preparing a *karyotype.* This procedure involves halting cell division (using chemicals), photographing, and then enlarging the image of the chromosomes.

Homologous chromosomes are then cut out of the photograph, displayed, and numbered. In Figure 12–10, showing a human karyotype, you can see that human body cells contain 22 pairs of homologous chromosomes, called *autosomes,* and one pair of *sex chromosomes.* In females, the sex chromosomes are alike; in males, they are different.

Karyotypes are useful in detecting genetic abnormalities such as *Down syndrome,* or *trisomy 21*, which is caused by an extra copy of chromosome 21. A pregnant mother can undergo genetic testing and discover whether or not her unborn child is afflicted with this disorder.

The following section of this chapter will discuss the process of mitosis in detail. It is important to remember that mitosis is successful when (1) the poles at each end of a cell and a microtubule spindle connecting the poles are formed, and (2) the shortening of spindle fibers causes the chromatids to separate and move into daughter cells.

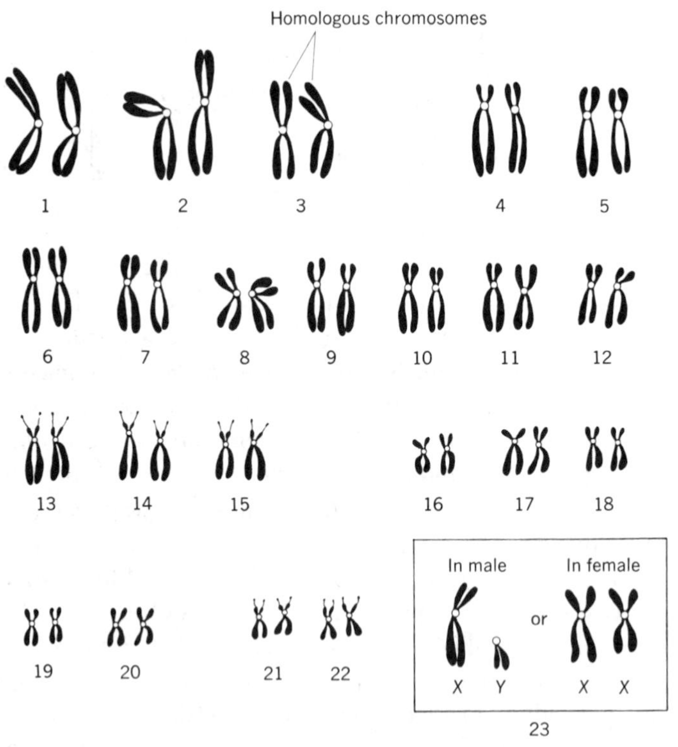

Figure 12–10 Human chromosomes (23 pairs)

Mitosis

The life cycle of a cell consists of two phases: (1) *mitosis,* called *phase M,* in which nuclear and cytoplasmic components divide; and (2) *interphase,* composed of subphases represented as G_1, S, and G_2 (see Figure 12–11). The following summarizes the events of interphase:

- G_1 is the period, or first *gap*, between DNA synthesis and mitosis.
- S is the period in which a cell grows, increases in size, and synthesizes DNA.
- G_2 is the period, or second *gap,* before mitosis; new DNA has been prepared and the cell is ready to divide.

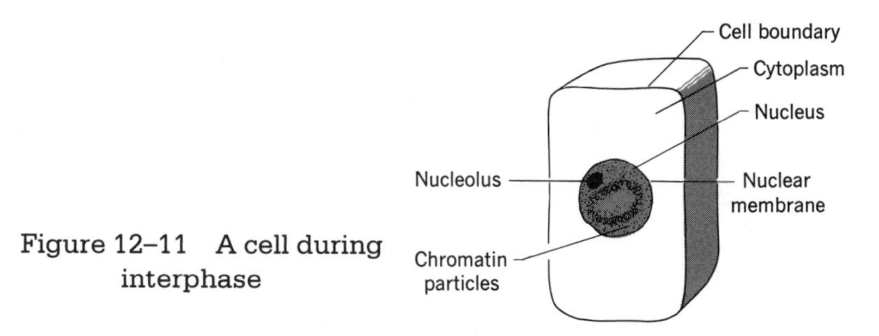

Figure 12–11 A cell during interphase

G_1 represents the beginning of interphase, G_2 represents the end, and S is the period between G_1 and G_2. The amount of time required for completion of the three subphases of interphase varies in different species. For example, a human skin cell takes about 24 hours to complete its life cycle, that is, to go from G_1 through actual mitosis. A yeast cell takes about two hours to complete its life cycle. Some cells in woody plants take 10 to 20 years to complete their life cycles.

Actual cell division, or mitosis, has four phases: *prophase, metaphase, anaphase,* and *telophase*. The phases are similar, but not identical, in plant and animal cells.

Mitosis in a plant cell. Refer to Figure 12–12 as you read about the stages of mitosis in a plant cell.

1. *Prophase.* During prophase, DNA strands condense into short, stubby rods, or chromosomes. These replicate, forming sister chromatids, which are connected by the centromere. The nuclear membrane and one or more nucleoli disintegrate, and fibers of cytoplasm radiate from opposite poles to form a *spindle.* The disintegrated components of the nuclear membrane are assimilated into the endoplasmic reticulum.

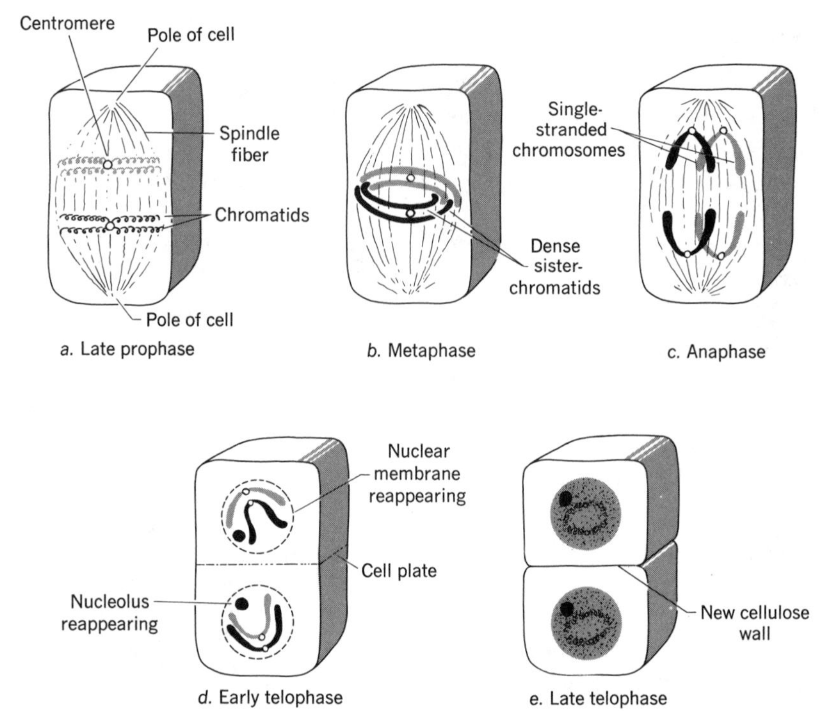

Figure 12–12 Stages in mitosis in a plant cell having two chromosomes

2. *Metaphase.* This phase of mitosis begins when all chromatid pairs are aligned in the center of the cell by spindle fibers. Each chromatid possesses a centromere and a *kinetochore* that attaches the chromatid to a microtubule. The chromatids line up in a circle perpendicular to the axis of the spindle, or *spindle equator.* By this time, spindle fibers extend from pole to pole. As a result, each chromatid is attached to one pole. At the end of metaphase, centromeres split, allowing the chromatids to move apart.

3. *Anaphase.* Now the chromatids are free (not in pairs), but still equipped with centromeres and kinetochores. Spindle fibers shorten, slide past each other, and push the poles apart. This causes the chromatids to move toward the poles. Research has shown that as spindle fibers shorten, they decompose at the poles. To demonstrate this, let a frankfurter represent a chromatid and a 30-cm length of string represent a spindle fiber. Make a knot to attach the "chromatid" to the "spindle fiber." Then attach the free end of the spindle fiber to a thumbtack, which represents a pole. Cut off a 3-cm length of spindle fiber at the pole. Again, attach the free end of the spindle fiber to the pole. The chromatid is now 3 cm closer to the pole. Repeat this procedure several times. You will no-

tice that the chromatid will draw closer and closer to the pole. This represents activity inside a cell during anaphase.

4. *Telophase.* At telophase, the main objective of mitosis has been achieved—replication of the cell's genetic material. Mitochondria, chloroplasts, and other organelles reappear and will be included in the daughter cells. After the chromatids separate from each other, they are called chromosomes. Cytoplasmic division, or *cytokinesis,* features the formation of a *cell plate* from vesicles produced by the Golgi apparatus. The vesicles migrate to the area of cell division and fuse, forming the cell plate. Then, a new plasma membrane and cell wall form next to the cell plate, which completes the formation of each daughter cell. The space between the daughter cells is called the *middle lamella.* Mitosis can be symbolized by the following: $G_1 \rightarrow S \rightarrow G_2 \rightarrow M \rightarrow C$, where C equals cytokinesis.

Mitosis in an animal cell. As you read about the stages of mitosis in an animal cell, refer to Figure 12–13. Unlike a plant cell, an animal cell lacks a cell wall and possesses a centrosome that encloses two centrioles. The structure of a centriole is discussed in Chapter 2. Mitosis in an animal cell repeats the $G_1 \rightarrow S \rightarrow G_2 \rightarrow M \rightarrow C$ cycle, but differs as centrioles divide

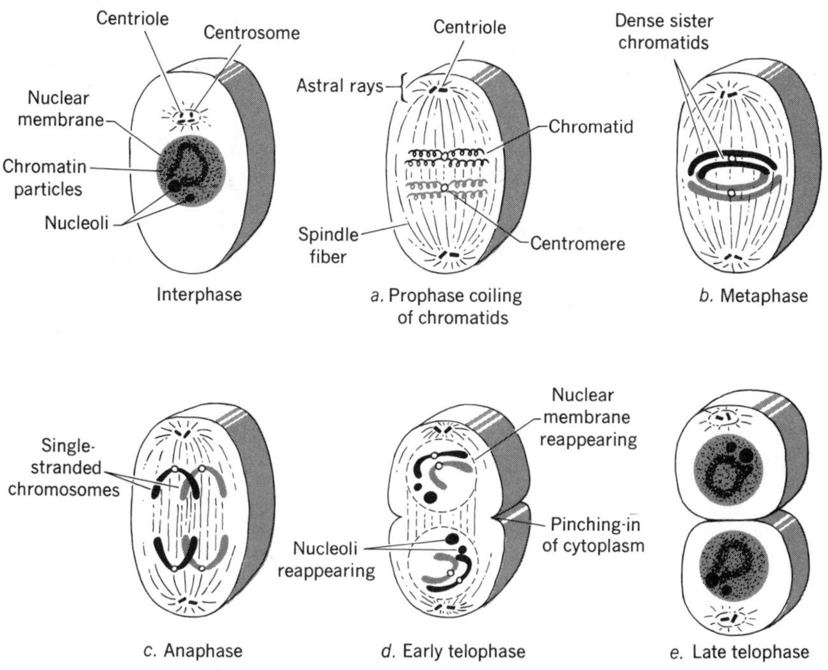

Figure 12–13 Stages in mitosis in an animal cell having two chromosomes

at the poles and serve as attachments for spindle fibers. Metaphase, anaphase, and telophase are similar in plant and animal cell mitosis. Cytokinesis in an animal cell involves a pinching-in of the cell membrane from opposite sides. This continues until complete separation of the cytoplasm takes place, resulting in two similar daughter cells.

The result of mitosis in plant and animal cells is the production and distribution of identical amounts of DNA to the daughter cells. In a unicellular organism, such as the paramecium, the division of a cell into daughter cells is the same process as reproduction. As in multicellular organisms, cell division includes mitosis and division of the cytoplasm. This type of reproduction is called asexual reproduction because only one parent and no specialized sex cells are involved. Asexual reproduction is discussed in Chapter 13.

Many organisms reproduce by sexual reproduction, in which two parents and specialized sex cells are involved. A different type of cell division—*meiosis*—is involved in the formation of sex cells.

Section Quiz

1. In plant cell mitosis, vesicles produced by the Golgi apparatus are building blocks of (*a*) spindle fibers (*b*) a cell plate (*c*) a new cell membrane (*d*) mitochondria.
2. Each chromatid is attached to the mitotic spindle at the (*a*) centrosome (*b*) centromere (*c*) kinetochore (*d*) cell plate.
3. The series of events $G_1 \rightarrow S \rightarrow G_2 \rightarrow M$ is followed by (*a*) cytokinesis (*b*) migration of centrioles (*c*) disappearance of nucleoli (*d*) DNA replication.
4. A chromatid becomes a chromosome at the (*a*) G_2 phase (*b*) G_1 phase (*c*) S phase (*d*) C phase.
5. Spindle fibers are formed from (*a*) microtubules (*b*) kinetochores (*c*) genomes (*d*) centrosomes.

Meiosis

Prokaryotic organisms such as bacteria reproduce by binary fission, a process that includes mitosis. In most cases, a new bacterium is identical to its parent cell. In effect, bacteria are usually *clones,* or a group of genetically identical cells descended from a single cell. Clones result from *asexual reproduction,* involving only one parent.

In sexual reproduction, a new organism forms from the union of a male *gamete*, or sperm cell, and a female gamete, or egg cell. The result of the union of two gametes is a fertilized egg cell called a *zygote*. The zygote undergoes a series of mitotic divisions, which results in the growth and development of an independent offspring. The normal number of chromosomes is called the *diploid* number, represented as $2n$. Gametes contain *one-half* the diploid number of chromosomes found in *somatic*, or body, cells. That number is called the *haploid*, or *monoploid*, number of chromosomes, represented as n. For example, the body cells of a dog contain 78 chromosomes; the sperm and egg cells of a dog contain 39 chromosomes each. In humans, the diploid number is 46; the monoploid number is 23. And in a pine tree, the diploid number is 24; the monoploid number is 12.

Meiosis is a process that extends over two cell divisions and reduces the diploid number of chromosomes to the monoploid number ($2n$ to n). Meiosis takes place in certain cells of the sex glands (ovaries and testes) of organisms that reproduce sexually. Meiosis maintains the species number of chromosomes by *reduction division*, or by producing monoploid (n) sex cells that unite in sexual reproduction to restore the diploid ($2n$) number of chromosomes characteristic of a species. The events of the two stages, called *meiosis I* and *meiosis II*, are presented in outline form. As you read on, refer to Figure 12–14.

Meiosis I

Prophase 1

1. During interphase and early prophase, each chromosome is duplicated as a result of DNA replication. Each chromosome is attached to its duplicate by a centromere; as in mitosis, the chromosomes in each pair are called *sister chromatids*.
2. Special molecules "cement" the paired chromatids, keeping them very close to each other as the result of a process called *synapsis*.
3. The DNA arrangement of one chromatid faces its corresponding DNA arrangement of the sister chromatid with little space between them.
4. *Nonsister* chromatids of homologous chromosomes, organized in a parallel arrangement called a *tetrad*, take part in *crossing-over* (Figure 12–15). That results in breaks at one or more sites along the length of the chromatids and corresponding parts are exchanged. Crossing-over produces new DNA combinations (genes) in chromatids that differ from those of the parent cells.

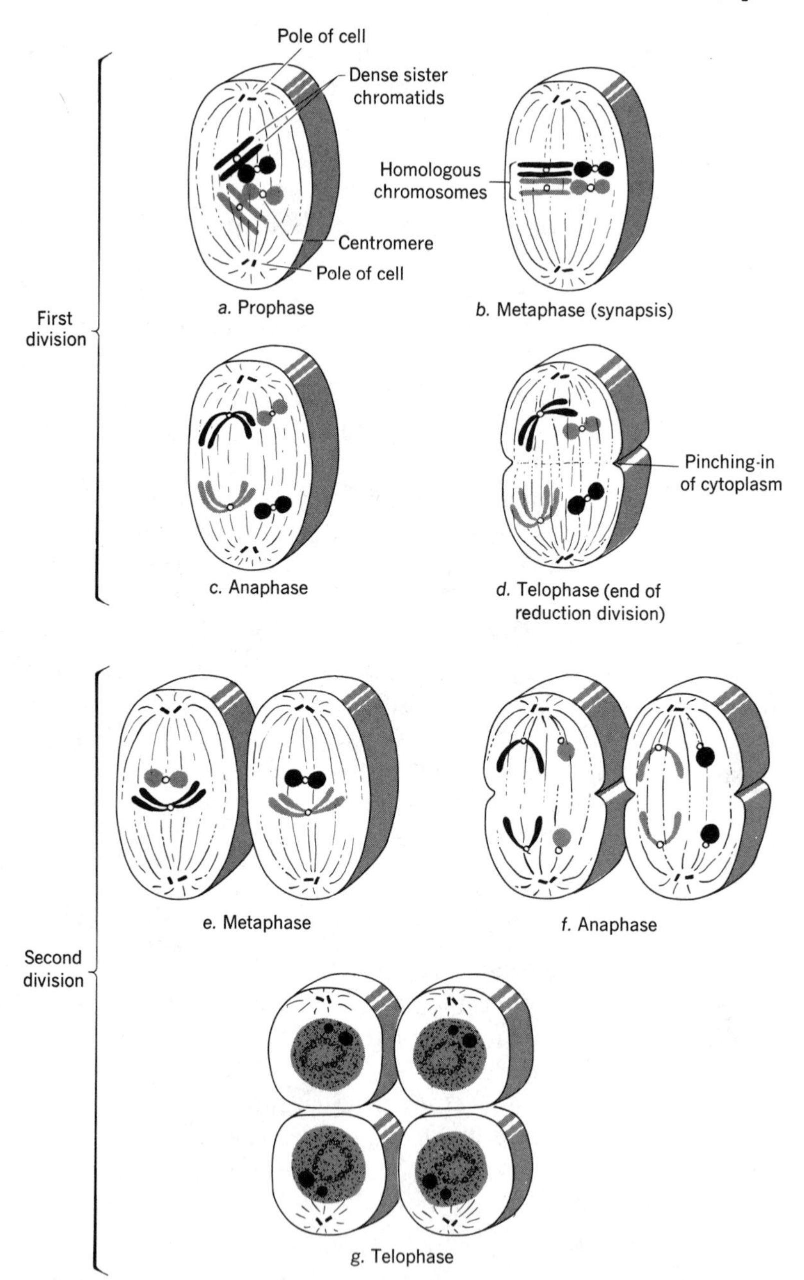

Figure 12–14 Stages in meiosis in an animal cell having four chromosomes

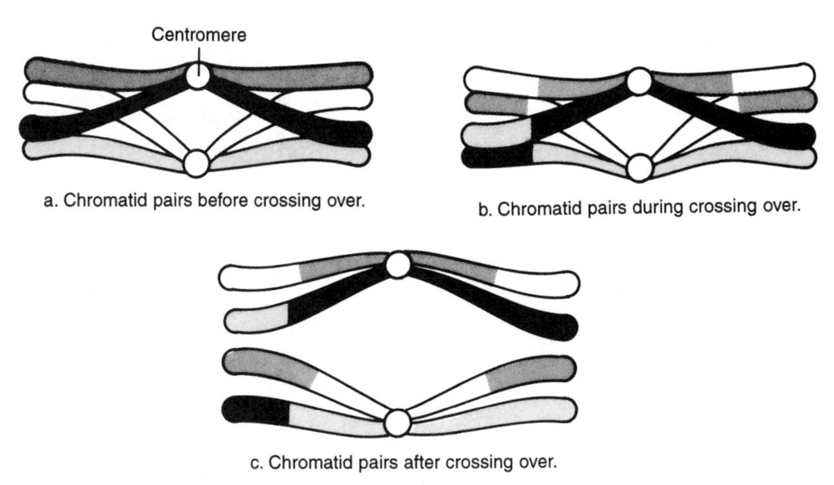

Figure 12–15 Crossing-over in chromatid pairs

Metaphase 1

1. The nuclear membrane and one or more nucleoli disappear; microtubules form a spindle extending to the poles; and centrioles, if present, divide and travel to the poles (as in mitosis).
2. Homologous chromatids are attached randomly to spindle fibers, which move them to align at the spindle equator.

Anaphase 1

1. Each homologue is pulled away from its partner and travels toward an opposite pole. Since each pair of homologues is positioned randomly at the spindle equator and their movements are random, one member of each pair will travel in a direction opposite to the other. As long as homologues move in opposite directions, it doesn't matter which member travels to which pole. This shuffling process, called *independent assortment*, is significant genetically because tetrad formation, crossing-over, and shuffling into position at the metaphase followed by separation at the anaphase produces an enormous number of possible gene recombinations. For example, with 23 pairs of human homologous chromosomes, it is possible to obtain millions of different combinations of genes in either a sperm or an egg. At anaphase, a pair of chromatids may fail to separate, leaving one daughter cell with an extra chromosome. This abnormal separation is called *nondisjunction*. Some effects of nondisjunction are discussed in Chapter 14.

Telophase 1

1. The first meiotic division is complete; reduction division has produced two monoploid (n) cells containing sister chromatids with DNA arrangements different from those of the parent cell.
2. The stage of cytokinesis, or separation of a cell into daughter cells, is similar in plant and animal cells.
3. The sister chromatids must be separated in meiosis II before they are called chromosomes.

Meiosis II

Prophase 2

1. DNA replication does not occur between the end of meiosis I and the onset of meiosis II.
2. Sister chromatids of each chromosome remain attached at a centromere. Spindle fibers and a spindle reform in the two monoploid daughter cells.

Metaphase 2

1. Chromatids are attached to spindle fibers, which move them to the spindle equator. Each chromatid is aligned with its partner, in preparation for separation.

Anaphase 2

1. Sister chromatids split, forming chromosomes.
2. Spindle fibers move the newly formed chromosomes to opposite poles.
3. Each potential daughter cell carries the monoploid number (n) of chromosomes characteristic of the species.

Telophase 2

1. Cytokinesis forms *four* daughter cells from the two cells produced by meiosis I.
2. Nuclear membranes and nucleoli reappear; spindle fibers disappear.
3. A gamete is formed—either a sperm cell or an egg cell.

Comparison of Mitosis and Meiosis

Figure 12–16 compares mitosis and meiosis. The differences between mitosis and meiosis are summarized in Table 12–4.

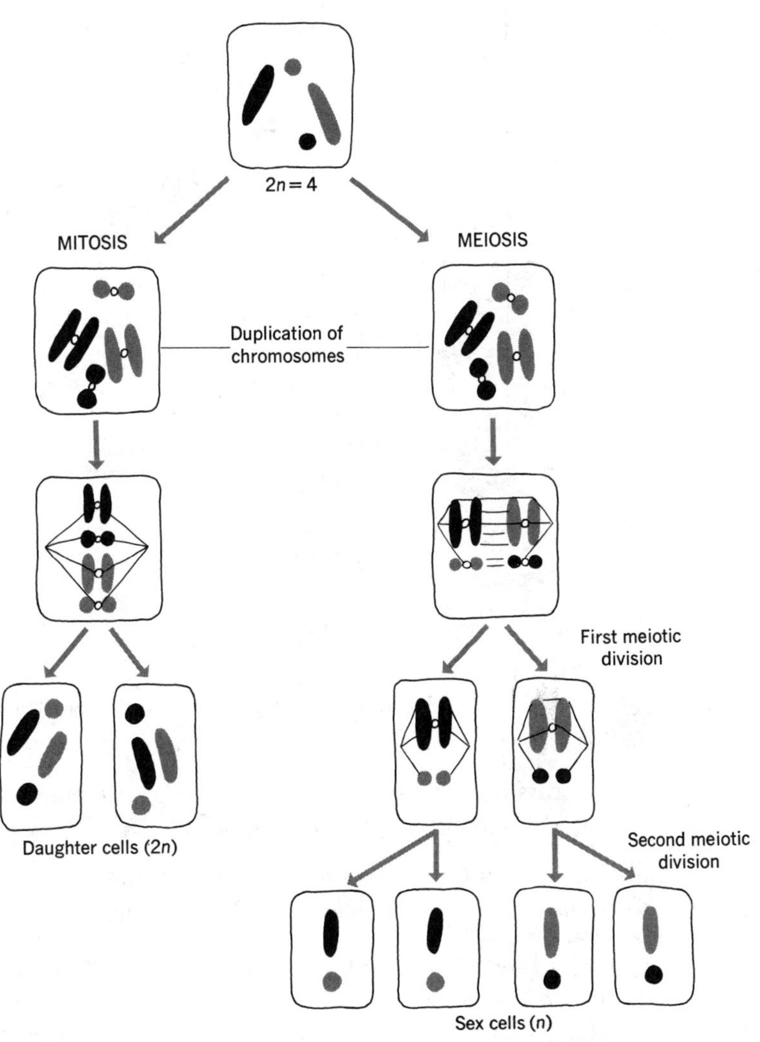

Figure 12–16 Comparison of mitosis and meiosis

Table 12-4 Summary of Differences Between Mitosis and Meiosis

Point of Comparison	*Mitosis*	*Meiosis*
Chromosomes in metaphase	Double-stranded chromosomes line up at equator of cell in *single* file	Double-stranded chromosomes line up at equator of cell in *double* file
Chromosomes at end of process	Species number (diploid) maintained in cells	One-half species number (monoploid) in cells
Number of cells formed from one cell	Two daughter cells	Four sperm cells in the male (Figure 12–17) One egg cell in the female (Figure 12–18)
Where the process takes place	In all organs	Only in certain cells of sex organs
Possibility of variations	Small, because the process maintains the identity of the chromosomes	Many, because chromosomes tend to break and change at the synapsis stage

Section Quiz

1. Which term best describes daughter cells formed by binary fission? (*a*) chromatids (*b*) gametes (*c*) clones (*d*) monoploid.
2. Synapsis and crossing-over usually occur during (*a*) meiosis II (*b*) prophase 1 (*c*) metaphase 2 (*d*) anaphase 1.
3. Independent assortment of genetic material occurs during (*a*) meiosis II (*b*) meiosis I (*c*) interphase (*d*) DNA replication.
4. A tetrad consists of four (*a*) chromosomes attached to each other by a kinetochore (*b*) monoploid cells formed by meiosis (*c*) chromatids attached to spindle fibers (*d*) chromatids attached to each other by a centromere.
5. In which stage of mitosis and meiosis is a spindle equator present? (*a*) metaphase (*b*) telophase (*c*) anaphase (*d*) prophase.

GAMETOGENESIS—FORMATION OF SEX CELLS

Meiosis produces the monoploid gametes in the sex organs. The general name for the process that produces mature sex cells is called *gametogenesis*. In males, sperm cells are produced by *spermatogenesis*; and in females, the process that produces *ova*, or egg cells, is called *oogenesis*.

Spermatogenesis

As you read about spermatogenesis, or sperm cell formation, refer to Figure 12–17, which outlines the process.

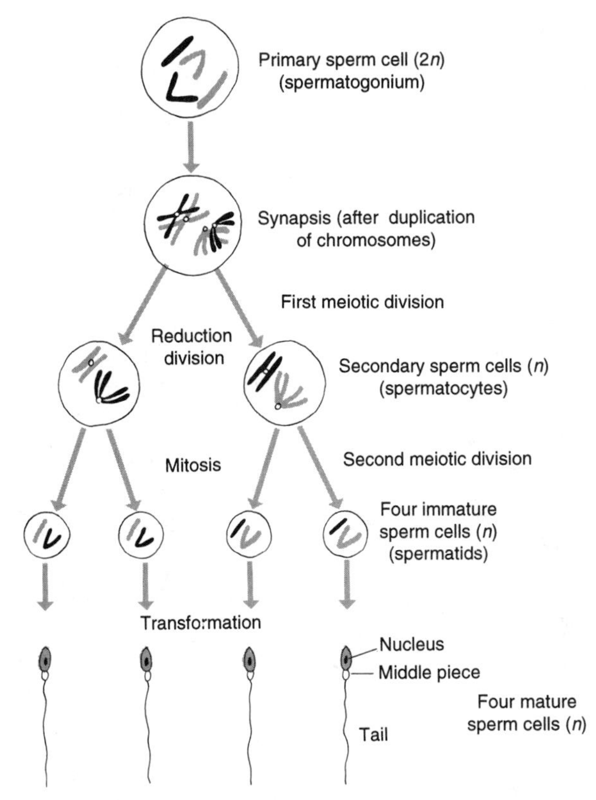

Figure 12–17 Schematic diagram of spermatogenesis

Spermatogenesis in most organisms follows the pattern of divisions discussed in the section on *meiosis*. In a human testis (testes, plural), cells called *spermatogonia* (spermatogonium, singular) divide many times, producing huge numbers of diploid *primary spermatocytes*. At maturity, these cells begin to divide, chromosomes pair off in synapsis, and then complete the first meiotic division (*reduction*). The resulting cells are called *secondary spermatocytes,* which contain the monoploid number of chromosomes. Then, a second meiotic division (*mitosis*) occurs, changing secondary spermatocytes to immature sperm cells, or *spermatids.* A process called *transformation* changes spermatids to mature, functional sperm cells.

Four sperm cells are formed from one primary spermatocyte. Spermatogenesis in humans takes about 10 weeks to be completed; different sections of a testis are involved and undergo spermatogenesis in rotation, which provides a continual supply of sperm cells.

Oogenesis

As you read about oogenesis, or egg cell formation, refer to Figure 12–18, which outlines the process.

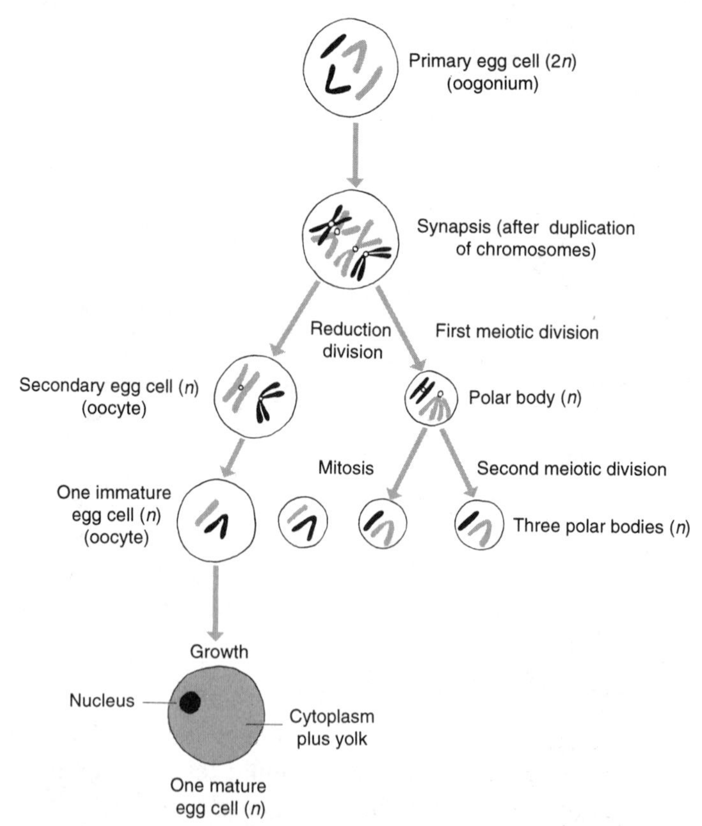

Figure 12–18 Schematic diagram of oogenesis

Oogenesis in most organisms follows the pattern of cell divisions outlined in *meiosis.* In humans, oogenesis includes the first meiotic division (*reduction*). Then, entrance by the head of a sperm cell triggers the egg cell's second meiotic division (*mitosis*).

At birth, a boy's testes are immature, which delays spermatogenesis for years. A baby girl, however, is born with about 400,000 primary oocytes! The immature sex cells develop from *oogonia* (oogonium, singular) within the ovary of a fetus *before birth.* About 500 *primary oocytes* normally develop into egg cells during a female's reproductive lifetime.

Primary oocytes remain diploid within an ovary until a girl attains sexual maturity, signaled by the onset of the first menstrual cycle. Menstruation triggers the first meiotic division (*reduction*), producing a monoploid *secondary oocyte* and a smaller cell called a *polar body.* Reduction division is followed by prophase of the second meiotic division (*mitosis*). The secondary oocyte remains in prophase even after it is released from the ovary. *Ovulation,* or release of an immature egg cell, occurs about every 28 days as part of the menstrual cycle.

Outside the ovary, the secondary oocyte remains in the prophase stage of mitosis unless it is penetrated by a sperm cell. The polar body completes mitosis without the need of a sperm cell; two smaller polar bodies are formed and accompany the secondary oocyte. If penetration by a sperm cell does not occur, the secondary oocyte and its two polar bodies die and disintegrate. If penetration does occur, the secondary oocyte resumes mitotic division, forming a larger cell and a third polar body. At that stage, a mature egg cell is ready to fuse its nuclear (monoploid) material with that of the sperm.

Comparison of Spermatogenesis and Oogenesis

1. In spermatogenesis, four functional sperm cells are formed from one primary sperm cell. In oogenesis, one functional egg cell is formed from one primary egg cell.
2. In spermatogenesis, polar bodies are not formed. In oogenesis, either two or three polar bodies are formed. The polar bodies do not take part in reproduction and usually disappear.
3. In both spermatogenesis and oogenesis, the diploid number of chromosomes is reduced to the monoploid number, which is the number characteristic of the species' sex cells.

4. Sperm cells formed in spermatogenesis move by a flagellum and lack stored food. Egg cells formed in oogenesis contain stored food and cannot move independently.

The information in this chapter will help you understand what happens to chromosomes (with their genes and DNA) during asexual and sexual reproduction, discussed in the following chapter.

Section Quiz

1. Sexual reproduction may result in variations among offspring because of (*a*) mitosis (*b*) meiosis (*c*) adaptations (*d*) cloning.
2. Polar bodies usually are present during (*a*) spermatogenesis (*b*) synapsis (*c*) oogenesis (*d*) reduction division.
3. In humans, the second meiotic division of oogenesis does not take place until (*a*) penetration by a sperm cell (*b*) a third polar body is formed (*c*) synapsis occurs (*d*) cytokinesis occurs.
4. In spermatogenesis, secondary spermatocytes develop into (*a*) sperm cells (*b*) spermatids (*c*) primary spermatocytes (*d*) monoploid spermatogonia.
5. Which statement is true? (*a*) An oocyte is a mature ovum. (*b*) An oocyte possesses the $2n$ number of chromosomes. (*c*) A spermatogonium is a mature sperm cell. (*d*) A spermatogonium possesses the $2n$ number of chromosomes.

Chapter Review Questions

The following questions will help you check your understanding of the material presented in the chapter.

1. The success of cells in passing on their characteristics to new cells depends upon the ability of (*a*) cells to make cytoplasm (*b*) cells to respond to changes in their environment (*c*) cells to grow (*d*) cell chromosomes to replicate.
2. The genetic code is most directly related to the sequence of (*a*) ribose units (*b*) deoxyribose units (*c*) nitrogenous bases (*d*) phosphates.

3. The sequence of purines and pyrimidines in a segment of a DNA chain are cytosine, guanine, adenine, thymine, adenine. The sequence in a complementary strand of newly made messenger RNA would be (*a*) cytosine, uracil, adenine, guanine, uracil (*b*) guanine, cytosine, uracil, adenine, uracil (*c*) uracil, adenine, cytosine, uracil, guanine (*d*) guanine, cytosine, thymine, adenine, thymine.
4. When a molecule of DNA replicates, it unwinds and "unzips" along the (*a*) bonds between the deoxyribose and phosphate units (*b*) bonds between a phosphate group and a nitrogenous base (*c*) phosphate-to-phosphate bonds (*d*) hydrogen bonds between the base pairs.
5. The bases that are linked together in a double-chain DNA molecule are (*a*) always the same as those in RNA (*b*) held together by strong covalent bonds (*c*) held together by weak hydrogen bonds (*d*) bonded to phosphate groups.
6. Which nitrogenous base is found in RNA but not in DNA? (*a*) thymine (*b*) uracil (*c*) adenine (*d*) guanine.
7. Between which types of compounds in a double-chain DNA molecule must the bonds break before replication can take place? (*a*) sugar–phosphate (*b*) sugar–base (*c*) purine base–pyrimidine base (*d*) phosphate–base.

Base your answers to questions 8 *and* 9 on your knowledge of biology and on the diagram, which shows part of a DNA molecule about to replicate.

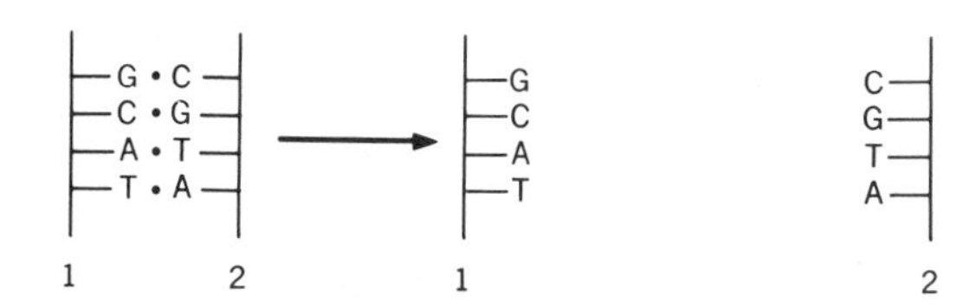

8. The two strands of DNA split because (*a*) hydrogen bonds are broken (*b*) they are separated by liquid currents (*c*) they are pulled apart by spindle fibers (*d*) the strands exchange material.
9. Strand 1 will attract free nucleotides with the sequence of bases represented by (*a*) TGAC (*b*) AGTA (*c*) CGTA (*d*) GCAT.
10. Within an organism, one cell has twice as much DNA as another cell. This is probably because the cell is (*a*) using energy (*b*) respiring (*c*) moving about (*d*) reproducing.

11. If the DNA code for the amino acid phenylalanine is AAA, the messenger RNA code for phenylalanine is (*a*) C-C-C (*b*) U-U-U (*c*) T-T-T (*d*) A-A-A.

Base your answers to questions *12 through 16* on the diagram below and on your knowledge of biology.

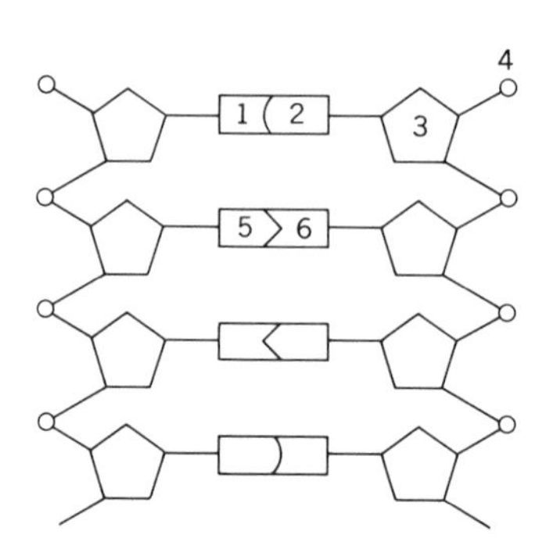

12. During mitosis, the molecule represented splits between (*a*) 1 and 2 (*b*) 2 and 3 (*c*) 3 and 4 (*d*) 1 and 5.
13. If nucleotide base 5 is adenine, nucleotide base 6 must be (*a*) adenine (*b*) guanine (*c*) cytosine (*d*) thymine.
14. The unit that is labeled 3 is probably a (*a*) ribose sugar (*b*) phosphate group (*c*) deoxyribose sugar (*d*) nitrogen compound.
15. A nucleotide contains the units numbered (*a*) 1, 2, and 3 (*b*) 1, 2, and 4 (*c*) 3, 4, and 5 (*d*) 2, 3, and 4.
16. If nucleotide base 5 was involved in the formation of an RNA molecule and base 5 was adenine, which base unit would be attracted to the site? (*a*) thymine (*b*) uracil (*c*) guanine (*d*) cytosine.
17. In a DNA molecule, the amount of guanine equals the amount of cytosine. This provides evidence in support of the fact that guanine and cytosine (*a*) are both nitrogenous bases (*b*) are paired in the nucleus (*c*) must neutralize each other (*d*) have the same chemical composition.
18. The coded message of a messenger RNA molecule is determined by the (*a*) amino acid sequence of an enzyme molecule (*b*) base sequence of a DNA molecule (*c*) monosaccharide sequence of a polysaccharide molecule (*d*) fatty acid sequence of a lipid molecule.

19. Mitosis in the cell of a hydra can be distinguished from mitosis in the cell of a bean plant because the bean plant cell lacks (*a*) cytoplasm (*b*) chromosomes (*c*) a nucleus (*d*) centrioles.

Base your answers to questions *20 and 21* on the cell division diagrams below.

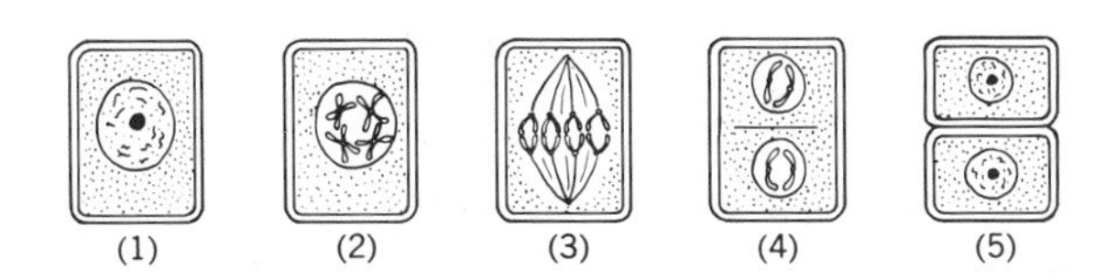

20. Which phrase best describes the cell division illustrated? (*a*) plant cell division, because of the presence of a centrosome and tetrads (*b*) animal cell division, because of the presence of centrosomes (*c*) plant cell division, because of the absence of centrioles and the presence of cell walls (*d*) animal cell division, due to the presence of asters and a temporary triploid condition.
21. Which would probably result if the cell were treated with colchicine, a chemical that induces polyploidy, during stage 1? (*a*) a doubling of the thickness of the cell membrane (*b*) a doubling of the chromosome number of the cell (*c*) the formation of a male gamete (*d*) the formation of a centrosome.
22. Which structure is observed during mitotic cell division in vascular plants but not in animals? (*a*) cell plate (*b*) centrosome (*c*) chromatid (*d*) spindle fiber.
23. After mitotic cell division, the number of chromosomes in each daughter cell compared to that of a parent cell is (*a*) the same (*b*) one-fourth as many (*c*) one-half as many (*d*) twice as many.
24. After the replication of chromosomes during mitosis, the chromosomes appear to be double-stranded. Each strand is known as a (*a*) centriole (*b*) centrosome (*c*) chromatid (*d*) spindle.
25. The process of crossing-over best explains how (*a*) linked genes become separated (*b*) albinism is maintained in a population (*c*) DNA is replicated (*d*) polyploidy in plants is controlled.
26. The diploid chromosome number in the cells of a certain plant is 14. How many chromosomes would a sperm nucleus in the pollen from this plant normally have? (*a*) 28 (*b*) 14 (*c*) 7 (*d*) 4.

27. How many sperm cells normally result when a single primary sex cell in the testis undergoes meiosis? (*a*) 1 (*b*) 2 (*c*) 3 (*d*) 4.
28. In humans, the nucleus of which cell would contain the same amount of DNA as the nucleus of a somatic cell? (*a*) mature egg (*b*) polar body (*c*) sperm (*d*) zygote.
29. If a species has 24 chromosomes in each muscle cell, how many chromosomes will a sperm cell contain? (*a*) 6 (*b*) 12 (*c*) 24 (*d*) 48.
30. If meiosis fails to occur during gamete formation in an organism normally containing 18 chromosomes in its body cells, the resulting polyploid cells would contain (*a*) 4.5 chromosomes (*b*) 9 chromosomes (*c*) 18 chromosomes (*d*) 36 chromosomes.
31. During synapsis in the first meiotic division, chromatids of homologous chromosomes overlap at various points. When disjunction occurs, portions of two chromatids may be exchanged with one another. This event is known as (*a*) crossing-over (*b*) polyploidy (*c*) nondisjunction (*d*) segregation.
32. In humans, meiosis is followed by an unequal division of the cytoplasm. This statement describes the process of (*a*) cleavage (*b*) pinocytosis (*c*) sperm cell formation (*d*) egg cell formation.
33. Meiosis is most closely associated with (*a*) sexual reproduction (*b*) vegetative propagation (*c*) budding (*d*) fission.
34. An organism that can successfully reproduce without undergoing meiosis is (*a*) an ameba (*b*) a dog (*c*) a grasshopper (*d*) a bird.

Biology Challenge

The following questions will provide practice in answering SAT II-type questions.

Part I

Base your answers to questions *1 through 5* on the following diagram, which shows a portion of a DNA molecule. In the diagram, numbers have been substituted for the letter symbols commonly used to represent different nitrogenous bases.

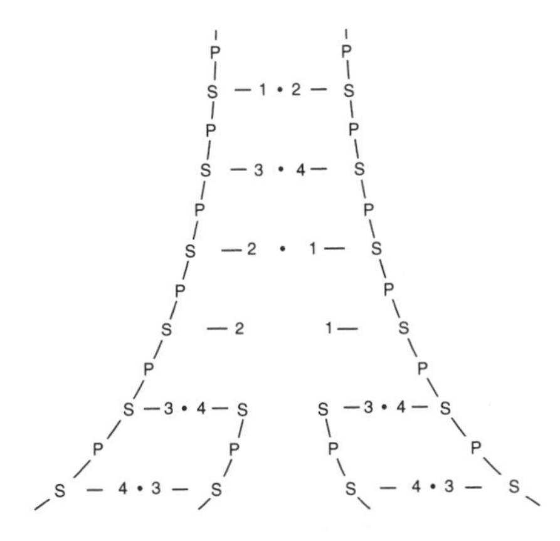

1. The process represented by the diagram is (*a*) deletion (*b*) replication (*c*) crossing-over (*d*) meiosis (*e*) nondisjunction.
2. Nitrogenous bases 1 and 2 are held together by (*a*) hydrogen bonds (*b*) peptide bonds (*c*) phosphate bonds (*d*) gene linkage (*e*) a centromere.
3. The process illustrated in the diagram normally results in the formation of (*a*) two identical molecules of mRNA (*b*) two dissimilar molecules of tRNA (*c*) one molecule of mRNA and one molecule of tRNA (*d*) two identical molecules of DNA (*e*) one molecule of DNA and one molecule of mRNA.
4. If a sequence of bases in this molecule is adenine, thymine, and cytosine, respectively, then the resulting sequence in a messenger RNA molecule will be (*a*) cytosine, thymine, adenine (*b*) guanine, adenine, thymine (*c*) uracil, adenine, guanine (*d*) cytosine, uracil, guanine (*e*) thymine, adenine, uracil.
5. The process represented by the diagram normally occurs during which stage in a cell's life cycle? (*a*) telophase (*b*) anaphase (*c*) prophase (*d*) interphase (*e*) synapsis.

Part II

A biology class studying cell division performed a laboratory investigation involving the following procedures:

1. A cotton-plugged test tube containing a nutrient broth culture of *Bacillus tumefaciens* was left standing for 48 hours to allow the bacteria to multiply and form a sediment.

2. The tube was unplugged and most of the fluid contents decanted into a jar of a germicidal solution, such as carbolic acid.
3. A chemical detergent was added to the bacterial sediment and mixed thoroughly.
4. Two droppers of 95 percent isopropyl alcohol were added to the detergent-bacteria mixture. A stirring rod was gently twisted in the mixture to gather coagulated material.
5. Two drops of the dye acridine-orange were added to the mixture and the stirring rod was twisted again in the mixture. The stained material clinging to the rod was then subjected to ultraviolet radiation. As a result, the material turned a characteristic color.

Use your knowledge of biology and the procedure described above to answer questions *1 through* 5.

1. The carbolic acid solution functions as a (an) (*a*) safety precaution (*b*) storage solution for nutrient broth (*c*) deodorant (*d*) emulsifier (*e*) pH adjustment.
2. An effect of the chemical detergent on bacteria is (*a*) removing wax from bacterial cell walls (*b*) breaking up bacterial cell walls and releasing cytoplasm and enzymes (*c*) breaking up bacterial cell walls and releasing DNA (*d*) inhibiting bacterial growth and multiplication (*e*) removing bacterial flagella.
3. The effect of adding alcohol solution to the mixture (*a*) dissolved lipids from the cytoplasm (*b*) removed phospholipids from the plasma (cell) membrane (*c*) neutralized bacterial antigens (*d*) coagulated strands of DNA (*e*) coagulated cytoplasmic proteins.
4. The stained stringy material clinging to the end of the stirring rod probably was (*a*) plasmids (*b*) DNA strands (*c*) bits of mitochondria (*d*) pieces of cell walls (*e*) neutralized antigens.
5. When stained with the dye acridine-orange and irradiated with ultraviolet light, DNA fluoresces (*a*) orange (*b*) green (*c*) blue (*d*) red (*e*) violet.

13

Reproduction: Asexual and Sexual

Learning Objectives

When you have completed this chapter, you should be able to:

- **Describe** methods of asexual reproduction in monerans, protists, fungi, and plants.
- **Describe** methods of asexual reproduction in animals.
- **Describe** methods of sexual reproduction in monerans and protists.
- **Discuss** the process of sexual reproduction in animals.
- **Identify** tissues and organs that are derivatives of the primary germ layers.
- **Describe** adaptations shown by mammals that contribute to the survival of the embryo.
- **Compare** the life cycles of representative animals.
- **Describe** the structure and function of the male and female human reproductive systems.
- **Describe** the process of sexual reproduction in plants.
- **Identify** ways in which seeds are dispersed.

OVERVIEW

In this chapter, you will learn how different organisms ensure the continuation of their species—generation after generation. You will learn about asexual and sexual reproduction, the two ways in which organisms reproduce—ways that contribute to a species' ability to survive under different environmental conditions.

ASEXUAL AND SEXUAL REPRODUCTION

In *asexual reproduction* only one parent is involved and no sex cells are required. Only mitotic cell divisions occur in asexual reproduction. As a result, a parent and its offspring usually have the same genetic (DNA) makeup.

In *sexual reproduction* two parents are involved, each contributing a sex cell, or *gamete.* The two gametes unite and form a single cell called a *zygote.* During meiosis, members of a pair of chromosomes tend to exchange genetic material before they separate. As a result, sex cells contain different sets of chromosomes and genes. The union of a sperm cell and an egg cell will produce gene combinations that differ from those of the parents. Sexual reproduction, more than asexual reproduction, enables a species to survive under changing conditions because the offspring usually have genetic makeups different from those of their parents and each other.

ASEXUAL REPRODUCTION IN MONERANS, PROTISTS, FUNGI, AND PLANTS

Methods of asexual reproduction in these groups include binary fission, budding, spore formation, and vegetative propagation.

Binary Fission

In *binary fission,* a parent cell divides to form two daughter cells, each of which is the same size and contains the same kind and quantity of DNA. Binary fission occurs among protozoans, bacteria, and many species of algae.

As you read about binary fission, refer to Figure 13–1, which shows this process in an ameba and a paramecium. First, the shape of the parent ameba becomes spherical. After the nucleus divides by mitosis, the cytoplasm pinches, which separates the nuclear material and the cytoplasm into two daughter amebas. Most species of paramecium have at least two nuclei, a larger nucleus called the *macronucleus* and a smaller nucleus called the *micronucleus.* When binary fission occurs, the micronucleus divides by mitosis, but the macronucleus merely lengthens and then separates into two pieces. Like an ameba, pinching in of the cytoplasm separates a parent paramecium into two daughter parameciums.

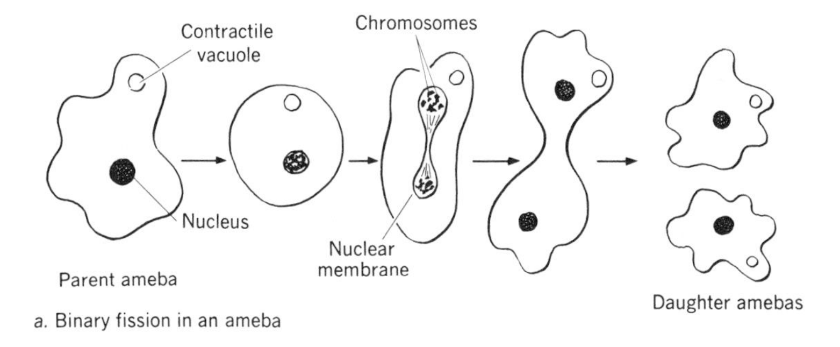

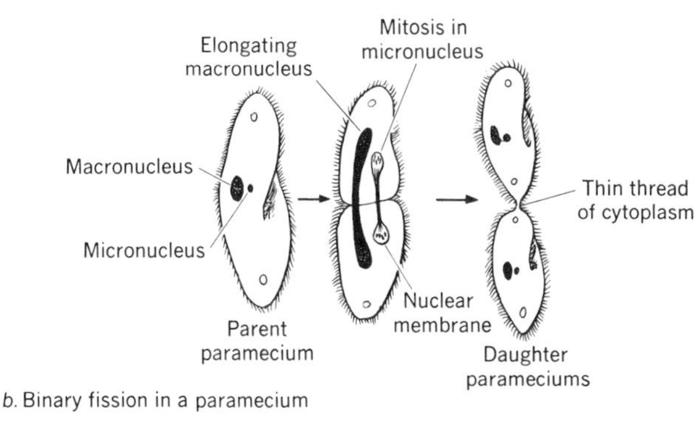

Figure 13–1 Binary fission

In bacteria and algae, there is no pinching in of the cytoplasm after the chromosomes have separated during mitosis. Instead, a new cell wall is formed in the middle of the parent cell, between the two masses of nuclear material formed by mitosis. This results in two daughter cells of approximately equal size and the same genetic makeup.

After binary fission, the parent cell no longer exists. Instead, there are two daughter cells, each about half the size of the parent cell.

Budding

In *budding,* mitosis provides each new daughter cell with the same kind and quantity of DNA as the parent cell. However, the cytoplasm divides unequally (Figure 13–2). The small cell, or daughter cell, is called a *bud.* It temporarily remains attached to the larger, parent cell.

Spore Formation

Spore formation is a common type of asexual reproduction in many fungi and plants, including yeast, bread mold, mushrooms, mosses, and

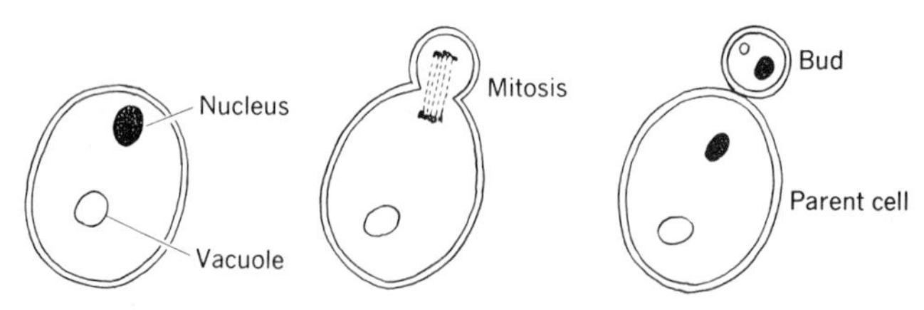

Figure 13–2 Budding in yeast

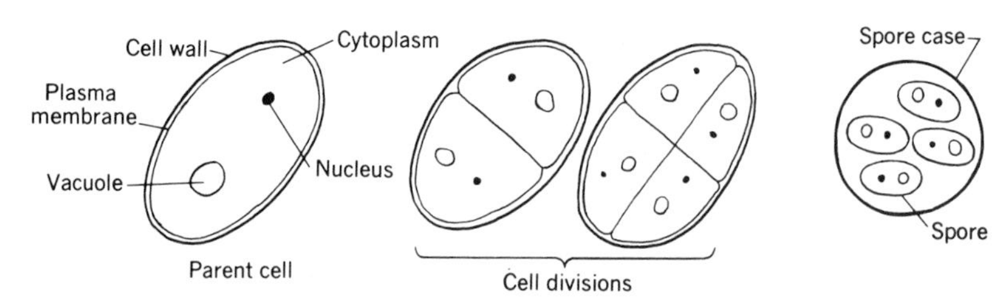

Figure 13–3 Stages of spore formation in yeast

ferns. A *spore* is capable of growing into a new organism. In spore formation, the nucleus and cytoplasm within a cell undergo a series of cell divisions, which produces many spores within the cell. The cell, now called a *spore case* (Figure 13–3), develops thick walls that are resistant to unfavorable conditions, such as lack of moisture and extreme heat and cold. Eventually the spores are liberated and scattered. Upon reaching a favorable environment, each spore grows into a new individual organism.

Vegetative Propagation

A structure that helps nourish a plant is called a *vegetative structure.* When an entire leaf, stem, or root—or a piece of one of these vegetative structures—develops into a whole plant, the process is called *vegetative propagation,* or *vegetative reproduction.* In this process, certain cells in either a leaf, stem, or root divide by mitosis and form cells that eventually develop into the tissues of a complete plant. The major types of vegetative propagation are illustrated and summarized in Figures 13–4 and 13–5.

Runners, rhizomes, layering, tubers, corms, and bulbs are mainly natural methods of vegetative propagation. Cutting and grafting, on the other hand, are artificial methods.

In *grafting,* a part (twig or bud) of a tree or shrub is placed in contact with a part (stem or root) of a different but related plant (Figure 13–5). The rooted plant is called the *stock.* The part that is attached to the stock is called the *scion.* In grafting, the cambium layer (growing layer) of the

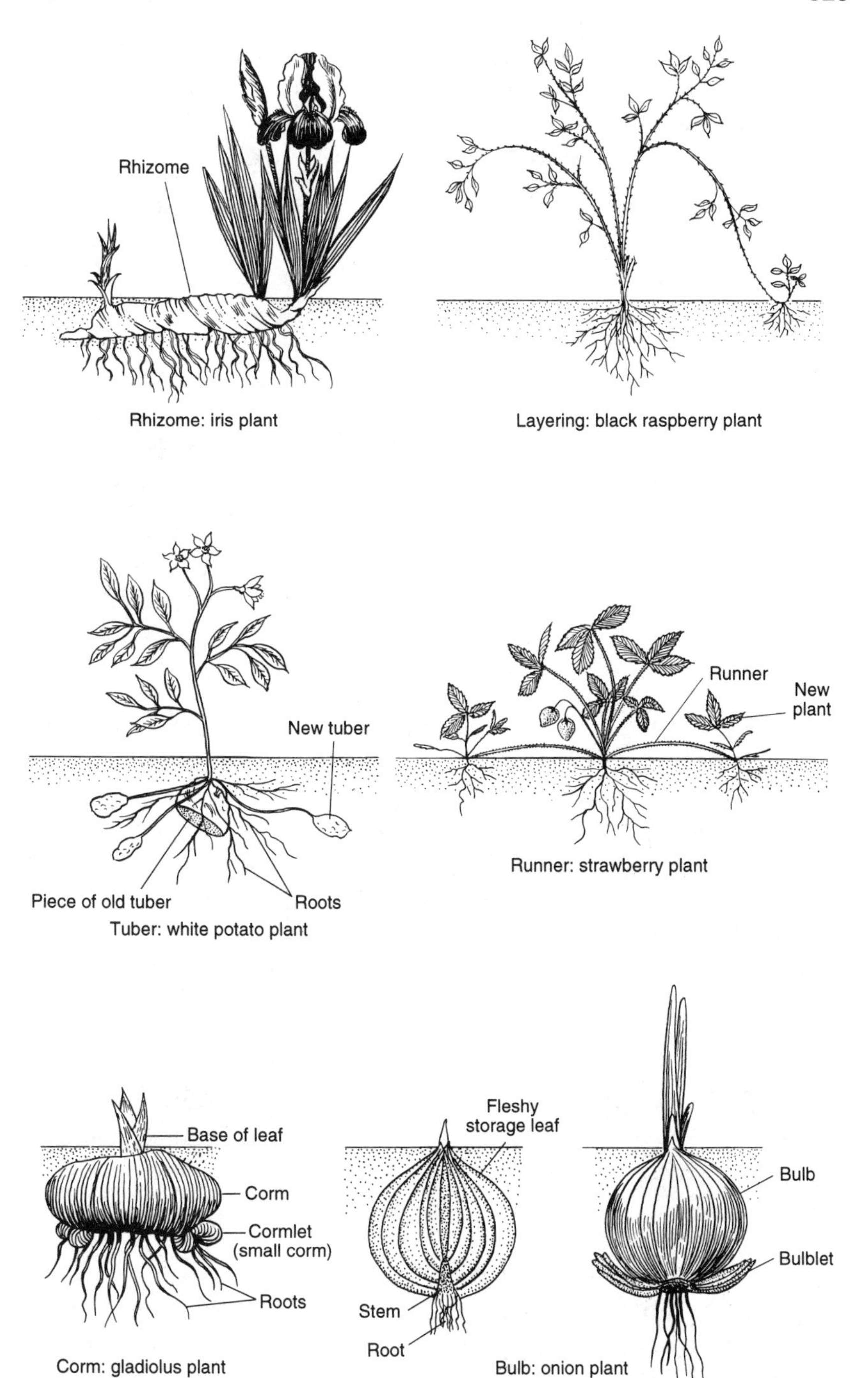

Figure 13–4 Vegetative propagation

scion is placed in contact with the cambium layer of the stock. As the cambiums grow, the joint heals and continuity is established in the xylem, phloem, and cambium of the scion and stock. Thus, if a twig from a Bartlett pear tree (scion) is grafted to a branch of an Anjou pear tree (stock), the scion will produce Bartlett pears. Any fruit that develops from the stock will be Anjou pears.

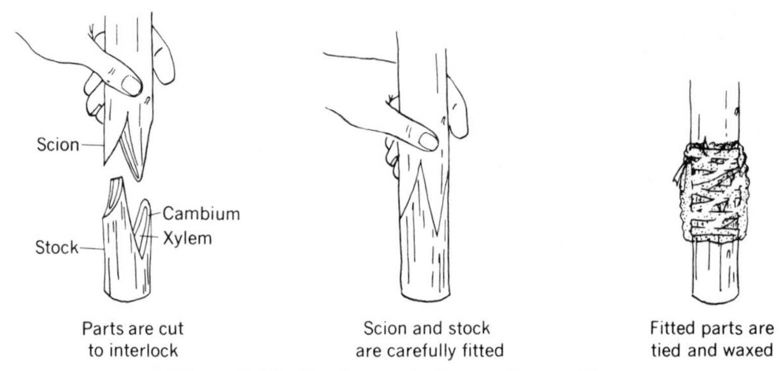

Figure 13–5 Steps in branch grafting

Grafting does not produce new varieties of plants. The main advantages of grafting are:

1. A desirable type of flower or fruit can be maintained because each member of a graft displays the characteristics of its own type. The stock functions only as support and as a source of nourishment for the scion.
2. Desirable seedless varieties can be reproduced.
3. Scions produce mature fruit sooner than trees that are grown from seeds.

ASEXUAL REPRODUCTION IN ANIMALS

Methods of asexual reproduction in animals include binary fission, budding, and regeneration.

Binary Fission and Budding

Figure 13–6 shows a planaria reproducing by binary fission and a hydra reproducing by budding. A planaria divides into two equal parts as a result of the pinching in of its body. The tail section grows a new head and the head section grows a new tail.

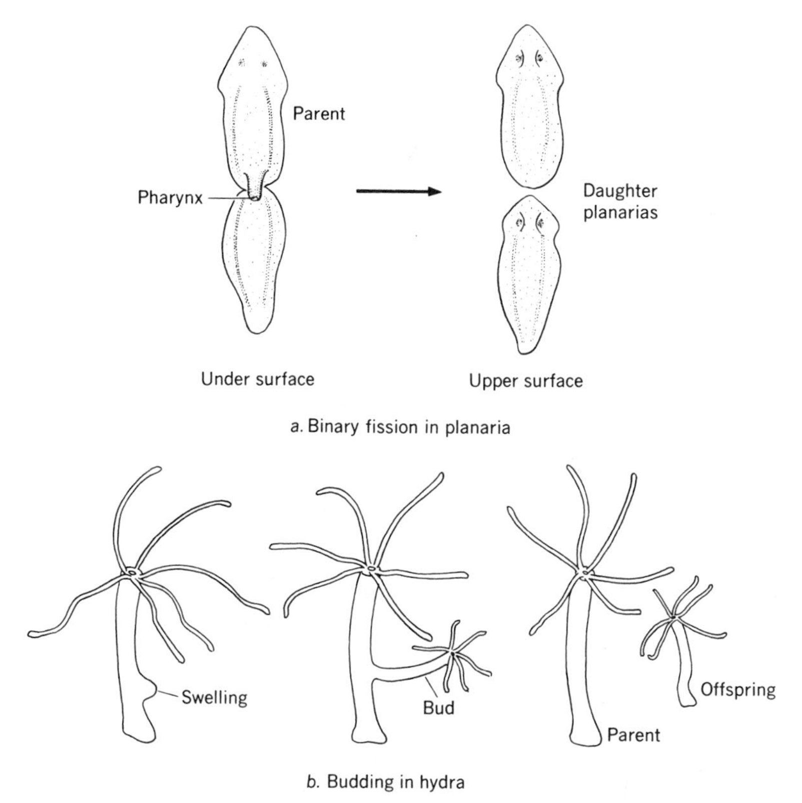

Figure 13–6 Asexual reproduction in some animals

In hydra, a bud begins as a group of cells that undergo repeated mitotic divisions and then increase in size. A mouth and tentacles soon develop at the free end of the bud. The gastrovascular cavity of the parent is continuous with that of the bud. When the bud is mature, it separates from the parent and lives as an independent organism.

Regeneration

The regrowth of a part of an animal is called *regeneration*. Some animals, such as hydra, planaria, starfishes, and crabs, can regenerate large sections of their bodies (Figure 13–7). Regeneration in animals is similar to vegetative propagation in plants.

The major advantage of asexual reproduction is that many identical offspring are formed in a short time. A species whose members produce large numbers of offspring with the same genetic makeup has a better chance to survive, provided the environment remains the same.

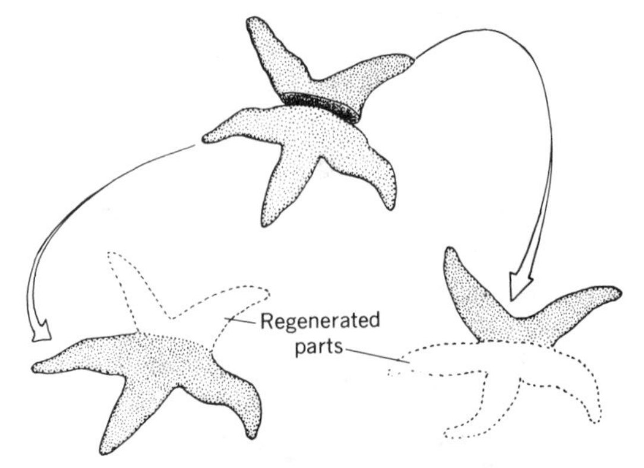

Figure 13–7 Regeneration in the starfish

SEXUAL REPRODUCTION IN MONERANS (BACTERIA)

Bacteria possess a single chromosome consisting of a circular DNA molecule. Bacterial DNA is relatively simple, containing a number of genes that suit the metabolic needs of a bacterium. Many bacteria also possess *plasmids*, small circular molecules of DNA or RNA that contain a few genes. The plasmids, called *F plasmids*, permit *conjugation*, in which a bacterium transfers DNA (or RNA) to another bacterium. The donor bacterium, called F^+, transfers DNA to a recipient, called F^-, which lacks a plasmid. The transfer of plasmid DNA or RNA is accomplished by a cytoplasmic bridge, called an *F filus*. Following transfer, the F^- bacterium (recipient) replicates new F^+ DNA (or RNA). Donor and recipient are now F^+ bacterial cells; the cells separate (Figure 13–8). Although conjugation in bacteria is a form of sexual reproduction, it is much less complicated than the joining of specialized sex cells of eukaryocytes. In Chapter 15, you will learn about the role of plasmids in genetic recombination techniques.

SEXUAL REPRODUCTION IN A PROTOZOAN

Paramecium is a ciliated protozoan commonly found in stagnant pools of water. It is a heterotroph that uses its thousands of cilia for swimming and sweeping food into its gullet. Paramecium reproduces asexually by binary fission and sexually by conjugation.

Figure 13–9, on page 334, shows the steps in conjugation of paramecium. Note that the two parameciums look alike; however, they are different chemically. Because of this chemical difference, the two

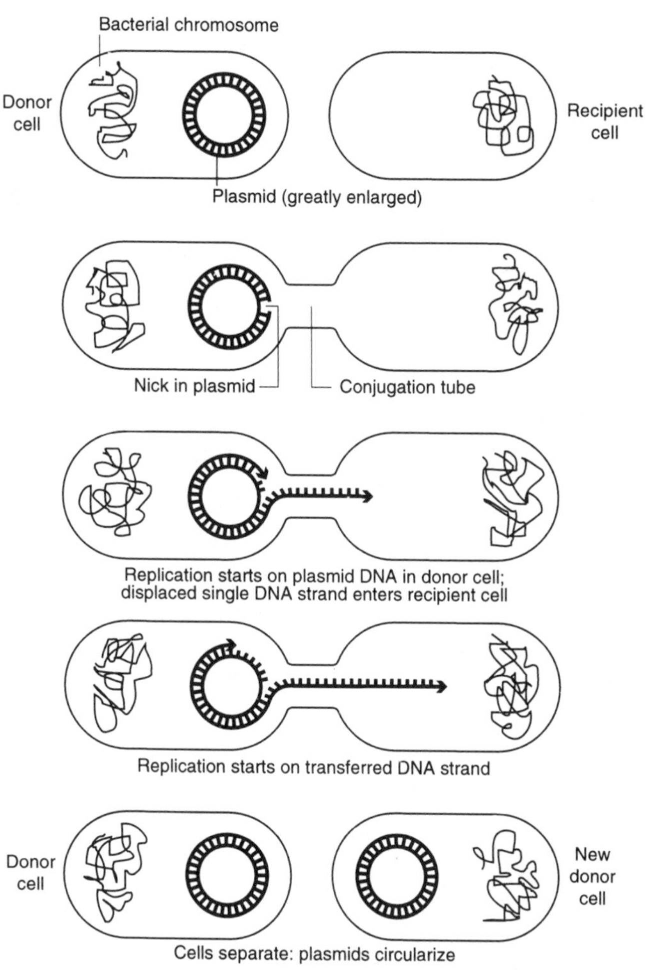

Figure 13–8 Transfer of plasmid DNA

parameciums are designated as *plus* (+) and *minus* (−) *strains*. The union of similar cells of opposite strains is called *conjugation*.

When conjugation occurs, two parameciums exchange portions of micronuclei that result from meiosis. The nuclear material exchanged has the monoploid number of chromosomes. After the exchange, the two micronuclei present in each paramecium fuse and form a zygote (diploid) micronucleus. Then the parameciums separate, macronuclei reform, and each paramecium undergoes a rapid series of binary fissions that result in many small offspring.

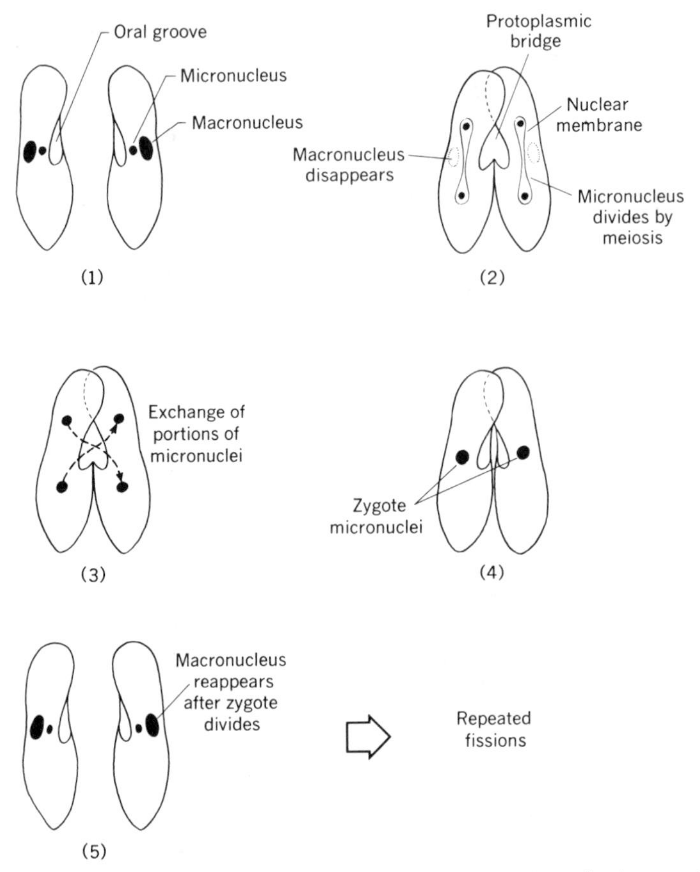

Figure 13–9 Steps in conjugation of paramecium (schematic)

SEXUAL REPRODUCTION IN AN ALGA

Spirogyra, a green alga, is a member of the protist kingdom together with red and brown algae. The body of spirogyra consists of a series of cylindrical cells forming a hairlike thread. Spirogyra is commonly called "pond scum" because it resembles a green mat floating on the surface. This alga reproduces both asexually by binary fission and sexually by conjugation.

Cells of spirogyra conjugate when colonies of opposite strains lie near each other (Figure 13–10). In conjugation, the contents of one cell passes through a tube into another cell and a zygote is formed. The zygote has the diploid number of chromosomes ($2n$), which later is reduced to the monoploid number (n) as a result of meiosis. Each new cell produced by the zygote has the monoploid number of chromosomes.

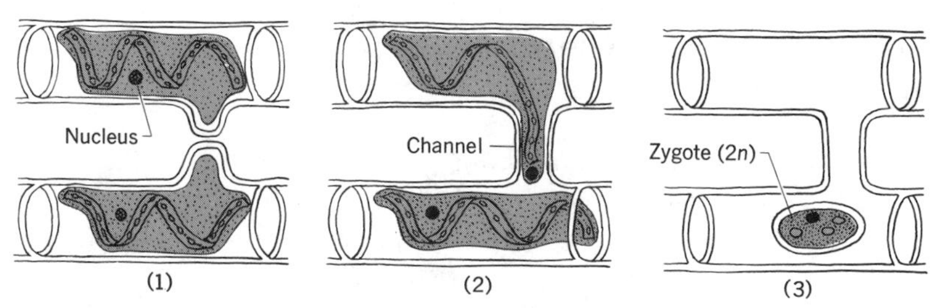

Figure 13–10 Steps in conjugation of spirogyra (schematic)

Section Quiz

1. A common form of *asexual* reproduction in fungi such as yeast and mushrooms is (*a*) vegetative propagation (*b*) budding (*c*) spore formation (*d*) binary fission.
2. Runners, rhizomes, tubers, and bulbs are natural methods of (*a*) vegetative propagation (*b*) budding (*c*) spore formation (*d*) grafting.
3. The regrowth of part of an animal that has been lost is called (*a*) budding (*b*) regeneration (*c*) binary fission (*d*) vegetative propagation.
4. Small circular molecules of DNA or RNA that may be transferred from one bacterium to another are (*a*) buds (*b*) spores (*c*) plasmids (*d*) corms.
5. Paramecium (a protozoan) and spirogyra (an alga) both can reproduce (*a*) asexually by binary fission only (*b*) sexually by conjugation only (*c*) asexually by budding and fission (*d*) asexually by binary fission and sexually by conjugation.

SEXUAL REPRODUCTION IN ANIMALS

Cells of the reproductive system ensure the next generation. The ability to transfer a line of hereditary characteristics from one generation to another prompted the 19th-century biologist August Weismann to call reproductive tissue *germplasm* instead of *somatoplasm*, or body tissue.

In animals, gametes, or sex cells, are produced in the *gonads*, or sex glands. Sperm cells result from spermatogenesis in a male gonad called the *testis* (plural, testes). Ova (singular, ovum), or egg cells, result from oogenesis in a female gonad called the *ovary*. Both types of sex cells are monoploid (n); when united by *fertilization*, a *zygote* is formed, and the diploid number ($2n$) of the species is restored ($n + n = 2n$).

Comparison of a Sperm Cell and an Egg Cell

Structure of a sperm cell. In humans, a spermatid develops a *head*, a *midpiece*, and a *flagellum* (tail). These structures adapt a sperm cell for swimming toward an egg cell and for penetrating it (Figure 13–11).

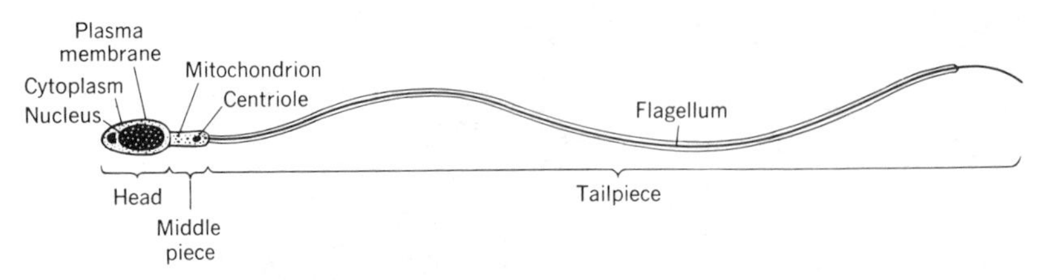

Figure 13–11 The parts of a sperm cell

Head. The head contains a monoploid load of genetic material and an *acrosome*, which is a vesicle that contains an enzyme called *hyaluronidase* and several proteinases. Hyaluronidase helps a sperm cell penetrate an egg cell by dissolving the egg's cell membrane. The cell membrane of an ovum contains *hyaluronic acid*, which strengthens the membrane. Hyaluronidase degrades hyaluronic acid, thus removing the cementing material between cells. The remainder of the head consists of condensed nuclear material.

Midpiece. Mitochondria within the midpiece provide energy for sperm cell locomotion. Energy in ATP generates the metabolic power that enables a flagellum to lash back and forth. A centriole in the midpiece functions as a basal body for the flagellum. A *basal body* is a self-reproducing organelle that produces flagella or cilia and functions in animal cell division.

Flagellum (tail). The tail is a typical flagellum that consists of an arrangement of microtubules with a characteristic 9 + 2 structure (see Chapter 2).

Structure of an egg cell. An egg cell is much larger than a sperm cell (Figure 13–12). As a result of the unequal cell divisions of meiosis, the cytoplasmic material is concentrated in one egg cell. Cytoplasm undergoes chemical changes that result in the production of *yolk*, a rich food for the developing embryo. The egg cell is monoploid (n). Egg cells of most animals are relatively large; mammalian egg cells, however, are microscopic in size and contain little yolk. Unlike a sperm cell, an egg cell cannot

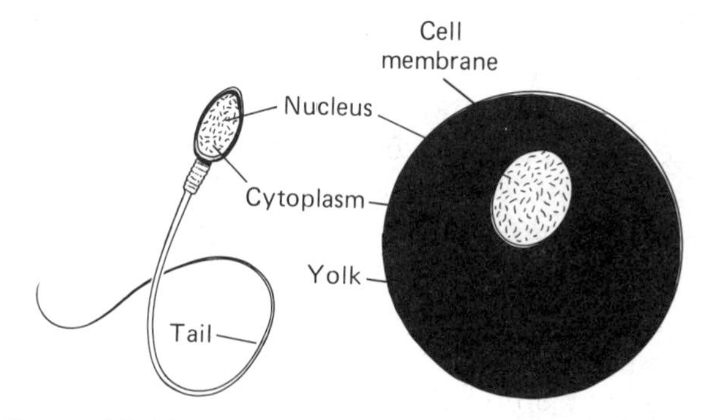

Figure 13–12 An egg cell (compared with a sperm cell)

move by itself. In most cases, it awaits its more mobile reproductive partner for fertilization.

Fertilization

Fertilization may take place inside the body of a female or it may take place outside the body. Internal fertilization occurs primarily in insects, reptiles, birds, and mammals. External fertilization occurs in most fishes and all amphibians.

Figure 13–13 shows fertilization of an egg cell by a sperm cell. When the head of a sperm cell touches the plasma membrane of an ovum, a small external cone is formed. The head and midpiece are drawn into the egg

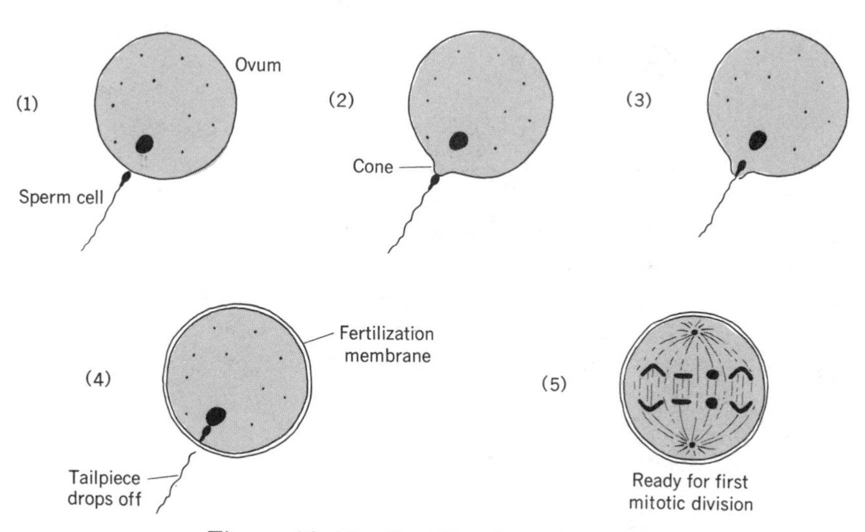

Figure 13–13 Fertilization of an egg

cell and the tail drops off. Then, a fertilization membrane is formed around the zygote. The *fertilization membrane* prevents other sperm cells from entering. Soon after formation of the fertilization membrane, centrioles and spindle fibers appear. Chromosomes of the sperm and egg line up and become attached by their centromeres to the spindle fibers. The diploid *zygote* is now ready for the first mitotic division that leads to the formation of an *embryo*.

Cleavage

The series of rapid cell divisions that change a zygote into a many-celled embryo is called *cleavage*. As you read on, refer to Figure 13–14, which shows the stages of cleavage. The three distinctive stages of cleavage are the morula, blastula, and gastrula. As the result of mitosis, every cell in each stage has the diploid number of chromosomes.

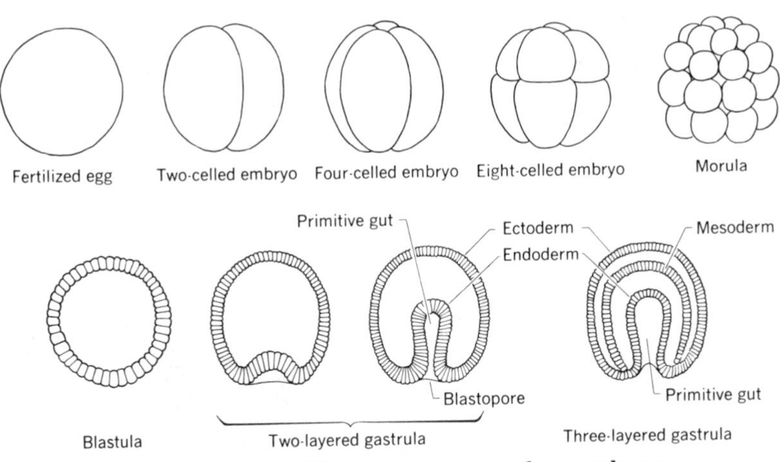

Figure 13–14 Cleavage stages of an embryo

A *morula* is a group of small cells that resembles a mulberry or blackberry fruit. A *blastula* is a hollow ball of many smaller, yet similar, cells; the single layer of cells encloses a space filled with fluid. A two-layered *gastrula* is formed mainly by the inward movement of some cells on one side of the blastula. The outer layer of cells of a gastrula is called *ectoderm;* the inner layer of cells is called *endoderm*. As the endoderm grows inward, it forms a deep pocket called the *primitive gut*. The opening to the primitive gut is called the *blastopore*. The primitive gut develops into the alimentary canal; the blastopore forms the anus; and a mouth forms at the opposite end of the anus.

A third layer of cells, called *mesoderm,* forms between the ectoderm and endoderm in the embryos of higher animals. The mesoderm produces a *three-layered gastrula.* The three layers of cells are called *primary germ layers* because they produce all of the body tissues and organs (see Table 13–1).

Table 13–1 Derivatives of the Primary Germ Layers

Primary Germ Layer	*Derivatives*
Ectoderm	Epidermis; sweat glands; hair; nails; skin; nervous system; parts of eye, ear, and skin receptors; glands (part of pituitary and adrenal glands)
Mesoderm	Skeletal system; muscle system; reproductive system; excretory system; circulatory system; dermis of skin; connective tissue
Endoderm	Inner lining of alimentary canal and respiratory tract; inner lining of ducts leading from liver; pancreas; salivary, thyroid, parathyroid, and thymus glands

Where Embryos Develop

The embryos of many animals develop externally, in water or on land. The embryos of most mammals, however, develop internally, within the mother's body.

External development in water. Egg cells fertilized externally usually develop into embryos outside of the female's body. External development occurs in aquatic animals such as mollusks, most fishes, and amphibians. The embryos receive food from yolk in the egg and obtain dissolved oxygen and water by diffusion.

External development on land. In many terrestrial (land) animals, eggs are fertilized inside the female's body. She then lays the eggs, which develop externally. Internal fertilization and external development are characteristic of terrestrial animals such as insects, reptiles, birds, and some mammals. In most cases, the zygote is encased in a hard or leathery shell that is porous enough for air to diffuse through it.

Shell-covered eggs possess membranes that enable the embryos inside to obtain nourishment and oxygen and to remove metabolic wastes. Embryonic membranes include the amnion, yolk sac, allantois, and chorion (Figure 13–15).

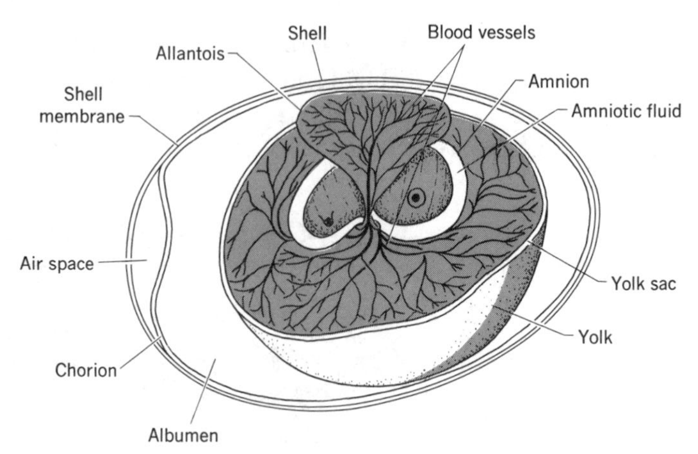

Figure 13–15 Embryonic membranes in a bird's egg (contents tilted)

The amnion. The amnion is a membrane that surrounds the embryo and is filled with *amniotic fluid.* The sac formed by the amnion and amniotic fluid protect the embryo by functioning as a shock absorber. The amniotic fluid also provides a favorable watery environment for development of the embryo.

The yolk sac. The yolk sac is a membrane that grows around the yolk of an egg. In the walls of the yolk sac are many blood vessels that carry nourishment from the yolk to the developing embryo.

The allantois. The allantois is another saclike membrane that possesses numerous capillaries lying close to the inner surface of an egg shell. In that location, the allantois functions as a respiratory organ by removing carbon dioxide within the egg and acquiring oxygen from the atmosphere. The allantois also functions as an excretory storage organ, since nitrogenous wastes are stored there.

The chorion. The chorion, or outer membrane, is located just under the membranes that line the inner surface of the porous shell. The chorion helps exchange respiratory gases between the shell and the capillaries of the allantois.

Internal development in mammals. Internal fertilization followed by internal development of the embryo is characteristic of most mammals. In mammals, the embryo develops within the *uterus*, or womb, of a female.

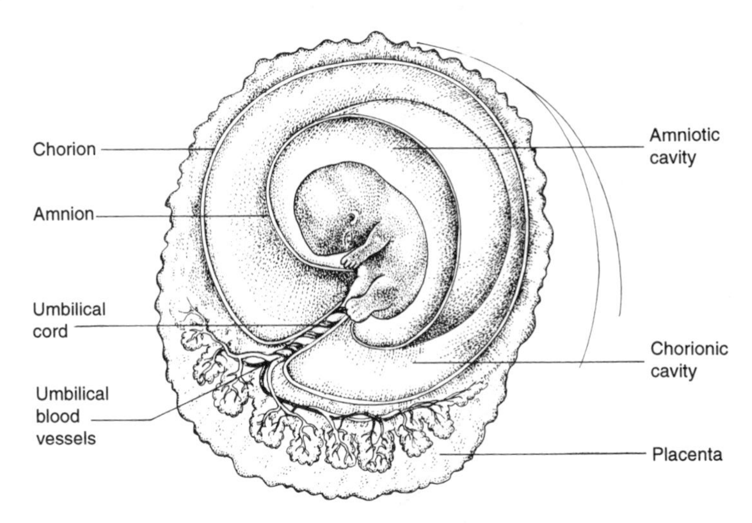

Figure 13–16 Embryo and placenta

In most mammals, the embryo obtains nourishment and oxygen, and gets rid of metabolic wastes, by means of a temporary organ called the *placenta* (see Figure 13–16). A placenta is formed partly from tissues of the uterus and partly from the chorion and allantois of the embryo. As a result, the embryo is attached to the inner wall of the uterus.

Note in Figure 13–17 that no actual connection exists between the blood vessels of the embryo and those of the mother. Instead, capillaries of the embryo in the placenta are bathed by the mother's blood flowing through blood sacs. Thus, digested nutrients, minerals, oxygen, hormones,

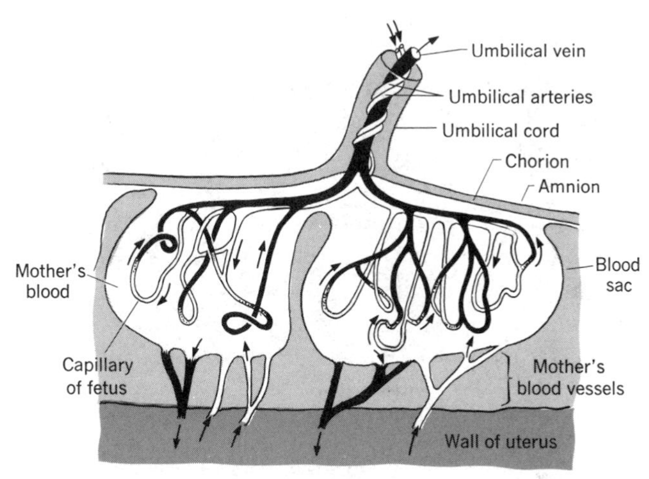

Figure 13–17 Part of a placenta (schematic)

water, and other substances pass into the blood of an embryo from the mother's blood by diffusion and osmosis. Wastes such as carbon dioxide, urea, minerals, and water pass from the embryo's blood to the mother's blood, also by diffusion and osmosis.

The embryos of most animals obtain nourishment from a large amount of yolk in the egg. However, the ova of mammals generally contain little yolk. Thus, a placenta is an evolutionary adaptation that enables a mammalian embryo to obtain continued nourishment after the yolk has been used up.

The blood vessels of the allantois develop into two arteries and one vein in the *umbilical cord,* the ropelike structure that connects the embryo (or fetus) to the placenta. Whereas in an adult most arteries carry oxygenated blood while the veins carry deoxygenated blood, it is the reverse in the embryo's (or fetus's) umbilical cord. The two umbilical arteries carry wastes and deoxygenated blood from the embryo (fetus) to the mother (through the placenta), who eliminates the wastes and oxygenates the blood. The umbilical vein carries nourishment and oxygenated blood from the mother (through the placenta), to the embryo (fetus). At birth, the umbilical cord is cut, thus separating the newborn baby from the placenta. The placenta and the part of the umbilical cord that is attached to the placenta are expelled from the uterus shortly after birth.

LIFE CYCLES OF REPRESENTATIVE INVERTEBRATES

Hydra

Under favorable conditions, hydra reproduce asexually by budding. Under unfavorable conditions, when the water temperature drops and the oxygen content in the water decreases in the fall, hydra develop testes and ovaries. Fertilization involves the fusion of a monoploid egg cell and a monoploid sperm cell within the ovary. The zygote remains in the ovary and develops into an embryo while still attached to the parent (Figure 13–18). Cleavage and development proceed from the two-cell stage through the morula, blastula, and two-layered gastrula stages. Mesoderm does not develop, and the gastrula consists of two germ layers, ectoderm and endoderm. The embryo becomes encased in a hard protective shell and then settles to the bottom of the lake or pond, where it remains all winter. When conditions are favorable again, the young hydra hatches from its shell and develops into an adult.

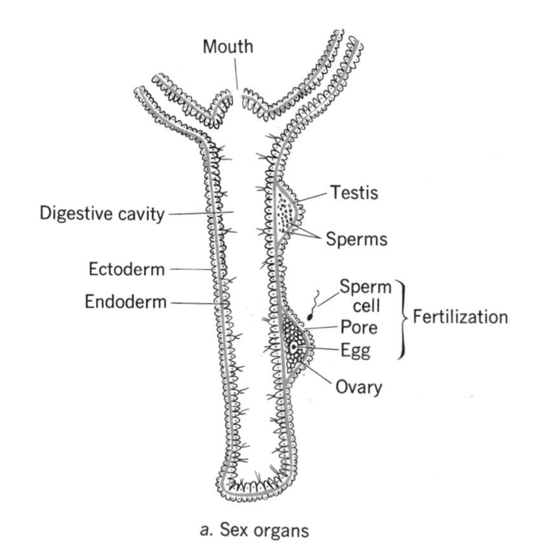

a. Sex organs

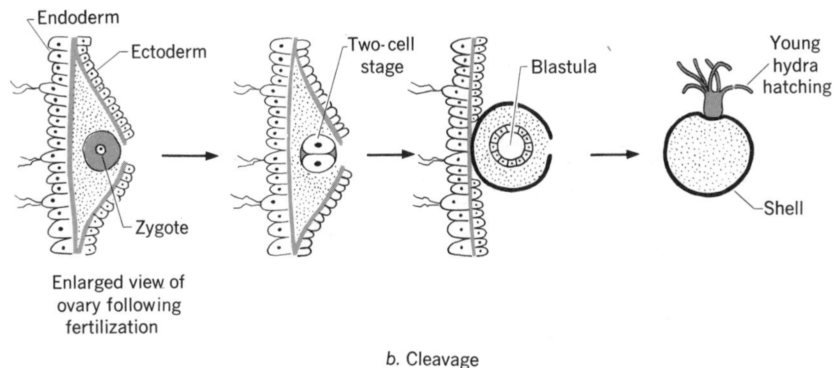

b. Cleavage

Figure 13–18 Sexual reproduction in hydra

Earthworm

An earthworm is a *hermaphrodite,* or an organism that has both male and female sex organs. Two earthworms, however, are necessary for sexual reproduction to occur. Figure 13–19a shows that the female reproductive organs include two ovaries, an oviduct leading from each ovary to an opening on the ventral surface, and two pairs of sperm sacs that receive sperm. The male reproductive system includes two pairs of testes, two sets of sperm-storage sacs, and two sperm ducts that lead to a pair of openings on the ventral surface. The earthworm's segments are numbered in the figure to help you locate the different structures.

Figure 13–19b shows two earthworms mating. Sperm cells from one earthworm are transferred to the sperm-receiving sacs of the other. After the earthworms separate, the *clitellum,* a swollen beltlike structure

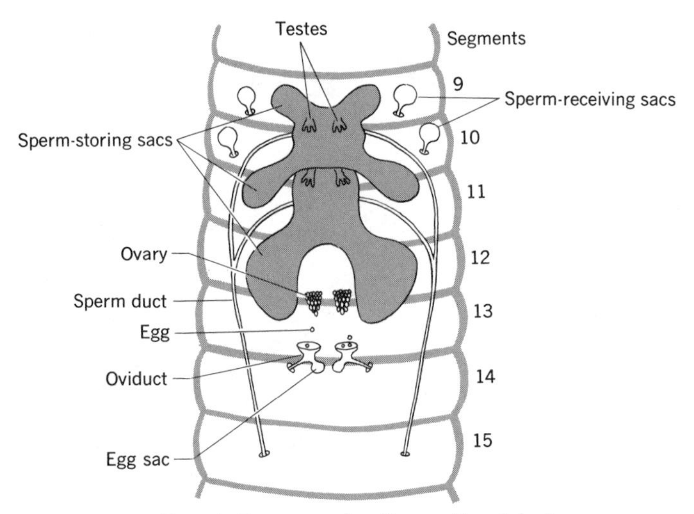

a. Reproductive organs of earthworm (dorsal view)

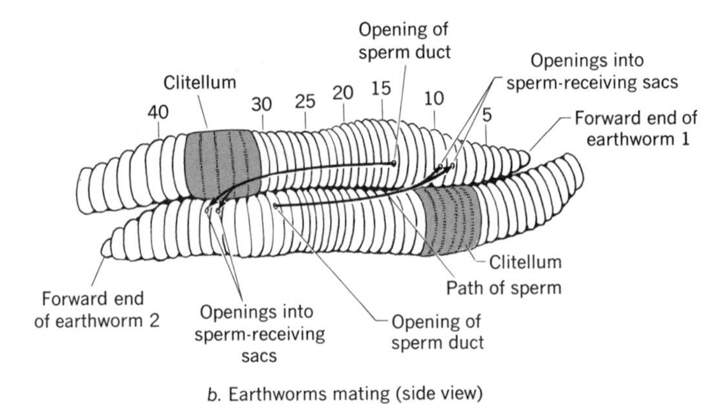

b. Earthworms mating (side view)

Figure 13–19 Earthworm reproduction

around the worm's body, secretes a thick mucus substance that partly coats each worm. Muscular contractions of the body spread the mucus coating over the openings of the oviducts. Eggs are then discharged from the oviduct openings into the mucus. As the coating spreads over the openings of the sperm-storage sacs, sperm cells are discharged into it. Then, after the coating is forced over the forward end of the worm, both ends of the coating close, forming a cocoon about 0.6 centimeter long. Fertilization, cleavage, formation of the gastrula, differentiation, and development occur within the cocoon. Young earthworms hatch from the cocoon in about three weeks.

Grasshopper

The reproductive system of a male grasshopper consists of a pair of testes and two sperm ducts that unite, forming a common duct that extends into a penis. The reproductive system of a female grasshopper consists of a pair of ovaries and a pair of oviducts that unite to form a vagina. After the penis transfers sperm cells into the vagina, the sperm cells are stored temporarily in a sperm-storage sac that branches off the vagina. When egg-laying takes place, shell-covered egg cells pass down the oviduct. As an egg passes the sperm sac, a sperm cell enters the egg through a tiny hole in the shell and fertilization occurs.

The female deposits fertilized eggs in the ground with an *ovipositor*, located at the tip of her abdomen. Cleavage, gastrula formation, differentiation, and partial growth take place before hatching. Then a tiny, wingless grasshopper called a *nymph* emerges from the egg (Figure 13–20). The nymph feeds on vegetation and grows by periodically molting (shedding) its exoskeleton. As the nymph becomes larger, differentiation continues and body structures such as wings grow and enlarge.

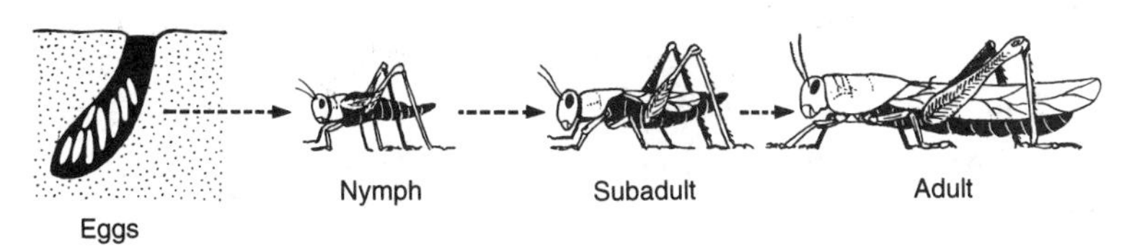

Figure 13–20 Grasshopper development

PARTHENOGENESIS

Parthenogenesis is the development of an embryo from an unfertilized egg cell. This process may be considered a type of asexual reproduction because an egg cell from a female parent produces new individuals without being penetrated by a sperm cell.

Individuals produced by parthenogenesis may be either diploid or monoploid, depending on the cell undergoing parthenogenesis. For example, zygotes of honeybees develop into either diploid female *workers* or diploid *queens*. In other cases, their unfertilized egg cells can develop directly into monoploid males called *drones*. Some fishes, amphibians, and lizards produce monoploid egg cells that undergo mitosis and produce some daughter cells with the diploid number of chromosomes, and some daughter cells with none. The resulting diploid cells divide as if they were zygotes, to form normal diploid individuals. The unfertilized egg cells of sea urchins, frogs, and rabbits may be stimulated artificially to develop into embryos. Methods that have been used include touching unfertilized eggs with needles, changing the water temperature and pH, and applying chemicals or hormones.

Section Quiz

1. Depending upon environmental conditions, hydra can reproduce (*a*) only asexually by budding (*b*) only sexually by producing sperm and egg cells (*c*) only asexually by binary fission (*d*) asexually by budding or sexually by sperm and egg cells.
2. Young earthworms develop within, and hatch from, a (*a*) clitellum (*b*) sperm-storage sac (*c*) cocoon (*d*) ovipositor.
3. The development of a grasshopper from fertilized egg to adult (*a*) includes a wingless nymph stage (*b*) includes a winged nymph stage (*c*) is parthenogenetic (*d*) is direct, without any body changes.
4. Two mature earthworms reproduce sexually by exchanging sperm cells, because they (*a*) are hermaphrodites (*b*) reproduce most often by regeneration (*c*) lack ovaries (*d*) possess monoploid body cells.
5. Parthenogenesis is the development of an offspring from (*a*) a fertilized egg cell (*b*) an unfertilized egg cell (*c*) a fertilized body cell (*d*) a nymph stage.

SEXUAL REPRODUCTION IN REPRESENTATIVE VERTEBRATES

Vertebrates, from fishes to mammals, have developed a variety of adaptations to help ensure survival. External fertilization followed by external development, found in some species, is accompanied by certain environmental dangers. Internal fertilization and internal development (and its variations), which evolved later in other species, generally ensures successful reproduction and survival of offspring.

Fishes

In fishes, the usual pattern of sexual reproduction is external fertilization and external development. A watery environment presents certain dangers for zygotes and developing embryos, including being eaten by parents or other predators, infection, temperature changes, and changes in chemical composition of dissolved substances. These dangers are overcome by females depositing great numbers of egg cells. Fertilization occurs shortly after when sperm cells are released by males. Not all egg cells are fertilized; those that are fertilized contain a plentiful supply of yolk to sustain a developing embryo. Most fish species grow and develop rapidly, from *fry* to *fingerling* to adult, without any major changes in physical appearance.

We see an evolutionary advance in some fishes to overcome environmental dangers. Instead of producing large numbers of eggs, these fish use internal fertilization and either internal or external development. For example, freshwater aquarium fishes, such as mollies and guppies, give birth to live offspring following internal fertilization and internal development. In some marine fishes, including various types of sharks, the embryos develop within the mother's body and are born alive.

Other fishes offer a degree of parental care to their young. For example, the male stickleback (a small freshwater fish) builds a protective nest; the male seahorse (a small marine fish) protects his developing young in a stomach pouch.

Amphibians

Frogs, toads, and salamanders mainly are terrestrial animals, but live close to water. In amphibians, the pattern of sexual reproduction is similar to that of most fishes: external fertilization and external development. In frogs, a male usually mounts a female, an act that stimulates release of egg cells from her cloaca. As egg cells pass out of the cloaca, sperm cells are released by the male. A fertilized frog egg, containing not much more yolk than a fish zygote, quickly develops a jellylike covering. This covering provides physical protection for the developing embryo and enables it to float on the surface of water, gaining warmth from the sun and oxygen from the atmosphere. A developing embryo shortly emerges from its jellylike home as a fishlike *larva,* or tadpole. A *tadpole* is a self-sufficient organism that ingests algae and organic debris. Tadpoles show evidence of fish ancestry by using a tail for locomotion and gills for breathing. A series of physical changes, called *metamorphosis,* controlled by the hormone thyroxine, transforms a tadpole into an adult frog. During metamorphosis, legs emerge, gills are replaced by lungs, and the tail is resorbed into the body. (See Figure 13–21, which shows the frog life cycle.)

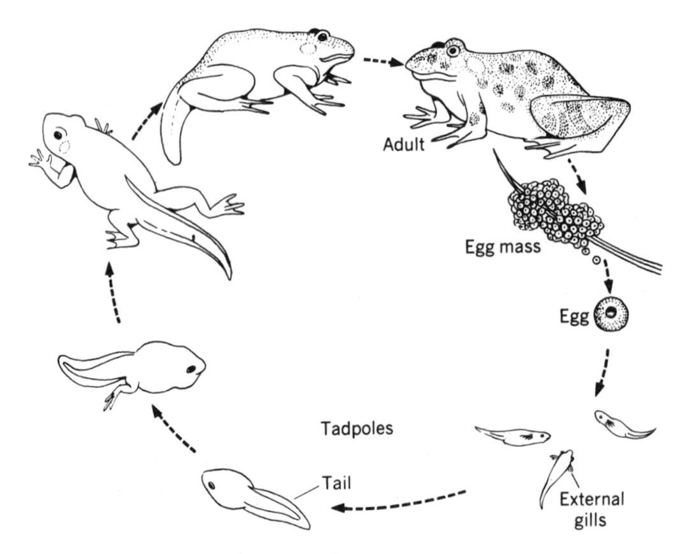

Figure 13–21 Frog development

Reptiles

The leathery skin of reptiles is an adaptation that reduces water loss by evaporation, enabling them to live on land successfully. In reptiles, sexual reproduction features internal fertilization and either external or internal development. A male uses a *penis*, or erectile tube, to introduce sperm into a female's reproductive passageway. The egg cells of a reptile are rich in yolk, enough to sustain a developing embryo during its incubation period. After fertilization, a leathery shell similar in size and shape to a bird's egg is secreted, which encloses the embryo, yolk, and albumen. In some reptile species, eggs hatch outside a female's body; in other species, the embryos develop internally and are born alive. A garter snake is an example of a reptile having internal fertilization and internal development.

Birds

Male birds lack an external penis. Sexual reproduction usually includes a "cloacal kiss," which occurs when the male brushes sperm cells at the cloaca, or reproductive opening, of a female. In birds, sexual reproduction features internal fertilization and external development. A bird's eggs are encased in a hard, calcium-rich protective shell that is porous to air. Fishes, amphibians, and most reptiles give little or no parental care to offspring. Birds, however, usually provide a great deal of care to their offspring. They incubate their eggs, feed their young, teach them to fly and obtain food, and protect them against predators.

Mammals

Not all mammals undergo internal fertilization and internal development. *Monotremes*, such as the *echidna* (spiny anteater) and the *duckbill platypus*, lay eggs following internal fertilization. The existence of monotremes provides evidence that mammals evolved from reptiles. Monotreme eggs are guarded by the mother. Parental care includes nursing the young. *Marsupials*, or pouched mammals such as the kangaroo and opossum, also lack a placenta and give birth to immature offspring. The young are helpless and spend considerable time in the mother's pouch, where they nurse at her milk-producing glands. After about a year, a mother kanagroo will not permit her offspring to climb back into the pouch. Most mammals are *placental*; the developing fetus is protected and well-nourished within the uterus until ready for birth and subsequent nursing. Newborn placental mammals receive much parental care for an extended period after their birth.

The adaptations of vertebrates promote the continued reproductive success of species. These adaptations are seen clearly in the progression from external fertilization in water to internal fertilization on land; also from external development in water, with its dangers, to the comparative safety of internal development accompanied by a continuous food supply during development and after birth. Vertebrates reflect their watery ancestry by requiring a moist environment (amniotic fluid) for their developing embryos.

Section Quiz

1. Metamorphosis of a tadpole into an adult frog is mainly controlled by (*a*) vitamin E and folic acid (*b*) the hormone thyroxine (*c*) the growth hormone (*d*) the hormone cortisone.
2. Which statement is correct? (*a*) All male vertebrates possess a penis. (*b*) An opossum and spiny anteater are closely related because both possess pouches in which to nurture their young. (*c*) A "cloacal kiss" is a method used by birds to transfer sperm to the female. (*d*) In a placenta, the blood vessels of fetus and mother are joined.
3. Two groups of vertebrates that include *both* egg-layers and live-bearers are (*a*) birds and fish (*b*) fish and reptiles (*c*) birds and amphibians (*d*) amphibians and reptiles.

4. The three primary germ layers in an embryo are the (*a*) ectoderm, blastula, and gastrula (*b*) morula, blastula, and gastrula (*c*) ectoderm, mesoderm, and endoderm (*d*) ectoderm, morula, and blastopore.
5. Most mammal embryos obtain oxygen and nutrients and remove wastes through the (*a*) placenta (*b*) fallopian tube (*c*) uterus (*d*) birth canal.

HUMAN REPRODUCTION

Humans are mammals that conform to a basic mammalian pattern of reproduction: internal fertilization, internal development, and a great deal of parental care. Although the sex chromosomes determine sex, the development of an embryo into either a male or female is strongly influenced by sex hormones that affect the reproductive system. The following section discusses the male reproductive system and its adaptations.

Male Reproductive System

As you read on, refer to Figure 13–22, which shows a model of the male reproductive system. The penis and *scrotum* are the distinguishing sexual organs of males. Testes are located within the scrotum. The *testes* are paired oval glands that develop in the body cavity of an embryo and normally descend into the scrotum of the fetus about the seventh month of gestation. The testes function best at a temperature several degrees below normal body temperature (37°C). For this reason, the scrotum and testes are located outside the body cavity.

Each testis consists of about 1,000 coiled tubes called *seminiferous tubules,* in which spermatogenesis takes place. Between developing sperm cells in the seminiferous tubules are found groups of *Sertoli cells* and clusters of *interstitial cells,* or *Leydig cells.* The Sertoli cells nourish the sperm cells and the Leydig cells secrete the male hormone *testosterone.* Mature sperm cells travel from the seminiferous tubules through fine tubes called *vasa efferentia* and are stored in the *epididymis,* which is a single tube coiled many times. Sperm cells leave the epididymis and enter the *vas deferens,* which is a tube that passes from the scrotum through the *inguinal canal* (passageway between the scrotum and body cavity) into the abdominal cavity, and then over the urinary bladder to join the *urethra.* At different times, the urethra transfers urine and sperm cells through the penis to the outside.

Semen is a mixture of sperm cells and secretions from *Cowper's glands, seminal vesicles,* and the *prostate gland.* An ejaculation of semen

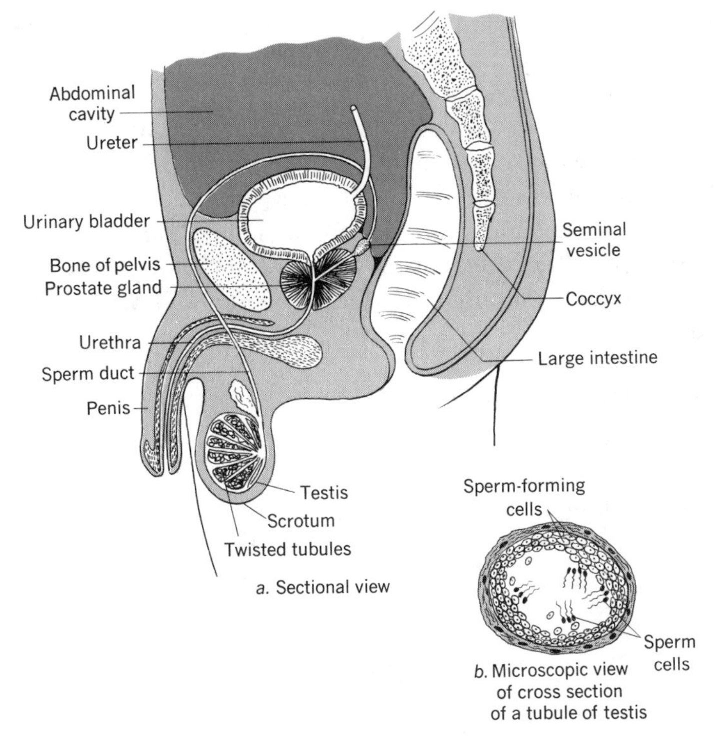

Figure 13–22 Reproductive system of the human male

contains about 50 to 150 million sperm cells per milliliter of fluid. Although only one sperm cell fertilizes an ovum, other sperm cells facilitate fertilization. By adding their hyaluronidase enzyme to that of the fertilizing sperm cell, other sperm cells ease penetration and fertilization.

The paired *Cowper's glands,* about the size of small peas, are located beneath and on either side of the prostate gland. The Cowper's glands secrete an alkaline fluid that neutralizes the acidic coating of urine that covers the lining of the urethra. The alkaline secretion protects sperm cells from chemical injury during ejaculation.

Seminal vesicles are paired structures about 5 cm long that are found between the vas deferens and prostate gland. Their secretion enters the prostate gland through tubes called *ejaculatory ducts.* The secretion of the seminal vesicles is sticky, alkaline, and contains fructose (simple sugar) and prostaglandins, which are discussed in Chapter 9. The alkalinity of this fluid helps neutralize the normal acidic secretions of the vaginal canal. Mitochondria in sperm cells use fructose as fuel for producing ATP, which enhance the movement of sperm cells.

The *prostate gland* is a chestnut-shaped gland that encloses the urethra as it emerges from the urinary bladder. This gland secretes a milky fluid that contains several enzymes which help sperm cells move. As a man ages, his prostate gland grows in size; at about age 45, further enlargement usually occurs, which can lead to medical problems.

The *penis* can enter a vaginal canal and introduce sperm cells to fertilize an egg. Ejaculation, or discharge of semen, is a reflex action that closes the sphincter (valve) at the base of the urethra. That action prevents semen from entering the urinary bladder and urine from mixing with semen.

Female Reproductive System

As you read on, refer to Figure 13–23, which shows a model of the female reproductive system. The ovaries, fallopian tubes, and uterus are the distinguishing sexual organs of females. *Ovaries,* which develop from the same embryonic tissues as do testes, are about the same size and shape as almonds. Ova develop from *oogonia* (singular, *oogonium*) in each ovary.

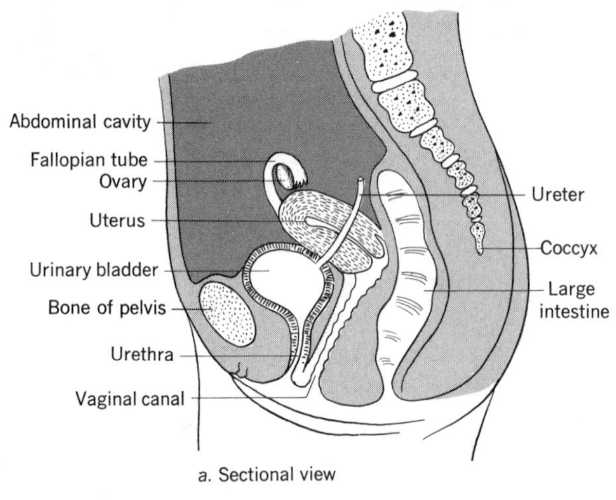

a. Sectional view

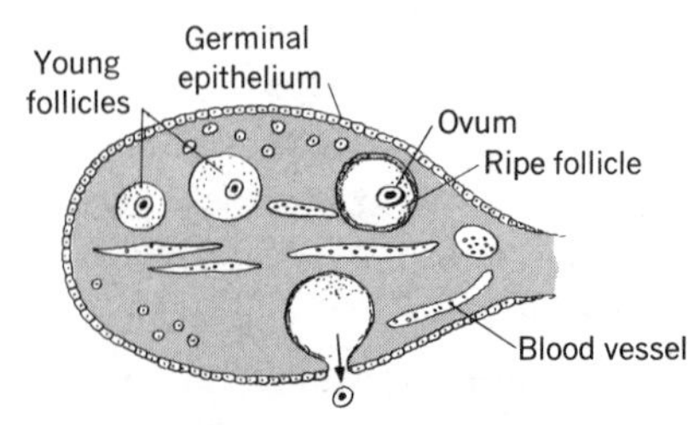

b. Ovary (microscopic view)

Figure 13–23 Reproductive system of the human female

Oogonia complete their mitotic divisions during fetal development. Consequently, the ovaries of a girl contain about 300,000 primary oocytes at birth. Each primary oocyte, covered by a layer of cells, is called a *primary follicle.* The primary follicles do not complete prophase 1 of meiosis until stimulated by FSH (*follicle-stimulating hormone*) from the pituitary gland. Stimulation of the primary follicles begins at puberty and continues, once about every 28 days, until a woman's reproductive life stops (menopause). As a result of FSH, a primary follicle and its primary oocyte enlarge. A secondary oocyte and a polar body are formed following the first meiotic division. The second meiotic division produces three more polar bodies and a mature egg, or ovum (see Figure 13–24). Discharge of an ovum from the ovary into a Fallopian tube takes place as a result of *ovulation,* a process that involves hormonal interactions.

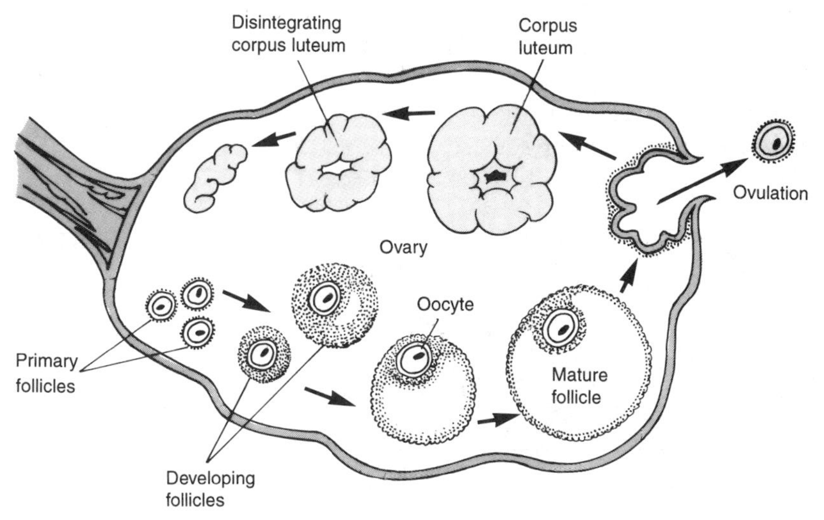

Figure 13–24 Ovulation

Fallopian tubes are the paired *oviducts* that transport egg cells. Human oviducts extend from each ovary for about 10 cm to the uterus. The opening of each oviduct is very near the ovary; when an ovum is discharged, it is swept into the oviduct's tubelike opening. Then, cilia from cells lining the oviduct and smooth muscle within the tubes transport the ovum by ciliary action and peristalsis. If fertilization occurs, a zygote is formed in the widest portion of a Fallopian tube. The zygote undergoes many mitotic divisions, from morula to blastula stages, and ending as a *blastocyst*. The blastocyst then travels to the uterus, where it is implanted. An unfertilized ovum lives for about two days and then dies.

The *uterus* is a muscular pear-shaped organ lined with a mucous membrane called the *endometrium.* A blastocyst is implanted in the endometrium, where it begins its nine-month period of development and growth, from embryo to full-term fetus. The uterus has a thick middle layer of smooth muscle, called the *myometrium.* During childbirth, the myometrium contracts, which expels a fetus from the uterus. The narrowest part of the uterus, called the *cervix*, or neck, opens into the vagina.

Menstrual Cycle

Women experience a sequence of changes in the ovaries and uterus during their reproductive years.

Menstruation is the periodic shedding of tissues that line the uterus together with blood and small blood vessels. After menstruation, the tissues that line the uterus begin to grow again. The menstrual cycle is a sequence of events that occurs about every 28 days. Sometime in the middle years, the menstrual cycle stops. This stopping is called *menopause.* The menstrual cycle consists of four stages: the follicle stage, the ovulation stage, the corpus luteum stage, and the menstruation stage. Figure 13–25 illustrates the events of the menstrual cycle.

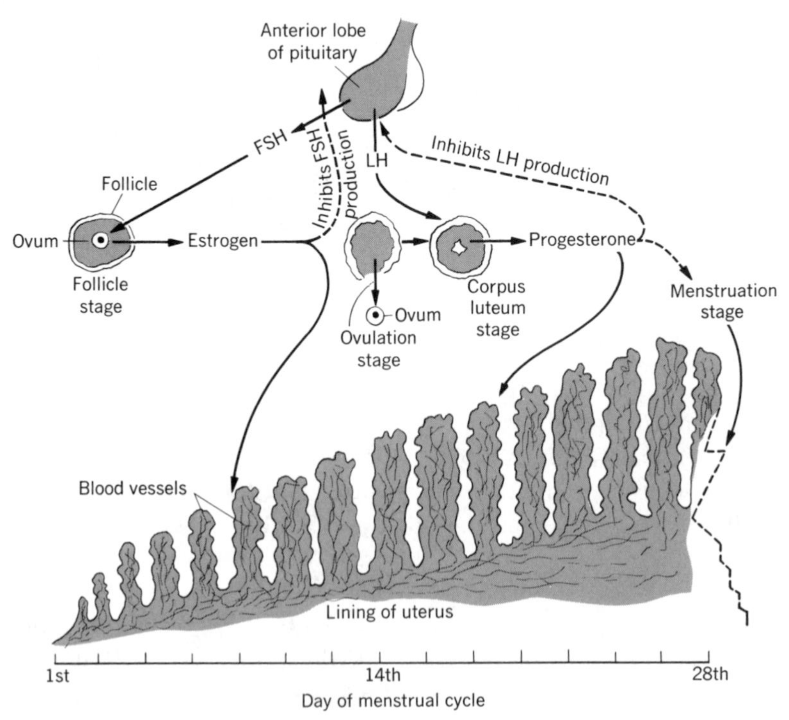

Figure 13–25 Menstrual cycle

The follicle stage. As discussed in Chapter 9, FSH stimulates follicle formation and maturation of the primary oocyte within the follicle. As the follicle enlarges, it fills with a fluid that contains *estrogens,* the female sex hormones. Estrogens cause the epithelial tissue lining the uterus to grow and thicken, and increases the number of blood vessels in the wall of the uterus. These changes in the uterus during the follicle stage prepare it for implantation of the developing embryo.

The ovulation stage. The high concentration of estrogens in the blood inhibits further secretion of FSH, which allows the *luteinizing hormone* (LH) of the pituitary gland to become active. During this stage, three different hormones—estrogen, FSH, and LH—are present in varying concentrations in the blood. This mixture of hormones triggers the follicle to burst, releasing the ovum (ovulation) at about the fourteenth day of the menstrual cycle.

The corpus luteum stage. Following ovulation, LH causes follicle cells to divide rapidly and produce a growth of pinkish-yellow cells (the corpus luteum) that fill the follicle. Under the influence of LH, the corpus luteum secretes the hormone *progesterone.* This hormone prevents further follicle formation, thickens the lining of the uterus, and causes an additional buildup of blood vessels within the wall of the uterus. At the end of this stage, the uterus is ready to receive an embryo, if one is present.

The menstruation stage. If an embryo is not present, the progesterone in the blood inhibits secretion of LH. In turn, the lowered concentration of LH inhibits secretion of progesterone. As the level of progesterone decreases, the tissue that lines the uterus becomes thinner and peels away from the wall of the uterus at about the twenty-eighth day of the menstrual cycle. As the lining peels away, many small blood vessels are ruptured, causing menstrual blood flow. Following menstruation, FSH once again stimulates follicle formation and the menstrual maturation of another primary oocyte.

Implantation of an Embryo

Cleavage of a zygote begins in the oviduct (Fallopian tube), and a blastocyst is formed. The blastocyst moves downward and implants itself in the thickened lining of the uterus (Figure 13–26). The continued secretion of progesterone from the corpus luteum helps maintain the thickness and rich blood supply of the lining of the uterus. As a result, the blastocyst undergoes *gastrulation,* and begins to differentiate into three *primary germ layers: ectoderm, endoderm,* and *mesoderm.* These embryonic membranes form all the tissues and organs of the body.

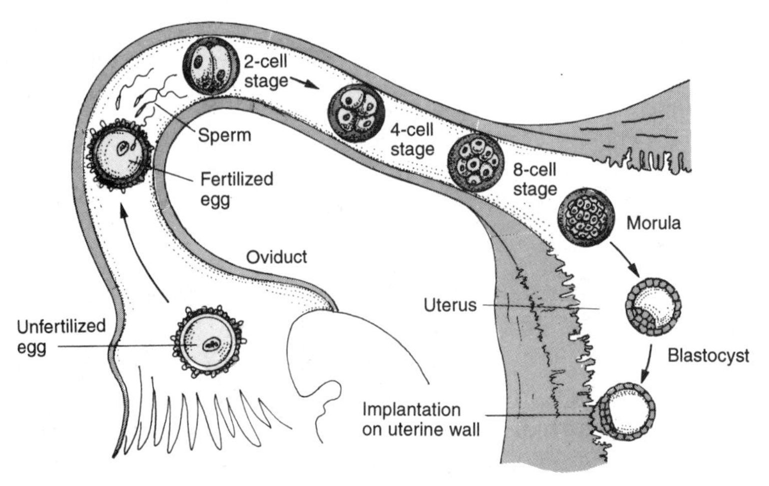

Figure 13–26 Development of the fertilized egg

After the embryonic membranes form, rootlike projections from the chorion join the tissue of the uterus, firmly attaching the embryo to the wall of the uterus. Here, a placenta is formed from tissues of the mother and the embryo. At this time, the placenta secretes additional progesterone and estrogen in addition to connecting an embryo to the uterus. Both hormones help stabilize the supply of blood vessels in the placenta while the embryo is implanted in the uterus.

Development after Implantation

Certain cells within the developing embryo multiply and develop along with the amnion, allantois, and the undeveloped yolk sac. When its major features are visible, the embryo is called a *fetus* (Figure 13–27).

Twinning. Twins are either fraternal or identical. *Fraternal twins* occur when two ova are fertilized by two different sperm cells. Fraternal twins may be the same sex or different sexes. *Identical twins* develop from the same zygote and thus are always the same sex. In this case, the cells of a two-celled embryo separate and each cell then develops independently. Identical twins usually share a placenta; fraternal twins usually have different placentas.

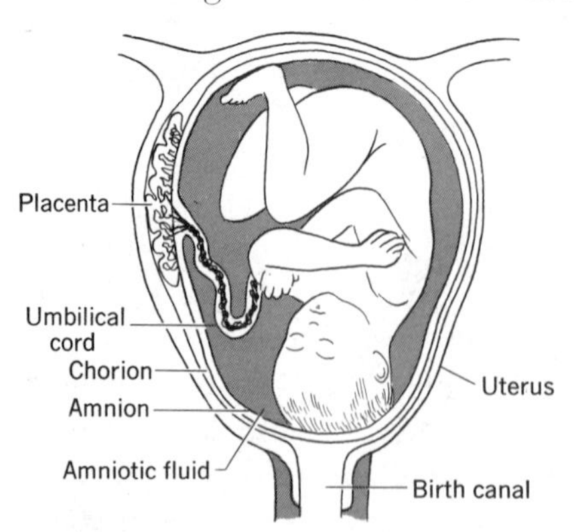

Figure 13–27 A developed fetus

Birth

In humans, the *gestation period,* or time from fertilization until birth, is about 280 days. Toward the end of this period, the progesterone level in the blood falls and the posterior lobe of the pituitary gland secretes the hormone *oxytocin.* Just before birth, oxytocin causes the muscular walls of the uterus to contract rhythmically and forcefully. As a result, the amnion usually bursts, spilling out its fluid. The baby is then forced out of the uterus into the vaginal canal and out of the mother's body. The newborn baby is still attached to the placenta by its umbilical cord, which is immediately tied and then cut. Further contractions of the uterus eventually dislodge the placenta with the remainder of the umbilical cord and expel it from the body. The expelled placenta and attached portion of the umbilical cord are called the *afterbirth.*

Toward the end of gestation, the pituitary gland secretes the hormone *prolactin,* which stimulates the mammary glands to produce milk. If a baby is being breast-fed, prolactin usually prevents follicle formation in the mother by inhibiting FSH production. When a baby is no longer breast-fed, prolactin secretion stops and secretion of FSH resumes the menstrual cycle.

Cloning

The cloning of mammals from an adult (differentiated) cell has been achieved by Scotish embryologist Dr. Jan Wilmut, who produced a lamb from an udder cell of an adult sheep. The lamb, named Dolly, developed from the DNA of a mammary gland cell that was fused with an egg cell from another sheep. The DNA from the mammary gland cell was incorporated into the egg cell after its own nucleus and DNA had been removed. A tiny electric shock stimulated division of the egg cell and its subsequent development into an embryo. After about one week, the embryo was implanted into a third sheep, which later gave birth to Dolly. Dolly is genetically identical to her biological mother, the one that supplied DNA from the udder cell. Although intended to help animal scientists produce better breeds of livestock, this breakthrough in genetic engineering has raised medical, ethical, and philosophical issues.

Section Quiz

1. Structures located within a scrotum include the (*a*) testis, epididymis, and vas deferens (*b*) testis, epididymis, and Cowper's gland (*c*) testis, seminal vesicles, and vas deferens (*d*) seminal vesicles, Cowper's gland, and epididymis.

2. Seminal fluid is composed of secretions from the seminal vesicles and (*a*) Cowper's glands and seminiferous tubules (*b*) prostate gland and Cowper's glands (*c*) prostate gland and seminiferous tubules (*d*) seminiferous tubules and vasa efferentia.
3. Ovulation in humans results from the interaction of the hormones (*a*) FSH, LH, and estrogens (*b*) FSH, LH, and progesterone (*c*) estrogens, progesterone, and LH (*d*) LH, estrogens, and progesterone.
4. Which hormones stabilize the blood supply in a placenta? (*a*) estrogens and FSH (*b*) LH and FSH (*c*) progesterone and estrogens (*d*) estrogens and prolactin.
5. A human zygote undergoes a number of mitotic divisions in a Fallopian tube and enters the uterus as a (*a*) gastrula (*b*) secondary oocyte (*c*) fetus (*d*) blastocyst.

SEXUAL REPRODUCTION IN PLANTS

Like animals, most plants reproduce sexually. Note the similarities and differences between the sexual process of plants and animals.

Flowering plants go through a pattern of alternating monoploid (n) and diploid ($2n$) generations. A *sporophyte,* or diploid plant, produces monoploid spores by meiosis; a *gametophyte,* or monoploid plant, produces monoploid sex cells by mitosis. At fertilization, sex cells unite to form a diploid zygote that grows into a new sporophyte. Both sporophyte and gametophyte are independent plants; of the two, the sporophyte is larger and more visible.

The gametophyte, or monoploid plant, represents the *sexual phase* of the reproductive process. The sporophyte, or diploid plant (the plant that usually is seen), represents the *asexual phase* (Figure 13–28).

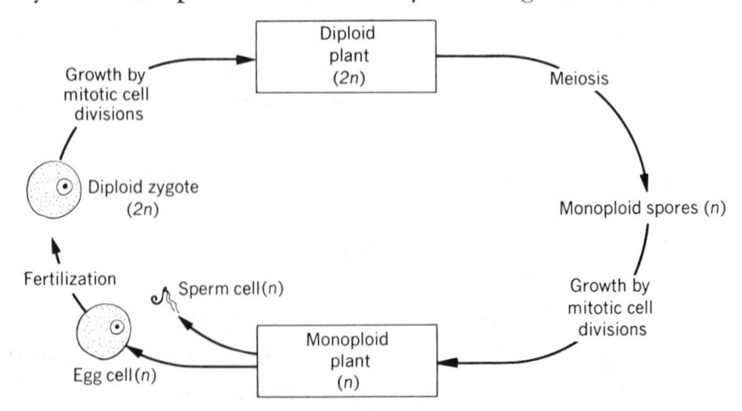

Figure 13–28 Reproductive cycle of plants (alternation of phases)

Life Cycle of Angiosperms (Seed Plants)

The spores of flowering plants usually are produced in a flower. Figure 13–29 shows the structure of a typical flower.

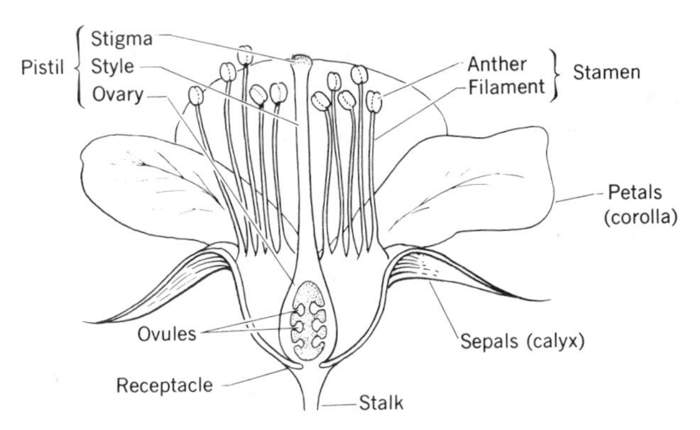

Figure 13–29 Structure of typical flower

The flower. A flower actually is composed of modified leaves. The components are: *sepals,* collectively called the *calyx*; *petals,* collectively called the *corolla*; *stamens*; and *carpels.*

Sepals. Sepals are small, green, leaflike structures that form an outer circle at the base of a flower. Sepals protect the flower bud and support the flower when the bud opens and unfolds.

Petals. Petals form a ring of structures inside the circle of sepals. Insect-pollinated flowers have petals that often are large and brightly colored; wind-pollinated flowers have petals that often are small and white, or they may not have any petals at all.

Stamens. Stamens form a ring of structures within the corolla. A stamen is a male reproductive organ that consists of an anther ("pollen box") in which pollen grains are formed, and a filament that supports the anther.

Carpels (pistils). Depending on the species, a flower may have one or more carpels, or *pistils.* These structures are usually located in the center of a flower. A carpel is a female reproductive organ that consists of three parts: a stigma, a style, and an ovary. A *stigma* usually is sticky or hairy, an adaptation for catching and holding pollen. Stigmas usually contain sugar, which helps pollen grains germinate and grow. The *style* is a tubelike connection between the stigma and ovary. An *ovary* is the enlarged part of the carpel that is connected directly to the *receptacle,* or stem tip, on which the flower rests. Many small, round, whitish structures called *ovules* are attached to the inside wall of the ovary. After fertilization, an ovule develops into a seed that contains an embryo plant.

Stamens and carpels are *essential organs* because they are necessary for sexual reproduction. Sepals and petals are *accessory organs* because they play a less important part in reproduction. The flower shown in Figure 13–29 possesses both male and female reproductive organs; other species possess flowers with only stamens or only carpels. Flowers with only stamens are considered male flowers, and flowers with only carpels are considered female flowers.

Pollination. The transfer of pollen from an anther to a stigma is called *pollination*. If pollen from a flower falls upon the stigma of the same flower or another flower on the same plant, the process is called *self-pollination*. The pea and violet plant regularly self-pollinate. If the pollen of a flower is carried to the stigma of a flower on another plant, the process is called *cross-pollination*. In most species, cross-pollination occurs more frequently than self-pollination. Sometimes a plant breeder transfers pollen from one flower to another using a soft brush. This method of transferring pollen is called *artificial pollination*.

Male functions. The meiotic divisions within an anther produce monoploid *microspores*. Each microspore develops into a pollen grain that grows a protective wall (Figure 13–30a). Each plant species constructs distinctive pollen walls. Thus, a plant can be identified by the pollen it produces. Shortly after a pollen grain is deposited upon a stigma, the pollen grain absorbs water and sugars from the stigma and begins to germinate. In this process, mitosis produces the generative nucleus and the tube nucleus. The *generative nucleus* divides to form two *sperm nuclei*, which fertilize the female gamete. The *tube nucleus*, forms a *pollen tube*, which transports the sperm nuclei to the female tissues of a flower. A growing pollen tube contains three nuclei—two sperm nuclei and one tube nucleus. It is the tiny male plant, or gametophyte stage of a plant life cycle (Figure 13–30b).

Female functions. The meiotic divisions within an ovule produce four monoploid *megaspores*. Three of these disappear, leaving one monoploid megaspore. The remaining monoploid megaspore undergoes three mitotic divisions resulting in eight nuclei. Four of the nuclei are located at one end of the cell and four at the opposite end. One nucleus from each group moves to the center of the cell and unites forming a diploid ($2n$) *fusion nucleus*. The three nuclei closest to the *micropyle* (opening) form an *egg nucleus* (n) and two short-lived *synergids*. The three nuclei at the

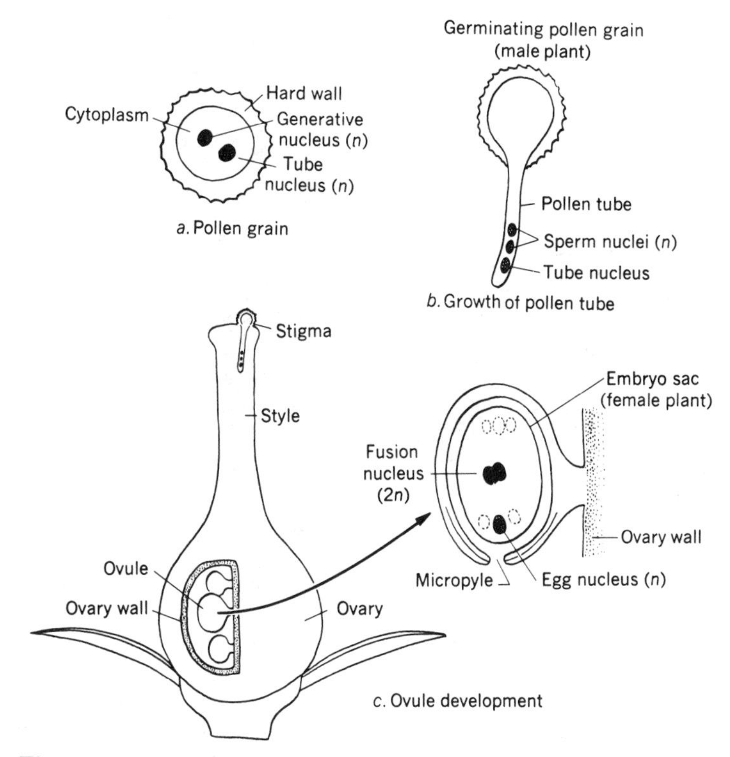

Figure 13–30 Development of a pollen grain and an ovule

opposite end are called *antipodals,* and also are short-lived. The entire structure containing seven nuclei (one diploid ($2n$) nucleus and six monoploid (n) nuclei, including the egg nucleus) is called an *embryo sac* (Figure 13–30c). An embryo sac within an ovule that contains the seven nuclei is the tiny female plant, or gametophyte stage of a plant's life cycle.

Fertilization. As you read on, refer to Figure 13–31, which shows fertilization in a flower. The tip of the growing pollen tube (the male plant) reaches the micropyle and enters the ovule. Then the tip of the pollen tube ruptures and its contents flow into the embryo sac. One sperm nucleus (n) unites with the egg nucleus (n) and forms the zygote ($2n$), which develops into the embryo. The other sperm nucleus (n) unites with the fusion nucleus ($2n$) and forms the *endosperm nucleus* ($3n$), which later nourishes the developing embryo. After fertilization has been accomplished, the tube nucleus disappears. The two unions that occur within the embryo sac are called *double fertilization.*

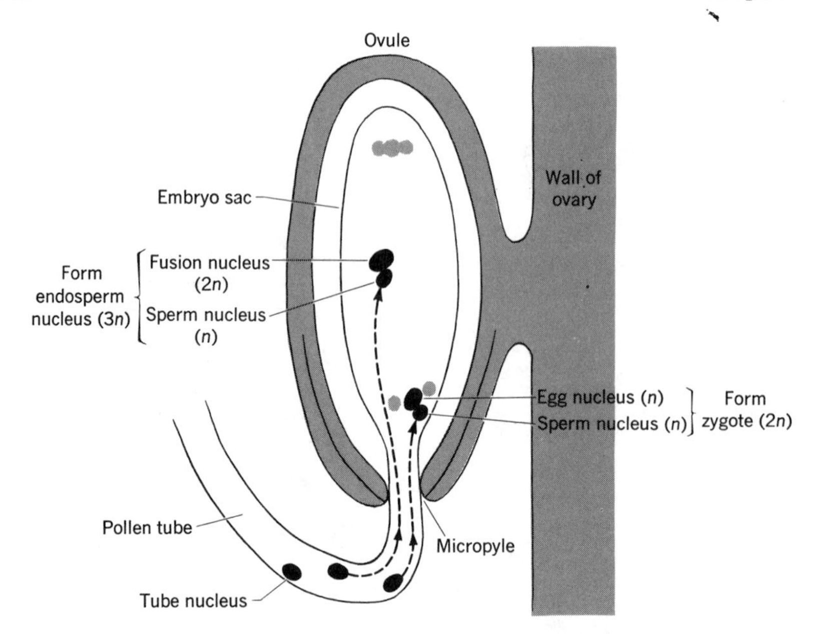

Figure 13–31 Fertilization in a flower

Seeds. After fertilization, the zygote divides by mitosis, forming an embryo that consists of a row of cells rather than a cluster of cells as in animal cleavage. These cells then differentiate and form the young asexual plant. The endosperm nucleus also divides by mitosis and forms *endosperm tissue,* which is an early source of nourishment for the embryo. In time, the double wall of the ovule dries up, forming a protective coat around the embryo. At maturity, the ovule is called a *seed.*

Flowering plants are divided into two main groups: dicots and monocots. In dicots, vascular bundles in the stem are arranged in a definite pattern; in monocots, vascular bundles are scattered throughout the stem (see Chapter 11). Seeds of dicot and monocot plants also show certain structural differences.

Dicot and monocot seeds. Study Figure 13–32, which compares a dicot seed (bean) with a monocot seed (corn). A dicot seed has the following structures:

1. Two fleshy *cotyledons* (seed leaves), which store food and enclose a young plant.
2. A *hilum,* or scar, which indicates where the seed was once attached by a short stalk to the ovary wall.

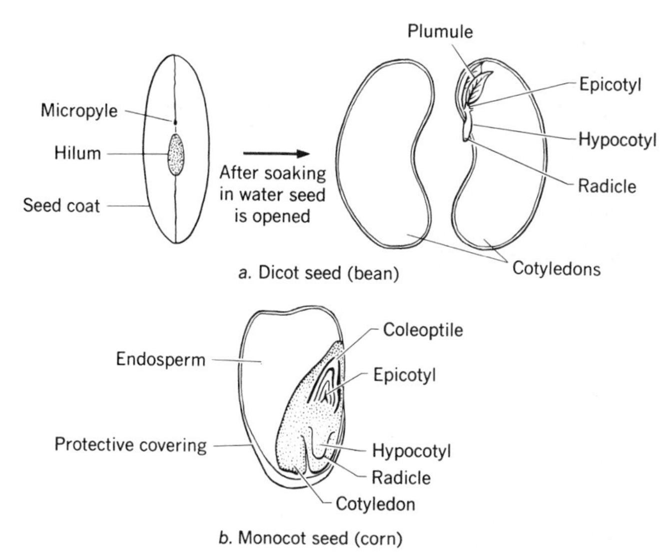

Figure 13–32 Seed structure

3. A young plant, which consists of an *epicotyl* and a *hypocotyl*. The upper part of the epicotyl consists of two miniature leaves—the *plumule*, or first bud of the young plant. The lower part of the epicotyl and most of the hypocotyl develop into the stem. The lower portion of the hypocotyl, called the *radicle*, develops into roots.

Figure 13–32 shows that a monocot seed has the following structures:

1. One small cotyledon that contains little stored food.
2. A large mass of endosperm tissue that covers the cotyledon and the young plant.
3. A coleoptile, which covers the tip of the epicotyl.

The major functions of the cotyledon are to produce enzymes that digest food stored in the endosperm and to transfer digested food to the young plant when it begins to sprout. As in the dicot seed, the radicle develops into roots. Unlike the dicot seed, the remainder of the hypocotyl also develops into roots. The epicotyl, under the influence of hormones (auxins) secreted by the coleoptile, develops into the stem and leaves.

Germination of seeds. The *germination*, or sprouting, of a seed requires adequate amounts of moisture and oxygen and a suitable temperature—between 18°C and 29°C (65°F and 85°F), depending upon the species. Under these conditions, germination begins with the digestion of stored nutrient material. Cellular respiration provides the energy required for cell growth, division, and development. The root system and the lower part of the stem develop first. The germination of a bean seed is shown in Figure 13–33a.

The germination of a monocot seed, such as corn, is similar to that of a dicot. However, an arch is not formed and the seed is not pulled out of the ground by the growing hypocotyl (Figure 13–33b).

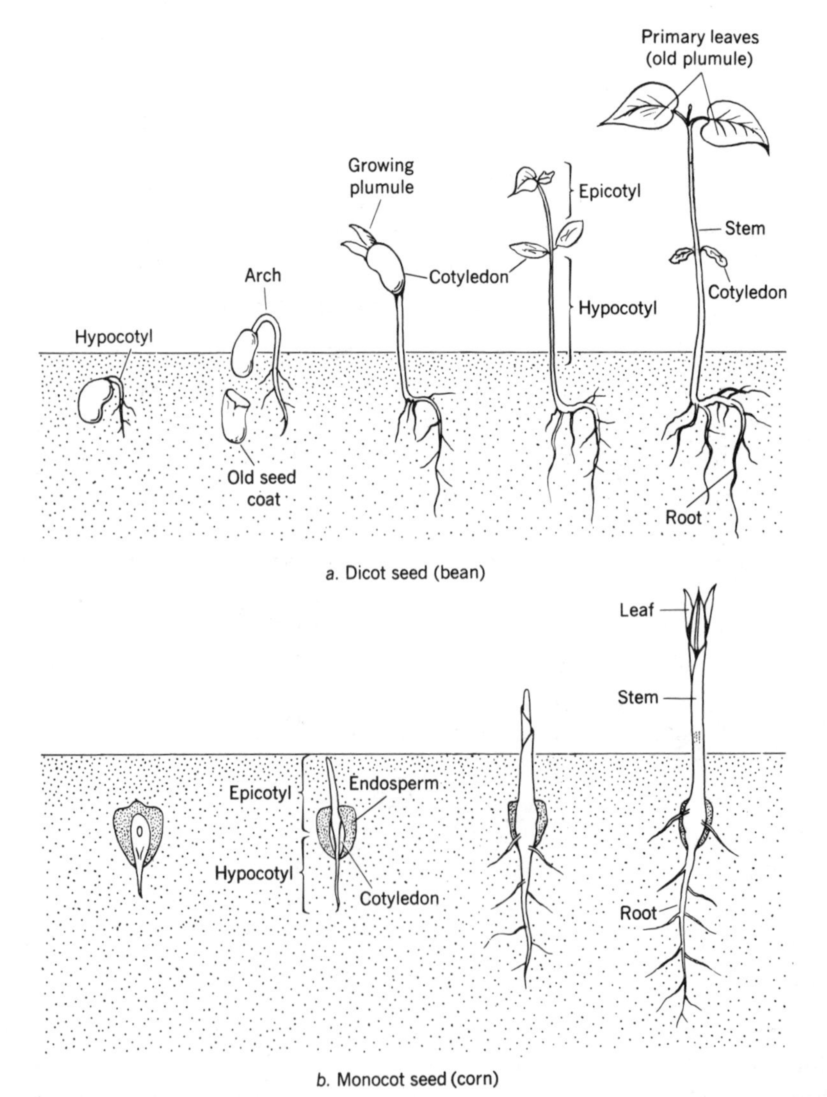

Figure 13–33 Germination of seeds

Fruits. As ovules ripen into seeds, the ovary of a flower enlarges and develops into a fruit. A *fruit* is a ripened ovary, or group of adjacent ripened ovaries, with any flower parts still attached. Some common vegetables, such as the tomato, cucumber, and pepper, are actually fruits. Some fruits, such as the pear and apple, are formed from the receptacle, which becomes the outer part of the fruit, and the ripened ovary, which becomes the core of the fruit.

Seed dispersal. In the course of evolution, plants developed different adaptations for dispersing (scattering) seeds (Figure 13–34). These adaptations enable young plants to obtain adequate water, minerals, light, and living space. As a result, the plants have a better chance to survive and reproduce.

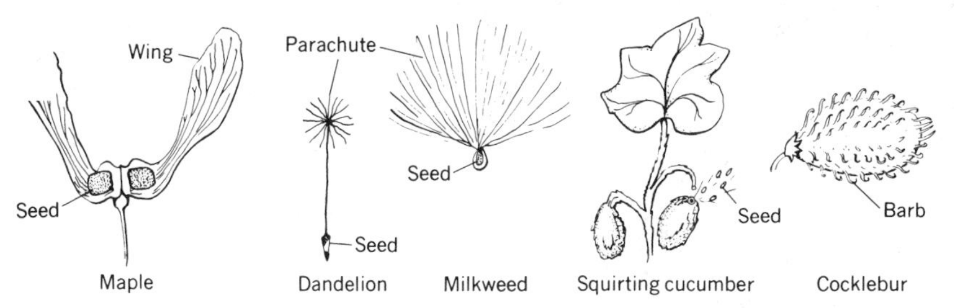

Figure 13–34 Adaptations for seed dispersal

The major methods of seed dispersal are wind, water, animals, and explosive fruits.

Dispersal by wind. Maple fruits bear winglike structures. Thus, when the fruit drops from a maple tree, the fruit whirls like a propeller and falls in an arc. This curved path carries the fruit away from the parent tree. Milkweed seeds and the fruits of dandelions have feathery structures shaped like a parachute. Air currents carry milkweed and dandelion fruits considerable distances from the parent plants.

Dispersal by water. Coconuts are adapted for floating in water. Water currents carry them far from the parent plants.

Dispersal by animals. Burdock and cocklebur fruits have hooks, or barbs, that become attached to the fur of passing animals. Eventually the seeds drop off the animals, landing in a distant location. Juicy fruits, such as raspberries and cherries, often are eaten by animals that cannot digest the seeds. When the animals eliminate wastes, the seeds fall on the ground far from the parent plant.

Dispersal by explosive fruits. The squirting cucumber and the touch-me-not have fruits that explode when they are touched. The fruits pop open with such force that the seeds are shot away from the parent plant.

The preceding survey of reproduction in animals, plants, and other organisms highlights the function of reproduction as a means of continuing the species. In both asexual and sexual reproduction, the species number of chromosomes is kept constant in each generation. Sexual reproduction is a major factor in evolution because it provides a source of genetic variations in offspring. In a changing environment, offspring that have favorable variations survive and continue the species. They may eventually develop into new species.

The next chapter discusses the transmission of hereditary traits from parent to offspring and explains how variations important to evolution may arise.

Section Quiz

1. Which structure develops into a male gametophyte? (*a*) anther (*b*) carpel (*c*) microspore (*d*) megaspore.
2. As a result of fertilization within an embryo sac, a fusion nucleus is (*a*) monoploid (*b*) diploid (*c*) triploid (*d*) tetraploid.
3. The carpel of a flower is also called a (an) (*a*) anther (*b*) pistil (*c*) sepal (*d*) stigma.
4. Which of the following is a fruit? (*a*) potato (*b*) tomato (*c*) carrot (*d*) cabbage.
5. The coleoptile of a corn seedling influences the growth of its (*a*) roots (*b*) endosperm (*c*) epicotyl (*d*) plumule.

Chapter Review Questions

The following questions will help you check your understanding of the material presented in the chapter.

1. Which method of asexual reproduction depends on wind dispersal of reproductive cells? (*a*) spore formation (*b*) fission (*c*) budding (*d*) vegetative propagation.

2. In a protist, which is true of reproduction by budding but not true of reproduction by fission? (*a*) equal cytoplasmic division (*b*) unequal cytoplasmic division (*c*) equal nuclear division (*d*) unequal nuclear division.

3. A peach tree bearing large sour peaches was crossed with another bearing small sweet peaches. One hundred seeds resulting from this cross were planted, and from this planting 20 trees that eventually produced large sweet peaches were obtained. By which process would it be possible to increase the number of peach trees that bear large sweet peaches? (*a*) hybridization (*b*) inbreeding (*c*) stem grafting (*d*) cross-pollination.

4. The ability to reproduce asexually is more often a characteristic of invertebrates than of vertebrates because invertebrates (*a*) possess more undifferentiated cells (*b*) cannot produce gametes (*c*) lack gonads (*d*) undergo maturation and cleavage.

5. Asexual reproduction in one-celled organisms never involves (*a*) replication of chromosomes (*b*) fusion of nuclei (*c*) formation of a spindle (*d*) formation of genetically identical daughter cells.

6. The development of an unfertilized egg is called (*a*) reduction division (*b*) spermatogenesis (*c*) parthenogenesis (*d*) hermaphroditism.

7. One difference between an egg and a sperm is that only the egg (*a*) contains mitochondria (*b*) is motile (*c*) can undergo parthenogenesis (*d*) can carry chromosomes.

8. Which group of animals stores the least amount of food in its eggs for developing embryos? (*a*) fish (*b*) birds (*c*) placental mammals (*d*) nonplacental mammals.

9. Sexual reproduction usually is characterized by (*a*) meiosis followed by fusion of gametes (*b*) meiosis followed by fission of gametes (*c*) mitosis followed by budding (*d*) mitosis followed by spore formation.

The series of diagrams on the next page represents stages in the reproduction and development of an invertebrate. For each term in questions *10 through 12*, select the number of the diagram, chosen from the series, that is most closely associated with that term. A number may be used more than once or not at all.

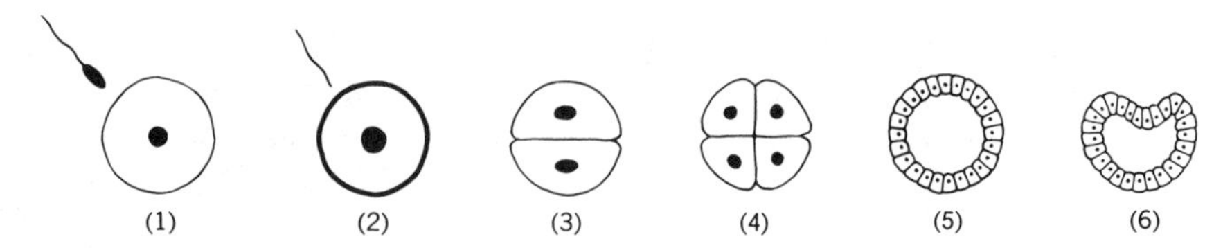

10. The zygote.
11. Formation of endoderm.
12. The blastula stage.
13. Which is a true statement concerning primary germ layers? (*a*) They develop into specialized tissues. (*b*) They are involved in fertilization. (*c*) They are found only in plants. (*d*) They protect the embryo against infection.
14. Which is usually associated with organisms whose reproduction involves external fertilization rather than internal fertilization? (*a*) parental care of the offspring (*b*) the production of large numbers of eggs (*c*) the embryo's direct nutritional dependence on the mother (*d*) a land environment.
15. Artificial parthenogenesis is the development of an egg cell that is (*a*) fertilized after being treated with chemicals (*b*) unfertilized, but has been treated by chemical or physical means (*c*) unfertilized and kept frozen until all enzymes are inactivated (*d*) fertilized before being treated with chemicals.

For each event associated with the reproductive cycle indicated in questions *16 through 20*, select the letter of the term, chosen from the following list, that best describes that event.

Terms

(*a*) Fertilization	(*e*) Oogenesis
(*b*) Mitosis	(*f*) Ovulation
(*c*) Menopause	(*g*) Puberty
(*d*) Menstruation	

16. The periodic breakdown and discharge of tissues lining the uterus occurs.
17. The reduction of chromosomes to the monoploid number during cell division is in the early stages of the cycle.
18. The completion of the cycle is delayed for a number of months.
19. The cycle stops permanently.

20. The rupture of the follicle results in the release of the egg cell.
21. The cells of an animal blastula are characterized by (*a*) frequent mitotic divisions (*b*) frequent meiotic divisions (*c*) monoploid chromosome numbers (*d*) many hormonal secretions.
22. External fertilization generally occurs in (*a*) mammals and birds (*b*) birds and reptiles (*c*) reptiles and amphibians (*d*) amphibians and fish.

For each phrase in questions *23 through 27*, select the letter of the structure, chosen from the list below, that best fits the description. A letter may be used more than once or not at all.

Structures

(*a*) Oviduct	(*d*) Uterus
(*b*) Scrotum	(*e*) Testis
(*c*) Ovary	(*f*) Cervix

23. Structure that produces sperm cells.
24. Narrowest part of the uterus.
25. Structure in which the fetus develops.
26. Outer pouch of the body that contains male sex glands.
27. Organ in which the corpus luteum develops.
28. In humans, how many monoploid sperms would be produced in the male gonads from each primary sex cell? (*a*) one (*b*) two (*c*) eight (*d*) four.
29. In most mammals, the diffusion of nutrients takes place between the capillaries of the mother and the capillaries of the embryo in the (*a*) navel (*b*) umbilical cord (*c*) yolk sac (*d*) placenta.
30. In the development of human embryos, gastrulation usually is followed immediately by (*a*) meiosis (*b*) differentiation (*c*) dispersal (*d*) parthenogenesis.
31. Identical twins have many similar characteristics primarily because they (*a*) develop from the division of a single zygote (*b*) develop from two fertilized egg cells (*c*) are the offspring of the same parents (*d*) are born at the same time.
32. The first development of secondary sex characteristics in a human female is most closely associated with (*a*) menopause (*b*) gestation (*c*) zygote formation (*d*) puberty.

33. Young kangaroos complete their development outside the mother's body. Therefore, which maternal organ of the kangaroo is most likely to be less developed? (*a*) mammary gland (*b*) uterus (*c*) pancreas (*d*) ovary.

34. As a result of meiosis in the male reproductive organ of a flower, monoploid nuclei will be found in the (*a*) pollen grain (*b*) cotyledon (*c*) epicotyl (*d*) filament.

35. Which is a function of the stigma? (*a*) It attracts insects. (*b*) It catches egg cells. (*c*) It catches pollen. (*d*) It ingests nutrients.

36. Where does germination of pollen grains normally occur? (*a*) on the anther (*b*) on the stigma (*c*) in the ovary (*d*) in the ovule.

37. Sperm nuclei must travel from the stigma of a flower to the ovule through the (*a*) oviduct (*b*) sepal (*c*) stomate (*d*) pollen tube.

38. The monoploid egg nucleus in flowering plants is produced in the (*a*) ovary (*b*) filament (*c*) stigma (*d*) anther.

39. The process in which dormant seeds develop into new plants when conditions are favorable is known as (*a*) pollination (*b*) germination (*c*) spontaneous generation (*d*) differentiation.

40. Following fertilization, which part of the flower develops into the seed? (*a*) ovule (*b*) ovary (*c*) receptacle (*d*) stigma.

Biology Challenge

The following questions will provide practice in answering SAT II-type questions.

Part I

Select the letter of the statement that best completes the sentence or answers the question. You may need to do library research.

1. Which group of organisms normally reproduces by forming clonal populations? (*a*) sea anemone, dragonfly, sponge (*b*) earthworm, yeast, bacterium (*c*) jellyfish, planaria, alga (*d*) sponge, yeast, sea anemone (*e*) bee, bacterium, butterfly.

2. In certain bacteria, F^+ and F^- plasmids enable a form of (*a*) sexual reproduction (*b*) asexual reproduction (*c*) anaerobic respiration (*d*) aerobic respiration (*e*) energy transformations.
3. Some plants possess a variety of adaptations, such as white night-blooming flowers, emitting odors, and producing flowers before leaves appear on stems of trees and shrubs. As a result, these plants survive by (*a*) ensuring self-pollination (*b*) ensuring cross-pollination (*c*) ensuring double fertilization (*d*) trapping and digesting insects (*e*) reverting to a heterotrophic nutrition.
4. A group of organisms that reproduces by parthenogenesis is (*a*) wasps, earthworms, fleas (*b*) ants, frogs, sea urchins (*c*) tapeworms, bees, starfishes (*d*) sponges, flatworms, oysters (*e*) ants, wasps, bees.
5. Which male animal lacks an external penis? (*a*) turtle (*b*) opossum (*c*) turkey (*d*) spiny anteater (*e*) rattlesnake.

Part II

Select the letter of the statement that best completes the sentence or answers the question.

1. Which process occurs in the uterus? (*a*) meiosis (*b*) follicle development (*c*) embryo development (*d*) fertilization (*e*) oogenesis.
2. The diploid number of chromosomes ($2n$) in the zygote results from (*a*) cleavage (*b*) meiosis (*c*) mitosis (*d*) fertilization (*e*) gastrulation.
3. Human egg cells and sperm cells are alike in that both (*a*) develop in gonads (*b*) possess adaptations for locomotion (*c*) possess the same relative amount of cytoplasm (*d*) are produced in about the same numbers (*e*) are about the same size.
4. The immediate result of meiosis is the formation of (*a*) embryos (*b*) cytoplasm (*c*) gonads (*d*) gametes (*e*) oogonia.
5. In humans, polar bodies are formed during (*a*) gastrulation (*b*) gestation (*c*) ovulation (*d*) spermatogenesis (*e*) oogenesis.

14

Heredity: Basic Principles

Learning Objectives

When you have completed this chapter, you should be able to:

- **Describe** Mendel's laws of heredity.
- **Discuss** the chromosome and gene theories of heredity.
- **Distinguish** between genotype and phenotype.
- **Use** a Punnett square to predict the results of a cross.
- **Apply** Mendel's laws to the inheritance of traits in various organisms.
- **Explain** how a testcross is used.
- **Describe** incomplete dominance of traits.

OVERVIEW

Heredity is the study of traits that are transmitted, or passed on, from one generation of living things to the next. Genetics is the branch of biology that explains how these traits are transmitted, or inherited. In this chapter, you will learn how traits are passed on from parent to offspring and about the contributions of Gregor Mendel to our understanding of patterns of inheritance.

MENDEL'S LAWS OF HEREDITY

The basic patterns of heredity were worked out in the 1850's by the Austrian monk *Gregor Mendel* when he studied the inheritance of traits in garden peas. Mendel selected garden peas for the following reasons:

- The flowers usually are self-pollinated.
- The flowers are large and easily pollinated by hand.
- Garden peas have easily recognizable contrasting traits, such as long and short stems, and purple and white flowers.

Biologists who studied the garden pea before Mendel failed to understand the basic patterns of inheritance. The main reason for their failure was that the biologists had studied many traits at a time. Mendel was successful because he studied a *single* trait at a time, and he kept careful records of the offspring that resulted from each cross.

Mendel's observations led him to conclude that organisms have many traits. Each trait is controlled by two characters, or determiners, one from each parent. When Mendel conducted his experiments, neither chromosomes nor genes had been discovered. Thus, his theory of determiners of heredity was truly remarkable. Mendel studied seven basic traits, which are listed in the first column of Table 14–1.

Law of Dominance

Mendel selected these seven traits because for each trait there were two contrasting types, or characters (listed in the second column of Table 14–1). He checked that each type was *pure* for a trait or that each plant with the pure trait arose from ancestors that always bred true for the trait.

Table 14–1 Cross-Pollination of Parents Having Contrasting Characters

Trait	*P_1 Generation (Pure Types)*	*F_1 Generation (Hybrids)*
Length of stem	Long stem X short stem	All long (100%)
Type of seed coat	Smooth seed X wrinkled seed	All smooth (100%)
Color of seed coat	Colored X white	All colored (100%)
Form of pea pod	Thick X thin	All thick (100%)
Position of flower	Side of stem X end of stem	All side of stem (100%)
Color of cotyledons	Yellow X green	All yellow (100%)
Color of pod	Green X yellow	All green (100%)

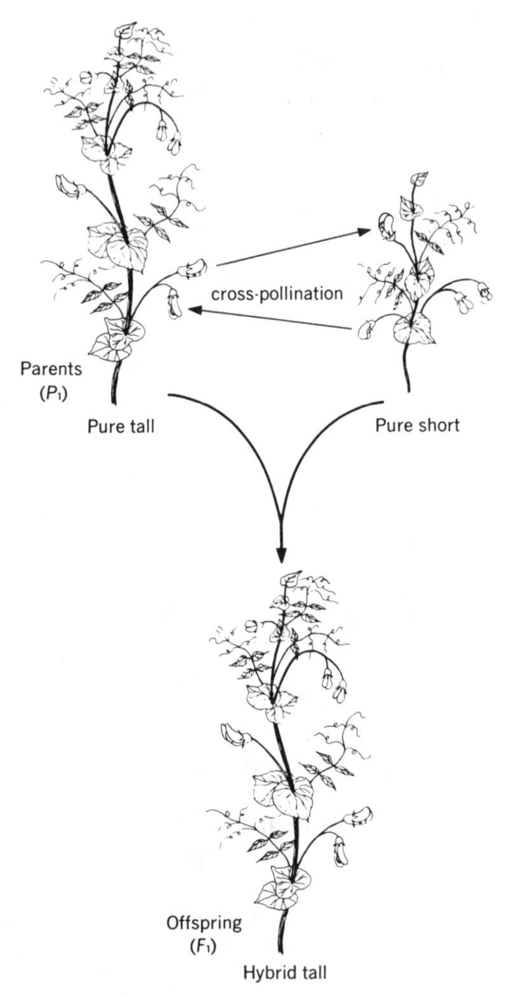

Figure 14–1 Mendel's experiment: cross-pollination of contrasting pure types

Mendel then cross-pollinated the flowers of plants with contrasting traits. He transferred pollen from the anthers of long-stemmed plants to the stigmas of short-stemmed plants (Figure 14–1). He also transferred pollen from short-stemmed plants to the stigmas of flowers of long-stemmed plants. Mendel named these plants the *first parental generation* (P_1) and the seeds he obtained from the matings the *first filial generation* (F_1).

Mendel performed similar crosses with plants possessing other contrasting pure traits, such as smooth seed and wrinkled seed (listed in the second column of Table 14–1). In the spring, Mendel planted the seeds produced from these crosses. In time, he observed that the offspring developed only one of the contrasting traits, listed in the third column of Table 14–1. These results led Mendel to formulate his *law of dominance*: When plants pure for contrasting traits are mated, all of the offspring of the F_1 show only one of the traits. The trait that appears is the *dominant* trait; the trait that does not appear, or that is hidden, is the *recessive* trait. All of the offspring of the F_1 were *hybrids*. Each hybrid received the two contrasting traits, although only one of the traits appeared.

Both a male and a female gamete are involved in fertilization. Thus, Mendel reasoned that one character for a trait is present in the male gamete, the other character for a trait is present in the female gamete. He also reasoned that a parent plant *pure* for a given trait has two like characters for the trait, and that all of the plant's sex cells possess the character for this trait.

Law of Segregation and Recombination

Mendel permitted the hybrids of the F_1 to undergo self-pollination to find out what happened to the hidden, or recessive, character in a hybrid (Figure 14–2). The offspring of the F_1, or first filial generation, that were produced by these matings he called the *second filial generation* (F_2). Table 14–2 summarizes his findings.

Mendel's analysis of the results indicated that most plants of the F_2 showed the dominant trait, and a smaller number showed the recessive trait. The ratio of plants showing the dominant trait to plants showing the recessive trait was about three to one (3:1), or 75 percent to 25 percent (fourth column of Table 14–2). From these observations Mendel formulated his second law of inheritance, called the *law of segregation and recombination:* When two hybrid plants are crossed, the contrasting characters for a given trait segregate (separate) when sex cells are formed. When fertilization occurs, it is possible for the recessive characters to recombine and appear in some of the offspring.

Probability. Mendel permitted plants of the F_2 generation to undergo self-pollination. After examining the offspring from these crosses, he observed that plants with a certain recessive trait, when mated to other plants with the same recessive trait, always produced offspring with the same recessive trait. In effect, the parent plants were pure for the recessive trait. Mendel also observed that when plants with a certain dominant trait were crossed to other plants with the same dominant trait, most of the offspring showed the dominant trait and others showed the recessive trait. After counting and analyzing the results, Mendel concluded that the

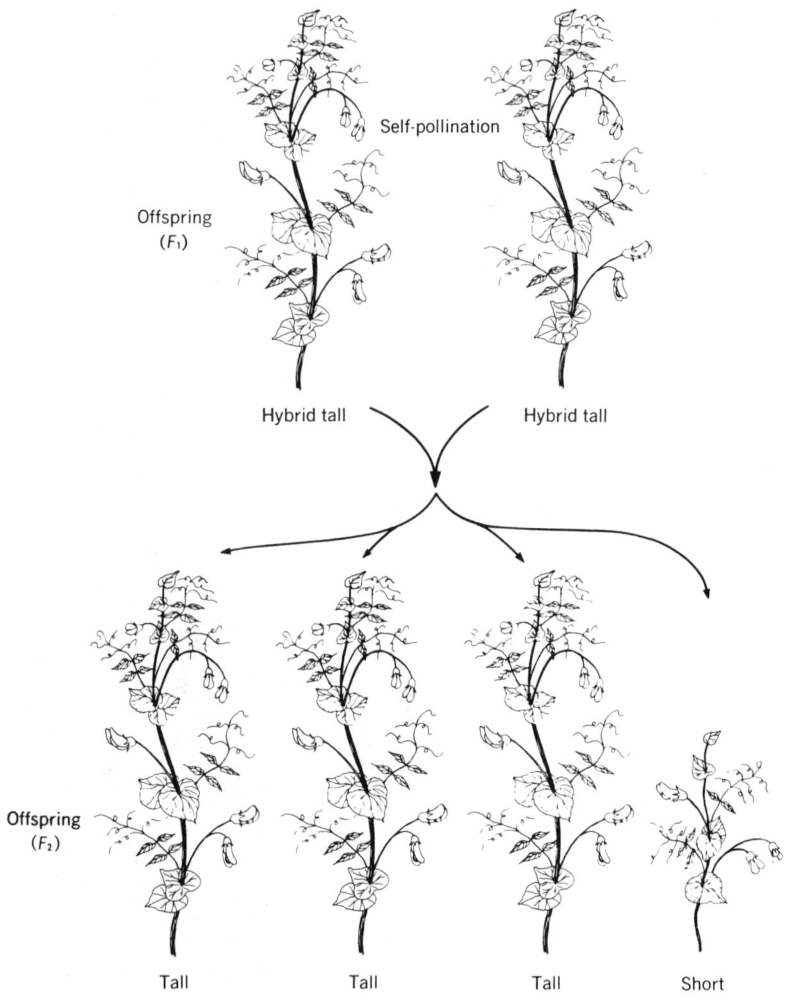

Figure 14–2 Mendel's experiment: self-pollination of hybrids

3/4:1/4 ratio (those plants showing the dominant trait to those plants showing the recessive trait) of the F_2 generation was actually 1/4 pure dominant: 2/4 hybrid dominant: 1/4 pure recessive, or a 1:2:1 ratio.

Mendel realized that these ratios could be attributed to the laws of probability, or chance. For example, if you toss two pennies at the same time for 100 times, you will get two heads about 25 percent of the time, a head and a tail about 50 percent of the time, and two tails about 25 percent of the time—a 1:2:1 ratio due to chance. Each penny represents a "hybrid" with two contrasting factors, a head and a tail. Each toss of the two coins represents a chance fertilization. Chance is involved in tossing coins

Table 14–2 Self-Pollination of Hybrids

Trait	*F_1 Hybrids*	*F_2*	*Ratio*
Length of stem	100% long	787 long (tall) 277 short	2.84 : 1
Type of seed coat	100% smooth	5,474 smooth 1,850 wrinkled	2.96 : 1
Color of seed coat	100% colored	705 colored 224 white	3.15 : 1
Form of seed pod	100% thick	882 thick 288 thin	2.95 : 1
Position of flower	100% side of stem	651 side of stem 207 end of stem	3.14 : 1
Color of cotyledons	100% yellow	6,022 yellow 2,001 green	3.01 : 1
Color of pod	100% green	428 green 152 yellow	2.82 : 1

as well as in the meeting of gametes during fertilization. Thus, a sex cell carrying the character for a dominant trait may be fertilized by a sex cell carrying a character for either the dominant or the recessive trait. Chance operating when gametes recombine during fertilization results in the 1:2:1 ratio of the F_2 generation.

Law of Independent Assortment

After studying the inheritance of two traits at once, Mendel formulated his third law of heredity, the *law of independent assortment:* Each character for a trait operates as a unit and is inherited independently of any other character. In one experiment, Mendel mated plants pure for long stems and yellow seeds with plants pure for short stems and green seeds. All the offspring had long stems and yellow seeds. He called the F_1 offspring *dihybrids* because they were hybrids for each trait (length of stem and color of seed). After Mendel mated the dihybrids, an analysis of the offspring of the F_2 generation did not show a 3:1 ratio. Instead, it showed four different types of plants, of which 9/16 had long stems and yellow seeds, 3/16 had long stems and green seeds, 3/16 had short stems and yellow seeds, and 1/16 had short stems and green seeds. He repeated the experiments with other dihybrid combinations and obtained the same results: a 9:3:3:1 ratio. The explanation for these results is discussed later in this chapter.

THE CHROMOSOME THEORY AND THE GENE THEORY OF INHERITANCE

Although Mendel's discovery was announced in 1865, it did not become widely known until 1900. Shortly after, modern concepts of heredity began to be developed by other scientists.

The Chromosome Theory of Inheritance

By 1900, many biologists had observed chromosomes and their behavior in mitosis and meiosis. *Walter Sutton*, a geneticist, compared the behavior of chromosomes with the behavior of the hereditary characters that Mendel had proposed. Sutton concluded that the characters that control heredity are located in the chromosomes because both the Mendelian characters and chromosomes (1) exist in pairs; (2) segregate during meiosis when gametes are formed; (3) recombine at fertilization; and (4) segregate independently of each other and pass from parents to offspring at fertilization.

The Gene Theory of Inheritance

When biologists compared the number of inherited traits of an organism with its species number of chromosomes, it was discovered that organisms have many more traits than they have chromosomes. For example, the American geneticist *Thomas Hunt Morgan* and his associates discovered hundreds of traits in the fruit fly, which only has four pairs of chromosomes. Morgan reasoned that only a small section of a chromosome, called a *gene*, determines a particular trait, and that a chromosome contains many genes.

Modern biologists view the gene as a specific part of a DNA molecule that possesses a code for a particular trait. The genes responsible for a particular trait occupy corresponding locations in each homologous (similar) chromosome of a pair. If the matching genes control contrasting characters, such as smooth seed and wrinkled seed in peas, the genes are called *alleles* (see Figure 14–3).

Phenotype and genotype. The physical appearance of a trait in an organism is called its *phenotype*. If an organism has two similar genes in homologous chromosomes, the organism is *pure* for the trait. The organism is *hybrid* for the trait if it possesses two *alleles*, or dissimilar genes. The *genotype* of a trait is the actual genetic composition of the trait. Thus, an organism pure for a given trait and one that is hybrid for the same trait have different genotypes. An organism whose genotype is pure for a particular trait is called *homozygous*. The genotype of an organism hybrid for

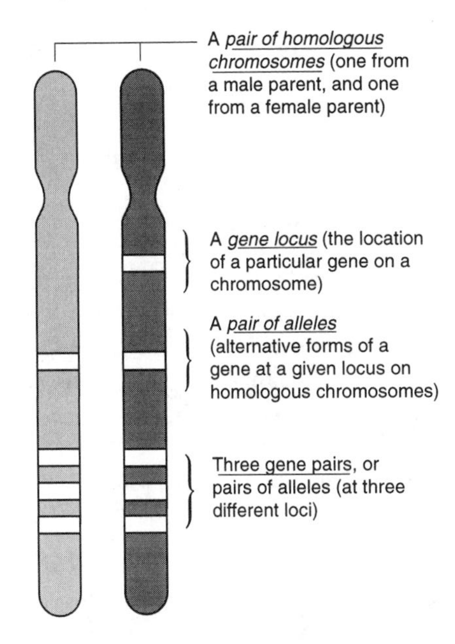

Figure 14–3 Homologous chromosomes

a particular trait is called *heterozygous*. Thus, two long-stemmed pea plants with the same phenotype have different genotypes when one is homozygous and the other heterozygous for the trait.

Symbols for genes. In Mendel's time, chromosomes and genes were unknown, so traits were referred to as characters. Today, the word *gene* is used instead of character or determiner. And, letters are used to designate traits; the letters represent the genes that determine the traits. In pea plants, for example, a capital *G* may be used to designate the dominant gene for green pod and a small g may be used to designate the recessive allele for yellow pod. Thus, a pea plant that bears pure, or homozygous, green pods can be designated as *GG;* the hybrid, or heterozygous, plant would then be designated as *Gg*; and the pure recessive, or plant bearing yellow pods, would be designated as *gg*. Note that in the case of a hybrid, the capital letter is written first.

EXPLAINING MENDEL'S EXPERIMENTS

To trace the inheritance of a particular trait, let us use the following guidelines:

1. Decide upon the symbols for each gene and show them in a key.
2. Show the phenotype of each parent.
3. Show the genotype of each parent.

4. Show the gene makeup in the gametes of each parent. Remember that each gamete contains only one gene, or allele, for a particular trait.
5. Show the possible fertilizations.
6. Express the results as ratios and percentages for the possible phenotypes and genotypes.

Tracing the Inheritance of One Trait at a Time

Now, let us apply these guidelines to study the inheritance of the color of pods in pea plants. Represented below are the results of a cross between a pure (homozygous) plant bearing green pods and a pure (homozygous) plant bearing yellow pods, thus:

KEY: Let *G* = gene for green pod
g = gene for yellow pod

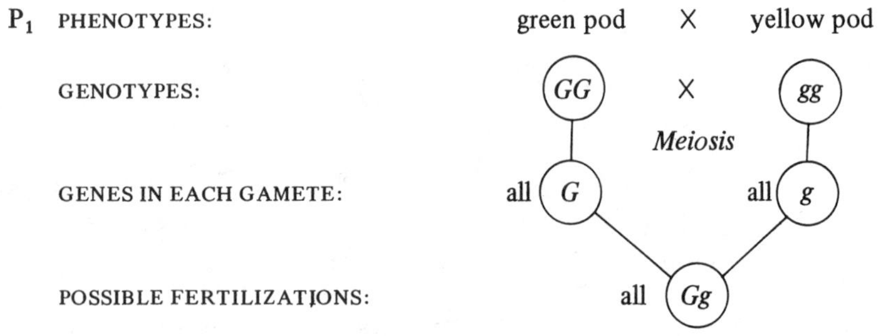

P_1 PHENOTYPES: green pod × yellow pod

GENOTYPES: (*GG*) × (*gg*)

Meiosis

GENES IN EACH GAMETE: all (*G*) all (*g*)

POSSIBLE FERTILIZATIONS: all (*Gg*)

RESULTS OF FERTILIZATIONS: all (100%) *Gg*—heterozygous green pods

The gametes formed by each parent contain the possible genes for the particular trait. Thus, the *GG* parent produces gametes all of which contain the dominant gene (*G*) for green pod color; the gg parent produces gametes all of which contain the recessive gene (*g*) for yellow pod color. The only possible combination of genes formed at fertilization is *Gg*. The effect of the recessive gene is not visible because the dominant gene allows only the effect of the dominant gene to be seen.

Using Punnett squares. The *Punnett square* is used to avoid overlooking any possible combination of gametes in a cross. A Punnett square is a diagram with as many vertical columns as there are types of gametes formed by one parent, and as many horizontal columns as there are types of gametes formed by the other parent. The genes (letters) are combined and written in the small squares formed by the intersecting columns, as shown below. Each small square represents both the genotype of the zygote and the percentage of this genotype. The large square represents 100 percent of the possible fertilizations.

Using the same key and the Punnett square, let us study the inheritance of the color of pods in pea plants again, thus:

P_1 PHENOTYPES: green pod × yellow pod

GENOTYPES:

GENES IN EACH GAMETE:

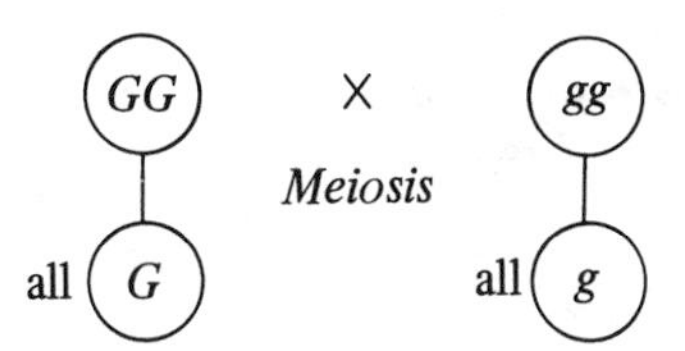

POSSIBLE FERTILIZATIONS:

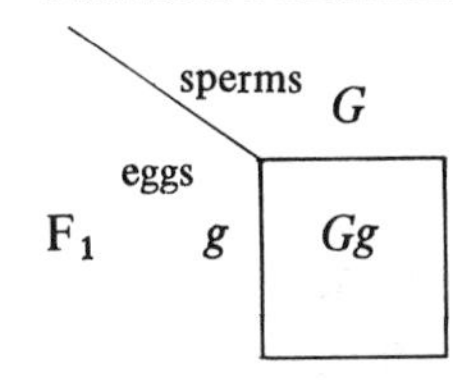

RATIOS FOR F_1 GENERATION:

Genotypes	*Phenotypes*
$\frac{1}{1}$ *Gg*	100% green pods

Now let us cross the heterozygous F_1 offspring (hybrid × hybrid).

F_1 PHENOTYPES: green pod × green pod

GENOTYPES:

GENES IN EACH GAMETE:

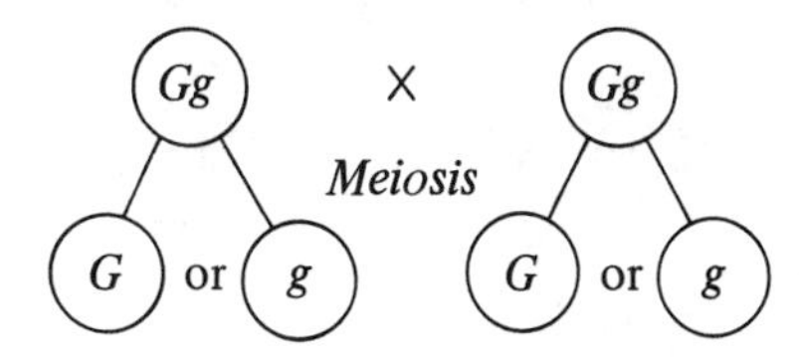

POSSIBLE FERTILIZATIONS:

eggs \ sperms	*G*	*g*
G	*GG*	*Gg*
g	*Gg*	*gg*

RATIOS FOR F_2 GENERATION:

Genotypes	*Phenotypes*
$\frac{1}{4}$ *GG*	25% green pods
$\frac{2}{4}$ *Gg*	50% green pods
$\frac{1}{4}$ *gg*	25% yellow pods

In this cross, the two types of gametes are *G* and *g*. Thus, a square is made with two vertical and two horizontal columns. Each small square within the large square contains the genotype of the individual formed by the fertilization of a particular sperm with a particular egg.

The table below shows the six possible crosses involving the inheritance of a specific trait. You can see that in crosses involving homozygous

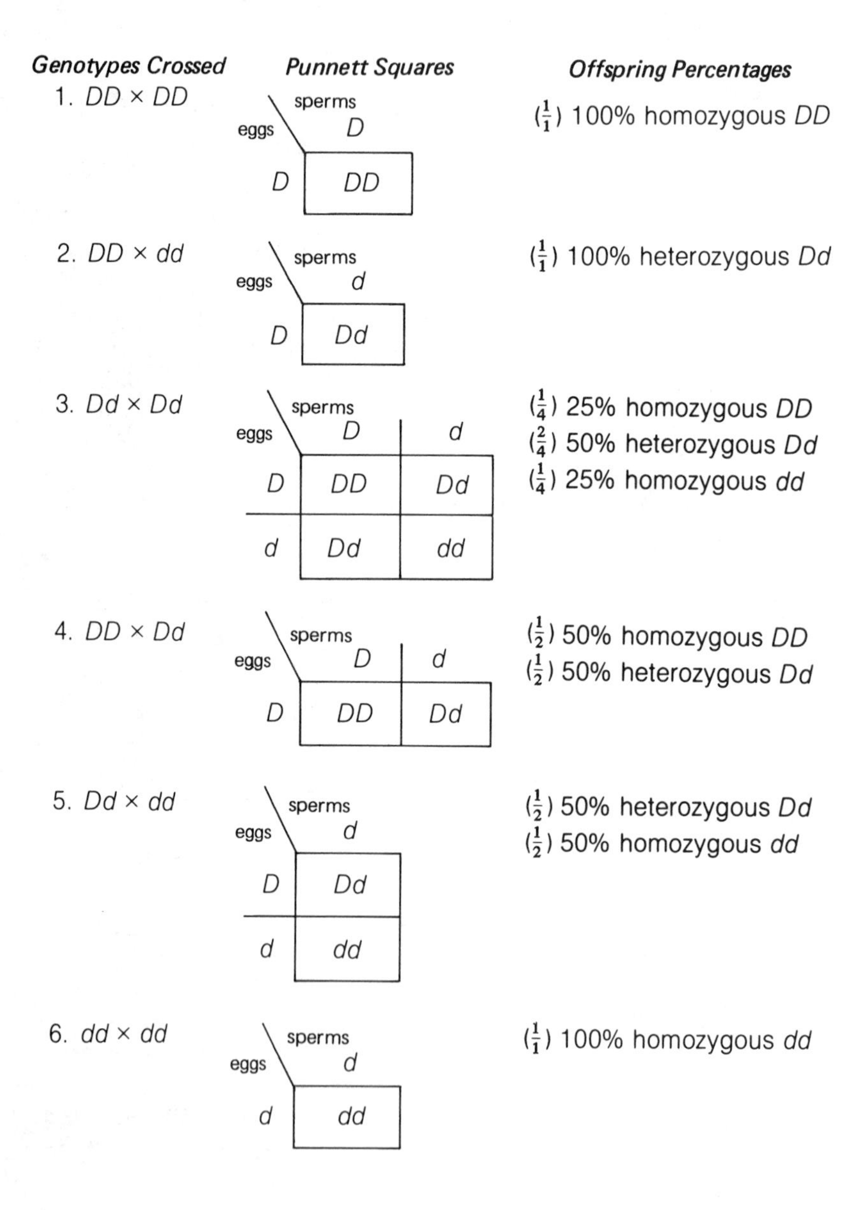

parents (crosses 1, 2, and 6), 100 percent of the offspring are alike, as the offspring of each cross has the same phenotype and genotype. If one parent is homozygous and the other parent is heterozygous (crosses 4 and 5), 50 percent of the offspring are homozygous and 50 percent are heterozygous. If both parents are heterozygous (cross 3), the ratio is 1:2:1, or 25 percent pure dominant, 50 percent hybrid dominant, and 25 percent recessive.

Tracing the Inheritance of Two Traits at a Time

Mendel's first two laws are related to the inheritance of only one trait. However, organisms are usually neither pure (homozygous) for all their traits, nor hybrids (heterozygous) for all their traits. Most organisms usually possess a combination of homozygous and heterozygous genotypes for hundreds or even thousands of traits. Mendel realized this and decided to investigate the inheritance of two traits at a time. He selected plants that were homozygous for long stems and yellow seed color (both homozygous dominant traits) and crossed them with pea plants that had short stems and green seeds (both homozygous recessive traits). All of the offspring had long stems and yellow seeds, thus:

KEY: Let L = gene for long stem (dominant)
l = gene for short stem (recessive)
Y = gene for yellow seed (dominant)
y = gene for green seed (recessive)

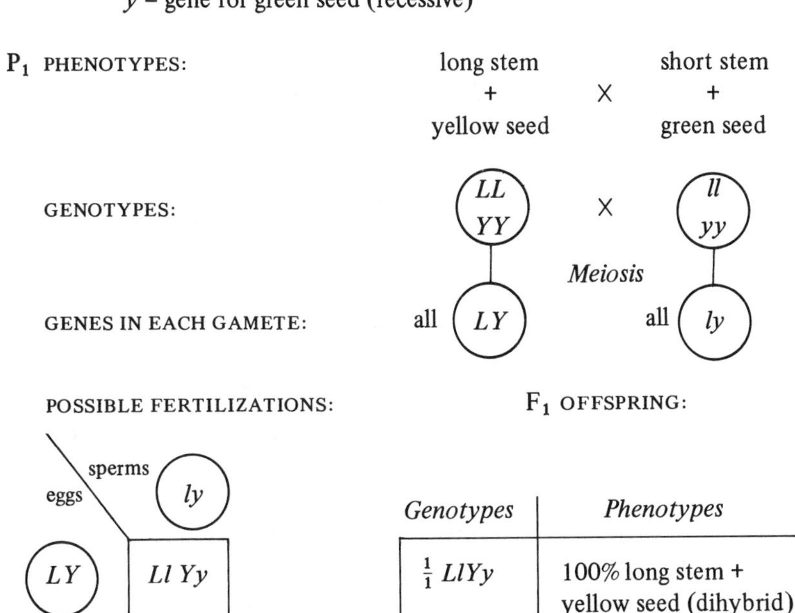

Genotypes	*Phenotypes*
$\frac{1}{1}$ *LlYy*	100% long stem + yellow seed (dihybrid)

You can see that the F_1 offspring are hybrid for both length of stem and seed color. An organism that is heterozygous for two traits is called a *dihybrid*; an organism that is hybrid for three traits is called a *trihybrid*; and so on.

Mendel then cross-pollinated the dihybrid plants of the F_1. The offspring of the F_2 consisted of four different types of plants in the approximate ratio of 9/16 long stems and yellow seeds, 3/16 long stems and green seeds, 3/16 short stems and yellow seeds, and 1/16 short stems and green seeds (or 9:3:3:1). Mendel repeated the experiment with dihybrids for other traits, such as pod color and flower position. Mendel noticed that the offspring were produced in a similar ratio. Keep in mind that the larger the number of offspring, the closer the results will approach the expected ratio of 9:3:3:1.

As a result of his experiments, Mendel formulated the *law of independent assortment.* This law states that the genes for height and the genes for seed color are located in different pairs of chromosomes. During meiosis, allelic genes separate independently of each other, as shown below in the gametes of the F_1. At fertilization, the genes unite in different combinations.

Using a Punnett square, you can see why Mendel obtained a 9:3:3:1 ratio in the F_2 of a cross between pea plants that were dihybrid for long stem and yellow seeds, thus:

F_1 PHENOTYPES:

long stem + yellow seed × long stem + yellow seed

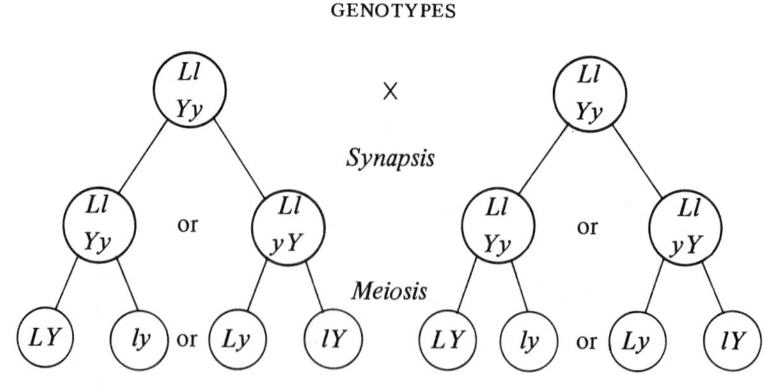

GENES IN EACH GAMETE

POSSIBLE FERTILIZATIONS:

eggs \ sperms	*LY*	*Ly*	*lY*	*ly*
LY	*LLYY*	*LLYy*	*LlYY*	*LlYy*
Ly	*LLYy*	*LLyy*	*LlYy*	*Llyy*
lY	*LlYY*	*LlYy*	*llYY*	*llYy*
ly	*LlYy*	*Llyy*	*llYy*	*llyy*

RATIO OF F_2 OFFSPRING:

Genotypes	*Phenotypes*		
$\frac{1}{16}$ *LLYY*	$\frac{1}{16}$ tall, yellow-seeded	6.25%	$\frac{9}{16}$ tall, yellow
$\frac{2}{16}$ *LLYy*	$\frac{2}{16}$ tall, yellow-seeded	12.5%	
$\frac{2}{16}$ *LlYY*	$\frac{2}{16}$ tall, yellow-seeded	12.5%	
$\frac{4}{16}$ *LlYy*	$\frac{4}{16}$ tall, yellow-seeded	25%	
$\frac{1}{16}$ *LLyy*	$\frac{1}{16}$ tall, green-seeded	6.25%	$\frac{3}{16}$ tall, green
$\frac{2}{16}$ *Llyy*	$\frac{2}{16}$ tall, green-seeded	12.5%	
$\frac{1}{16}$ *llYY*	$\frac{1}{16}$ short, yellow-seeded	6.25%	$\frac{3}{16}$ short, yellow
$\frac{2}{16}$ *llYy*	$\frac{2}{16}$ short, yellow-seeded	12.5%	
$\frac{1}{16}$ *llyy*	$\frac{1}{16}$ short, green-seeded	6.25%	$\frac{1}{16}$ short, green

An examination of the results of the dihybrid cross reveals that 12 plants have long stems and four have short stems—a ratio of 3:1. Regarding seed color, 12 plants are yellow and four are green—a separate ratio of 3:1. In other words, the genes for stem length are inherited separately from the genes for seed color (independent assortment). Also, each character is inherited as a unit—the inheritance of stem length has no effect upon the inheritance of seed color.

Section Quiz

1. One reason why Mendel selected garden peas is because they (*a*) germinate very rapidly (*b*) have small flowers (*c*) have large fruits (*d*) can be easily pollinated by hand.
2. The ratio of the F_2 generation when gametes recombine is (*a*) 1:2:1 (*b*) 1:3:1 (*c*) 1:4:1 (*d*) 3:1:1.

3. The physical appearance (gene expression) of a certain trait in an organism is called the (*a*) genotype (*b*) phenotype (*c*) allele (*d*) chromosome.
4. Mendel was able to formulate the law of segregation and recombination when he had (*a*) produced hybrids (*b*) produced mutations (*c*) recorded F_1 results (*d*) counted F_2 types.
5. Which cross illustrates Mendel's principle of segregation and recombination? (*a*) BB × bb (*b*) Bb × Bb (*c*) bb × bb (*d*) BB × BB.

MENDELIAN INHERITANCE IN OTHER ORGANISMS

Drosophila (*Fruit Fly*)

T. H. Morgan and his associates used the fruit fly (Figure 14–4) in their experiments for the following reasons:

1. The fruit fly has a short life cycle (10 to 12 days), which enabled Morgan to observe many generations in one year.
2. The fruit fly produces many offspring at one time.
3. The fruit fly is easily reared in small bottles.
4. The fruit fly's body cells contain only four pairs of chromosomes.

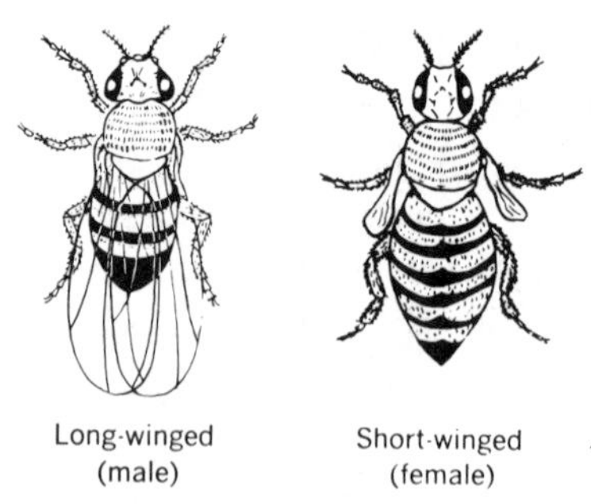

Figure 14–4 Fruit flies (enlarged)

In the fruit fly, long wings are dominant to short (vestigial) wings; red eyes are dominant to white eyes; and yellow body color is dominant to black body color. Morgan discovered that when homozygous long-winged flies were mated to homozygous short-winged flies, all the offspring of the crosses (F_1) had long wings. These results agreed with Mendel's law of dominance. To find out if the law of segregation and recombination worked for fruit flies as it did for Mendel's pea plants, Morgan allowed the males and females of the F_1 to mate, thus:

KEY: Let L = gene for long wing (dominant)
l = gene for short wing (recessive)

F_1 PHENOTYPES: long wing × long wing

GENOTYPES: Ll × Ll

Meiosis

GENES IN EACH GAMETE: L or l L or l

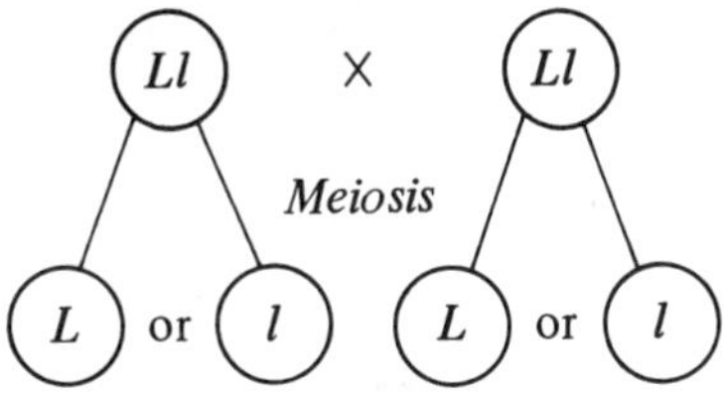

POSSIBLE FERTILIZATIONS:

eggs \ sperms	L	l
L	LL	Ll
l	Ll	ll

F_2 OFFSPRING:

Genotypes	*Phenotypes*
$\frac{1}{4}$ LL	25% long wing
$\frac{1}{2}$ Ll	50% long wing
$\frac{1}{4}$ ll	25% short wing

These results, the 1:2:1 ratio, or 25 percent pure dominant, 50 percent hybrid dominant, and 25 percent recessive, agreed with Mendel's second law of heredity.

Guinea Pig

In the guinea pig, black fur is dominant to white fur, and shaggy fur is dominant to smooth fur. The inheritance of fur color in the guinea pig is shown in Figure 14–5.

Corn

Corn is frequently used to study inheritance because each ear of corn contains many kernels, or seeds, each of which is the result of a separate fertilization. In corn, purple seed color is dominant to yellow, and smooth seed is dominant to wrinkled seed. If an ear of corn contains 308 purple kernels and 96 yellow kernels, which generation does it represent? If your answer is "F_2 generation," you are correct because only when a hybrid is crossed with another hybrid (F_1) can you expect a phenotype ratio of 3:1.

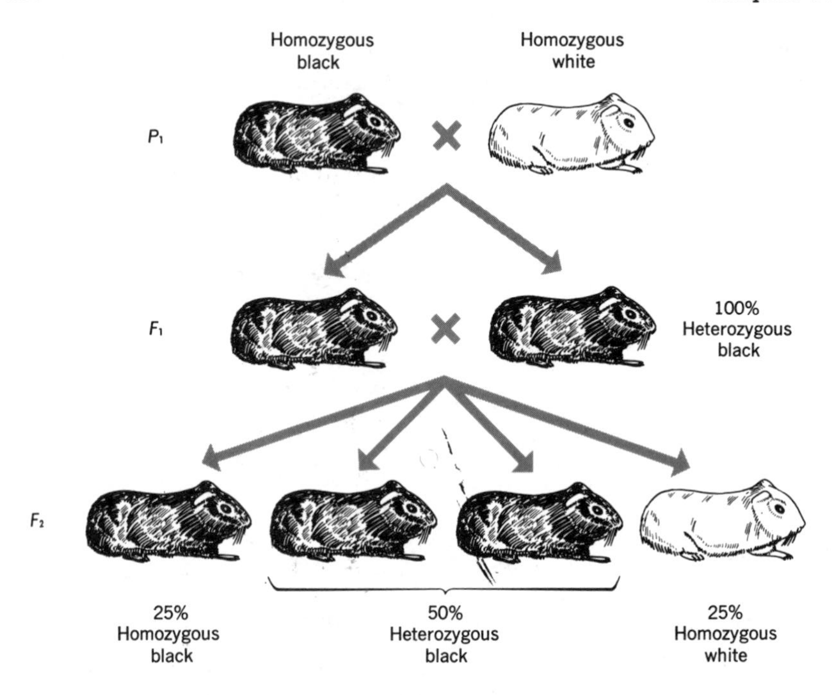

Figure 14–5 Inheritance of coat color in guinea pigs

Microorganisms

Bacteria and other microorganisms are favored for genetic studies for the following reasons:

1. They reproduce rapidly, producing many generations in several hours. This makes it easier to collect and analyze experimental data rapidly.
2. Large populations can be easily grown, stored, and managed. They do not involve problems of feeding, waste removal, and temperature regulation. By selectively varying the temperature, composition of nutrient material, water availability, and other metabolic factors, abnormal forms or new adaptations may arise. In each case, the DNA is studied and analyzed and, finally, related to an organism's metabolism or change in metabolism.
3. Some microorganisms, such as certain fungi, undergo a type of alternation of generations in which their monoploid stage (n) is conspicuous and long-lived. The monoploid generation is expressed phenotypically by a *single set* of either dominant or recessive genes. There is no need to testcross to determine genotype, as with diploid ($2n$) organisms.

Thus, chromosome and gene studies are undertaken more often at the cellular level than at the multicellular level. The reason is that genetic research is easier using a small amount of simple DNA. In Chapter 15, you will learn about recombinant DNA techniques that utilize bacteria as vectors.

THE TESTCROSS

Organisms that are homozygous dominant for a trait have the same phenotype as organisms that are heterozygous dominant for the trait. Such individuals are crossed with a homozygous recessive to determine the genotypes. This experimental procedure is called a *testcross* or *back-cross*.

For example, to determine whether a black guinea pig is pure (homozygous) or hybrid (heterozygous) for black color, the guinea pig is mated with several white (recessive) guinea pigs. If all the offspring are black, the genotype of the black parent must be homozygous, thus:

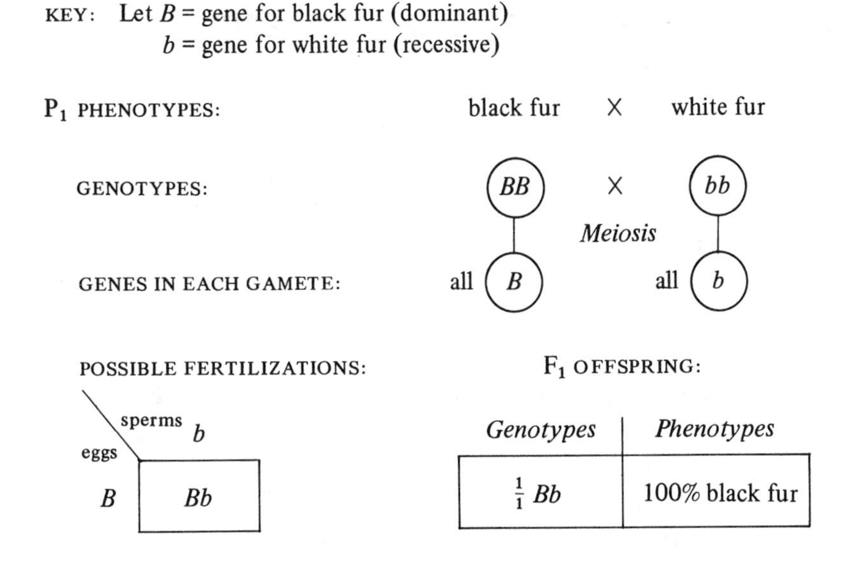

Genotypes	*Phenotypes*
$\frac{1}{1}$ *Bb*	100% black fur

If some offspring are black and some are white, the genotype of the black parent must be heterozygous, thus:

P_1 PHENOTYPES: black fur X white fur

GENOTYPES:

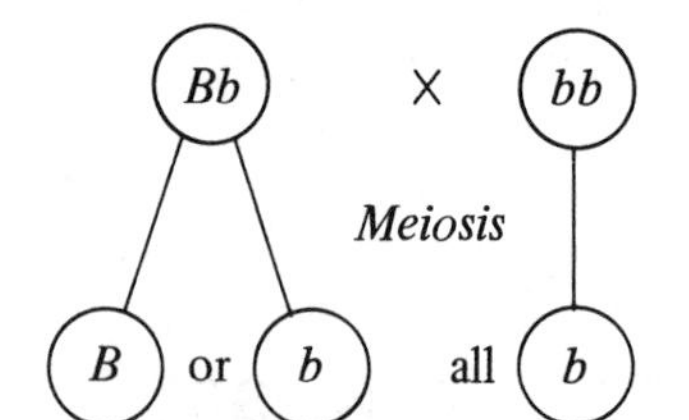

GENES IN EACH GAMETE:

POSSIBLE FERTILIZATIONS:

eggs \ sperms	*b*
B	*Bb*
b	*bb*

F_1 OFFSPRING:

Genotypes	*Phenotypes*
$\frac{1}{2}$ *Bb*	50% black fur
$\frac{1}{2}$ *bb*	50% white fur

INCOMPLETE DOMINANCE (BLENDING INHERITANCE)

When inheritance was studied in other organisms, biologists discovered that for some traits Mendel's law of dominance did not apply because the F_1 offspring did not resemble either parent. Instead, the appearance of the offspring was a blend of the contrasting traits of both parents. This lack of dominance in traits is called *incomplete dominance* or *blending inheritance*.

Flower Color in Japanese Four O'Clock Plants

When red-flowered four o'clock plants are crossed with white-flowered four o'clock plants, all of the F_1 offspring are pink (see Figure 14–6). Since neither the allele for red flower color nor that for white flower color is dominant, the letters for both alleles are capitalized.

You can see that in the F_2 generation the phenotype ratio is 1:2:1, not 3:1. Thus, 25 percent of the offspring are red *(RR)*, 50 percent are pink *(RW)*, and 25 percent are white *(WW)*. In this case, Mendel's law of segregation and recombination applies rather than his law of dominance. Since the hybrids are pink, the phenotype indicates the genotype.

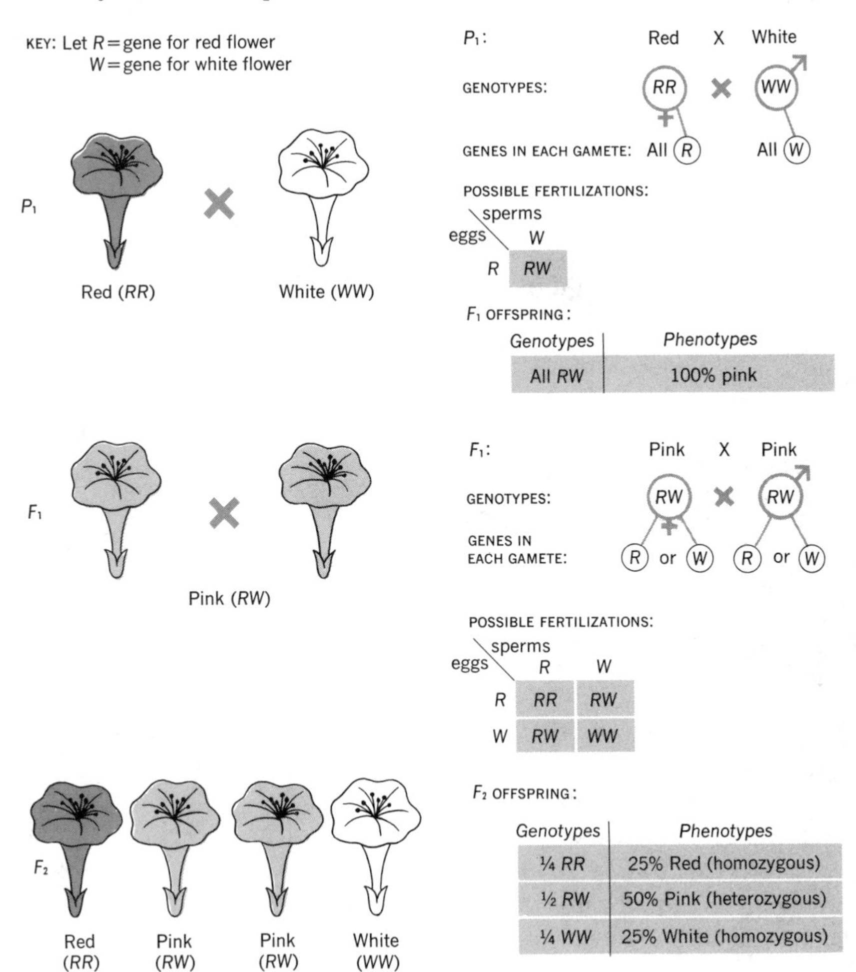

Figure 14–6 Blending inheritance in Japanese four o'clocks

Feather Color in Andalusian Chickens

Figure 14–7 shows that when black Andalusian chickens are crossed with white Andalusian chickens, all of the F_1 offspring are blue. The blue appearance of the feathers is due to the way light is reflected from the sequence of black and white markings on the feathers. As with four o'clock plants, neither the allele for black feathers nor the allele for white feathers is dominant.

In the F_2 generation the phenotype ratio is 1:2:1, not 3:1. Thus, 25 percent of the offspring are black (*BB*), 50 percent are blue (*BW*), and 25 percent are white (*WW*).

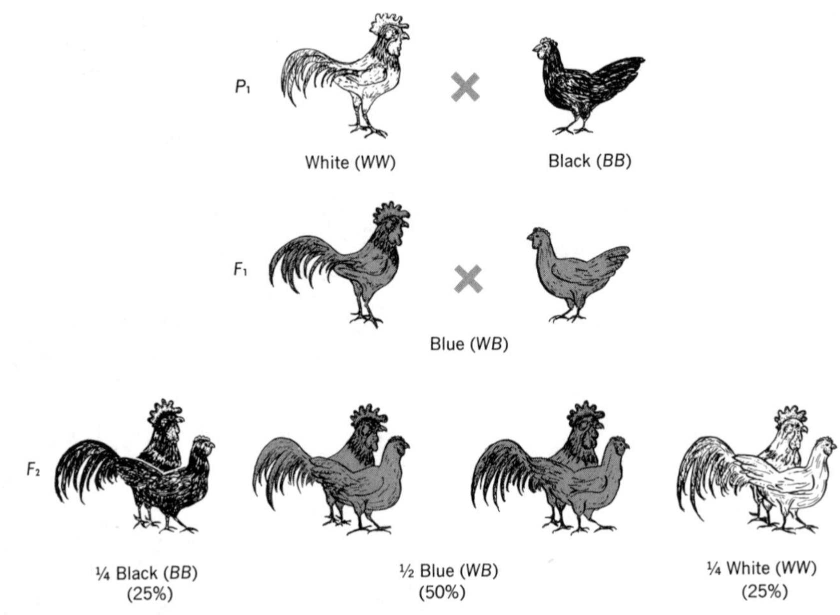

Figure 14–7 Blending inheritance in Andalusian chickens

Mendel showed that the inheritance of traits from generation to generation can be explained by his laws of heredity. Other biologists, including Sutton and Morgan, modified and extended Mendel's ideas by developing the idea of chromosomes and genes and explaining certain exceptions to the law of dominance. In the next chapter we will consider principles of heredity unknown to Mendel and also consider the changes in heredity that affect evolution.

Section Quiz

1. The body cells of a worm contain only two chromosomes. The probability that any two traits will be on the same chromosome is about (*a*) 100 percent (*b*) 75 percent (*c*) 50 percent (*d*) 25 percent.
2. Mendel's characters, or factors, are comparable to modern (*a*) chromosomes (*b*) genomes (*c*) genes (*d*) genotypes.
3. The scientist who first presented clear evidence that Sutton's chromosome idea was correct and that Mendel's characters were located on chromosomes is (*a*) Brown (*b*) Morgan (*c*) Watson (*d*) Punnett.

4. An organism that is homozygous recessive for a trait is crossed with an organism that is heterozygous dominant for the same trait. This is called (*a*) recombination (*b*) a testcross (*c*) segregation (*d*) a phenotype.
5. When offspring appear as a blend of the contrasting traits of both parents, this illustrates (*a*) complete dominance (*b*) incomplete dominance (*c*) blending inheritance (*d*) answers b and c.

Chapter Review Questions

The following questions will help you check your understanding of the material presented in the chapter.

1. Genetic material is generally thought to consist of phosphate combined with (*a*) proteins and sugar (*b*) nitrogen bases and sugar (*c*) proteins and lipids (*d*) sugars and amino acid.
2. Which scientist is credited with stating the principle that traits among offspring result from combinations of dominant and recessive unit characters? (*a*) Darwin (*b*) Watson (*c*) Lamarck (*d*) Mendel.
3. With reference to any single pair of alleles, if an organism produces two types of gametes, that organism is best described as (*a*) polyploid (*b*) heterozygous (*c*) homozygous (*d*) mutagenic.
4. The genotypes of organisms mainly are determined by the (*a*) environment they inhabit (*b*) genotypes of their parents (*c*) ratio of genes to chromosomes (*d*) phenotypes of their grandparents.
5. Normally it would be incorrect to use the term heterozygous to describe any (*a*) gamete (*b*) zygote (*c*) hybrid (*d*) genotype.
6. Because of a short life cycle and few chromosomes, an ideal species for experimental studies in genetics is the (*a*) garden pea (*b*) human (*c*) fruit fly (*d*) maple tree.
7. According to Mendel, when a pea plant homozygous for tallness is crossed with a pea plant homozygous for shortness, their offspring will resemble (*a*) neither parent (*b*) both parents (*c*) the hybrid parent only (*d*) the dominant parent only.
8. The phenotype of a pea plant can be determined most easily by (*a*) looking at it (*b*) crossing it with a recessive plant (*c*) crossing it with a similar plant (*d*) looking at the parents.

9. The letter T represents the gene for the dominant trait (tall) and the letter t represents the gene for the recessive trait (short). Which statement concerning the offspring is most likely correct if a cross is made between Tt and Tt? (Assume a large number of offspring.) (*a*) All the offspring will be short. (*b*) All the offspring will be tall. (*c*) 50 percent of the offspring will be short. (*d*) 75 percent of the offspring will be tall.

10. In pea plants, tallness is dominant over shortness. If 50 percent of one generation of pea plants are short, what were the probable genotypes of the parents? (*a*) $TT \times tt$ (*b*) $Tt \times tt$ (*c*) $Tt \times Tt$ (*d*) $tt \times tt$.

11. An investigator destroys all the white-eyed fruit flies in a culture and maintains only the red-eyed flies. How can the production of white-eyed flies in future generations of the cultures be explained? (*a*) The white-eyed flies are homozygous dominant. (*b*) The gene for white eyes is incompletely dominant. (*c*) The red-eyed flies were pure recessive. (*d*) Some of the red- eyed flies were heterozygous.

12. In humans, right-handedness is dominant over left-handedness. What are the most probable genotypes of two right-handed parents who have a left-handed child? (*a*) $RR \times RR$ (*b*) $RR \times rr$ (*c*) $Rr \times RR$ (*d*) $Rr \times Rr$.

13. Although albino plants die before producing seeds, the albino trait does not disappear because (*a*) homozygous normal plants become albinos when deprived of light (*b*) albinism prevents the synthesis of chlorophyll (*c*) darkness activates the gene for albinism (*d*) many green plants are heterozygous.

14. Which of the crosses below best illustrates Mendel's principle of segregation? (*a*) $BB \times bb$ (*b*) $Bb \times Bb$ (*c*) $bb \times bb$ (*d*) $BB \times BB$.

15. If a homozygous dominant individual is crossed with a heterozygous dominant individual, the probability that any offspring will exhibit the recessive trait is (*a*) 0 (*b*) 1/8 (*c*) 1/4 (*d*) 1/2.

16. A genetic cross that may yield recessive offspring is best represented by (*a*) $TT \times tt$ (*b*) $TT \times Tt$ (*c*) $TT \times TT$ (*d*) $Tt \times Tt$.

17. Curly hair is dominant over straight hair. What is the probability of a heterozygous curly-haired father and a homozygous curly-haired mother having a child with straight hair? (*a*) 100 percent (*b*) 50 percent (*c*) 25 percent (*d*) 0 percent.

18. If half the offspring of four o'clock plants are pink and half are white, the parents' phenotypes probably were (*a*) a red and a white (*b*) a pink and a red (*c*) both pink (*d*) a pink and a white.

19. A black guinea pig that has a smooth coat is mated with a white guinea pig that has a rough coat. The production of offspring that have smooth, white fur is an example of (*a*) autosomal linkage (*b*) sex linkage (*c*) nondisjunction (*d*) independent assortment.
20. To determine whether an organism is homozygous dominant, it should be crossed with an organism that is (*a*) recessive (*b*) heterozygous (*c*) hybrid (*d*) homozygous dominant.
21. A phenotype that appears for the first time and differs from both parents is directly related to a missing enzyme. This probably was caused by a change in the (*a*) environment (*b*) nucleotide sequence (*c*) reproductive method (*d*) centrosome structure.
22. Four o'clock flowers exhibit incomplete dominance, so that a crossing of red- and white-flowered plants produces plants that bear pink flowers. If the gene for red is represented by *R*, then the pink flower's allele pattern is (*a*) *PP* (*b*) *Pw* (*c*) *RW* (*d*) *RR*.
23. A cross between a white rooster and a black hen results in 100 percent blue Andalusian offspring. When two of these blue Andalusians are crossed the offspring probably will be (*a*) 100 percent white (*b*) 75 percent black, 25 percent white (*c*) 75 percent blue, 25 percent white (*d*) 25 percent white, 25 percent black, 50 percent blue.
24. When round squash are crossed with long squash, all offspring are oval in shape. How many different genotypes are expected when round squash are crossed with oval squash? (*a*) one (*b*) two (*c*) three (*d*) four.
25. When large numbers of roan cattle are interbred, percentages occur as follows: 25 percent red, 50 percent roan, and 25 percent white. These results illustrate (*a*) independent assortment (*b*) blending inheritance (*c*) dominance (*d*) natural selection.
26. When long radish plants were pollinated with pollen from round radish plants, the plants obtained produced oval radishes. This illustrates the principle of (*a*) dominance (*b*) segregation (*c*) independent assortment (*d*) incomplete dominance.
27. In drosophila, the allele for vestigial wings (*w*) is recessive to the normal long wing allele (*W*). The allele for white eye (*r*) is recessive to the allele for normal red eye (*R*). In a mating *WWRr* × *Wwrr* (assuming large numbers of offspring), what proportion of offspring will show normal traits (long wings and red eyes)? (*a*) 50 percent (*b*) 25 percent (*c*) zero percent (*d*) not enough information given to answer the problem.

Biology Challenge

Part I

The following questions will provide practice in answering SAT II-type questions.

1. The ears of women usually are hairless. It is reasonable to infer that (*a*) the sex chromosomes of females are homozygous for hairless ears (*b*) the Y chromosomes of males contains a dominant gene for hairy ears (*c*) hairless ears in females is a sex-linked trait (*d*) the X chromosome of males contains a recessive gene for hairy ears (*e*) genes play no part in determining hairlessness in either sex.
2. John and Mary can roll their tongues. Their six-year-old son cannot. A reasonable hypothesis for the son's inability to roll his tongue is (*a*) tongue rolling develops as a result of training and practice (*b*) both parents are homozygous dominant for tongue rolling (*c*) both parents are homozygous recessive for tongue rolling (*d*) both parents are heterozygous for the trait (*e*) during meiosis, crossing-over took place.
3. Red cattle can be produced by breeding a roan cow with a white bull. Which statement is *correct*? (*a*) 100 percent of the offspring resulting from a number of matings will be roan. (*b*) 100 percent of the offspring resulting from a number of matings will be white. (*c*) 50 percent of the offspring resulting from a number of matings will be red. (*d*) To establish a pure line of red cattle, breed red cattle with roan cattle. (*e*) To establish a pure line of red cattle, interbreed them.
4. Which statement is *incorrect*? (*a*) An organism whose normal number of chromosomes is 32, has 64 chromatids in its secondary oocyte. (*b*) In humans, black hair is dominant over red hair and having six fingers is dominant over five fingers. (*c*) In grasshoppers, an egg cell fertilized by a sperm cell lacking an X chromosome develops into a male. (*d*) If an allele is not dominant, it may be recessive. (*e*) An organism whose genotype is *CC* and *EE* for two traits is homozygous dominant for both traits.
5. Some scientists claim that luck played an important role in Mendel formulating his basic laws of inheritance. His luck was in selecting traits that were (*a*) not linked (*b*) different and linked (*c*) studied together (*d*) not pure (*e*) common to the garden pea and other legumes.

Part II

Questions *1 through 3* describe certain procedures associated with the objectives of different investigations. For each procedure, select the letter of the statement that best describes the objective.

1. Two long-winged fruit flies were mated. The offspring produced were 48 long-winged and 14 short-winged flies. The procedure was performed (*a*) to determine if long wings is sex-linked (*b*) to determine if short wings is dominant or recessive (*c*) to repeat Mendel's law of dominance (*d*) to initiate a population study (*e*) to study the relationship between ratios and sex of offspring.
2. A student removed the salivary glands of a caterpillar, squashed and mounted them on a microscope slide, stained the tissue, covered the slide with a cover-slip, and then observed the slide under low and high power. This enabled the student (*a*) to test for the presence of DNA (*b*) to study the structure of secretory cells (*c*) to observe the large chromosomes (*d*) to discover crossing-over locations (*e*) to identify different alleles.
3. In corn, the gene *R* for red color of seeds is dominant over gene *r* for green color. Gene *N* for normal seed is dominant over gene *n* for abnormal seed. The results of crossing a heterozygous red plant with normal seeds with a green plant with abnormal seeds were: 98 red, normal; 51 red, abnormal; 104 green, abnormal; 46 green, normal. The plants were crossed (*a*) to determine if linkage is involved (*b*) to determine parental genotypes (*c*) to study phenotypic ratios (*d*) to study crossover ratios (*e*) to perform a testcross.
4. Which statement is *correct*? (*a*) A grasshopper egg fertilized by a sperm cell lacking an X chromosome will develop into a female. (*b*) The A and B alleles of human blood types show that genes are not necessarily either dominant or recessive. (*c*) In peas, yellow seed color is dominant to green. A homozygous yellow seed plant crossed with a green seed plant will produce the following phenotypic ratio: 75 percent yellow, 25 percent green, 25 percent yellow, 50 percent green. (*d*) An organism with the genotype *Dd Ee Ff* may form six different types of gametes. (*e*) A testcross must involve a homozygous dominant for a particular trait.

15

Heredity: Further Principles

Learning Objectives

When you have completed this chapter, you should be able to:

- **Describe** ways in which the work of Morgan contributed to our understanding of sex determination and sex linkage.
- **Explain** how crossing-over causes an exchange of genetic material between chromosomes.
- **Discuss** ways in which multiple alleles affect the appearance of certain traits.
- **Describe** ways in which the environment affects gene expression.
- **Associate** changes in genetic material with gene mutations.
- **Discuss** ways in which genes control the chemical reactions that occur in cells.
- **Identify** some of the agents that cause gene mutations.
- **Discuss** ways in which genes influence the inheritance of some human traits.
- **Describe** ways in which scientists use recombinant DNA.

OVERVIEW

After the rediscovery of Mendel's laws of heredity in 1900, the development of the chromosome and gene principles of inheritance led to the discovery of other patterns of inheritance. Some of these patterns have influenced our ideas about evolution. In this chapter, you will learn about various patterns of inheritance in humans and other organisms.

MODERN PRINCIPLES OF HEREDITY

Major patterns of inheritance that were discovered after 1900 include linkage, sex determination, sex linkage, nondisjunction, crossing-over, multiple alleles, and the effect of the environment on genes.

Linkage

When Morgan mated fruit flies that were dihybrids, he observed a 3:1 ratio in the offspring of some of the crosses, not the 9:3:3:1 ratio he usually observed in the F_2 offspring of other species. For example, when he mated fruit flies that were dihybrid for yellow body and long wings, he obtained F_2 offspring 75 percent of which had a yellow body and long wings and 25 percent of which had a black body and short wings. Morgan reasoned that this ratio occurs because the genes for the two traits are located on the same chromosome, or are *linked.* Thus, the genes for yellow body and long wings are on one chromosome and the genes for black body and short wings are on the homologous chromosome. During meiosis, the genes for body color and wing length do not separate, but stay together and pass as a unit into a specific gamete. *Linkage* means that specific genes tend to stay together, to segregate together during meiosis, and to be inherited together independently of other gene groups in different chromosomes.

Sex Determination

In the fruit fly, the body cells of both males and females possess three pairs of homologous chromosomes that are alike. In the male, however, the fourth pair of chromosomes consist of two unmatched chromosomes; in the female, the fourth pair of chromosomes are matched (Figure 15–1). The fourth pair of chromosomes, called the *sex chromosomes,* determine the fruit fly's sex. In the female, the matched pair of sex chromosomes are called *X chromosomes*; in the male, the unmatched sex chromosomes are called the *X and Y chromosomes*. Thus, a male fruit fly is designated as XY and a female is designated as XX. The other three pairs of chromosomes, called *autosomes,* have no direct connection with *sex determination.*

Each body cell of a human male contains 22 pairs of autosomes plus a pair of sex chromosomes—the X and Y chromosomes. Each body cell of a human female contains 22 pairs of autosomes plus a pair of sex chromosomes—two X chromosomes.

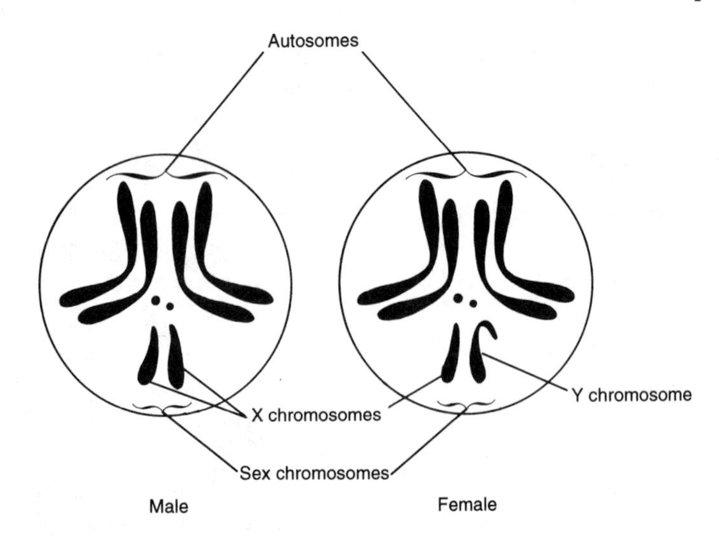

Figure 15–1 Chromosomes of the fruit fly

Sex determination in fruit flies and humans. Let us trace the inheritance of sex in fruit flies and humans by observing what happens to the sex chromosomes as they pass from one generation to the next. (Autosomes, which are not involved in sex determination, will not be shown.) The symbol ♂ will designate a male and the symbol ♀ will designate a female.

KEY: Let XX = sex chromosomes of female
XY = sex chromosomes of male

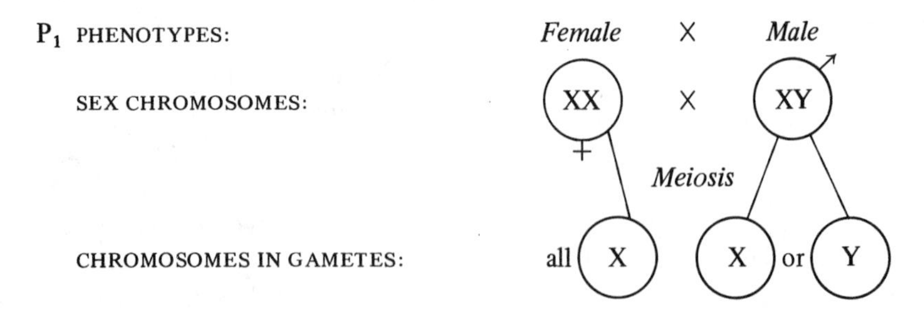

POSSIBLE FERTILIZATIONS:

eggs \ sperms	X	Y
X	XX	XY

F_1 OFFSPRING:

Sex Chromosomes	*Phenotypes*
$\frac{1}{2}$XX	50% females
$\frac{1}{2}$XY	50% males

You can see that the female forms only one type of gamete with respect to sex: all the egg cells carry an X chromosome. The male forms two types of gametes with respect to sex: 50 percent of sperm cells carry the X chromosome and the remaining 50 percent carry the Y chromosome. The sex of an individual organism is determined at fertilization. If there are many fertilizations, the chances are that 50 percent of the offspring will be males and 50 percent will be females, or a ratio of 1 XY : 1 XX. According to the laws of probability, the chance of obtaining either a male offspring or a female offspring is 1/2.

Sex determination in other organisms. Each body cell of the male grasshopper possesses 23 chromosomes—11 pairs of autosomes plus an X chromosome. Thus, the sex chromosomes of a male grasshopper are XO. Each body cell of the female grasshopper possesses 24 chromosomes—11 pairs of autosomes plus two X chromosomes. Thus, the sex chromosomes of the female are XX. The male forms two types of sperm cells: 50 percent carry the X chromosome and 50 percent do not carry the sex chromosome. All the female's egg cells carry an X chromosome. Therefore, at fertilization, egg cells fertilized by sperm cells carrying an X chromosome develop into female offspring. Egg cells fertilized by sperm cells lacking the sex chromosome develop into male offspring.

In birds, butterflies, and moths, the males possess two matched sex chromosomes called ZZ and the females possess two unmatched sex chromosomes called ZW. All the male's sperm cells, therefore, possess a Z chromosome. The female forms two types of egg cells, some that carry a Z chromosome and others that carry a W chromosome. Again, the sex of the offspring is determined at fertilization.

Sex Linkage

When Morgan mated homozygous red-eyed female fruit flies to white-eyed males, all the F_1 offspring were red-eyed, indicating that red eye is dominant to white eye. When he mated F_1 red-eyed females to red-eyed males, the offspring in the F_2 generation showed a ratio of three red-eyed flies to one white-eyed fly. Morgan examined the white-eyed flies and found that all of them were males. These white-eyed males made up one-half of the male population, the other half were red-eyed. There were no white-eyed females. Apparently eye color is associated, or linked, with the chromosomes that determine sex. In other words, eye color is *sex-linked.* Let us use Morgan's method to trace the inheritance of eye color in fruit flies.

When the inheritance of a sex-linked trait is traced using a Punnett square, the key includes the sex chromosomes and their linked genes.

1. Heterozygous red-eyed female × red-eyed male

KEY: X^R = chromosome bearing the dominant gene for red eye
X^r = chromosome bearing the recessive gene for white eye
Y^o = chromosome bearing no gene for eye color

F_1 PHENOTYPES: red-eyed female X red-eyed male

GENOTYPES:

GENES IN EACH GAMETE:

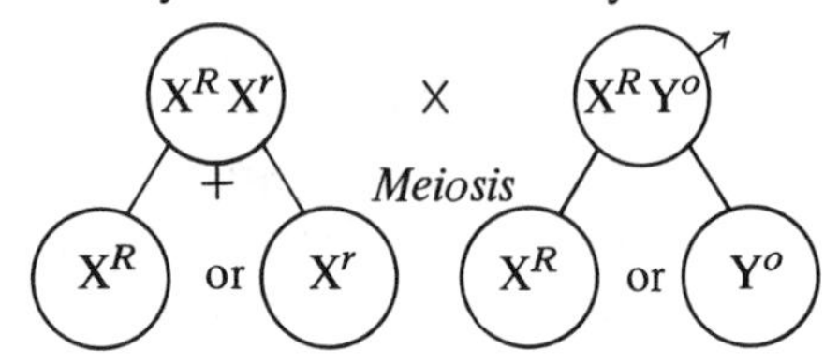

POSSIBLE FERTILIZATIONS:

eggs \ sperms	X^R	Y^o
X^R	$X^R X^R$	$X^R Y^o$
X^r	$X^R X^r$	$X^r Y^o$

F_2 OFFSPRING:

Genotypes	*Phenotypes*
$\frac{1}{4} X^R X^R$ $\frac{1}{4} X^R X^r$	50% red-eyed females
$\frac{1}{4} X^R Y^o$	25% red-eyed males
$\frac{1}{4} X^r Y^o$	25% white-eyed males

This cross shows that when a heterozygous red-eyed female mates with a red-eyed male, all the females are red-eyed, half the males are red-eyed, and the other half are white-eyed. In the white-eyed males, the recessive gene on the X chromosomes is not overshadowed by any other gene for eye color. In the red-eyed males, the single dominant gene on the X chromosome is enough to cause the development of red eyes.

2. Heterozygous red-eyed female × white-eyed male

KEY: Let X^R = sex chromosome bearing the dominant gene for red eye
X^r = sex chromosome bearing the recessive gene for white eye
Y^o = sex chromosome bearing no gene for eye color
O^o = lack of sex chromosome and gene for eye color due to nondisjunction in the female

F_2 PHENOTYPES: heterozygous red-eyed female × white-eyed male

GENOTYPES: $X^R X^r$ (♀) × $X^r Y^o$ (♂)

Meiosis

GENES IN EACH GAMETE: X^R or X^r ; X^r or Y^o

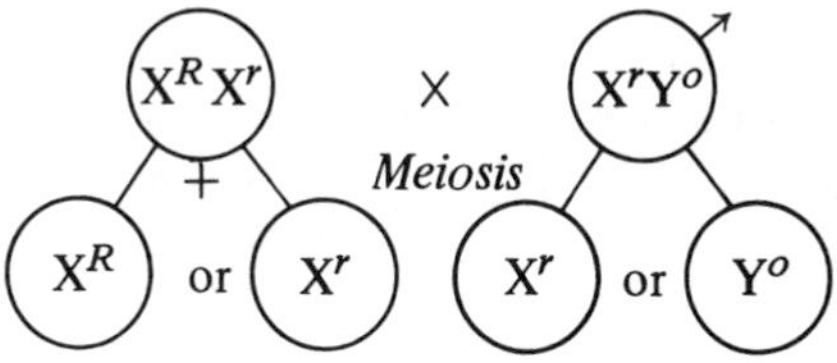

POSSIBLE FERTILIZATIONS:

eggs \ sperms	X^r	Y^o
X^R	$X^R X^r$	$X^R Y^o$
X^r	$X^r X^r$	$X^r Y^o$

F_3 OFFSPRING:

Genotypes	*Phenotypes*
$\frac{1}{4} X^R X^r$	25% red-eyed females
$\frac{1}{4} X^r X^r$	25% white-eyed females
$\frac{1}{4} X^R Y^o$	25% red-eyed males
$\frac{1}{4} X^r Y^o$	25% white-eyed males

This cross shows that when a heterozygous red-eyed female is mated with a white-eyed male, two types of offspring result in both sexes: half of each sex has red eyes and the other half has white eyes. All the white-eyed females possess two recessive genes; all the white-eyed males possess only one recessive gene. This led Morgan to conclude that only red-eyed females could be heterozygous for the trait and be carriers of the recessive gene.

From these investigations, Morgan discovered that (*a*) allelic genes for red eyes and white eyes are located on the X chromosomes, and (*b*) the Y chromosome lacks either allele for eye color. In the male, a single recessive gene for white eyes can express itself. In the female, two recessive genes are required for the white-eyed condition to express itself. The association of a gene with a sex chromosome is called *sex-linkage*. Studies have revealed that there are about 100 sex-linked traits in the fruit fly.

Nondisjunction

One of Morgan's co-workers, *Calvin Bridges*, mated white-eyed females to red-eyed males. Instead of obtaining all red-eyed females, Bridges occasionally noticed that some white-eyed females were produced. Bridges knew that two recessive genes are necessary for a female to have white eyes. He explained the appearance of white-eyed females by assuming that during meiosis, the paired XX chromosomes in a female parent failed to separate. As a result, two abnormal egg cells were formed: one egg cell having both X chromosomes (XX), and the other egg cell lacking sex chromosomes. The failure of homologous chromosomes to separate, or disjoin, after synapsis is called *nondisjunction.* Using a Punnett square, let us test Bridges' explanation:

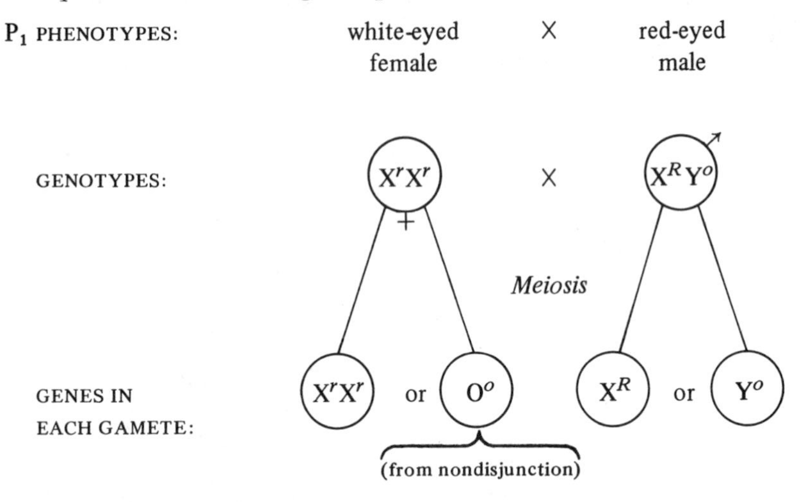

POSSIBLE FERTILIZATIONS:

eggs \ sperms	X^R	Y^o
X^rX^r	$X^RX^rX^r$	$X^rX^rY^o$
0°	X^R0^o	0^oY^o

F_1 OFFSPRING:

Genotypes	*Phenotypes*
$\frac{1}{4}X^RX^rX^r$	25% zygotes die (expected females)
$\frac{1}{4}X^rX^rY^o$	25% white-eyed females
$\frac{1}{4}X^R0^o$	25% sterile red-eyed males
$\frac{1}{4}0^oY^o$	25% zygotes die (expected males)

Although fruit flies possessing the genotype $X^r X^r Y^o$ have a Y chromosome, they are, nevertheless, females. Also, flies having the genotype X^R0^o lack a Y chromosome but are, nevertheless, males. Bridges verified his reasoning by examining the cells of the offspring. He found that the cells of some offspring had an extra chromosome; the cells of others lacked one chromosome. The presence of extra chromosomes was additional proof that chromosomes carry hereditary traits.

Crossing-over

During the synapsis phase of meiosis, the chromatids of paired homologous chromosomes may intertwine, exchange segments, and then separate (Figure 15–2). As a result, new linkage groups are formed. The chromatids that have crossed over and exchanged segments consist of gene sequences different from the original gene sequences. The process of breaking linkage groups in chromatids and exchanging segments of chromosomes is called *crossing-over.*

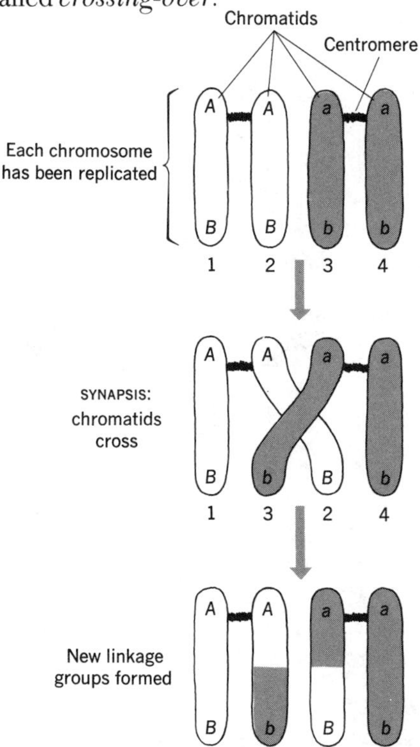

Figure 15–2 Crossing-over

As you read on, refer to Figure 15–3, which shows the possible result of crossing-over in a dihybrid fruit fly. In this case, crossing-over occurs only in the female. According to the genotypes of the parents, if crossing-over did not occur, we could expect the usual ratio of 1 *YyLl* : 1 *yyll*. If crossing-over occurred, as shown in the diagram, the new gene sequences should produce four types of offspring in the ratio of 1 *YyLl*:1 *Yyll*:1 *yyLl*:1 *yyll*. However, the actual results of this mating showed that of all the off-

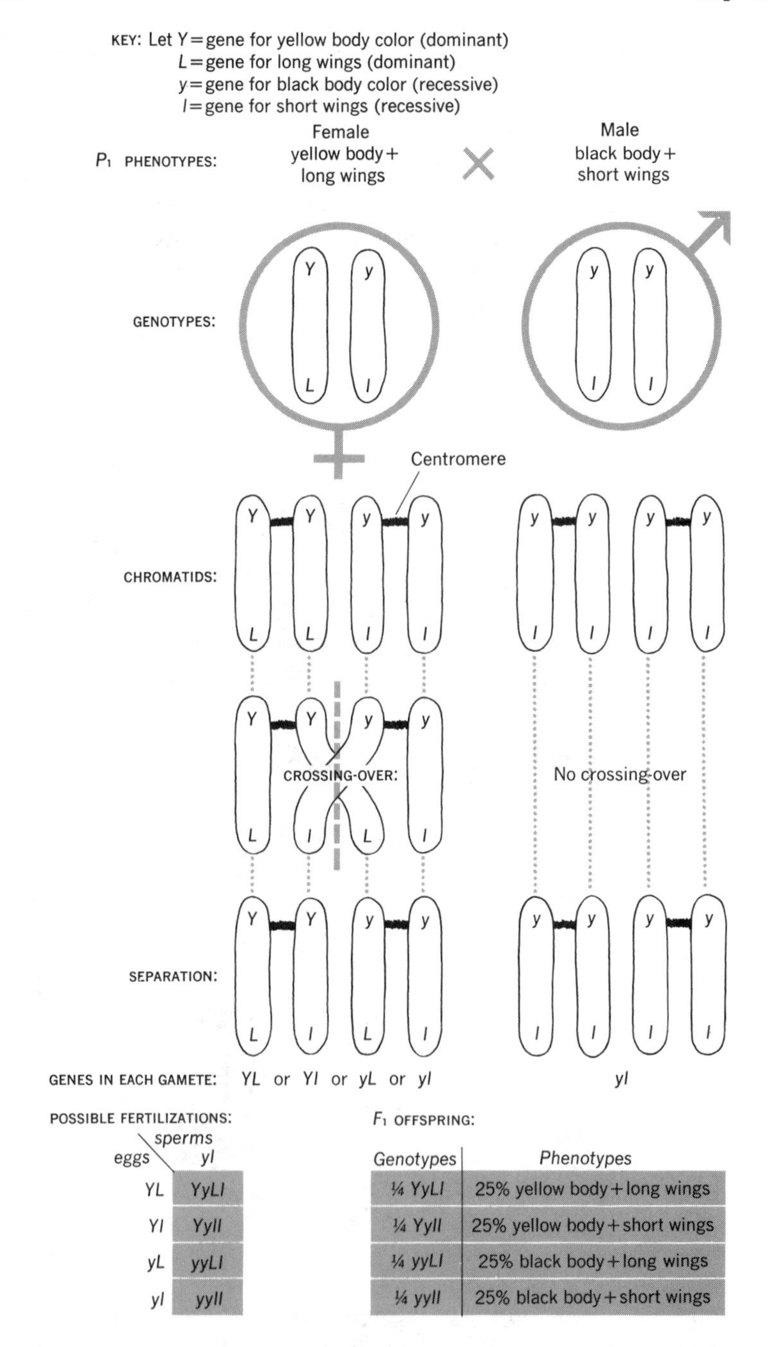

eggs \ sperms	yl
YL	YyLl
Yl	Yyll
yL	yyLl
yl	yyll

Genotypes	Phenotypes
¼ YyLl	25% yellow body + long wings
¼ Yyll	25% yellow body + short wings
¼ yyLl	25% black body + long wings
¼ yyll	25% black body + short wings

Figure 15–3 Possible results of crossing-over in a dihybrid fruit fly

spring, 41 percent had yellow bodies and long wings, 41 percent had black bodies and long wings, 9 percent had yellow bodies and short wings, and 9 percent had black bodies and short wings. Thus, breaks in linkage can produce variations in offspring and unexpected ratios. The inheritance of variations by offspring is a basis for the evolution of new species.

Multiple Alleles

In time, biologists studying inheritance realized that certain results can be explained only if more than two alleles, or *multiple alleles,* are present for a particular trait. The fruit fly has at least ten alleles for eye color; the snapdragon plant has nine alleles for flower color; mice have at least four alleles for fur color; and humans have three alleles for blood type. In each case, only two alleles for a given trait can be present in an individual and determine the appearance of the trait.

Effect of the Environment

How an organism actually develops depends upon the interactions between its genes and the environment. For example, a certain variety of fruit fly will develop curly wings when reared at a temperature of 25°C. When reared at 16°C, the fruit flies develop straight wings. If the offspring of these straight-winged fruit flies are reared at 25°C, all the offspring develop curly wings (Figure 15–4). In this case, a difference of 9°C determines the expression of the gene for wing type.

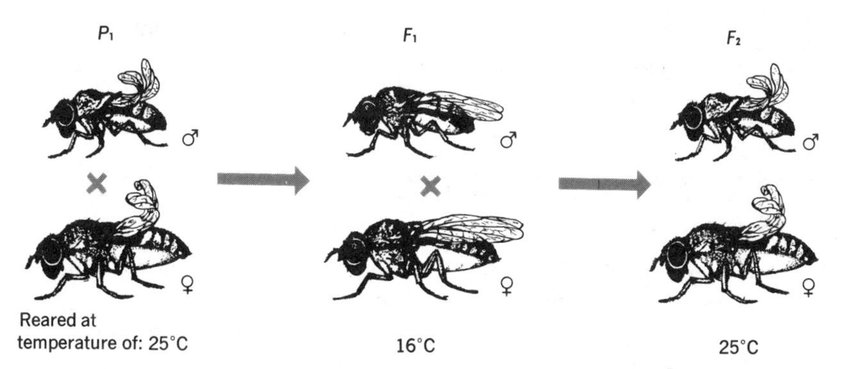

Figure 15–4 Effect of temperature on wing development of fruit fly

A Himalayan rabbit is a variety of domestic rabbit that has a white body and black ears, nose, tail, and feet. The genes for black fur color in a Himalayan rabbit do not express themselves unless the air temperature is near freezing. Thus, only the extremities that reach freezing temperatures produce black fur. When a patch of white fur is plucked from the rabbit's skin, black fur will grow back in the same region if the rabbit is reared at a temperature of about 0°C.

Studies of identical twins that were separated at birth and reared in different environments indicate that for some traits, such as eye color and blood type, the environment does not affect the expression of genes. However, the expression of genes for traits such as height and ability to learn are affected by environmental factors.

Section Quiz

1. The organelles that possess the greatest quantity of DNA are the (*a*) ribosomes (*b*) chloroplasts (*c*) chromosomes (*d*) mitochondria.
2. Which male animal possesses a pair of matched sex chromosomes? (*a*) moth (*b*) fruit fly (*c*) grasshopper (*d*) earthworm.
3. Chromosomes that do not have a direct connection with sex determination are the (*a*) X chromosomes (*b*) Y chromosomes (*c*) autosomes (*d*) alleles.
4. An exchange of chromatid segments producing a change in gene sequences may occur as a result of (*a*) mitosis (*b*) translocation (*c*) nondisjunction (*d*) crossing-over.
5. Linkage helps explain (*a*) the one gene–one enzyme concept (*b*) certain phenotypic results of segregation (*c*) the phenotypic results of translocation (*d*) the appearance of a hairless mouse in a litter of normal mice.

CHANGES IN HEREDITY

Under normal conditions, genes are remarkably stable. At times, however, changes occur in genes and the traits they control. Changes in genes or chromosomes and the corresponding changes in inherited traits are called *mutations*, a word first used by the Dutch botanist *Hugo De Vries*. Examples of mutations include white-eyed fruit flies, hornless (polled) cattle, and seedless oranges, grapefruits, and grapes. Most mutations are recessive and harmful to the organism in which they appear. However, a recessive mutant gene usually is overshadowed by the normal gene on the homologous chromosome. Some mutations are dominant and are noticeable in the offspring.

Mutations and DNA

Mutations of either genes or chromosomes are actually changes in DNA. This was discovered as the result of a series of experiments involving pneumonia bacteria that underwent the process of transformation. In genetics, *transformation* is an inherited change in a cell caused by the transfer of DNA from a cell of another species.

Griffith's experiments. As you read on, refer to Figure 15–5, which illustrates the experiments of *Frederick Griffith,* an English bacteriologist. In 1928, Griffith conducted an experiment in which he injected mice with certain live pneumococci that possessed a jellylike capsule. Griffith observed that all of the mice contracted pneumonia and died. He also observed that when he injected mice with certain live pneumococci that lacked a capsule, the mice did not contract pneumonia. Thus, he concluded that the presence of a capsule on living pneumococci caused pneumonia. When Griffith injected mice with a mixture of dead bacteria possessing capsules and live bacteria lacking capsules, the mice contracted pneumonia and died. Griffith could not explain the effects of the

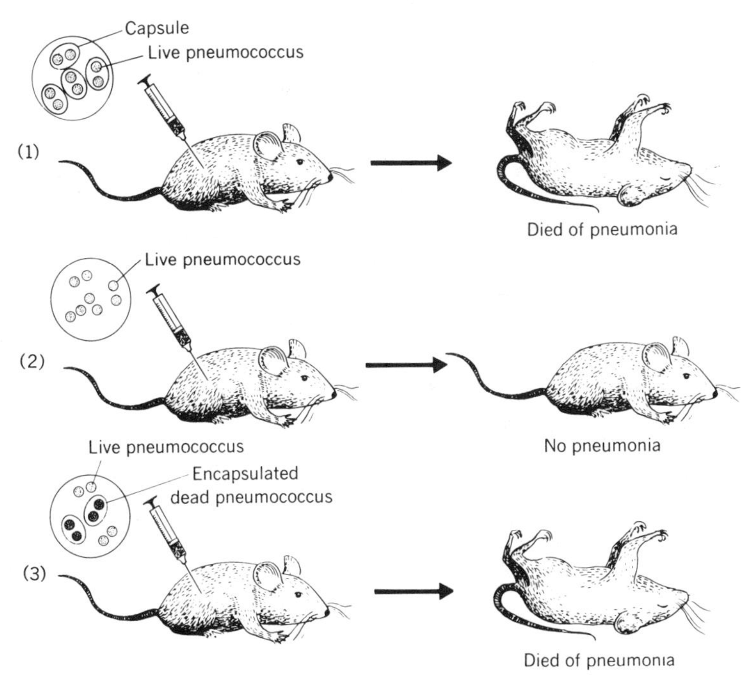

Figure 15–5 Griffith's experiment

mixture on the mice. The explanation was formulated 16 years later, as a result of the investigations of *Oswald T. Avery, Colin Macleod,* and *Maclyn McCarty.*

Avery's experiments. As you read on, refer to Figure 15–6, which illustrates the concept of transformation. Avery mixed dead pneumococci possessing capsules with live pneumococci lacking capsules. The researchers discovered that the live bacteria incorporated into their bodies the DNA of the dead bacteria with capsules. Apparently, the DNA responsible for capsule formation became part of the genetic code of the live bacteria. This experiment showed that the genetic trait of capsule formation coded by one bacterium could be transferred to another.

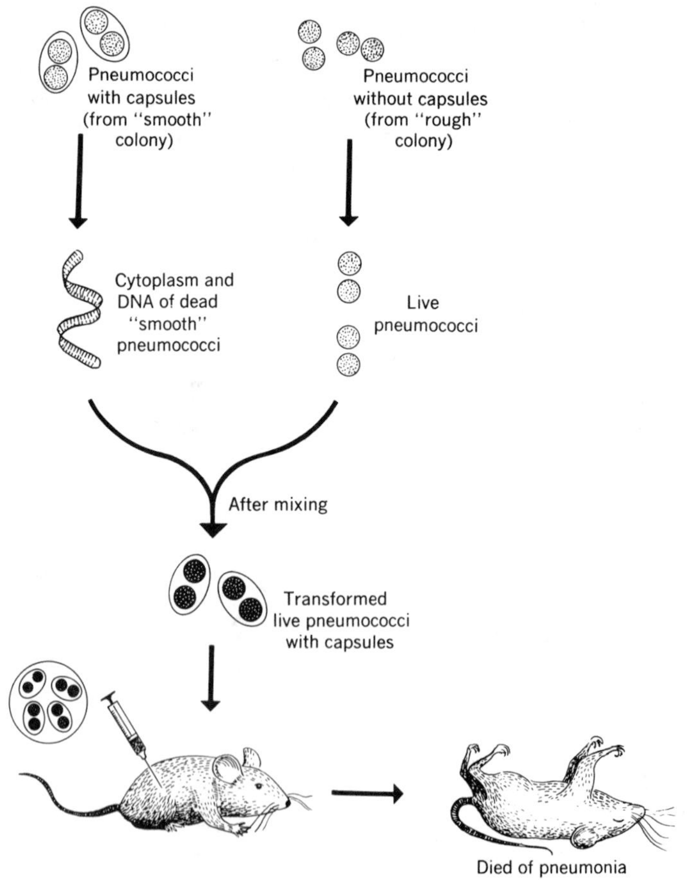

Figure 15–6 Avery's experiment

In 1952, *Joshua Lederberg* and *Norton Zinder* discovered the process of transduction while investigating a type of virus called a bacteriophage. A *bacteriophage* is a virus that is parasitic on bacteria. *Transduction* is the transfer by a virus of DNA from one bacterium to another bacterium.

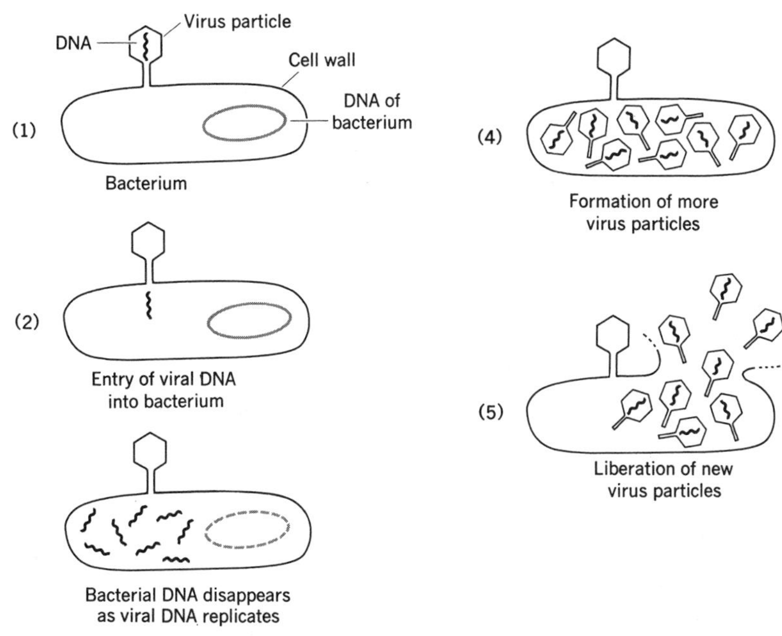

Figure 15–7 Reproduction of bacteriophage

Experiments of Lederberg and Zinder. Figure 15–7 shows how a bacteriophage infects bacteria and how the virus reproduces. Figure 15–8 illustrates what happens during transduction. In (1), a susceptible bacterium is attacked by the virus; in (2) the bacterium is destroyed and its DNA becomes part of the DNA of the new viruses. When a virus with modified DNA attacks a bacterium resistant to attack (3), the virus discharges its DNA into the bacterium (4) but the resistant bacterium is not destroyed. Instead, the viral DNA becomes part of the DNA of the resistant bacterium and (5) part of its genetic code. Thus, the resistant bacterium has acquired a new gene. This gene expresses itself as a flagellum in the next generation. When reproduction occurs by binary fission, (6) each of the offspring develops the new trait, a flagellum.

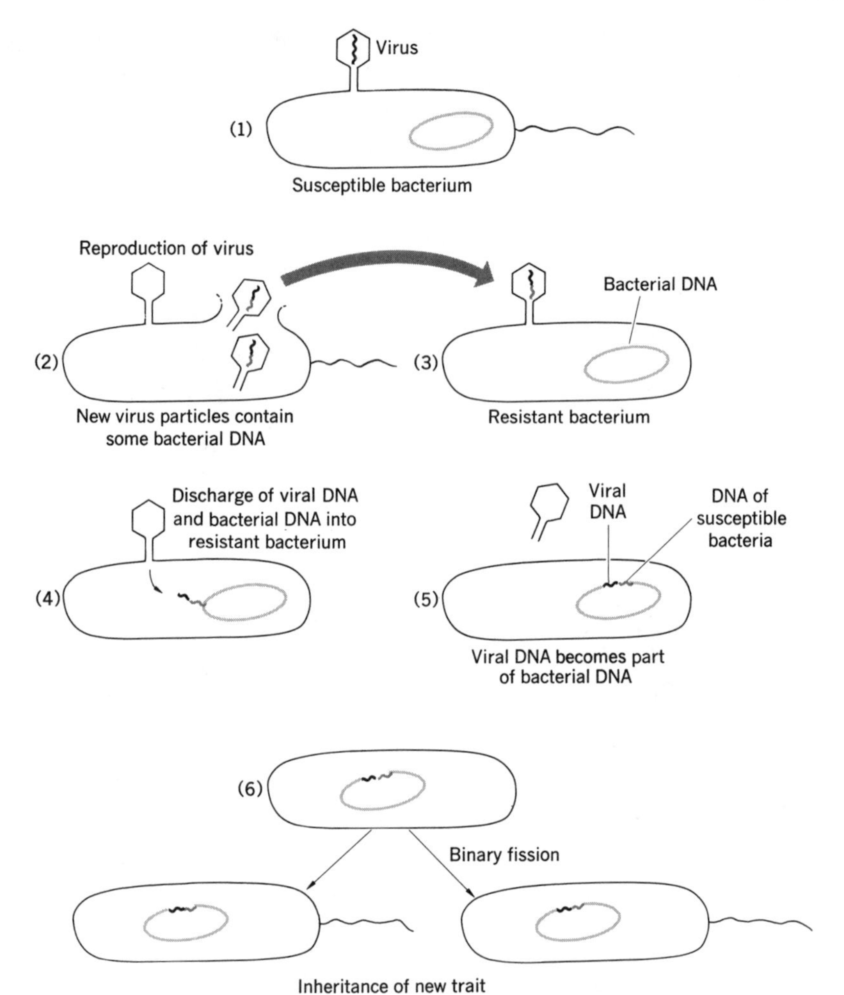

Figure 15–8 Transduction of viral DNA

Genes Control Chemical Reactions

In the early 1940's *George Beadle* and *Edward Tatum,* studying the pink mold *Neurospora* (Figure 15–9), hypothesized that genes control the production of enzymes, which control the chemical reactions that occur in living cells. You can see that *Neurospora* forms spore sacs that contain eight spores. When the spores are grown on a minimal culture medium containing sucrose, nitrates, and biotin (a B-complex vitamin), the spores germinate and grow because they can manufacture other essential substances needed for growth.

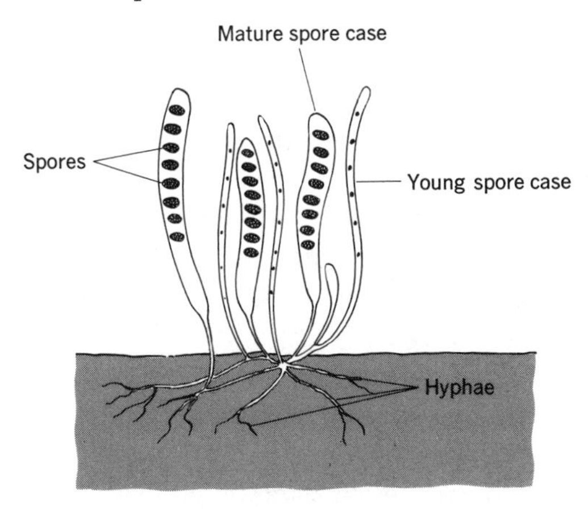

Figure 15–9 Neurospora mold

Beadle and Tatum exposed the spore cases of *Neurospora* to X rays and then transferred the irradiated spores to a minimal culture medium (Figure 15–10). Some spores germinated and grew into molds; other spores did not germinate. These nongerminating spores were then transferred to an enriched medium that contained all the substances required for normal growth. In a few days the spores germinated and grew into molds. Beadle and Tatum concluded that the X rays had caused a mutation in some genes, and that the mutations prevented the manufacture of substances the spores needed to germinate.

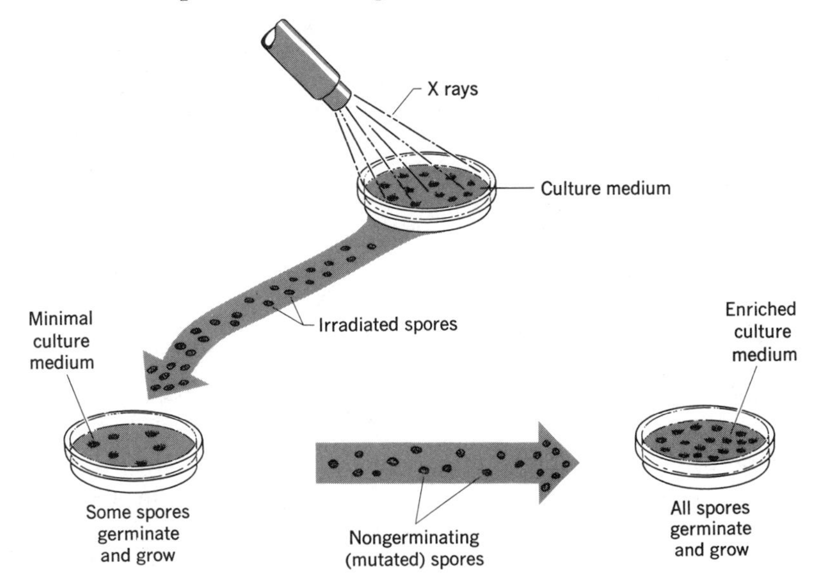

Figure 15–10 Experiment of Beadle and Tatum

Then Beadle and Tatum attempted to find out which essential substances were not manufactured. They prepared a series of minimal culture media each of which contained a different amino acid or vitamin. Individual irradiated spores were transferred to each different medium. By observing in which medium germination occurred, Beadle and Tatum could identify the amino acid or vitamin that could not be manufactured by the irradiated spore. The main ideas of the *one gene–one enzyme theory* of Beadle and Tatum are:

1. The manufacture of a specific substance by a cell depends upon the presence of a specific enzyme.
2. The presence of a specific enzyme depends upon the presence of a specific gene.
3. Either an artificially induced or a natural mutation in a gene can prevent the manufacture of an essential substance.
4. Accordingly, the relationship between genes and enzymes appears to be that one gene is responsible for the formation of one enzyme.

The manufacture of a specific substance may involve a series of enzymes. Therefore, a change in any one of the genes that control the production of one of the required enzymes usually prevents the formation of an end product. In the series of reactions shown in Figure 15–11, if gene 1, 2, or 3 is changed, product C is not formed. If gene 3 is changed, only products A and B are formed.

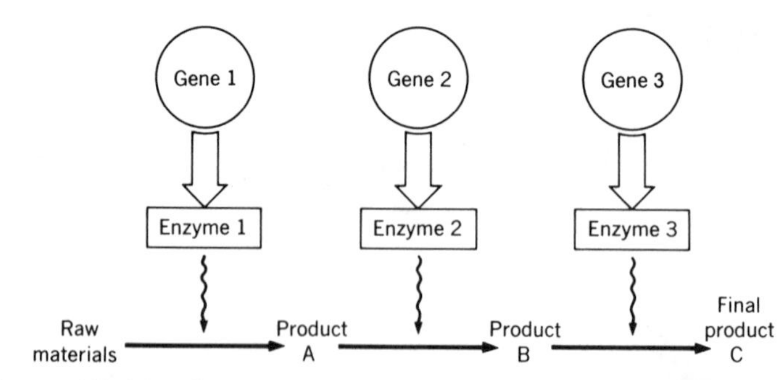

Figure 15–11 Genes, enzymes, and the formation of products

Gene mutations

Gene mutations are changes in the DNA code. They can occur under one or more of the following conditions:

Substitution. In substitution, the sequence of nitrogenous bases is changed by the replacement of one base by another. For example, if part of an original DNA sequence is normally TAG and T is replaced by G, then the sequence is changed to GAG. This change in the original DNA sequence produces new codes in the messenger RNA.

Addition. In *addition*, an extra one of the four bases is added to the sequence. For example, if part of an original DNA sequence is normally TAG CCA and a T is added between G and C, the modified DNA sequence becomes TAG TCC, thus leaving out the A. Again, such a change in the original DNA sequence produces new codes in messenger RNA.

Removal. In *removal*, a nitrogenous base drops out of the original DNA sequence. For example, if part of an original DNA sequence is TAG CCA and the G drops out, the modified sequence becomes TAC CA (incomplete). This rearrangement of the original DNA sequence also tends to result in a new code in messenger RNA.

The changes, or *genetic mutations*, in DNA codes caused by substitution, addition, or removal of genes prevent the synthesis of normal proteins at the ribosomes. In turn, this can produce a changed trait. Evidently, gene mutations resulting from changes in the DNA code are significant factors in the evolution of new species.

Chromosome Mutations

Mutations usually occur when the structure of a chromosome is changed or the number of chromosomes is changed. These changes, called *chromosome mutations*, are caused by crossing-over, nondisjunction (both discussed previously), deletion, translocation, inversion, and duplication. Figure 15–12 shows how these changes affect the structure of chromosomes.

1. *Deletion* is the permanent loss of a small piece of chromosome from a cell.
2. *Translocation* is the transfer of a chromosome segment to a nonhomologous chromosome. (In crossing-over, an exchange of chromosome segments occurs between homologous chromosomes only.)
3. *Inversion* is the detachment and subsequent reorientation of a chromosome segment. The inverted segment then becomes reattached to the chromosome.
4. *Duplication* occurs when a segment of a chromosome breaks away and becomes attached to an entire homologous chromosome.

Thus, chromosome mutations can also produce new traits because deletion, translocation, inversion, and duplication of chromosomes form new DNA codes.

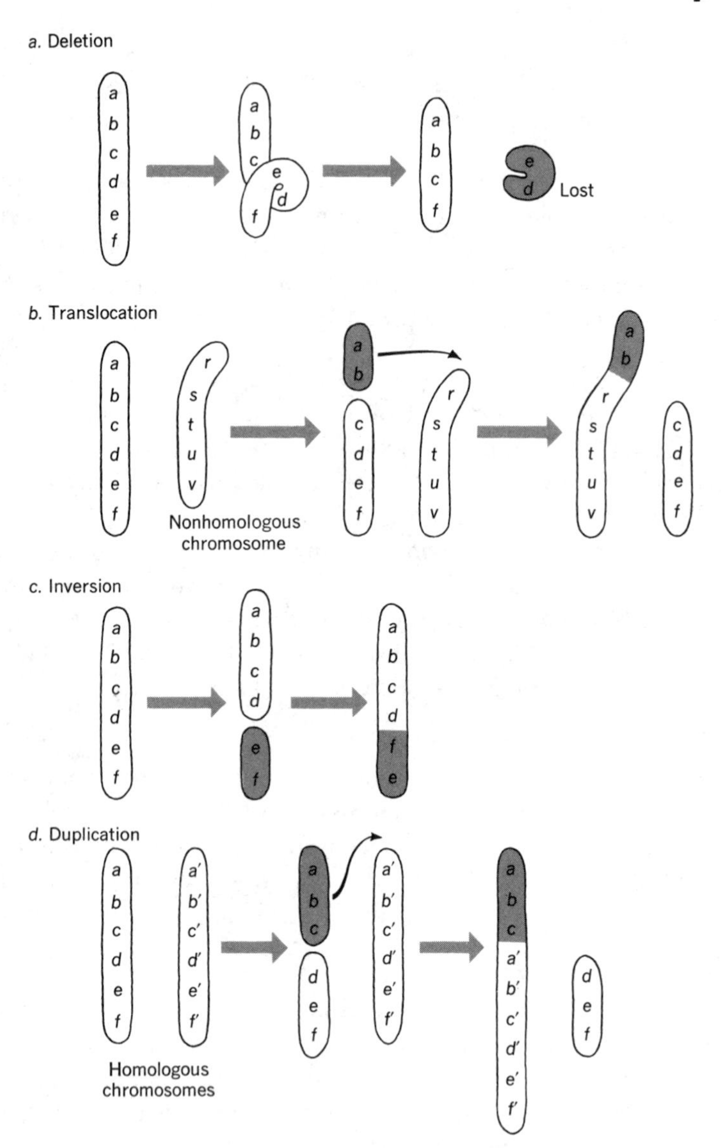

Figure 15–12 Chromosome mutations

AGENTS CAUSING MUTATIONS

Chemical or physical environmental factors that may cause either a gene or a chromosome mutation are called *mutagenic agents*.

Mutagenic Chemicals

Certain substances can interfere with the normal functioning of genes and chromosomes and thus cause mutations. The following are some examples of mutagenic chemicals:

Colchicine. This organic chemical is a plant extract that stops mitosis by preventing the separation of chromosomes during anaphase. As a result, the cell in which mitosis began contains two diploid sets of chromosomes, or $2n + 2n = 4n$. When the monoploid number of chromosomes has been multiplied more than twice, the condition is called *polyploidy*. The $4n$ condition is described as *tetraploid*; the $3n$ condition ($2n + n$) is described as *triploid*. Gametes containing an abnormal number of chromosomes are formed when polyploid cells undergo meiosis. As a result of fertilization, zygotes may be formed with $3n$, $4n$, $5n$, $6n$, $8n$, and other polyploid numbers. Plants with polyploid cells usually are larger and sturdier than normal diploid plants. The embryos of animals with polyploid cells usually do not survive. Many types of cancer cells are polyploid.

Antibiotics. Actinomycin D, an antibiotic, combines with the nitrogen bases in the DNA of organisms such as bacteria. This interferes with the formation of messenger RNA.

Mustard gas. This poisonous gas was used as a weapon during World War I. Mustard gas interferes with nucleotide formation in cells, which prevents the normal replication of DNA. For this reason, mustard gas occasionally is used in the treatment of cancer as a way of interrupting uncontrolled cell divisions.

Other mutagens. Toxic chemical wastes and hallucinogenic drugs such as LSD may produce breaks in chromosomes. Damaged chromosomes cause mutations, which may be expressed in the offspring as physical or mental defects.

Penetrating Radiations

Ultraviolet rays and radioactive elements such as uranium can alter genes and chromosomes by producing abnormal ions in cells. These ions can interfere with the pairing of nitrogen bases during DNA replication. For example, penetrating radiations may cause the base adenine to join with the base cytosine instead of thymine. The resulting change in the sequence of bases causes a mutation.

In 1927 *Herman J. Muller*, a student of T.H. Morgan, used X rays to irradiate fruit flies and produce mutations. Other forms of radiation, especially from radioisotopes such as cobalt-60, have been used to produce mutations in plants and animals since Muller's use of X rays. One important reason for banning nuclear bomb tests in the atmosphere is that the radioactive fallout from these tests may produce harmful mutations and cause an increase in the number of cases of cancer.

Section Quiz

1. A condition in which cells have some multiple of the normal chromosome number is (*a*) translocation (*b*) deletion (*c*) inversion (*d*) polyploidy.
2. Which describes an inherited change in a cell caused by the transfer of DNA from a cell of another species? (*a*) transduction (*b*) transformation (*c*) substitution (*d*) addition.
3. Which geneticist used X rays as a way of producing mutations? (*a*) Avery (*b*) Muller (*c*) Mendel (*d*) Morgan.
4. Most mutations in organisms are both (*a*) recessive and harmful (*b*) recessive and helpful (*c*) dominant and harmful (*d*) dominant and helpful.
5. The one gene–one enzyme theory resulted from the research of (*a*) Griffith (*b*) De Vries (*c*) Beadle and Tatum (*d*) Lederberg and Zinder.

SUMMARY OF PRINCIPLES OF GENETICS

The first principles of genetics were Mendel's laws of dominance, segregation and recombination, and independent assortment. Sutton added the concept that hereditary units were carried by the chromosomes. Then Morgan and subsequent geneticists made it clear that these hereditary units were genes.

The gene principle of heredity includes the following concepts:

- Genes are the units that control heredity.
- Genes control the structure and the functioning of organisms.
- Genes in a chromosome are linked, somewhat like beads on a string.

- Breaks in linkage and crossing-over of genes occur.
- In humans, sex is determined at fertilization by the combination of X and Y chromosomes.
- The X chromosomes bear the genes associated with sex-linked traits.
- Abnormal disjunction of chromosomes during meiosis can produce unexpected types of offspring.
- Genes control chemical reactions by coding for the synthesis of specific enzymes.
- Gene changes, or mutations, are changes in the DNA code.
- The environment influences the expression of a trait by affecting the genes that control the trait.

The following section discusses the part played by chromosomes and genes in human heredity.

HUMAN HEREDITY

It is difficult to research human heredity for the following reasons:

- Humans cannot be used as subjects in controlled breeding experiments.
- Completely pure strains of individuals do not exist.
- It takes a long time for humans to mature and have offspring.
- Compared to other organisms, humans have few offspring.
- Humans rarely keep complete and accurate records of births, deaths, the appearance of a new trait, or the disappearance of an old trait.
- People often are reluctant to reveal information about family members.
- Human body cells contain 46 chromosomes, each of which is small and not easily studied.

Pedigrees

Information about human heredity has been obtained by tracing the inheritance of some traits using family trees, or pedigrees. For example, in Figure 15–13, the blue-eyed offspring shown in line 2 indicates that the mother in line 1 was hybrid for brown eyes. Was the man who married her daughter (3 on line 2) pure or hybrid for brown eyes? If you examine the children resulting from this marriage, you can see that one child is brown-eyed and the other is blue-eyed. Therefore the man must be hybrid for brown eyes. In this way we can reason out the genotypes of individuals from their phenotypes.

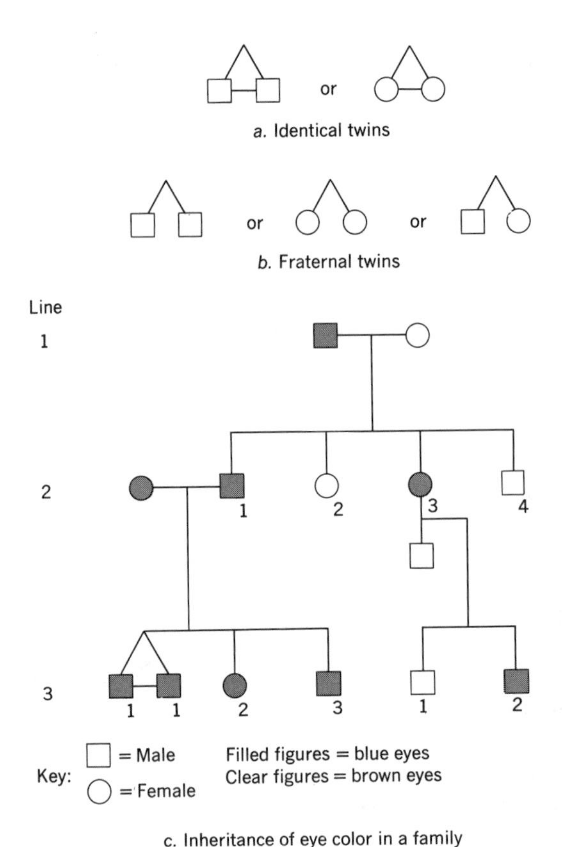

Figure 15–13 Human pedigree chart

The Inheritance of Some Human Traits

Table 15–1 lists some human traits and some pertinent facts about each trait.

Eye color. In humans, eye color is controlled by several pairs of non-allelic genes. One pair determines whether the brown pigment *melanin* will develop. An individual who is homozygous recessive for color is an albino. An albino lacks pigment in any part of the body, including the iris of the eyes. Thus, the eyes of an albino appear pink because the color of the capillaries can be seen through the colorless cells of the iris. Other pairs of genes control the location, concentration, and intensity of the melanin particles seen in the iris. As a result, eye color varies from dark brown to blue.

Table 15–1 Human Inheritance

Trait	*Dominant*	*Recessive*
Skin color	Normal (pigmented)	No pigment (albino)
Eye color (some colors not well understood)	Brown (some types may be recessive)	Blue (some types may be dominant)
Hair color	Dark	Light
Hair type	Curly (or wavy)	Straight
Ear lobe	Free	Attached
Nose size	Large	Small
Ability to taste PTC (phenylthiocarbamide)	Tasting	Nontasting
Number of fingers	Six fingers (a mutation)	Five fingers
Blood types	Type A Type B Type AB is an example of blending inheritance	Type O Type O
Rh blood factor	Rh^+	Rh^-
Color vision	Normal	Red-green colorblindness (sex-linked)
Arch of foot	Flat	Arched
Blood-clotting speed	Normal	Hemophilia (sex-linked)
Mental ability	Normal	Feeblemindedness (some types)
Tongue movement	Ability to roll sides	Inability to roll sides
Thickness of lips	Thick	Thin

Blood types. Your blood type (see Chapter 6) depends upon the inheritance of multiple alleles that express themselves as antigen A, antigen B, and O (no antigen). Analysis of family pedigrees indicates that the genes for antigen A and antigen B are dominant over the gene that does not produce an antigen. However, the gene for antigen A is not dominant over the gene for antigen B. When these two genes are present together, their effect is combined, similar to blending inheritance.

These alleles usually are designated as I^A, I^B, and i. However, the letters *A*, *B*, and *O*, respectively will be used to simplify tracing the inheritance of blood types. Table 15–2 shows blood phenotypes and their possible genotypes.

Table 15–2 Blood Phenotypes and Genotypes

Blood Phenotype	*Genotype (Possible Alleles)*
A	*AA* (homozygous) or *Ao* (heterozygous)
B	*BB* (homozygous) or *Bo* (heterozygous)
AB	*AB* (heterozygous—incomplete dominance)
O	*oo* (homozygous recessive)

A parent who is hybrid for type A and a parent who is hybrid for type B can produce four children with different blood types. A Punnett square shows how this can occur:

KEY: Let *A* = gene for antigen A
B = gene for antigen B
o = gene for no antigens

P₁ PHENOTYPES:

GENOTYPES:

GENES IN EACH GAMETE:

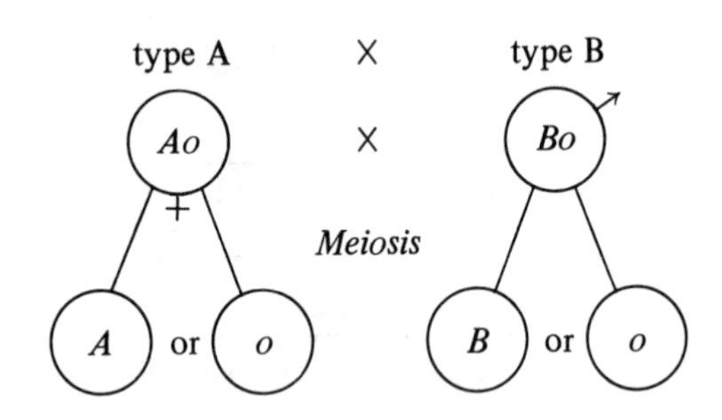

POSSIBLE FERTILIZATIONS:

eggs \ sperms	*B*	*o*
A	*AB*	*Ao*
o	*Bo*	*oo*

F_1 OFFSPRING:

Genotypes	*Phenotypes*
$\frac{1}{4}$ *AB*	25% type AB
$\frac{1}{4}$ *Ao*	25% type A
$\frac{1}{4}$ *Bo*	25% type B
$\frac{1}{4}$ *oo*	25% type O

The only family member who can donate blood to all others is type O, the universal donor. The only family member who can receive blood from all others is type AB, the universal receiver.

Rh factor. Rh antigen is *not* produced (Rh^-) in the presence of a pair of recessive genes ($rh^- rh^-$). Rh antigen *is* produced (Rh^+) in the presence of either two dominant alleles or one dominant allele ($Rh^+ Rh^+$ or $Rh^+ rh^-$). The different blood types and the Rh factor are inherited independently of each other because the genes for blood types are located in another chromosome pair. Thus, a person may have type A blood and be either Rh^+ or Rh^-.

Skin color. Skin color, height, and body form are inherited traits that result from the interaction of a group of several pairs of genes. Each gene produces a certain effect. Dominance or recessiveness does not exist among the genes of the group. Rather, the effect of one gene adds to the effect of another gene. The final effect is the sum of the effects of all genes in the group.

In humans, skin color is caused by the cumulative effect of at least three pairs of genes. Suppose P is used to represent the gene for much pigment, p to represent the gene for little pigment, and the numbers 1, 2, and 3 to represent the gene pairs. Then the genotype of a dark-skinned person could be represented by $P_1P_1P_2P_2P_3P_3$. Similarly, the genotype of a light-skinned person could be $p_1p_1p_2p_2p_3p_3$. A child born of two such people could have the genotype $P_1p_1P_2p_2P_3p_3$ and have a phenotype that was intermediate in skin color.

Sex-linked traits. In humans there are about 150 disorders that are sex-linked, including color blindness, hemophilia, and some forms of muscular dystrophy. Many color-blind people see red or green colors as shades of yellow or gray. Hemophiliacs have blood that clots slowly because the blood lacks a protein called the *antihemophilic factor.* Thus, a person with this disorder bleeds excessively, even from minor cuts or scrapes. The major symptom of muscular dystrophy is progressive wasting away of muscles. The genes for sex-linked disorders are recessive and are located in the X chromosome.

Let us trace the inheritance of hemophilia in a marriage between a female carrier (normal, but carrying one recessive gene) and a normal male. (The Y chromosome does not carry an allele for the antihemophilic factor.) Thus:

KEY: Let X^H = chromosome bearing the gene for normal clotting
Let X^h = chromosome bearing the gene for hemophilia
Y^o = chromosome bearing neither gene

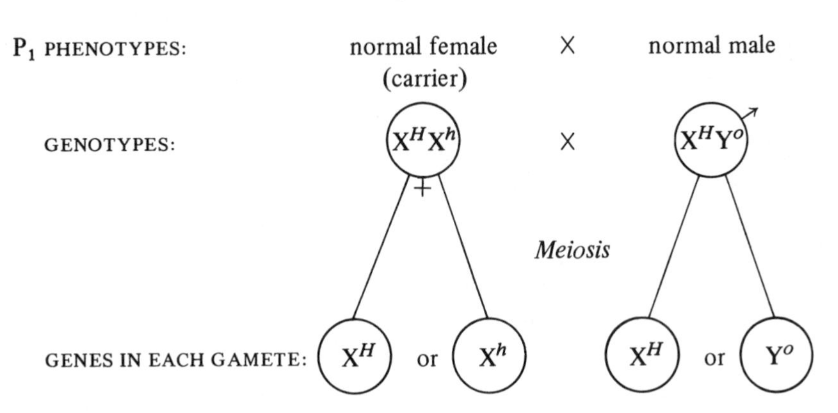

POSSIBLE FERTILIZATIONS:

eggs \ sperms	X^H	Y^o
X^H	X^HX^H	X^HY^o
X^h	X^HX^h	X^hY^o

F_1 OFFSPRING:

Genotypes	*Phenotypes*
$\frac{1}{4}$ X^HX^H	25% normal females
$\frac{1}{4}$ X^HX^h	25% normal females (carriers)
$\frac{1}{4}$ X^HY^o	25% normal males
$\frac{1}{4}$ X^hY^o	25% hemophiliac males

You can see that half the daughters are carriers and that half the sons are hemophiliacs.

Traits involving chromosome disorders. The number of chromosomes in a cell can change as a result of *nondisjunction,* or abnormal separation of homologous chromosomes during meiosis. This genotypic change usually produces a noticeable phenotypic change, manifested as a physical and/or mental disorder such as Down syndrome, Turner syndrome, and Klinefelter syndrome. A *syndrome* is a combination of observable symptoms characteristic of a specific disorder.

Down syndrome. Possibly due to aging of their eggs, women in their forties are, statistically, more likely than women in their twenties to produce two copies of chromosome 21 in their egg cells. Fertilization of such an abnormal egg cell by a normal sperm cell results in a total of 47 chromosomes (45 autosomes + 2 sex chromosomes) in each cell of the developing embryo. Down syndrome also is called *Trisomy 21* because there are three copies of chromosome 21 in every cell. This disorder, which may affect about one infant in 800, is characterized by varying degrees of mental retardation, short stature, stubby hands and feet, and eyelid folds. (Refer back to Figure 12–10 to see a normal human karyotype of 46 chromosomes.)

Turner syndrome. About one female infant in 5,000 female births has an XO sex chromosome genotype (44 autosomes + only one X sex chromosome = 45 chromosomes). This female infant develops into a short woman with undeveloped breasts and ovaries. Most women suffering from Turner syndrome have short thick necks and are sterile. Some also suffer from a defective aorta, misshapen kidneys, double ureters, and premature old age.

Klinefelter syndrome. About one male infant in 500 male births has a trisomy XXY sex chromosome genotype (44 autosomes + XXY sex chromosomes = 47 chromosomes). This male infant develops into an unusually tall man with fat deposited in female patterns. Klinefelter syndrome males are also characterized by undeveloped testes, enlarged breasts, and sterility.

Traits involving gene defects. In humans there are more than 700 recessively inherited disorders and more than 900 dominantly inherited disorders. Some recessively inherited disorders are phenylketonuria, albinism, sickle-cell anemia, cystic fibrosis, galactosemia, and Tay-Sachs disease. Some dominantly inherited disorders are glaucoma, achondroplasia, Huntington disease, and hypercholesterolemia.

Phenylketonuria. The complete metabolism of the amino acid *phenylalanine* into the pigment melanin is regulated by several enzymes acting in series. The incomplete metabolism of phenylalanine is caused by the inheritance of a gene defect that prevents the production of one of these enzymes. This causes the accumulation of a poisonous substance called *phenylpyruvic acid* in the blood and urine. Phenylpyruvic acid causes brain damage in the developing embryo, a condition called phenylketonuria (PKU). It is a common practice in most hospitals to test newborn babies to determine whether the baby has inherited this disorder. If the result is positive, the baby is fed a diet that contains no phenylalanine to prevent further brain damage.

Albinism. In albinism, the pigment melanin is not produced. Thus, an albino individual lacks normal pigmentation. In this disorder, the enzyme that normally metabolizes the amino acid *tyrosine* to melanin is absent because of the absence of the dominant gene.

Sickle-cell anemia. In sickle-cell anemia, the red blood cells are crescent- or sickle-shaped because they contain an abnormal type of hemoglobin. A normal hemoglobin molecule consists of several hundred amino acid units arranged in a definite sequence. A person afflicted with sickle-cell anemia is homozygous recessive. The recessive genes change the normal sequence of amino acid units by placing the amino acid *valine* in the position normally occupied by the amino acid *glutamic acid.* This difference between normal hemoglobin and sickle-cell hemoglobin causes some red blood cells to collapse and become sickle-shaped when the blood cells lose oxygen to the tissues. The distorted cells then clog the blood vessels, break up, and are destroyed. This results in a condition in which the blood is unable to carry sufficient oxygen to the tissues.

Heterozygous individuals are normal, but still carry the recessive trait. They often are immune to malaria. Immunity to malaria has survival value because it enables people living in regions where malaria is prevalent to survive, reproduce, and thus pass along their genes.

Cystic fibrosis. In cystic fibrosis, the cells that line the respiratory and digestive systems produce large amounts of mucus. As a result, the lungs and glands such as the pancreas may become plugged. Cystic fibrosis is caused by a recessive allele on chomosome 7. With treatment, people suffering from this disorder can live to early adulthood.

Galactosemia. In galactosemia, milk sugar is not metabolized normally. As a result, galactose accumulates in the brain and spleen, causing damage to these organs.

Tay-Sachs disease. Most victims of this incurable disease are of Eastern European Jewish ancestry. Tay-Sachs babies are born brain-damaged and die in early childhood. Tay-Sachs disease is caused by the lack of a specific enzyme for the breakdown of lipids in the brain.

Glaucoma. In glaucoma, pressure increases in the eyeball, eventually causing damage to the retina. Glaucoma is a major cause of blindness.

Achondroplasia. This condition is a form of dwarfism.

Huntington disease. In this disease, there is a progressive wasting of the nervous system.

Hypercholesterolemia. In this disorder, high levels of cholesterol accumulate in the blood, increasing the risk of heart disease.

Genetic Counseling

Parents at risk of having children with genetic defects often seek *genetic counseling*, which informs parents of the probability of producing a child with genetic abnormalities. Genetic counseling is a procedure that involves examining family pedigrees and following up with either *amniocentesis* or *chorionic villi sampling.*

Amniocentesis. If pedigrees of both parents reveal the possibility of carrying a defective allele for a genetic disorder, the pregnancy is called *high-risk.* In that case, it is wise for a pregnant woman to undergo amniocentesis. In this procedure, ultrasound techniques are used to locate the position of the fetus within the amniotic sac; then a long needle is used to draw out a small amount of amniotic fluid from the sac. Epithelial cells in the amniotic fluid are tissue-cultured and a karyotype is made and examined. Chromosomal abnormalities can be detected and revealed to the prospective parents. (Refer back to Figure 13–27, which shows a fetus in the amniotic sac.)

Chorionic villi sampling. In this recently developed procedure, which is less invasive than amniocentesis, cells are removed from the chorion, an embryonic structure that is part of the placenta. The cells are tissue-cultured, a karyotype made, and then examined for evidence of chromosomal abnormalities. Ethics, family values, religion, societal values, and the health of the fetus and mother are considered in the decision whether to continue the pregnancy. (Refer back to Figure 13–16, which shows an embryo and the chorionic structure.)

Section Quiz

1. The sex genotype of an individual with Turner syndrome is (*a*) XXY (*b*) XX (*c*) XY (*d*) XO.
2. The condition called Trisomy 21 often is called (*a*) Klinefelter syndrome (*b*) Down syndrome (*c*) Turner syndrome (*d*) none of these.
3. Which chromosomal aberration at meiosis most often produces defective genotypes in offspring? (*a*) nondisjunction (*b*) deletion (*c*) inversion (*d*) translocation.

4. A karyotype is a visual representation of (*a*) a diploid set of chromosomes (*b*) a monoploid set of chromosomes (*c*) organelles inside a cell (*d*) a DNA molecule.
5. An embryo with the sex genotype XXY develops into a (*a*) normal male (*b*) sterile male (*c*) normal female (*d*) sterile female.

MOLECULAR GENETICS

A clear understanding of molecular genetics, including genetic engineering, depends on a knowledge of the basic facts about DNA. These are presented below in outline form:

- Four "letters" of the DNA alphabet (A, T, G, and C) can be used to "write" numerous messages or instructions. (A, U, G, and C are the "letters" of the RNA alphabet.)
- The instructions are delivered by triplets of nucleotides called *codons*. Collectively, codons represent a genetic code or blueprint.
- A sequence of codons along a DNA molecule is, in effect, a series of *genes*. The codons instruct the ribosomes in how to assemble specific proteins.
- Thus, codons may be considered "words" that are "read" by the ribosomes. Codons correspond to any one of 20 amino acids found in the cytoplasm. Different codons may correspond to the same amino acid. (Refer back to Table 12–2.)
- Not all codons are involved in the actual process of protein synthesis. Certain codons either start the process or stop it at completion.
- In some animals, including humans, instruction codons are interrupted by other codons called *introns*. Introns are noncoding sequences within a gene; thus they do not code for protein synthesis. The coding regions of genes are called *exons*, which code the amino acid sequences of proteins that are to be synthesized, or *expressed*.
- When DNA splits and transcribes instructions to messenger RNA (mRNA), the introns are removed. Lacking introns, messenger RNA leaves the nuclear area and enters the cytoplasm, eventually joining the ribosomes.
- At the ribosomes, the "language," or instructions encoded in mRNA, directs the joining of selected amino acids into chains, or proteins. The process in which amino acids are joined to produce proteins is called *translation*.

Recombinant DNA

Scientists involved in genetic engineering make frequent use of *recombinant DNA*, which represents an assortment of genes from two or more sources combined in one cell. To study and analyze genes of different organisms, scientists had to develop ways to cut DNA strands at specific regions. That represented a challenging problem for three reasons:

1. DNA strands are relatively long. For example, there are about two meters of DNA strands in each human cell.
2. Using a human cell as a model, there are about six billion base pairs in two meters of DNA.
3. All of the genetic material is tightly packaged in 46 chromosomes.

The problem was solved by *Dr. Daniel Nathans*, an American geneticist who cut out DNA fragments using *restriction enzymes* produced by bacteria. (In 1978, Nathans won the Nobel Prize for Physiology or Medicine for his accomplishments in genetics research.) The restriction enzymes cut DNA at points where particular nucleotide sequences, or codons, are specific for a protein. Nathans' work led to the production of more than 200 different restriction enzymes—all developed from bacteria. Thus, more than 200 different nucleotide sequences can be cut, or separated, from a DNA strand. When a DNA fragment is cut out, a space is left with ends that can be spliced together, making the DNA strand whole again. A DNA fragment taken from one or more other organisms can be used to fill in this space, thus introducing a set of "foreign" codons into the DNA strand. An enzyme called *DNA ligase* is used to splice, or recombine, ends.

This basic recombinant technique was refined to include bacterial plasmids. A *plasmid* is a tiny ringlike DNA molecule containing several thousand base pairs (see Figure 15–14). In the laboratory, scientists

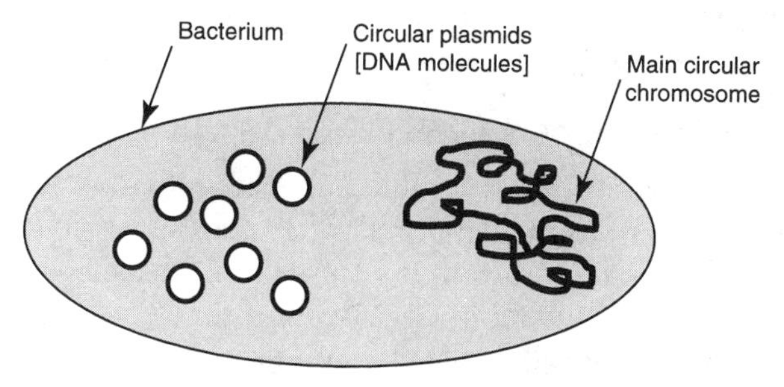

Figure 15–14 A bacterium's DNA plasmids

remove a plasmid from a bacterium, use restriction enzymes to cut a space, and then splice in foreign DNA fragments using DNA ligase (see Figure 15–15). The altered plasmid can be considered a "hybrid" because it contains foreign genetic material. The hybrid plasmid with foreign DNA can be transferred back into the bacterium from which it was removed. Then, more bacteria containing the hybrid DNA can be replicated.

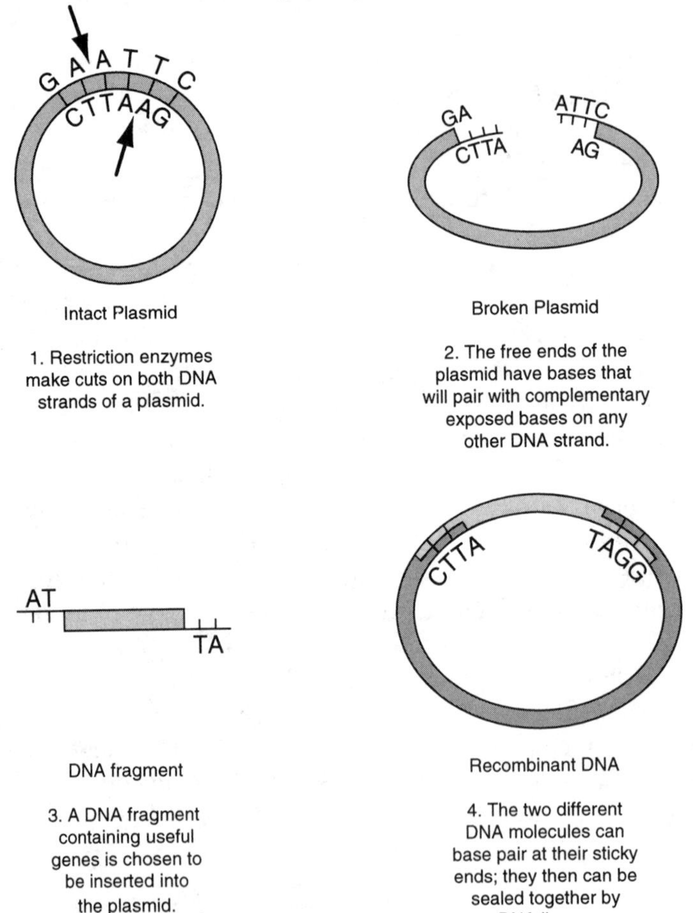

Figure 15–15 Recombinant DNA technique

At present, scientists use recombinant techniques to develop bacterial colonies that produce useful biological substances such as insulin, human growth hormone, and interferon. *Interferon* is a substance, produced by virus-infected cells, which inhibits viral replication.

Scientists have inserted foreign genes into plants and into the eggs of certain fishes, amphibians, reptiles, birds, and mammals. Organisms whose genotypes have been artificially changed by such genetic engineering are called *transgenic organisms.* Examples of useful transgenic organisms are transgenic pigs, which produce meatier pork chops, and transgenic soybeans, which are herbicide-resistant.

Should gene therapy, or recombinant DNA techniques, be used in human cells to replace defective alleles with normal ones? For example, should a gene that codes for human growth hormone be given to a short person? Should a gene that codes for human insulin be given to a diabetic? The complexity of human gene functions, coupled with the fact that the human genome is not completely understood, raises ethical questions concerning the joining of foreign genes with human genes.

Section Quiz

1. A nucleotide triplet that apparently has no function, such as UAC, is called a (an) (*a*) exon (*b*) intron (*c*) codon (*d*) axon.
2. Plasmids containing nucleotides are usually present in (*a*) bacteria (*b*) human cells (*c*) plant cells (*d*) viruses.
3. Restriction enzymes are used by geneticists to (*a*) inhibit defective genes (*b*) cut pieces of DNA out of plasmids (*c*) connect nucleotides (*d*) stop DNA replication.
4. The joining of amino acids at ribosomes to form proteins is a function of (*a*) tRNA (*b*) mRNA (*c*) specific introns (*d*) DNA ligase.
5. Recombinant DNA technology uses plasmids (*a*) as sources of purified genes (*b*) to locate specific exons (*c*) to copy DNA molecules (*d*) as cloning vectors.

Chapter Review Questions

The following questions will help you check your understanding of the material presented in the chapter.

1. According to the gene-chromosome theory, a pair of alleles for a trait (*a*) occupy corresponding positions on homologous chromosomes (*b*) must be different from each other (*c*) occupy different positions on homologous chromosomes (*d*) are located on the same chromosome.

2. If genes for two different traits are located on the same chromosome, they are (*a*) allelic (*b*) linked (*c*) segregated (*d*) intermediate.
3. The chromosomes that are not involved in sex determination are called (*a*) centrosomes (*b*) centrioles (*c*) ribosomes (*d*) autosomes.
4. In fruit flies, black body color and normal wings frequently are inherited together. One possible conclusion is that genes for these traits (*a*) are allelic pairs (*b*) are located on different chromosome pairs (*c*) are examples of the principle of independent assortment (*d*) are located on the same chromosome.
5. The gene for sex-linked traits such as color blindness is carried on the (*a*) Y chromosome (*b*) X chromosome (*c*) homologous chromosome (*d*) autosomes.

Base your answers to questions 6 *through* 8 on the representation below depicting a change in a portion of the base sequence in a DNA molecule.

$$\text{A T C G T A} \xrightarrow{\text{X ray}} \text{A A C G T A}$$

6. The change may be interpreted as a (*a*) gene mutation (*b*) nucleic acid replication (*c*) nucleotide synthesis (*d*) gene replication.
7. In humans, a change similar to the one shown may cause an individual to be afflicted with (*a*) Down syndrome (*b*) polyploidy (*c*) phagocytosis (*d*) sickle-cell anemia.
8. An important characteristic of this type of change is that frequently it (*a*) involves many chromosomes at once (*b*) is an advantage to an organism (*c*) may be passed on to offspring (*d*) requires a change in the environment.
9. Occasionally, a person may have either three X chromosomes, or two X chromosomes and a Y chromosome, in each body cell. This abnormal condition is caused by (*a*) multiple alleles (*b*) nondisjunction (*c*) artificial selection (*d*) crossing-over.
10. Which is caused by nondisjunction? (*a*) hemophilia (*b*) phenylketonuria (*c*) nearsightedness (*d*) Down syndrome.
11. Crossing-over normally occurs during (*a*) mitosis (*b*) meiosis (*c*) fertilization (*d*) cleavage.
12. There are many shades of skin coloring in humans. From this fact it may be assumed that the trait for skin coloring in humans is deter-

mined by (*a*) a single pair of alleles (*b*) cytoplasmic alleles (*c*) more than one pair of alleles (*d*) linked alleles.

13. In sickle-cell anemia, the abnormal hemoglobin differs from normal hemoglobin by (*a*) a single amino acid (*b*) the amount of coenzymes present (*c*) the number of iron atoms (*d*) the number of genes present.
14. In a human male, which sex chromosome makeup indicates that nondisjunction occurred during the formation of his mother's gametes? (*a*) XXX (*b*) X (*c*) XY (*d*) XXY
15. If a defective gene occurs on the X chromosome, the gene will normally be transmitted to male offspring only by (*a*) the mother (*b*) the father (*c*) segregation (*d*) mutation.
16. A plant has 18 chromosomes, although its species normally has 12 chromosomes. This mutation is an example of (*a*) recombination (*b*) polyploidy (*c*) crossing-over (*d*) incomplete dominance.
17. The fact that DNA has been found in the chloroplasts and mitochondria of some cells is possible evidence of (*a*) cytoplasmic as well as chromosomal inheritance (*b*) bacterial transformation (*c*) spontaneous mutations (*d*) the presence of a mutagenic agent.
18. The dissimilarities among living things mainly are caused by (*a*) their physiological activities (*b*) their DNA and RNA codes (*c*) the type of bonding in their compounds (*d*) the chemical elements in their protoplasm.
19. If dead type *A* bacteria are added to cultures of living type *B* bacteria, living type *A* forms appear. This may best be explained by (*a*) a change in the nutrition of the bacteria (*b*) the spontaneous generation of type *A* bacteria (*c*) the transfer of DNA from the dead type *A* bacteria to the living type *B* bacteria (*d*) the induction of a mutation in the dead type *A* bacteria.
20. The inheritance of the ABO blood groups in humans may be explained by a model using (*a*) linkage (*b*) multiple alleles (*c*) crossing-over (*d*) independent assortment.
21. Four children of the same parents have different blood types. Which is true about the genotypes of the parents? (*a*) Both parents are homozygous. (*b*) Both parents are heterozygous. (*c*) The father is homozygous and the mother is heterozygous. (*d*) The father is heterozygous and the mother is homozygous.

22. Which parents are most likely to produce a color-blind daughter? (*a*) a mother who is a carrier and a father who is normal (*b*) a mother who is color-blind and a father who is normal (*c*) a mother who is neither a carrier nor color-blind and a father who is color-blind (*d*) a mother who is a carrier and a father who is color-blind.

Base your answers to questions 23 *and* 24 on the information below. In humans, red-green color blindness is due to a sex-linked recessive gene.

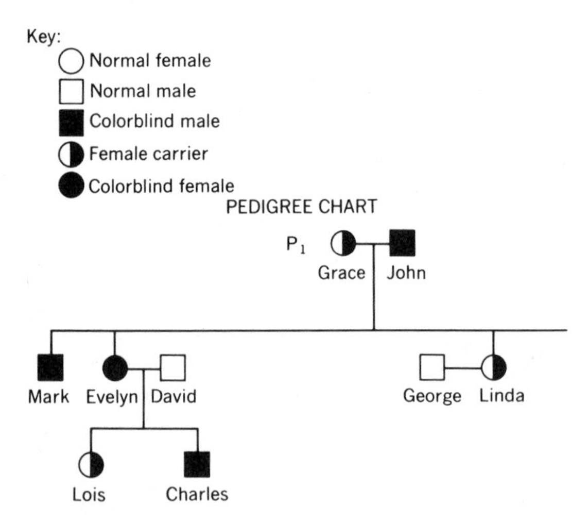

23. Which assumption about the P_1 generation is true? (*a*) John had one gene for color blindness. (*b*) John had two genes for color blindness. (*c*) Grace carried two genes for color blindness. (*d*) Grace was color-blind.

24. If Linda and George have a son, what is the probability that he will be color-blind? (*a*) 0 percent (*b*) 25 percent (*c*) 50 percent (*d*) 75 percent

25. A hemophiliac man is married to a woman who does not have genes for hemophilia. What is the probability that their first child will be a boy who is a hemophiliac? (*a*) 0 percent (*b*) 25 percent (*c*) 50 percent (*d*) 100 percent

26. To explain why hemophilia is more frequent in males than in females, it was suggested that the gene for hemophilia might be located on the Y chromosome. Which observation contradicts this suggestion? (*a*) Hemophilia may occur in females. (*b*) Y chromosomes are smaller than X chromosomes. (*c*) Males have XY chromosomes. (*d*) Few males are hemophiliacs.

27. Suppose an orange has been artificially bred using recombinant DNA to have a much greater tolerance to cold temperatures than other normal oranges have. This is an example of a (an) (*a*) pedigree (*b*) codon (*c*) intron (*d*) transgenic organism.
28. One reason why the manipulation of genes is so difficult is because (*a*) DNA strands are relatively short (*b*) genes cannot be seen by a microscope (*c*) chromosomes are changing constantly (*d*) there are billions of base pairs in a strand of DNA.
29. People may benefit from recent advances in genetic engineering if they suffer from (*a*) diabetes (*b*) dwarfism (*c*) viral infections (*d*) all of these.

Biology Challenge

The following questions will provide practice in answering SAT II-type questions.

Part I

Answer questions *1 through* 5 based on library research, your knowledge of biology, and the schematic shown below. The schematic shows a metabolic pathway starting with the amino acid phenylalanine.

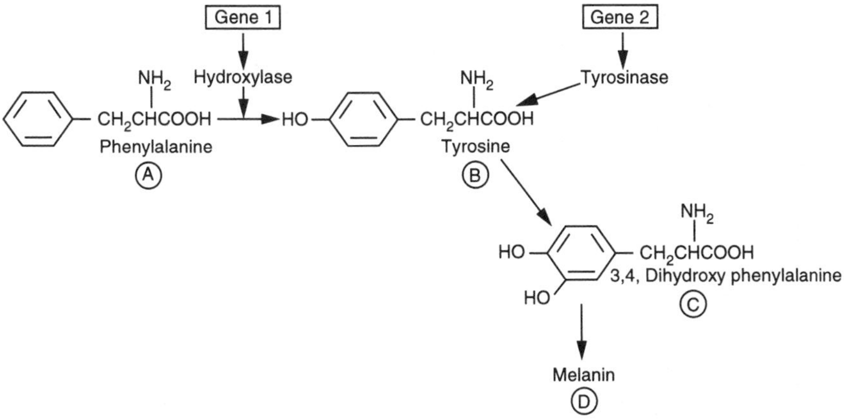

Gene 1
Hydroxylase
NH_2
$CH_2CHCOOH$
Phenylalanine
(A)
Gene 2
Tyrosinase
HO
NH_2
$CH_2CHCOOH$
Tyrosine
(B)
HO
HO
NH_2
$CH_2CHCOOH$
3,4, Dihydroxy phenylalanine
(C)
Melanin
(D)

1. The reactions of the pathway occur in a series: A → B → C → D. The product of one reaction serves as the substrate for the next reaction. The substrate for melanin is (*a*) phenylalanine (*b*) hydroxylase (*c*) tyrosine (*d*) dihydroxy-phenylalanine (*e*) tyrosinase.

2. A gene 2 allele, when homozygous, is the probable cause of an inborn error of metabolism called (*a*) Down syndrome (*b*) albinism (*c*) Turner syndrome (*d*) sickle-cell anemia (*e*) phenylketonuria.
3. Tyrosine is a substrate for products other than melanin, for example, homogentisic acid. This acid is normally converted to carbon dioxide and water with the aid of an oxidase enzyme. *Alcaptonurics* are individuals who lack oxidase. Such individuals (*a*) are hemophiliacs (*b*) suffer from achondroplasia (*c*) do not metabolize milk sugar (*d*) cannot secrete mucus (*e*) eliminate black urine.
4. Should gene 1 be faulty in a child and phenylalanine accumulate in the blood, the child usually is (*a*) mentally retarded (*b*) anemic (*c*) six-fingered (*d*) an albino (*e*) a dwarf.
5. Cretinism, a type of mental retardation, results when an enzyme fails to convert tyrosine into (*a*) thyroxine (*b*) epinephrine (*c*) hemoglobin (*d*) phenylthiocarbamide (*e*) growth hormone.

Part II

Select the letter of the statement that best completes questions *1 and 2*.

1. Frederick Griffith, a biologist, observed that when he injected mice with certain live pneumococci that possessed a jellylike capsule, the mice contracted pneumonia and died. He also observed that when he injected mice with certain live pneumococci that lacked a capsule, the mice did not contract pneumonia. When he injected mice with a mixture of dead bacteria possessing capsules and live bacteria lacking capsules, the mice contracted pneumonia and died. An explanation for the death of these mice is that (*a*) all dead bacteria can cause pneumonia (*b*) capsules have no connection with pneumonia (*c*) capsule formation became part of the genetic code of the live bacteria (*d*) toxins released by the dead bacteria caused pneumonia (*e*) transduction took place.
2. In 1950, scientists Maurice Wilkins and Rosalind Franklin, at King's College in England, extracted DNA from calf thymus cells. Wilkins observed: "A fiber of DNA (was) almost invisible, like a filament of a spider web." Wilkins asked a graduate student in X-ray crystallography to take diffraction pictures of DNA threads in order to discover their structure. The name of that student was (*a*) Linus Pauling (*b*) James Watson (*c*) Marshall W. Nirenberg (*d*) Severo Ochoa (*e*) Max Perutz.

Answer questions 3 *through* 5 based on the following paragraph and your knowledge of biology. You may have to do library research.

Certain genes, called *oncogenes*, are related to the growth of cancer cells. To date, 20 human oncogenes have been discovered. Oncogenes appear to be necessary, but not solely responsible for the growth of cancer cells. It is believed that two or more different oncogenes, working together, are necessary to change normal cells to malignant cells. Most oncogenes were discovered after scientists determined that certain viruses, moving in and out of cells, pick up bits of DNA.

Dr. John Cairns, professor of microbiology at the Harvard School of Public Health, stated: "Out of the millions of times this happens, an occasional virus picks up genes that make a recipient cell go crazy. If you can find this virus, it hands you the cancer gene on a plate." Mammals, fishes, and fruit flies possess normal genes that may become activated as oncogenes by association with a retrovirus (containing RNA).

3. The above paragraph may lead us to infer that (*a*) a common set of genes, persisting over millions of years of evolution, may be the key to many cancers (*b*) one important cause of cancer is a retrovirus (*c*) bits of DNA in cells that code for tumors always are present in mammals, fishes, and fruit flies (*d*) a chicken with cancer could not possibly possess oncogenes (*e*) it is impossible to interfere with oncogene activity and thereby stop or reverse cancer development.
4. Chromosomal breaks, translocations, deletions, and point mutations (*a*) are significiant because they always provide normal genes that turn into oncogenes (*b*) are not significant because the changes usually are mended by self-repair mechanisms (*c*) result from retrovirus invasion (*d*) result from non-disjunction (*e*) result from transformation.
5. Dr. Cairn's recipient cell may "go crazy" due to (*a*) translocation (*b*) spontaneous mutation (*c*) duplication (*d*) transduction (*e*) restriction enzymes.

16

History of Life on Earth

Learning Objectives

When you have completed this chapter, you should be able to:

- **Describe** methods scientists use to estimate the age of Earth.
- **Identify** ways in which Earth has changed over time.
- **Discuss** the importance of fossil evidence to explain the changes that have occurred on Earth over time.
- **Identify** characteristics of certain geologic eras and periods.
- **Describe** stages in the evolution of humans.
- **Discuss** some results of the evolutionary process.

OVERVIEW

How did the Earth originate? How did life begin on our planet? Scientists currently think that Earth developed from gaseous matter, and that life later developed from simple organic compounds. In this chapter, you will learn how species of protists, fungi, plants, and animals gradually evolved from primitive forms of life. Most of these early species are extinct; many others still exist. During the course of evolution, simple organisms gave rise to the more complex types that now inhabit Earth.

THE ORIGIN OF EARTH AND ITS ROCKS

Scientists know from evidence in the Earth's rocks that life has existed for billions of years. The history of life on Earth reveals that great changes have occurred in living things during different periods of time. Most of the evidence supporting the hypothesis that simple organisms gave rise to more complex organisms is obtained from fossils and other sources. This evidence has led scientists to conclude that present-day organisms evolved from a common ancestor. This is the basis for the modern theory of evolution.

In order to understand the modern theory of evolution, first learn about the formation and age of Earth and its rocks, and learn how rocks reveal evidence of past life.

Scientists think that during the formation of the solar system, Earth may have developed from gases that condensed into hot molten matter. This matter gradually cooled and solidified. Rocks that form from the solidification of molten matter, such as granite and basalt, are called *igneous rock*. In time, water accumulated in rock basins and formed Earth's oceans. Weathering and erosion of igneous rocks broke them down into smaller particles that eventually became soil. In many regions rock fragments and soil were transported by glacier, wind, and running water and deposited elsewhere. The final destination for most of the soil was in quiet bodies of water such as lakes and bays. Such deposited materials are called *sediments*. As layers of sediments accumulated, they were compressed by the weight of the overlying layers and the water above them. Compressed particles, cemented together by minerals in the water, formed *sedimentary rock*. In this manner, sand became sandstone, clay became shale, and chalky shell fragments became limestone. As a result of pressure and heat, igneous and sedimentary rocks were sometimes changed into harder, denser *metamorphic rock*. And in this manner, sandstone became quartzite, shale became slate, and limestone became marble.

THE AGE OF EARTH

Several methods have been used to estimate the age of Earth's features. One method is based on the assumption that it takes about 1,000 years for 30 cm of sedimentary rock to form. Thus, if you know the thickness of a sedimentary rock layer, you can determine the approximate age of the rock layer. A second method is based on the assumption that the oceans were originally bodies of fresh water. If you divide the average concentration of dissolved salts in the sea by the average amount of salt added in a year, you can obtain a rough estimate of the age of the oceans.

The most accurate method of determining the approximate age of Earth is to find out the age of the original igneous rock by using the radioisotopes uranium-238 and potassium-40. Radioisotopes decay at a definite rate, and they eventually change into nonradioactive elements.

The amount of time for half of a given quantity of a radioisotope to decay to a stable element is called its *half-life*.

Uranium-238, which occurs naturally in many igneous rocks, has a half-life of about 4.5 billion years and decays to form lead-206. By determining the ratio of lead to uranium in an igneous rock, it is possible to calculate the age of the igneous rock. Using this method, rock samples have indicated ages that range from 4 to 5 billion years. The half-life of potassium-40 is about 1.4 billion years; potassium-40 eventually decays into stable argon gas. The potassium → argon method confirms the estimates obtained by the uranium → lead method. If the oldest igneous rocks are about four billion years old, then the material from which these rocks were formed must be older, thus making the age of Earth about 4.5 billion years old.

The radioisotope carbon-14, a normal constituent of some carbon dioxide molecules in the atmosphere, is used to estimate the age of wood and bone specimens less than 40,000 years old. Carbon-14 has a half-life of about 5,700 years and decays to form stable nitrogen-14. Nonradioactive carbon-12 is the ordinary isotope of carbon present in most carbon dioxide molecules. As a result of respiration and photosynthesis, all living things accumulate both carbon-12 and carbon-14 in a fixed ratio. At death, cells no longer accumulate carbon-14; the carbon-14 content in the cells decreases as the radioisotope decays to form stable nitrogen-14. As this occurs, the ratio between the two carbon isotopes changes. The new ratio can be used to calculate, or *date*, the age of the remains of many organisms or materials made from them.

THE ORIGIN OF LIFE

About 300 years ago, people believed that new generations of living things arose from nonliving matter. This belief, called *spontaneous generation*, was disproved by the experiments of *Francesco Redi* (1668), *Lazzaro Spallanzani* (about 1780), and *Louis Pasteur* (1864). However, the problem of how life evolved remains unsolved. Under the conditions that existed several billion years ago, life may have evolved from nonliving matter.

Most scientists agree that life originated in warm, shallow seas. Some scientists thought that autotrophs were the first living things because they make their own food. Other scientists, however, claimed that the chlorophyll molecule found in all autotrophs is too complex to have appeared so early in Earth's history. Accordingly, most scientists now believe that simple heterotrophs were the first living things, based on the *heterotroph hypothesis* developed by A. I. Oparin in 1936.

The Heterotroph Hypothesis (Oparin Hypothesis)

The heterotroph hypothesis is based on the following inferences about Earth's environment:

Early atmosphere. Earth's original atmosphere consisted of a mixture of the gases ammonia (NH_3), methane (CH_4), hydrogen (H_2), and water vapor (H_2O). In contrast, the atmosphere today consists mainly of a mixture of the gases nitrogen (79 percent), oxygen (20 percent), carbon dioxide (0.04 percent), water vapor (variable amounts), and small amounts of the rare gases, such as neon, argon, and xenon.

Early temperature. The temperature of Earth and its atmosphere was much higher during our planet's early history than it is today. Also, Earth and its atmosphere received more energy from the sun four billion years ago than today. This energy is *radiant energy*—heat, ultraviolet rays, gamma rays, and cosmic rays.

Formation of organic compounds. As the Earth cooled, water vapor condensed and precipitated to fill rock basins, thus forming the ancient seas. Rain that fell on the land dissolved minerals from the igneous rocks and carried the minerals to the sea. Lightning from storms and radiant energy from the sun caused the gas molecules of the original atmosphere to react and form different organic compounds, such as amino acids, glucose, fatty acids, glycerol, and nucleotides. These complex molecules combined with rain and formed an organic "soup" in the shallow marine environments that had formed. In time, the "soup" became thicker as new organic compounds such as proteins (including enzymes), nucleic acids, and ADP and ATP were formed. Carbon atoms tend to combine with other carbon atoms and with hydrogen atoms. Thus, it is probable that the ancient seas were like tremendous vats in which chemical combinations occurred continuously.

Formation of coacervates. In the ancient seas, complex organic compounds interacted to form clusters, or aggregates, of molecules called *coacervates* (Figure 16–1). The availability of glucose, enzymes, and some energy from molecules of ATP enabled coacervates, which consist of inorganic and organic compounds, to undergo anaerobic respiration. And, the availability of DNA and nucleotides enabled the coacervates to reproduce. In effect, the coacervates were the first heterotrophs, the original forms of life.

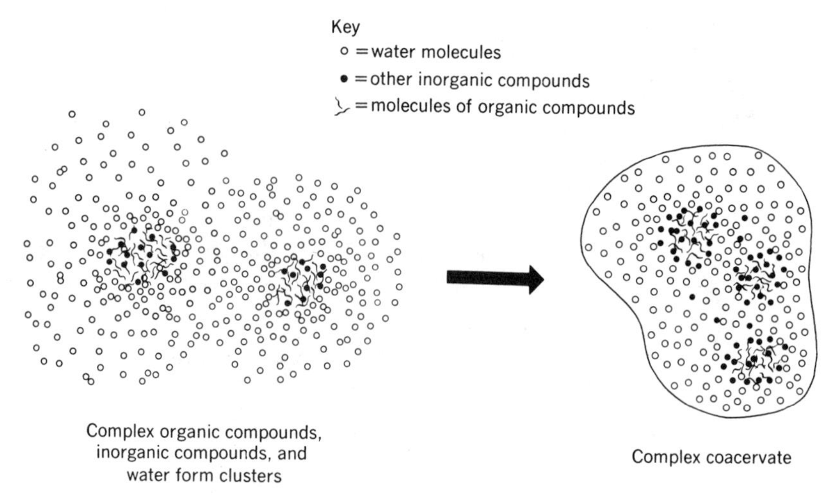

Figure 16–1 Formation of a coacervate

Autotrophs evolved from heterotrophs. The heterotrophs carried out anaerobic respiration, which released carbon dioxide to the seas and atmosphere. The carbon dioxide in the atmosphere screened out the high-energy radiation that until then caused molecules to combine. In time, as the food supply of the primitive organisms grew scarce, mutations in DNA molecules eventually produced new enzymes and chlorophyll. In this manner, some heterotrophs evolved into the first autotrophs—organisms that could produce organic substances from inorganic materials by photosynthesis. The newly formed autotrophs not only provided food for themselves and for the heterotrophs, but also released oxygen to the seas and the atmosphere.

Development of aerobic respiration. As oxygen accumulated in the atmosphere, the remaining methane reacted with oxygen, forming carbon dioxide and water. The remaining ammonia and other compounds reacted with oxygen, forming nitrogen, carbon dioxide, and more water, all of which became part of the lower atmosphere.

The presence of dissolved oxygen in seawater enabled aquatic organisms to carry out aerobic respiration. Aerobic respiration made more energy available than anaerobic respiration because the aerobic process releases more ATP molecules. This enabled the evolution of larger and more active organisms. It is estimated that it took about two billion years for organisms to develop the ability to carry out all the life processes.

The heterotroph hypothesis differs from the idea of spontaneous generation in the following ways:

1. Life did not evolve suddenly, but gradually, over hundreds of millions of years.
2. The environment in which life originated was very different from the environment today.

Evidence for the Heterotroph Hypothesis

In 1953, *Stanley Miller*, a graduate student at the University of Chicago, constructed the apparatus shown in Figure 16–2, which simulates the conditions of the atmosphere that existed about four billion years ago. After operating the apparatus for a week, Miller analyzed the material that had accumulated at its base. He discovered that the material contained amino acids and other types of organic compounds.

Several years later, *Sidney Fox* formed simple proteins by heating a mixture of amino acids. This confirmed the idea that the first proteins

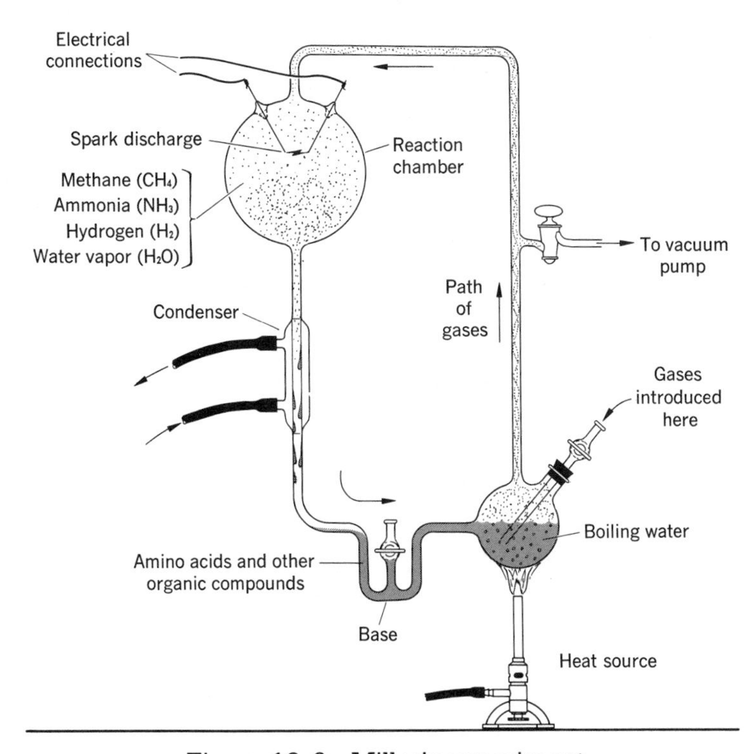

Figure 16–2 Miller's experiment

could have been naturally formed from amino acid units. Successful synthesis of some kinds of RNA and DNA in the laboratory made it easier for scientists to accept the idea of the natural development of self-duplicating molecules. To date, a living cell has not been created from a mixture of inorganic and organic compounds.

FOSSILS AND EVOLUTION

Fossils are the remains of prehistoric organisms. The remains include bones, teeth, shells, as well as the other hard parts of animals, plants, and unicellular organisms preserved in rock. Traces and impressions of organisms are also types of fossils. *Paleontology* is the branch of biology involved with the discovery, classification, and interpretation of the great diversity of past life forms. Paleontologists use fossils as evidence to support their ideas about the evolution of different forms of life on Earth. In effect, fossils are pieces of an evolutionary jigsaw puzzle that, interpreted properly, can provide some understanding of the biological and geological events that took place over long periods of time.

A change involving major extinctions and appearances of new life forms is called *macroevolution*. For example, the appearance of mammals, about 230 million years ago, from reptilian ancestors that subsequently became extinct is an example of macroevolution. In contrast, *microevolution* involves changes within a population that result in a modified life form. An example of microevolution is the development of a modern one-toed horse from a three-toed horselike ancestor. Natural selection, or Darwinian evolution, is the process by which all macroevolutionary and microevolutionary changes occur.

Fossil Remains

The actual remains of organisms were preserved as fossils in the following ways:

In ice. Remains of the woolly mammoth, an extinct relative of the elephant, have been discovered in the frozen soil of Siberia and Alaska. The mammoth remains were complete, including the flesh and skin. Woolly mammoths lived until about 25,000 years ago.

In amber. *Amber* is hardened resin that once oozed as sap from ancient trees. Insects, such as extinct flies and ants, sometimes became trapped in the fresh, sticky resin. The insects became fossils as the resin hardened into amber.

In tar pits. A tar pit is a pool of viscous asphalt tar. The La Brea Tar Pits in Los Angeles, California, contain the fossil bones of many extinct mammals including the saber-toothed cat and the giant sloth. Animals that tried to walk or run across the surface of the tar pits became trapped, sank into the pools, and their remains were preserved by the tar.

In coal. Swampy tropical forests made up mainly of giant tree ferns once covered parts of Siberia and the United States, including Alaska. As the tree ferns died, they fell into water, became submerged, and then were covered by particles of sand, gravel, clay, and mud. As a result of chemical reactions, hydrogen and oxygen gas in the organic compounds were released, leaving mainly solid carbon behind. In time the remaining plant materials solidified and changed to *bituminous* (*soft*) *coal*, a type of sedimentary rock. Many beds of soft coal contained impressions of leaves and the bark of stems.

Indicated Fossil Remains

The indicated remains of organisms were also preserved as fossils in the following ways:

Impressions and imprints. Organisms often left impressions of feet and other body parts in soft sediment (mud or clay). As a result of baking and hardening by the sun, the *impressions*, or *imprints*, were preserved. Later, additional layers of sediment covered the imprints. The imprints became permanently preserved as fossils when the mud or clay and the layers of sediment changed into sedimentary rock. Footprints of dinosaurs found in sedimentary rock are examples of imprints.

Casts and molds. As dead organisms (or parts of organisms) became embedded in clay and decomposed, cavities were left where the organisms had been. The cavity walls functioned as *molds* that showed external features of the organisms. Natural *casts* formed later, when the cavities filled with mineral matter containing salts of iron and calcium, and hardened. Many casts and molds of brachiopods and other extinct aquatic animals have been found.

Petrification. In some regions, dead organisms were covered by water that contained large amounts of dissolved minerals. This prevented rapid decay of the remains. As the remains slowly decayed, mineral matter gradually replaced the organic matter in the organism, thus turning it into "stone." This process is called *petrification*. Trilobites, bones of extinct fishes, and wood have been found as petrified remains.

Determining the Age of Fossils

The approximate age of a fossil can be determined from the sedimentary rock layer in which the fossil is located. The deeper the rock layer, the older the fossil. Once the age of a particular fossil has been determined, it can be used as a *key*, or *index fossil*, to calculate the age of other organisms in similar layers of sedimentary rock.

Section Quiz

1. The appearance of mammals about 230 million years ago is an example of (*a*) catastrophism (*b*) microevolution (*c*) macroevolution (*d*) divergent evolution.
2. The age of Earth is closest to (*a*) one million years (*b*) one billion years (*c*) four million years (*d*) four and one-half billion years.
3. A prehistoric animal whose frozen body was preserved for thousands of years was the (*a*) eohippus (*b*) coelacanth (*c*) saber-toothed cat (*d*) woolly mammoth.
4. The replacement of bone or wood by minerals is called (*a*) amber formation (*b*) transmutation (*c*) petrification (*d*) vulcanization.
5. The heterotroph hypothesis was proposed by (*a*) Oparin (*b*) Darwin (*c*) Miller (*d*) Fox.

The Fossil Timetable

An examination of fossils in sedimentary rock presents a picture-book story of the geologic evolution of Earth and many of the organisms that have lived on it. For convenience, the time involved, about 2.5 billion years, is divided by scientists into large intervals called *eras*; these are divided into shorter intervals called *periods*. The major eras and the dominant life forms during these eras are shown in Table 16–1.

Archean Era. Inorganic and organic molecules were present when the Archean Era began more than four billion years ago. At some point, life forms originated from these organic molecules; the oldest fossils are the traces of 3.8 billion-year-old bacteria found in sedimentary rocks.

However, the earliest and most significant era, regarding the evolution of life, is the *Proterozoic Era*.

Proterozoic Era. This era began about 2.5 billion years ago, shortly after the origin of life and evolution of prokaryotic unicellular organisms. The

Table 16–1 The Fossil Record

Era	*When Era Began (millions of years ago)*	*Dominant Plants*	*Dominant Animals*
Cenozoic	65	Modern (flowering plants)	Humans and modern mammals; birds; insects
Mesozoic	225	Conifers and cycads	"Age of Reptiles" (Dinosaurs)
Paleozoic	570	Ferns and mosses	Amphibians; fishes; and higher invertebrates
Proterozoic	2,500	Algae (plantlike protists)	Protozoans (animallike protists); sponges
Archean	More than 4,000	Inorganic and organic molecules; origin of life	

Proterozoic Era featured the rise of eukaryotic cells, land masses concentrated at Earth's equator, and changes in Earth's climate, including an increase of atmospheric oxygen. These changes helped set the stage for the development and successful evolution of multicellular organisms. An ice age at the end of the Proterozoic Era wiped out vast numbers of *plankton* (aquatic communities of microorganisms floating on surface). The ice age was followed by a gradual warming of Earth. These events triggered the evolution of multicellular, aerobic organisms; fossils of these organisms have been found in the Ediacaran Hills of Australia. Scientists think that the land areas on Earth at this time did not feature any kind of life. About 600 million years ago, the Proterozoic ended and the Paleozoic Era began.

Paleozoic Era. This era is divided into a number of periods, including the *Cambrian*, *Ordovician*, *Silurian*, *Devonian*, *Mississippian*, *Pennsylvanian*, and *Permian*.

Cambrian Period. This period featured many significant events, including the following:

- The movement of land masses away from the equator, resulting in new coastlines, enclosed bodies of water, and new habitats.

- An increase in the concentration of atmospheric oxygen to the present-day level of about 20 percent.
- The development of multicellular organisms with more than two tissue layers.
- The specialization of tissues, which led to the development of organs and organ systems.
- Microevolutionary and macroevolutionary processes flourished: species disappeared and new species appeared. The species that persisted are ancestors of present-day organisms.

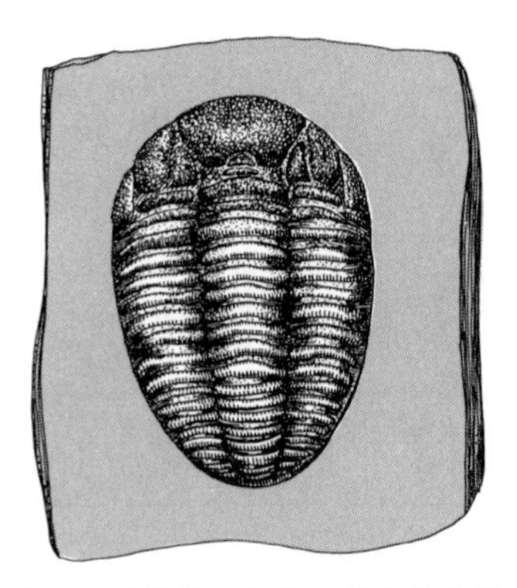

Figure 16–3 A fossil trilobite

Different invertebrate species evolved during the Cambrian Period, including sponges, jellyfishes, sea anemones, corals, flatworms, roundworms, mollusks, annelid worms, arthropods, and echinoderms. The most abundant Cambrian fossils are the trilobites (Figure 16–3). *Trilobites* are ancestors of the phylum *Arthropoda*, or "jointed-leg" animals, consisting of about 850,000 different species. Arthropods now include such organisms as insects, spiders, and crustaceans. Fossils of diatoms and seaweeds also have been found in Cambrian rocks.

Ordovician Period. During this period, trilobite species diminished, as corals, starfishes, and fishlike species evolved and flourished. *Graptolites* are characteristic fossils of this period, among many others. Graptolites currently are thought to be related to primitive *chordates* (animals with a notochord at a stage of their life cycle). The evolution of diverse, multicellular marine algae, including the chlorophytes (green algae), rhodophytes (red algae), and phaeophytes (brown algae) also occurred during the Ordovician Period.

Silurian Period. The most notable animal of the Silurian Period is a marine invertebrate called the sea scorpion, which resembled gigantic

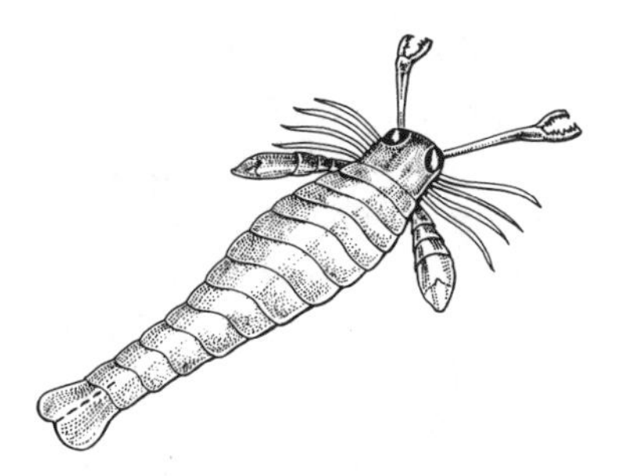

Figure 16–4 A reconstruction of a fossil sea scorpion

shrimp nearly three meters long (Figure 16–4). Coral and fishes with movable jaws were common. The first air-breathing animals, including land scorpions and millipedes, appeared. Plants, such as club mosses and *psilophytes*, also flourished during the Silurian Period. Psilophytes had few branches and no leaves. Paleontologists think that the psilophytes were transitional plants—with some characteristics of marine algae, and some characteristics of land plants.

Devonian Period. This period is called the *Age of Fishes* because large numbers of cartilaginous (such as sharks and rays) and bony fishes inhabited the seas. On land, amphibians (Figure 16–5), mosses, and true vascular plants became common. By the end of the Devonian Period, a revolutionary adaptation had evolved in plants—the *seed*. Plants that produced seeds were better able to survive on land. A seed is adapted for survival on land because it resists drying, contains stored food for the plant embryo, and germinates only under the proper temperature and water availability. Toward the end of the Devonian Period, alternation of generations, including a conspicuous sporophyte generation and a reduced gametophyte generation, represented a new method of reproduction in plants.

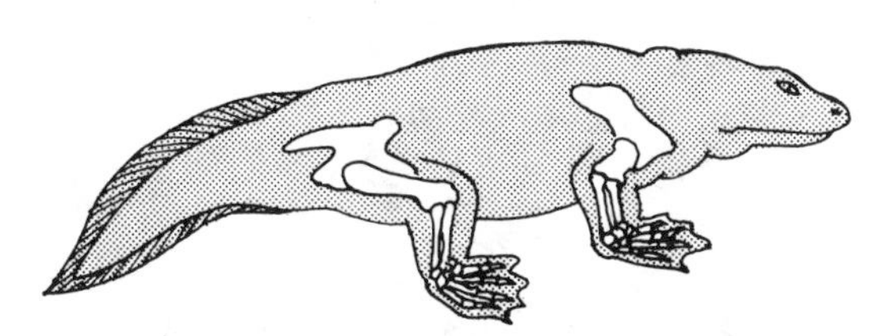

Figure 16–5 A primitive amphibian

Mississippian Period. During this period, filter-feeding, flowerlike marine animals called *crinoids* flourished. Crinoids lived on the bottom of shallow seas, attached to the seafloor by "stems." In addition, unicellular marine animals called *foraminiferans* were so numerous during the Mississippian Period that entire beds of limestone were composed of their calcium-rich skeletons. Both terrestrial plants and animals are noticeably missing from the fossil record of the Mississippian Period. Perhaps conditions were unfavorable for life on land or unsuitable for the preservation of fossils.

Pennsylvanian Period. During this period, swamps and forests formed and accumulated abundant plant remains, which subsequently were buried and compressed into coal beds. Giant tree ferns and insects were numerous, especially cockroaches. Cockroach fossils are so abundant from the Pennsylvanian Period that some scientists jokingly call this period the *Age of Cockroaches*. By the close of this period, amphibians with some reptilian features had begun to appear.

Permian Period. The Permian Period featured the rise of the Appalachian Mountains; active volcanoes in present-day California and Oregon; formation of the Ural Mountains in Europe; and the accumulation of snow and ice in the Southern Hemisphere—ushering in an ice age for most of Earth south of the equator. In the Northern Hemisphere, conebearing trees replaced tree ferns, and reptiles appeared in greater numbers.

Driven by a cooler climate and a variety of new habitats, animal evolution was characterized by an increase in diversity—providing the basis for the evolution of dinosaurs, modern reptiles, birds, and mammals. At the close of the Permian Period, early mammals first appeared. The transitional forms were *therapsids*, terrestrial animals with combined reptilian and mammalian features (Figure 16–6). Therapsids were different from reptiles, in that, among other features, they maintained a high internal temperature. Modern reptiles (and amphibians) have body temperatures that vary with the temperature of the external environment.

Figure 16–6 A fossil therapsid

Mesozoic Era. The Mesozoic Era (composed of the Triassic, Jurassic, and Cretaceous periods) began about 225 million years ago. It is called the *Age of Reptiles* because dinosaurs and other reptiles were the most abundant vertebrates. (See Figure 16–7.) Mesozoic reptiles inhabited all environments—land, water, and air. During this era, the climate was stable and warm. Macroevolution occurred, which resulted in the extermination of many species, including swimming reptiles, called *icthyosaurs*, plant-eating dinosaurs, such as *Diplodocus*, carnivorous dinosaurs, such as *Tyrannosaurus rex*, and the *pterodactyl*, a flying reptile with batlike wings.

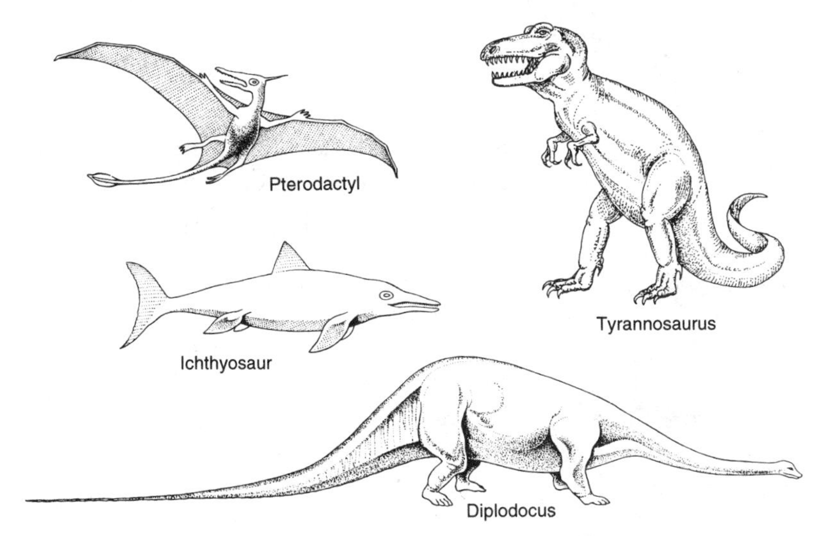

Figure 16–7 Mesozoic reptiles and dinosaurs

Dinosaurs ranged in size from that of a chicken to giant flesheaters and huge herbivores. Currently, some paleontologists think that dinosaurs may have maintained a high internal temperature, as do birds and mammals. If true, that may explain the ability of dinosaurs to migrate and inhabit regions that varied in temperature; in addition, it is thought that dinosaurs gave rise to warmblooded birds.

Archeopteryx was a primitive reptilelike bird that possessed scales and feathers covering parts of its body, teeth, and a pair of claws on its wings and on its feet (Figure 16–8). Most paleontologists cite archeopteryx as an example of an intermediate, or in-between, form to illustrate modern

Figure 16–8 Archeopteryx—a "birdlike" animal

evolutionary theory. During the Triassic Period, therapsids gave rise to the first true mammals; then the therapsids became extinct. The first true mammals were small, mouselike, warm-blooded, furry, and they fed primarily on insects. These mouselike mammals were active at night, or nocturnal, avoiding the larger carnivorous dinosaurs that were active during the day. The dominant plants of the Mesozoic Era were conebearing trees. During the Cretaceous Period, angiosperms, or true flowering plants, spread on land, populating the landscape with shrubs, trees, and herbaceous plants. At the same time, new insect species evolved, such as bees, wasps, ants, termites, flies, and mosquitoes. Some of these insects depended on angiosperm plants for their survival. As a result, paleontologists think that angiosperms must have coevolved with insects. The Mesozoic Era ended with the mass extinction of dinosaurs and other creatures, paving the way for other life forms in the Cenozoic Era.

Cenozoic Era. The Cenozoic Era, commonly called the *Age of Mammals*, covers the past 65 million years. The continents were in approximately the same position they are today. New ocean currents developed, which changed climates in different parts of Earth. These changed climates influenced the evolution of plants and animals. Angiosperms became the dominant plant form. Grasses, which helped feed herbivorous mammals, became common. In the early Cenozoic, small mammals, such as rodents and lemurs, became more numerous, and modern species of reptiles and birds appeared. *Hyracotherium* (ancestor of modern horses), whalelike mammals, primitive monkeys and apes, cats, and early dogs appeared. Camels, giraffes, antelope, deer, wolves, and hyenas became abundant in what is now Europe, Asia, and North America. In the North-

Figure 16–9 A saber-toothed cat and a woolly mammoth

ern Hemisphere, large deposits of snow and ice accumulated, which led to periodic movements of continental glaciers southward in all directions. The last continental glacier receded from the area that is now called Long Island, New York, about 10,000 years ago. Saber-toothed cats and woolly mammoths inhabited North America before this last glacial retreat (see Figure 16–9).

Living Fossils

Living plants and animals that closely resemble prehistoric ancestors are called *living fossils*. They include the lungfish, coelacanth fish (Figure 16–10), platypus, horseshoe crab, and ginkgo tree. The lungfish resembles members of an ancestral group related to an animal connecting fishes and amphibians. The lobe-finned coelacanth fish, which has fins that possess leglike bones, was believed to be extinct, but was discovered alive by a South African fisherman in 1938. The platypus is an egg-laying mammal. Today's horseshoe crab has the same basic structure as the fossils of horseshoe crabs found in ancient rocks. And, the ginkgo, or maidenhair, tree has persisted for millions of years without changing. It is frequently seen on city streets and in parks.

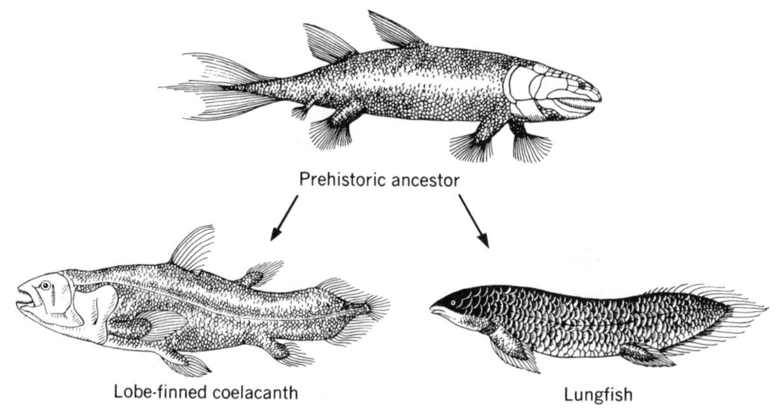

Figure 16–10 Two "living fossil" fishes and their ancestor

Why Species Became Extinct

The fossil record indicates that many species became extinct, others remain unchanged, and still others evolved into modern species. The major reasons why species became extinct include changes in Earth's surface, changes in climate, changes in heredity, and destruction by other organisms.

Changes in Earth's surface. As a result of the emergence and submergence of landmasses, many species of aquatic and terrestrial organisms became extinct. Some organisms became fossils, others left no trace of their existence. Only those organisms possessing suitable adaptations to the new environmental conditions survived. The extensive coal beds of Germany, North America, and Siberia, and the fossils of fishes, clams, and oysters on mountaintops in Wyoming attest to widespread movements of Earth's crust in the past. Localized movements of Earth's crust, such as earthquakes and volcanoes, also produced the extinction of some species.

Changes in climate. Plant fossils found in coal beds in cold regions are related to modern plants found only in warm climates. Thus, the extensive coal beds in present-day cold regions indicate that a tropical climate was once characteristic of these regions. Organisms that were not adapted to survive in a cold climate died; organisms that survived were either adapted to live and reproduce in a cold climate or were able to migrate to warmer regions. For example, bears migrated and adapted to different climates in different regions. The adaptability of bears to climate change may explain why bears are found in arctic as well as temperate zones of the world.

Changes in heredity. Organisms that inherited structures and functions that enabled them to adapt to a changing environment survived. Organisms that inherited traits that did not have survival value became extinct. For example, if a mutation causing short legs appeared in a running animal, the animal would become easy prey for its predators; these short-legged running animals would probably become extinct.

Destruction by other organisms. In the struggle for survival, organisms usually prey on each other. A possible reason for the extinction of dinosaurs toward the end of the Mesozoic Era was the appearance of small mammals. Some of these small mammals, such as rodents, may have eaten the eggs of many dinosaurs. In more recent times, humans have overhunted the passenger pigeon, causing its extinction.

The Direction of Evolution

The fossil record generally indicates that complex life forms evolved from simpler forms. In some cases, an organism lost an adaptation and gained another. The tapeworm, for example, lost its digestive system and gained the ability to absorb digested food from a host. These changes enable the tapeworm to survive within the digestive system of other organisms. The fossil record is incomplete for most organisms. However, the fossil records of the horse, camel, elephant, and human are detailed enough to indicate the direction of their evolution.

Evolution of the horse. Figure 16–11 shows how the modern horse (genus *Equus*) evolved from *Hyracotherium*, or *Eohippus*, the ancestor of several horselike species. A study of the bones of *Hyracotherium* indicates that the animal was about the size of a collie dog, had four toes on its front feet and three toes on its hind feet, and had teeth adapted to chewing leafy plants. Over a period of 60 million years, ancestral horses and their descendants evolved into the modern horse, which has a single toe (hoof) on each foot and high molar teeth adapted to grinding grasses. Attached to a modern horse's main toe bone are two slender bones, called *splints*, which are vestiges, or remains, of the toes lost during its evolution.

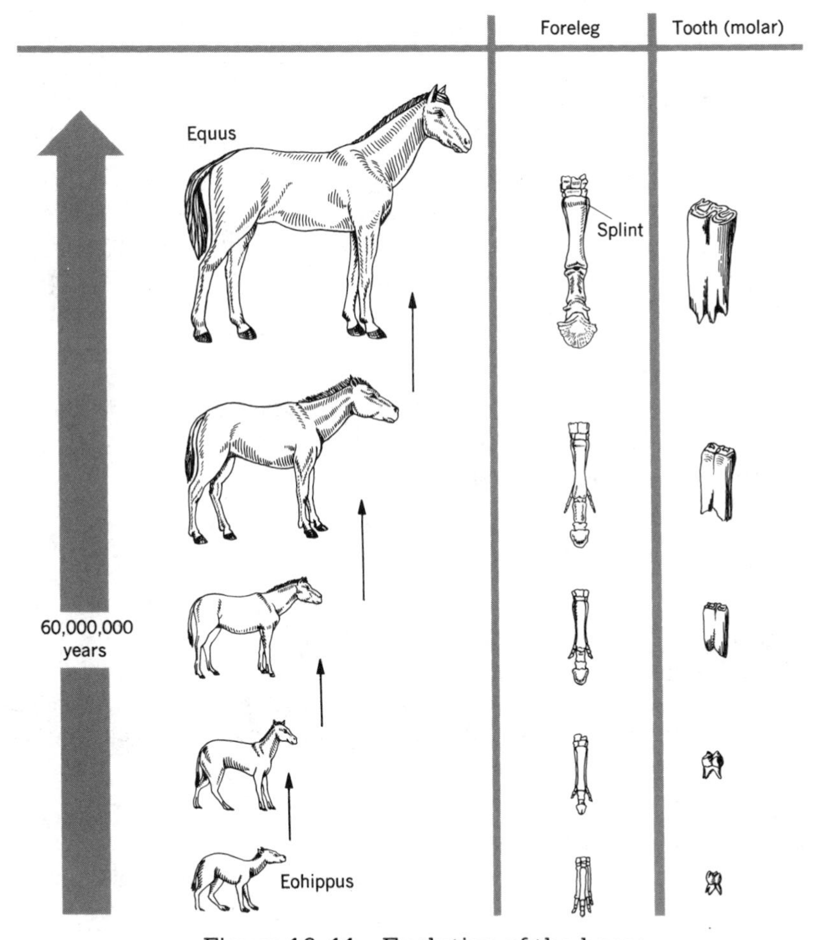

Figure 16–11 Evolution of the horse

Section Quiz

1. Multicellularity and tissue specialization are characteristics of organisms that first appeared in the (*a*) Cambrian Period (*b*) Silurian Period (*c*) Permian Period (*d*) Devonian Period.
2. Green, red, and brown algae first appeared in the (*a*) Cenozoic Era (*b*) Paleozoic Era (*c*) Mesozoic Era (*d*) Proterozoic Era.
3. Most paleontologists think that mammals probably evolved from (*a*) birds (*b*) dinosaurs (*c*) therapsids (*d*) diplodocus.
4. The remains of *Archaeopteryx* indicate that birds are most closely related to (*a*) flying insects (*b*) reptiles (*c*) flying mammals (*d*) amphibians.
5. A form of animal life that probably arose after fishes was (*a*) amphibians (*b*) mollusks (*c*) trilobites (*d*) crinoids.

Evolution of Humans

Humans are *primates*, members of an order of mammals that includes monkeys and apes (chimpanzees, gibbons, orangutans, and gorillas)—our closest animal relatives. After primates made their appearance, a split occurred, separating two prosimian (pre-monkey) lines from a third *anthropoid* ("humanlike") line. *Prosimians* are a relatively primitive group of primates that includes tarsiers and lemurs. Anthropoid primates feature the following characteristics:

- A large brain encased in a protective brain case (cranium).
- Opposable thumbs and flexible hands.
- Prolonged parental care of young.
- Well-developed binocular vision.
- A highly complex nervous system.

Apes, ancestral humans, and modern humans are called *hominoids*. All humans and their most humanlike ancestors are called *hominids*. Studies of hominoid fossils show that the hominid line separated from the line leading to chimpanzees and gorillas about six to eight million years ago. Yet, DNA studies indicate that chimpanzees are still more closely related to humans than they are to gorillas.

The genus Australopithecus. The reasons for the emergence of a separate hominid line from the main hominoid group are unknown. It is clear, however, that the first real hominid group, called *Australopithecus*, arose in Africa about four million years ago, and disappeared about one million years ago. In 1974, a team of paleontologists found an almost complete 3 to 3.5 million-year-old skeleton of an early australopithecine woman. Dubbed *Lucy,* her skeleton reveals that she walked erect on two legs (*bipedal*) and had rounded (not square) jaws—both are human features. Lucy and other early australopithecines were short (about 1.3 meters tall), and weighed about 21 kilograms. Their brain was about the same size as that of a present-day gorilla (about 400 cubic centimeters). Australopithecines ate both plants and animals. These features of the genus *Australopithecus* are ancestral to features of our genus *Homo*.

Lucy, and other fossils like it, are classified in the species called *Australopithecus afarensis*. Fossils of three larger, more recent species of australopithecines have been discovered: *A. boisei*, *A. africanus*, and *A. robustus* (see Figure 16–12). This group of hominids existed between one and three million years ago. At present, much controversy exists among paleontologists about the naming of certain australopithecines, their initial appearance, and which ones evolved from the others or lived at the

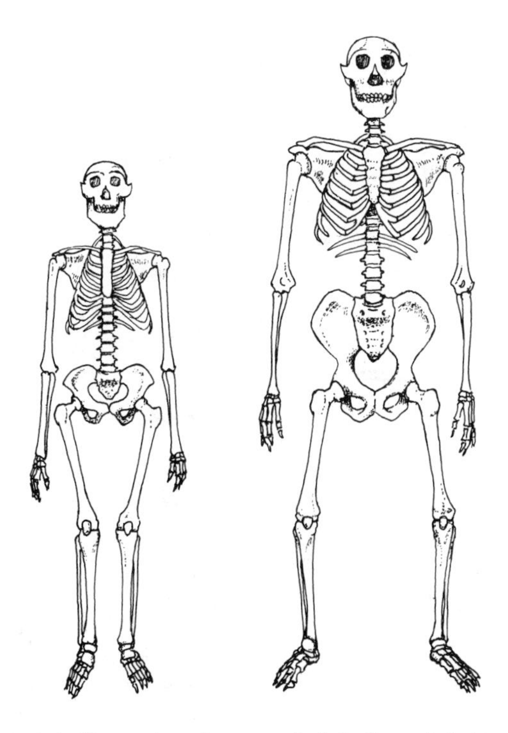

Figure 16–12 A comparison of *A. afarensis* and *A. boisei*

same time (see Figure 16–13). However, most paleontologists agree that early hominids shared special traits that enabled the development of more advanced hominids.

The genus *Australopithecus* flourished most likely for the following reasons:

1. Bipedalism enhanced by modifications of bone structure and arrangement of legs, pelvis, and lower spinal column added greater mobility and smoothness of movement.
2. The teeth and jaw structure of these early hominids enabled them to eat both plants and animals. In effect, *Australopithecus* represents a trend in hominids away from a strictly herbivorous diet to an omnivorous one.
3. Most important, hominids had larger brains than hominoids. The cranial capacity of early hominids ranged over time from 400 to 600 cubic centimeters—quite a difference from the brain size of a chimpanzee, which ranges from 280 to 400 cubic centimeters. A larger brain size meant improved survival skills.

The genus Homo. The hominid *Homo habilis*, the first species in our genus, appeared about two million years ago and survived for about a half-million years. The species name *habilis* means handy, or *able*, and *Homo habilis* was able to make and use stone tools. Fossil bones and tools associated with this species were discovered in sedimentary rock in Olduvai Gorge, Tanzania. Hominid skeletons unearthed in 1987 show that *H. habilis* was relatively short, apelike, and resembled the australopithecines.

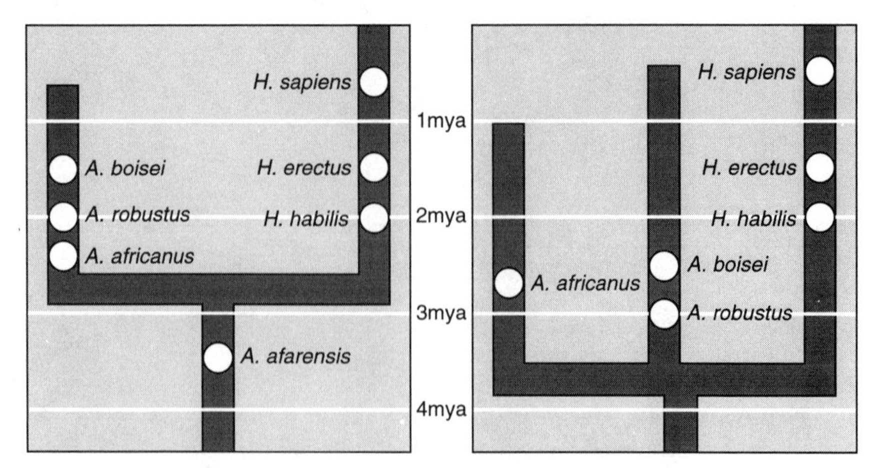

Figure 16–13 Two different models of hominid evolution

The next hominid to arise in this genus was *Homo erectus*, which probably evolved from *H. habilis* about 1.5 million years ago.

Homo erectus was an advanced hominid with survival skills that enabled this species to migrate from Africa into Asia and Europe. This hominid species lasted for about one million years. "*Peking man*" and "*Java man*" are examples of *H.erectus* (Figure 16–14). *Homo erectus* walked upright, had a thick skull and heavy brow ridge, and was probably as tall as modern humans. Importantly, *H. erectus* had a much larger brain capacity than did previous species of hominids. A skeleton of a young male *H. erectus* was found in Kenya by the well-known paleontologists Alan Walker and Richard Leakey. The specimen, named the *Nariokotome* boy, was about two meters tall, with a brain case that, in life, had contained an 880 cubic centimeter brain. In addition, artifacts indicate that *H. erectus* made and used fire and stone tools.

Figure 16–14 A reconstruction of *Homo erectus*

Homo sapiens neanderthalensis. Our species, *Homo sapiens*, evolved about a half-million years ago in Africa. *H. sapiens* spread from Africa to all parts of the world, taking over the territories and habitats of *H. erectus*. *Neanderthals*, classified as *Homo sapiens neanderthalensis*, arrived in Europe about 70,000 years ago. Fossils of these hominids were found in the

Neander Valley, Germany, in 1856, hence their name. Neanderthals were muscular, short, and powerfully built. They had massive skulls, bony ridges above the brows, and large brains (Figure 16–15). Neanderthals made and used tools from stone, such as hand axes, borers, scrapers, and spearheads. They wore skins of animals; buried their dead with flowers, food, and weapons; cared for their sick and injured; and lived a communal life. Neanderthals lasted for about 300,000 years, and then were replaced about 40,000 to 100,000 years ago by modern humans, probably also out of Africa. Some paleontologists think that Neanderthals may have interbred with modern humans about 40,000 to 50,000 years ago.

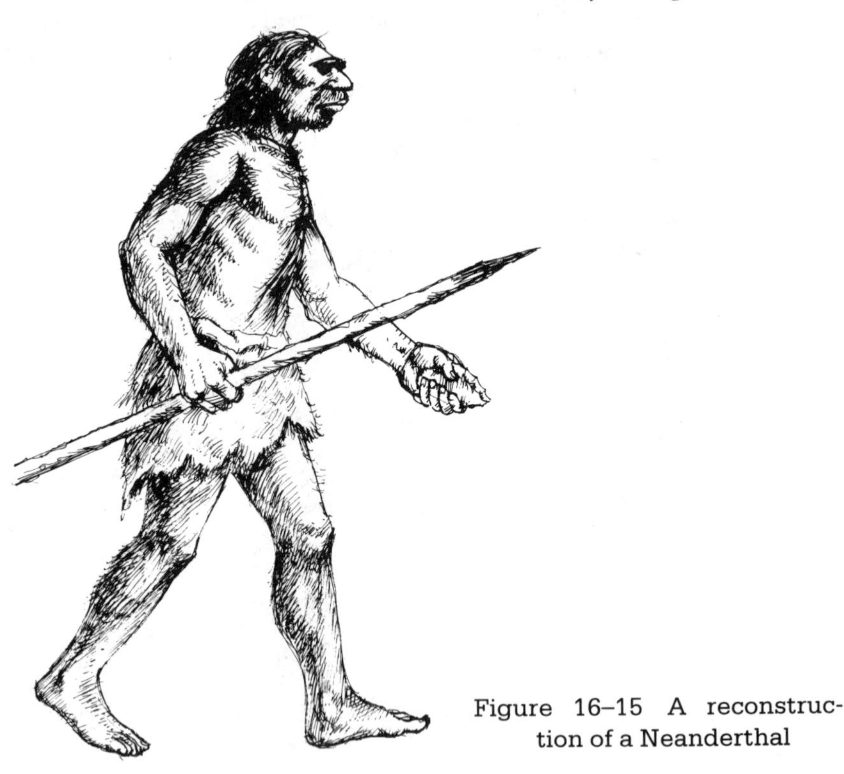

Figure 16–15 A reconstruction of a Neanderthal

Homo sapiens sapiens. Cro-Magnon humans, the first members of our subspecies *Homo sapiens sapiens*, were physically weaker than Neanderthals, but had a more sophisticated culture. In general, Cro-Magnon humans were tall, with a graceful body, featuring a head with finer features including a large, broad forehead, and a well-developed chin. They made sophisticated tools, weapons, pottery, and ornaments out of polished stone, bone, and ivory. Cro-Magnons also made clothing from animal skins. Other artifacts reflect artistic abilities and religious beliefs.

Early *Homo sapiens sapiens* lived in social groups, with males doing the hunting and fishing, and females performing agricultural and domestic tasks. According to one widely accepted theory, modern humans spread from Africa to Asia and Europe. Much later, people migrated to Siberia and the North American continent, arriving there about 14,000 years ago. At about that time, the last glacial period ended and the ice receded northward—melting, raising sea levels, and thus cutting off various land migration routes.

All humans belong to the same species, *Homo sapiens sapiens*, and usually are grouped into three major groups: Negroid, Caucasoid, and Mongoloid.

THE RESULTS OF EVOLUTION

Evolution is not a change in an individual organism during its lifetime. Rather, it represents the cumulative changes in a population over many generations. Evolution is *opportunistic*, taking advantage of major forces that drive the evolutionary process. Charles Darwin was aware of the results of evolution and considered them in formulating his theory of natural selection.

Similarities in structure. Charles Darwin and his colleagues already knew that certain body parts of different animals were similar in structure and arrangement. For example, Darwin observed that the arms of a human, legs of a horse, and wings of a bat are similarly constructed and arranged, although each function very differently. His evolutionary theory emphasized these similarities. Figure 16–16 shows similarities in the forelimbs of five representative vertebrates. The forelimbs are *homologous structures* because all five are derived from the same embryological

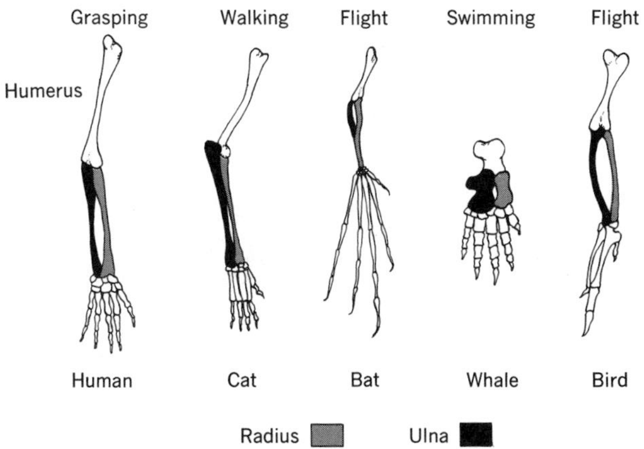

Figure 16–16 Comparison of forelimbs of vertebrates

tissue. Note how the bones correspond (are homologous), yet each forelimb functions differently.

The wings of a butterfly and those of a bird are examples of *analogous structures*, or structures that have the same function but differ in structure and embryological origin. A bat's wing and a human's arm are homologous structures. The bones of a bat wing and a human arm are similar in structure and origin, although a wing is used for flying and a human arm is used for performing various manipulations. The presence of homologous organs in different, but related, organisms is the result of descent from a common ancestor according to Darwin's evolutionary theory.

Similarities in development. Darwin also was interested in the embryological development of vertebrates. He observed that the embryos of different species had a similar shape, and at some point possessed gill pouches and a tail (see Figure 16–17). The presence of gill pouches, even in air-breathing animals, is understandable when we realize that land forms actually evolved from a common aquatic ancestor. In addition, the embryos of more closely related animals, such as the chicken and tortoise, appear to have a greater resemblance to each other than do the embryos of less closely related animals, such as the chicken and salamander.

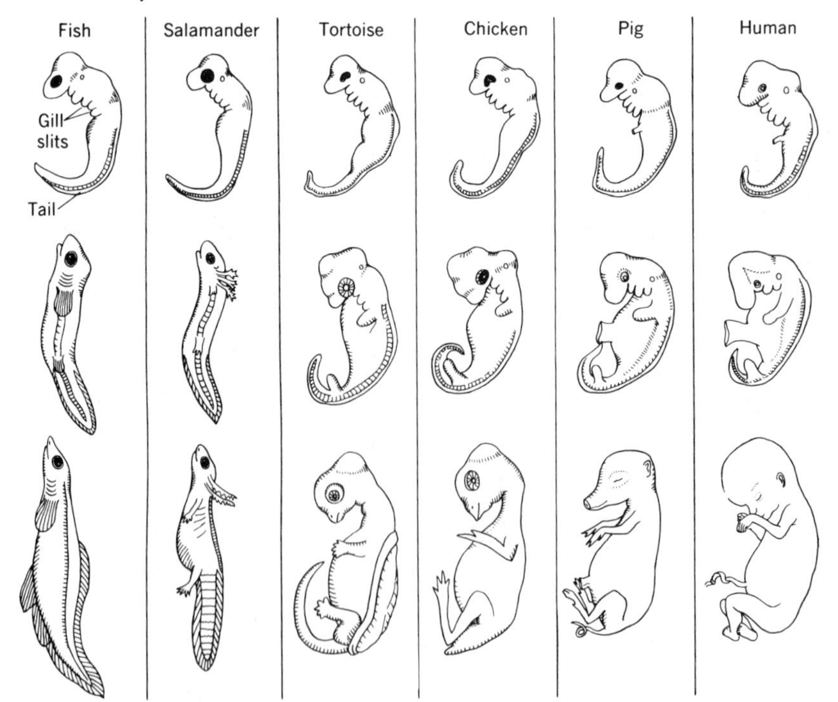

Figure 16–17 Comparison of embryos of vertebrates

Among invertebrates, the early embryos of mollusks and segmented worms are very similar; the early embryo of a starfish, however, bears little resemblance to the embryo of either a mollusk or a segmented worm. As a result, you can conclude that mollusks and segmented worms are closely related and probably arose from a common ancestor. The starfish, however, probably arose from another aquatic ancestor.

Similarities in biochemistry. Charles Darwin did not have any knowledge of biochemistry. He and the biologists of his time were unaware that living things resemble each other in their chemical composition and in the way they transform, store, and use energy. All living things possess DNA and RNA. Yet, organisms differ from each other because their DNA codes are not exactly the same. For example, many chemical reactions that occur during digestion, protein synthesis, cellular respiration, and DNA replication in a cat and in a frog are similar. Cats and frogs are different, however, because they acquired different sets of genes during the course of evolution.

In mammals, the chemical composition of hormones and enzymes are so similar that extracts from cattle and pigs are sometimes used to treat humans. The immunological reactions of mammals are similar. Consequently, antibodies taken from horses and sheep can be used to provide humans with immunity to certain diseases.

Vestigial structures. Darwin considered certain apparently useless body parts and organs, called *vestigial structures*, evidence for evolution. He reasoned that an apparently useless organ, such as the appendix in humans, must have been functional in an ancestral type. In humans, there are more than 100 vestigial structures, besides the appendix, such as certain ear muscles, skin muscles, and fused tail bones (coccyx). Other types of vestigial structures include hipbones in whales and splint bones in horses. Likewise, vestigial leg bones in a snake are evidence of an ancestor with legs. (See Figure 16–18.)

Coevolution. Many angiosperms, or flowering plants, coevolved with insects such as butterflies and bees. In *coevolution*, two different species evolve together in a reciprocal fashion; an evolutionary change in one species occurs as a response to a change in the other. For example, the benefits of flower pollination and food collection by bees evolved together: bees collect nectar and pollen for food, and plants complete the reproductive process (fertilization). Darwin had observed many examples of coevolution during his many years of field research. Thus, he included coevolution and its significance in his evolutionary theory.

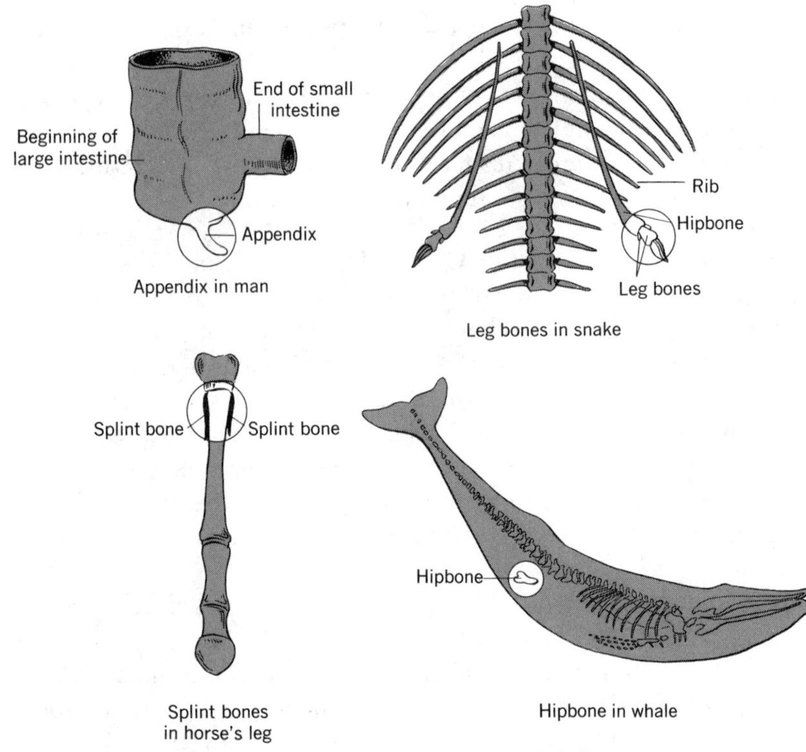

Figure 16–18 Examples of vestigial organs

Predation. One animal feeding on another is *predation*. A lion feeds on a zebra, a bear feeds on salmon, and a ladybird beetle feeds on aphids. Over time, predators evolved adaptations that enable them most efficiently to capture their prey. In response, prey species became better adapted to escape their predators. Thus, predators and prey tend to coevolve—this interaction between unrelated species, therefore, affects their mutual evolution.

Geographic Isolation

Some species are distributed throughout the world, while others only inhabit a particular area. For example, many species of placental mammals, such as rats, are widely found in many parts of the world. Nonplacental mammals (marsupials), however, are mainly native to Australia. (An exception is the opossum, a marsupial native to North America and South America.) The marsupials of Australia are isolated by a natural barrier that surrounds this continent—the ocean. The distribution of species over the Earth is the result of the geologic changes that have occurred on

Earth and the evolutionary changes that have occurred in the species of organisms.

The appearance of similar species in different regions of the world is best explained by considering the following factors in evolution: center of migration, competition and the struggle to survive, the appearance of barriers, and the inheritance of adaptive variations. Using these evolutionary factors, the distribution of marsupials can be explained as follows:

Center of migration. The particular region in which the ancestral type lived is called the *center of migration*. Fossil evidence indicates that at one time there were no placental mammals. Only marsupial mammals existed in Asia and Australia, which previously had been one large landmass. The more densely populated regions of this landmass became centers of migration as the marsupials competed for food and living space.

Competition and the struggle to survive. Competition for food and living space forced the marsupials to migrate to other regions of the landmass and eventually to interbreed. Under the conditions that were present in these regions, the marsupial population was stable.

Appearance of barriers. A natural obstacle that prevents organisms from migrating to other areas is called a *natural barrier*. Natural barriers include bodies of water, deserts, canyons, tall mountains, and very strong winds. If organisms become geographically isolated for a long period of time as the result of such barriers, they cannot interbreed with other members of their species, and will in time become different from them. The longer the period of separation, the greater the differences.

In the case of marsupials, the ancient Asian-Australian landmass divided into two continents separated by a natural barrier, the ocean, which prevented further migration. Consequently, the Asian and Australian marsupials became isolated into two separate populations. Because of the ocean separating them, interbreeding of these two populations became impossible.

Inheritance of adaptive variations. Occasional mutations and the results of independent assortment that arose in a population could not be passed on to the other population because of the natural barrier. This illustrates *speciation* as a result of geographic isolation. As different mutations accumulated and were inherited in each population, differences between the two populations became even greater. In Asia and elsewhere, the environmental conditions changed; marsupials eventually became extinct, and only placental mammals survived. In Australia, environmental conditions remained relatively stable and marsupials did not have to compete with placentals. As a result, marsupials such as the kangaroo, koala, and wombat evolved and flourished.

Classification

Taxonomy is the science of natural classification (see Chapter 18—Classification). Scientists recognize evolutionary relationships between organisms and arrange them in a taxonomic scheme that indicates these relationships. Organisms that are closely related have many similarities and some differences; organisms that are distantly related have fewer similarities and greater differences. By studying such a classification system, especially noting the presence of intermediate forms, you can see how different groups of organisms evolved from common ancestors. The internationally accepted system of classification groups organisms into a *Kingdom*, *Phylum*, *Class*, *Order*, *Family*, *Genus*, and *Species* (and *subspecies*). This classification scheme is useful because organisms are related structurally by evolutionary descent.

The process of evolution by natural selection not only explains the great diversity of living things on Earth, but also the resemblance in structures among the many different species. In the next chapter, the main theories of evolution will be discussed.

Section Quiz

1. Coevolution involves (*a*) microevolution within a given population (*b*) macroevolution within a given population (*c*) reciprocal evolution between two unrelated species (*d*) convergent evolution between two unrelated species.
2. *Hyracotherium* is considered to be the ancestor of (*a*) horses (*b*) camels (*c*) elephants (*d*) whales.
3. Anthropoid primates are distinguished from prosimians (lemurs and tarsiers) by possessing (*a*) opposable thumbs (*b*) color vision (*c*) vestigial structures (*d*) specialized canine teeth.
4. Apes, humans, and australopithecines are classified as (*a*) hominoids (*b*) hominids (*c*) anthropoids (*d*) prosimians.

5. "Lucy," the hominid fossil, is classified as a (an) (*a*) *Homo erectus* (*b*) *Australopithecus africanus* (*c*) *Australopithecus afarensis* (*d*) *Homo habilis*.

Chapter Review Questions

The following questions will help you check your understanding of the material presented in the chapter.

1. According to the heterotroph hypothesis, the amount of oxygen in the atmosphere today is related to the evolution of which process? (*a*) respiration (*b*) reproduction (*c*) hydrolysis (*d*) photosynthesis.
2. Conditions for the evolution of aerobic organisms may have been established only after a chemical reaction evolved that enabled some organisms to carry out (*a*) nondisjunction (*b*) photosynthesis (*c*) mutualism (*d*) heterotrophic nutrition.
3. According to the heterotroph hypothesis, living things evolved in a(n) (*a*) desert environment (*b*) forest environment (*c*) vacuum (*d*) ocean environment.
4. According to the heterotroph hypothesis, the first living things on Earth probably obtained food by (*a*) making organic food from inorganic raw materials (*b*) making inorganic food from organic raw materials (*c*) taking in previously formed organic food from the oceans (*d*) absorbing previously formed organic food from the atmosphere.
5. According to the heterotroph hypothesis, the earliest heterotrophs on Earth carried out (*a*) photosynthesis (*b*) chemosynthesis (*c*) anaerobic respiration (*d*) aerobic respiration.
6. The heterotroph hypothesis was supported when it was discovered that (*a*) chlorophyll is similar in structure to DNA (*b*) fermentation yields a large amount of ATP (*c*) enzymes are not necessary for photosynthesis (*d*) inorganic compounds may interact to form amino acids.
7. A heterotroph that is aerobic can best be described as an organism that (*a*) uses free O_2 for respiration but cannot manufacture food (*b*) uses free O_2 for respiration and can manufacture food (*c*) does not use free O_2 for respiration and cannot manufacture food (*d*) does not use free O_2 for respiration but can manufacture food.

8. Which group of animals would have nucleic acids most similar to the nucleic acids of amphibians? (*a*) reptiles (*b*) birds (*c*) mammals (*d*) arthropods.

9. The best means of discovering whether there is a close evolutionary relationship between animals is to compare (*a*) blood proteins (*b*) use of forelimbs (*c*) foods consumed (*d*) habitats occupied.

10. The fact that the Rh factor is found in the blood of rhesus monkeys as well as in human blood indicates that (*a*) human blood is identical to monkey blood (*b*) humans and monkeys may have a common ancestor (*c*) humans descended from monkeys (*d*) rhesus monkeys are related to humans but not to other monkeys.

11. The bony structure of a human's arm and hand most closely resembles the (*a*) wing of a grasshopper (*b*) wing of a butterfly (*c*) wing of a bat (*d*) fin of a shark.

12. The similarity between some mollusk larvae and annelids suggests that (*a*) annelids were mollusks at one time (*b*) mollusks were annelids at one time (*c*) life arose in the sea (*d*) annelids and mollusks had a common ancestor.

13. Antibodies produced by the blood of a horse can be used as a substitute for the antibodies that a human may fail to produce. This fact most strongly supports which statements? (*a*) All chemicals produced by living organisms are inorganic. (*b*) All chemicals produced by horses are antibodies. (*c*) Horses and humans occupy the same niche. (*d*) Horses and humans have a common ancestor.

14. Skeletal similarities between two animals of different species are probably due to the fact that the two species (*a*) live in the same environment (*b*) perform the same functions (*c*) are genetically related to a common ancestor (*d*) have survived until the present time.

15. Body structures that are fundamentally alike and develop in the same way are (*a*) embryological (*b*) homologous (*c*) heterozygous (*d*) allelic.

16. In humans, an example of a vestigial structure is the (*a*) cerebrum (*b*) nephron (*c*) pancreas (*d*) appendix.

17. The human embryo has gill pouches. A possible explanation for the appearance of these vestigial structures is that (*a*) the embryo is surrounded by amniotic fluid in the uterus, and gill pouches are

needed to carry out respiration (*b*) at one time, humans lived in the sea (*c*) humans probably evolved from fish alive today (*d*) humans and fish have a common ancestor.

18. In estimating the age of fossil specimens, an important concept is that (*a*) the flippers of whales and the arms of humans are homologous structures, which indicates that mammals have a common ancestor (*b*) the fossils of animals that have been found in Australia are quite different from those found on mainland Asia (*c*) similarities of vertebrate embryos in early development may suggest common ancestry among vertebrates (*d*) in undisturbed layers of Earth's crust, the older layers are the lowest ones, and each succeeding layer is younger than the layer below it.

19. Fossil evidence indicates that the modern camel is larger than its ancestors. The best explanation for this is (*a*) use and disuse (*b*) natural selection (*c*) homologous structures (*d*) crossbreeding with other species.

20. The fossil remains of two different animals are found at the same depth in undisturbed rock. As a result, you can conclude that these two animals (*a*) were members of the same phylum (*b*) died about the same time (*c*) occupied the same ecological niche (*d*) represent two groups that evolved from each other.

21. Many varieties of marsupials have evolved in Australia over a long period of time because of the (*a*) geographic isolation of the continent (*b*) internal development of the young marsupials (*c*) cold-blooded condition of marsupials in a warm environment (*d*) lack of deciduous trees on the continent.

22. The survival value of a trait is most likely determined by the (*a*) DNA code which produced it (*b*) type of mutation from which it resulted (*c*) environment in which it functions (*d*) sex of the individual possessing it.

23. The most probable sequence in evolution was (*a*) unicellular organisms → land-dwelling vertebrates → aquatic vertebrates → multicellular invertebrates (*b*) unicellular organisms → aquatic vertebrates → land-dwelling vertebrates → multicellular invertebrates (*c*) unicellular organisms → multicellular invertebrates → aquatic vertebrates → land-dwelling vertebrates (*d*) unicellular organisms → multicellular invertebrates → land-dwelling vertebrates → aquatic vertebrates.

Biology Challenge

The following questions will provide practice in answering SAT II-type questions.

Part I

Answer the following questions in two parts. First, find four items in each right-hand set (1 to 5) that are related to *one* item in the corresponding left-hand set (*A* to *C*). Check this item on the left and the four related items on the right. Next, underline the checked item in the left set and the one *unchecked* (unrelated) item in the right set.

Example:

X (*A*) vertebrates — (1) crayfish
(*B*) amphibians — X (2) toad
(*C*) mammals — X (3) rabbit
X (4) salamander
X (5) eagle

1. (*A*) Cenozoic Era
(*B*) Paleozoic Era
(*C*) Mesozoic Era

(1) amphibians
(2) dinosaurs
(3) sharks
(4) tribolites
(5) ferns

2. (*A*) Silurian Period
(*B*) Devonian Period
(*C*) Cambrian Period

(1) graptolites
(2) seeds
(3) mosses
(4) frogs
(5) sharks

3. (*A*) Cretaceous Period
(*B*) Pennsylvanian Period
(*C*) Permian Period

(1) mammals
(2) reptiles
(3) therapsids
(4) archeopteryx
(5) conifers

4. (*A*) Cenozoic Era (1) angiosperms
   (*B*) Paleozoic Era (2) *Hyracotherium*
   (*C*) Mesozoic Era (3) apes
   (4) ruminants
   (5) roaches

5. (*A*) hominid (1) *Homo habilis*
   (*B*) hominoid (2) *Homo erectus*
   (*C*) primate (3) Neanderthal
   (4) A. *afarensis*
   (5) gorilla

Part II

Base your answer to question *1* on the following paragraph and on your knowledge of biology.

It has been noted that the embryos of "higher" animals resemble embryos of "lower" animals; and that the embryo of a "higher" animal develops through a series of stages similar to those of "lower" animals.

1. Which statement is most closely in agreement with the above information? (*a*) A human embryo resembles a mature fish at a certain stage of development. (*b*) The blastula stage of a mouse embryo may be compared to a spherical multicellular embryo of a "lower" animal. (*c*) The theory may be expressed as "phylogeny recapitulates ontogeny." (*d*) Analogous structures are derived from a similar evolutionary origin. (*e*) The more remote the ancestry of two species, the greater the structural similarities.

2. A "tree" of ape and hominid divergence, based on immunological studies of amino acid sequences of three proteins, shows the primates branching off as follows: gibbons, orangutans, gorillas, chimpanzees, then hominids. You can infer from this that (*a*) amino acid sequences in humans and chimpanzees are very similar (*b*) amino acid sequences in chimpanzees and gorillas are identical (*c*) human and chimpanzee lines diverged 100 million years ago (*d*) early hominids were more related to gorillas than to humans (*e*) humans, chimpanzees, and gorillas can safely exchange blood.

17

Evolution: Change Over Time

Learning Objectives

When you have completed this chapter, you should be able to:

- **Describe** ways that Lamarck contributed to our understanding of evolution.
- **Discuss** strengths and weaknesses in Darwin's theory of natural selection.
- **Explain** how the modern theory of evolution was developed.
- **Discuss** various examples of evolution in modern times.
- **Apply** the Hardy-Weinberg Law to population genetics.

OVERVIEW

Fossil evidence supports the hypothesis that as Earth's environments changed, so did living things. Paleontologists have discovered that most present-day species differ from those of the past. Yet, all species apparently evolved from a common ancestor. In the course of evolution, many species became extinct, while others, adapting to changing environments, survived. In this chapter, you will learn about the different theories that have been proposed to explain how life on Earth has changed over time.

THEORIES OF EVOLUTION

Most paleontologists now support the idea that the rate of evolution is variable. For example, evolution has been very slow for cockroaches, which have remained virtually unchanged for almost 300 million years. Evolution has been more rapid for humans and extremely rapid for pesticide-resistant insects and antibiotic-resistant bacteria that have appeared during the past 25 years.

Several widely accepted theories of evolution have been proposed to explain how species change over time, and how certain adaptations may result in the production of a new species from an existing population.

Lamarck's Theory of Evolution

In 1801, Jean Lamarck proposed a theory based on his idea that if certain individuals often used an organ or part of body, that organ or body part became larger and stronger. Lamarck held that new species develop from ancestors as the result of *need, use and disuse*, and the *inheritance of acquired traits* that accumulate over a long period of time as follows:

Need. If necessary, an organism develops a structure that helps it survive. For example, the short-necked, short-legged ancestral giraffe needed to stretch its neck and legs to get food (tree leaves) when food near the ground (leafy shrubs) became scarce. Giraffes that could not compete with other animals for tree leaves starved to death.

Use and disuse. The giraffe stretched its neck and forelegs to obtain leaves in trees high above the ground. As a result, these body parts of the ancestral giraffe became longer.

Inheritance of acquired characteristics. The longer necks and legs acquired by stretching were passed on to each new generation of giraffes. Thus, according to Lamarck's theory, the modern giraffe is the result of the accumulated inheritance of these acquired traits.

Weaknesses in Lamarck's Theory

Lamarck's theory is unacceptable for the following reasons:

Evolution is not direct. The fossil record shows that evolution does not proceed directly from less adapted to better adapted organisms. Instead, evolution proceeds along many different lines, and in many different directions, as shown, for example, by the fossil records of horses and humans.

Acquired characteristics are not inherited. The experiments of *August Weismann* (1889) and those of *W. E. Castle* and *J. C. Phillips* (1909) disproved Lamarck's theory that acquired characteristics are inherited.

Weismann's experiments. Weismann tried to establish a breed of tailless mice to show that changes in the genetic material in reproductive cells are the only changes that can be transmitted to offspring. He cut off the tails of 20 generations of mice and observed that all of the offspring of each generation had tails just as long as the tails of the parents.

As a result, Weismann proposed the theory of *continuity of germplasm,* which claims that the *germplasm* (reproductive tissue) of an organism descends directly from the germplasm of its parents. Consequently, changes in the *somatoplasm* (nonreproductive tissue) of the parents, such as the loss of a tail or acquiring large muscles, are not transmitted to the next generation because the changed somatoplasm does not change the germplasm.

Experiments of Castle and Phillips. Castle and Phillips surgically removed the ovaries from a white (*bb*) guinea pig and replaced them with ovaries from a homozygous black (*BB*) guinea pig. Black fur is a dominant trait in guinea pigs. After the white pig bearing the "black" ovaries recovered from the operation, it was mated with a white male guinea pig (*bb*). The offspring of this mating were all black (*Bb*). This provided evidence that the "white" somatic environment in which the ovaries of the black guinea pig were placed had no effect upon the ability to produce egg cells that carried the dominant gene for black fur color.

The results of these experiments help explain somewhat similar situations in humans. For example, in China, it used to be considered beautiful for a woman to have small feet. Thus, for generations the Chinese had bound the feet of young girls so that they would grow up with small feet. Yet no decrease has been noted in the foot size of modern Chinese women whose feet have never been bound. These and other examples have led scientists to conclude that an acquired trait is not transferred to the next generation by way of the germplasm.

Darwin's Theory of Natural Selection

After a five-year tour (1831-1836) around the world as a naturalist on the British ship *Beagle*, Charles Darwin was convinced that new species evolved from a common ancestor.

Darwin's study of the animals on the Galápagos Islands, in the Pacific Ocean, almost 1,000 km west of Ecuador, puzzled him considerably. The islands are inhabited by 13 species of finches that differ from each other (Figure 17–1). However, since the similarities among the finches were greater than the differences, Darwin concluded that the different species descended from a common ancestor. Natural selection enabled finches with specific adaptations to survive in their particular environments.

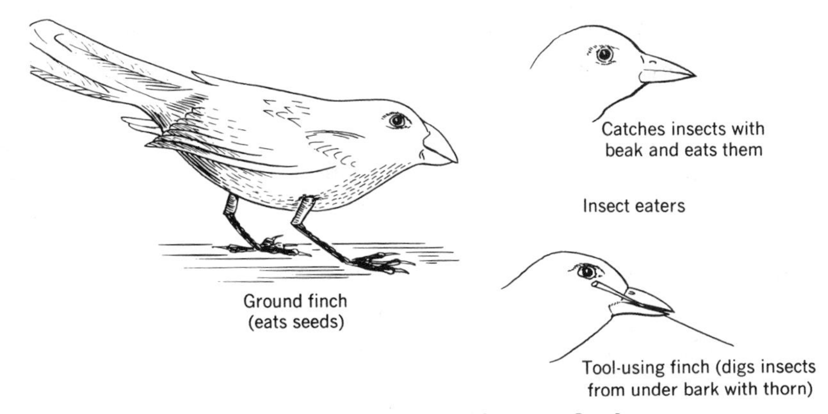

Figure 17–1 Some Galápagos finches

After collecting evidence on variations in species for about 20 years, Darwin described his findings in *On the Origin of Species by Means of Natural Selection*, published in 1859. About the same time another naturalist, *Alfred Wallace*, working independently of Darwin, came to the same conclusions. Instead of feuding over who should receive credit for the theory, each gave credit to the other. The theory of natural selection proposes the following ideas to explain evolutionary change:

Overproduction. All organisms tend to produce more offspring than can survive. For example, a female codfish lays over five million eggs at a time. If all these eggs were fertilized, and survived to adulthood, the sea would be filled with codfishes.

Struggle for existence. Overproduction results in competition for available food and living space between members of the same species and between members of different species. Unsuccessful organisms die; successful organisms survive. Over a long period of time, competition keeps the population relatively constant. Thus, many codfish eggs are fertilized but few develop. Most young codfish are eaten by other animals; only about two eggs in the more than five million codfish eggs laid by one female develop, grow to adulthood, and reproduce.

Variations. Although the offspring of a species are similar, variations are usually present. A variation may give an organism a particular advantage in the struggle for existence. For example, a codfish with larger fins than other codfish may escape its enemies by swimming more rapidly than a codfish with smaller fins.

Survival of the fittest. An organism with a variation that better adapts it to the environment may have a better chance to survive long enough to reproduce and pass on its genes. Organisms with less favorable variations do not compete successfully and do not live long enough to reproduce. In effect, the variation that enables an organism to adapt better to its particular environment is selected by one or more factors in the environment.

Inheritance. A favorable variation, or adaptation, is transmitted to the offspring. Over a long period of time, inherited adaptations accumulate in more and more offspring. New species may evolve.

In contrast to Lamarck, Darwin would have explained the evolution of the giraffe in this way:

1. The short-necked, short-legged giraffes produced young.
2. The competition for food and living space led to a *struggle for existence.*
3. Among the *variations* that arose among the giraffes' offspring were some with longer necks, some with longer legs, and others with both longer necks and longer legs.
4. As food closer to the ground became scarce, the giraffes with longer necks and longer legs were able to eat leaves from the higher branches of trees. The other giraffes had to compete for the decreasing amount of food on shrubs and on the lower branches of trees. Thus, the long-necked and long-legged giraffes survived better as the result of *natural selection*.
5. These giraffes reproduced and *passed the favorable adaptations to their offspring* (inheritance). After many generations, the modern giraffe evolved and the other giraffes became extinct.

Today, scientists accept most of Darwin's theory because it stresses the importance of the changing environment in relation to natural variations.

Strengths in Darwin's Theory

Most scientists support Darwin's theory that explains evolution for the following reasons:

1. Natural selection relies on the inheritance of variations, or differences from an established type. These inherited variations, or *adaptations*, are often favorable to survival in a changing environment.
2. Darwin understood that evolution best describes what occurs to pop-

ulations rather than major groups of organisms. Scientists define a population as a group of interbreeding organisms living in a specific area. In effect, Darwin provided the basis for relating inherited variations and the gene pool of a population. A *gene pool* constitutes the collective alleles and their frequencies in a given population.
3. Unlike Lamarck's theory, Darwin's theory of natural selection is not a purposeful but rather a random process that produces species.
4. Modern organisms have evolved from ancestral types, and are continually changing.

Weaknesses in Darwin's Theory

Modern scientists point out the following weaknesses in Darwin's theory:

1. Darwin did not explain how variations arose.
2. Darwin did not distinguish between variations caused by differences in heredity and those acquired as a result of environmental changes.
3. Darwin believed that favorable variations of both types were inherited. However, experimental evidence has shown that variations due to changes in the somatoplasm are not inherited.
4. Darwin believed that evolution occurred at a steady rate. However, evidence shows that evolution often proceeds at different rates for different organisms at different times. This idea is called *punctuated equilibria*, meaning that long, stable periods of relatively little change are interrupted by brief periods of great change. Although this idea of punctuated equilibria is controversial, it is clear that organisms have changed and continue to change over time.

De Vries' Theory of Mutations

In 1901, *Hugo De Vries,* a Dutch botanist, proposed that *mutations*, or gene alterations, are the major cause of variations. He said that mutations are either harmful, of no value, or adaptive. Only adaptive mutations enable organisms to survive. For example, the long neck and long legs of some giraffes arose as spontaneous mutations. These mutations were selected as favorable adaptations among competing giraffes for survival. Thus, a new species evolved—the modern giraffe.

De Vries' mutation theory strengthened Darwin's theory by showing that variations are mutations, but it did not explain how mutations occur. In addition, the mutation theory did not consider that individual mutations are seldom significant enough to produce a new species in just one generation.

Modern Theory of Evolution

The modern theory of evolution is a synthesis of Darwin's theory of natural selection, De Vries' mutation theory, and modern genetics. Using knowledge of modern genetics, scientists explain the appearance of mutations as changes in DNA, chromosomal aberrations, and recombination of traits resulting from independent assortment. The modern theory of evolution explains a giraffe's evolution in the following manner:

1. *Overproduction*. More short-necked, short-legged giraffes were produced than could survive as their food supply decreased.
2. *Struggle for existence*. The scarcity of shrublike vegetation near the ground caused the entire population of these animals to compete for leaves on the lower branches of trees.
3. *Variation*. Among the competing giraffes, favorable mutations arose, such as longer necks and longer forelegs. Giraffes with the longest necks and forelegs survived in greater numbers than the others. They reproduced and their favorable genes were passed on to their offspring in succeeding generations.
4. *Survival of the best-adapted*. Giraffes with longer necks and forelegs survived because they could feed upon leaves in the higher branches of trees. The short-necked, short-legged giraffes died. In succeeding generations, additional mutations occurred that further increased the length of the giraffe's neck and forelegs. In time, giraffes could feed upon leaves high above the ground without competition from other species.
5. *Inheritance*. Evolution of the modern giraffe resulted from the inheritance of these favorable mutations from generation to generation.

Section Quiz

1. Which scientist supported the idea that changes occurred because of a need? (*a*) De Vries (*b*) Mendel (*c*) Lamarck (*d*) Darwin.
2. Which statement describes a part of Darwin's theory of evolution? (*a*) Variation is not essential to evolution. (*b*) Too few offspring are produced in any population. (*c*) Organisms with favorable adaptations tend to survive. (*d*) Unfavorable mutations cause an increase in populations.
3. The main cause of variations according to De Vries' theory is (*a*) migration (*b*) mutation (*c*) selection (*d*) genetic drift.

4. Inherited variations that favor survival in a changing environment are called (*a*) spontaneous mutations (*b*) acquired characteristics (*c*) adaptations (*d*) gene alterations.
5. The modern theory of evolution does *not* include the following concept: (*a*) inheritance of acquired characteristics (*b*) natural selection (*c*) De Vries' mutation theory (*d*) overproduction of offspring.

EVOLUTION IN MODERN TIMES

In modern times, some species have evolved rapidly in response to drastic environmental changes caused by humans. Examples of rapid evolution include the peppered moth, and organisms resistant to chemicals such as DDT and penicillin.

The Peppered Moth

Peppered moths vary in color. Some are light-colored with dark markings; others are black with light markings. Before 1845 the only species of peppered moth common in the city of Manchester, England, was light-colored (Figure 17–2). The moth's color helped it blend with the light color of the bark of local trees, thus enabling the moth to escape being seen by predatory birds. As a result of the Industrial Revolution which began in 1845, soot from coal used to power factories settled everywhere.

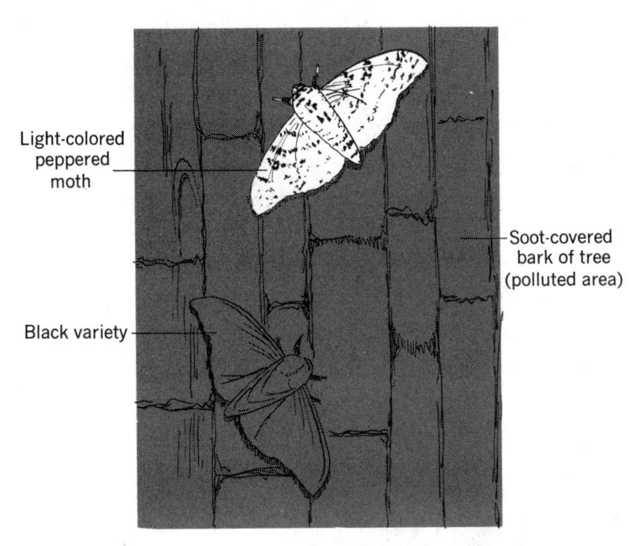

Figure 17–2 Peppered moth varieties

Some years later, a dark-colored peppered moth was found on the soot-covered bark of one of the trees. After that the dark peppered moth population increased rapidly. By the late nineteenth century, dark moths made up about 99 percent of the population of peppered moths. Light-colored peppered moths were very difficult to observe.

The dark moths survived and produced offspring like themselves because a chance mutation enabled the moths to blend with the darkened tree trunks. As a result, the light-colored moths were readily eaten by birds and most of the dark moths survived. Now that soot emission from factories is controlled, the light-colored moths are reappearing in increasing numbers.

EVOLUTION IN POPULATIONS

Modern theory recognizes that the factors of evolution act on the population, and not on individual organisms. A *population* is a group of individual organisms of the same species inhabiting a given region. All the buttercups living in a meadow, for example, make up a population.

Scientists have discovered that, in nature, variation is the rule for organisms whose sex cells result from genetic recombination. The genetic variability within a population of a species is the basis for evolutionary change. Natural selection "chooses" beneficial, or advantageous, alleles in a population, which may result in new species developing over time. *Population genetics* is the study of the changes in the genetic makeup of populations. Population geneticists investigate the behavior of alleles and how changes in allele frequencies lead to evolutionary change. For example, in a particular population of fruit flies, there might be individuals with long wings and others with short wings, but in different frequencies. Let us assume that the allelic genes *L* (for long wings) and *l* (for short wings) in the fruit fly are transmitted according to the recognized principles of heredity. As a result, some fruit flies would be genotype *LL*, some *Ll*, and some *ll*. A combination of all genes in the population for wing length make up the gene pool for this trait. The frequency with which each allele occurs usually is expressed as a percentage or a decimal, and is called the *gene frequency*.

The Hardy-Weinberg Law

In 1908, *G. H. Hardy*, an English mathematician, and *W. Weinberg*, a German physician, independently concluded that genetic variation is an evolutionary factor. They discovered that the frequency of an allele for a trait tends to remain constant from generation to generation in a stable population. In a large population in which (a) random mating occurs, and (b) the proportions of alleles remain unchanged, the original proportions of genotypes also remain constant in succeeding generations. Dominant alleles do not replace recessive alleles. The unchanged proportions of alleles making up genotypes are said to be in *Hardy-Weinberg equilibrium*. This concept, which states that sexual reproduction alone does not affect genetic equilibrium and cause evolution, is called the *Hardy-Weinberg Law*.

To see how the Hardy-Weinberg Law operates, let us apply it to a population of 200 fruit flies—100 males and 100 females. The fruit flies have the following genotypes:

Table 17–1 Wing Types in Fruit Flies

Males	*Females*
36 are *LL* (homozygous long-wing)	36 are *LL* (homozygous long-wing)
48 are *Ll* (heterozygous long-wing)	48 are *Ll* (heterozygous long-wing)
16 are *ll* (homozygous short-wing)	16 are *ll* (homozygous short-wing)

If all the males and females produce ten gametes each, you would calculate the gene frequency in the population as shown in Table 17–1.

In this case, the male fruit flies produce 1000 gametes; 600 are *L* and 400 are *l*. Thus, the *L* gametes make up 0.60 (60 percent) and the *l* gametes make up 0.40 (40 percent).

In this population, the gene frequency of the *L* allele in both the males and the females is 60 percent and the gene frequency of the *l* allele is 40 percent.

Because of random matings between the males and females, the following fertilizations and offspring genotypes are possible:

Sperm	+	*Egg*	=	*Genotype of Offspring*
L	+	*L*	=	*LL*
l	+	*l*	=	*ll*
L	+	*l*	=	*Ll*
l	+	*L*	=	*Ll*

The frequency with which any sperm will unite with any egg depends upon the relative abundance of the *L* and *l* gametes. You can show these matings with a Punnett square, using the gene frequencies of both parents in this population of fruit flies.

eggs \ sperms	L (0.60)	l (0.40)
L (0.60)	LL (0.36)	Ll (0.24)
l (0.40)	Ll (0.24)	ll (0.16)

You can see from the results of the matings that 36 percent (0.36) of the offspring are LL, 48 percent (0.48) are Ll, and 16 percent (0.16) are ll. This distribution of gene frequencies matches the distribution of the original population. Although there may be many more offspring than parents, the gene frequencies did not change. If matings between individuals of a population occur at random, if the population is large, and if mutations and migrations do not occur, the gene frequencies tend to remain constant from one generation to the next. This is the basic idea of the Hardy-Weinberg Law. The principle also can be expressed algebraically, as shown in the section below.

The Hardy-Weinberg Law Expressed Algebraically

Using the same fruit fly population, the frequency of the L alleles is 60 percent (0.60) and the frequency of the l alleles is 40 percent (0.40). The sum of the frequencies is 1 ($0.60 + 0.40 = 1.00$). Thus, the gene pool containing L and l genes is 100 percent, of which 60 percent are L and 40 percent are l.

By using p to represent the gene frequency of one allele and q to represent the gene frequency of the second allele, you can express the gene pool by means of the equation

$$p + q = 1$$

If p represents the dominant gene (L), then $p = 0.60$.
If q represents the recessive gene (l), then $q = 0.40$.
Then:

$$0.60 + 0.40 = 1$$

Since the Hardy-Weinberg Law assumes that the males and females of a population produce equal numbers of gametes, the gene pool for a set of alleles consists of p and q amounts of each allele. Let us show the results of random matings, again using the previous gene frequencies, but using p for the L allele and q for the l allele.

eggs \ sperms	p (0.60)	q (0.40)
p (0.60)	$p \times p$ (0.36)	$p \times q$ (0.24)
q (0.40)	$p \times q$ (0.24)	$q \times q$ (0.16)

These results may be simplified as follows:

$$p \times p = p^2 \qquad p \times q = pq$$
$$p \times q = pq \qquad q \times q = q^2$$

By combining like terms, the gene frequencies are $p^2 + 2pq + q^2$. Since these gene frequencies make up 100 percent of the gene pool of the offspring, the expression may be put into equation form as follows:

$$p^2 + 2pq + q^2 = 1$$

This equation states that in a particular population, the results of random mating will show a distribution of genotypes in the following frequencies:

p^2 = frequency of the homozygous dominant individuals
$2pq$ = frequency of the heterozygous individuals
q^2 = frequency of the homozygous recessive individuals

Now substitute the gene frequencies, $p = 0.60$ and $q = 0.40$, in the equation for the fruit fly population:

$$p^2 + 2pq + q^2 = 1$$
$$0.36 + 0.48 + 0.16 = 1$$

In summary, the offspring of the original population will be 36 percent homozygous dominant (*LL*), 48 percent heterozygous (*Ll*), and 16 percent homozygous recessive (*ll*).

Solving Gene Frequency Problems

In a population, if the frequency of one allele is known, the equation $p^2 + 2pq + q^2 = 1$ may be used to determine the frequency of the other gene as well as the genotypes and phenotypes. For example, in guinea pigs the gene for black fur (*B*) is dominant over the gene for white fur (*b*). If 9 percent of this population has white fur, what percentage of the guinea pigs is heterozygous black? What percentage is homozygous dominant?

Given: $q^2 = 9\%$ or 0.09 (gene frequency of the homozygous white recessive, *bb*)

$$q^2 = 0.09$$
$$q = \sqrt{0.09} = 0.3$$

Since $p + q = 1$, then:

$$p + 0.3 = 1$$
$$p = 1 - 0.3 = 0.7$$

Now that you know the value of both p and q, you can find the percentage of heterozygous black individuals, or $2pq$:

$$2pq = 2(0.7 \times 0.3) = 0.42, \text{ or } 42 \text{ percent}$$

Your calculations will show that 42 percent of the guinea pig population is heterozygous for black fur. If 9 percent is homozygous recessive (given) and 42 percent is heterozygous, the remaining 49 percent is homozygous dominant. Another way to find the percentage of homozygous dominant individuals is to calculate the value of p^2. Since $p = 0.7$, $p^2 = 0.49$, or 49 percent. The gene frequency of the homozygous black, *BB*, is 49 percent.

Despite the variety of genotypes within a population, the Hardy-Weinberg equation explains why a population may not change significantly or evolve for many generations. Although variations necessary for evolution in this population are present, evolutionary change is not taking place.

EVOLUTION AND THE HARDY-WEINBERG LAW

If the frequency of genes in a stable population tends to remain constant, evolution will not occur. If the frequency of genes changes, the rate of evolution will increase. The frequency of genes in a population is influenced by the following factors:

1. Selection. The genes for a trait that favors survival tend to increase in frequency; genes for the contrasting, or low-survival, trait tend to decrease in frequency. This occurs because individuals possessing the low-survival genes die, thus ending the transmission of these genes to the next generation. The microevolution of the dark-colored peppered moths in England illustrates this type of selection.

2. Population size. In a small population, the chance that the genes for a recessive mutation will combine during fertilization is greater than in a large population. The reason is that in a small population the probability of hybrids meeting and mating is increased. The chance incorporation of a mutant gene in a small population, called *genetic drift*, often explains the appearance of varieties having nonadaptive structures.

3. *Mutation rate.* Genes mutate at variable rates. If the mutation rate of any gene is more rapid than the rate at which natural selection eliminates its bearers, the frequency of that gene will increase and a shift in frequency will occur.

4. *Migration.* If an individual of one variety migrates to a new region and then successfully mates with related individuals, a new set of genes may be introduced into the gene pool of the native population. In time, the new genes may have greater survival value than those in the original gene pool of the native population.

5. *Isolation.* If parts of a population become isolated so that the separated groups cannot interbreed, then each group functions as a small population. In a small population, genetic drift may cause the gene frequency to shift. *Geographic isolation* by mountains, deserts, and large bodies of water can separate populations and prevent interbreeding. *Biological isolation* involves factors such as the failure of individuals to mate, and the failure of fertilized eggs to undergo cleavage.

Briefly, the basis of evolution is natural selection acting on adaptations that have either high or low survival value. The resulting change in gene frequencies in a population is the basis for the evolution of a new species. The following chapter will describe the diversity of species that, through evolution, are adapted to live in different regions on Earth.

Section Quiz

1. Which statement is *correct*? (*a*) In time, more favored genes in a population make up a greater proportion of the population's gene pool. (*b*) A characteristic of an organism, acquired during its life, may be transmitted to genes and passed on to offspring. (*c*) In England, the dark-colored form of the peppered moth is increasing in frequency as the country solves some of its air pollution problems. (*d*) Acquired traits and natural selection are used in conjunction to explain evolution.

2. Which statement is *incorrect*? (*a*) The frequency with which each allele occurs is called the *gene frequency*. (*b*) Dominant alleles do not replace recessive alleles. (*c*) The unchanged proportion of alleles

that make up genotypes are in *Hardy-Weinberg equilibrium.* (*d*) Sexual reproduction causes evolution.

3. In the algebraic expression $p^2 + 2pq + q^2$, the term $2pq$ refers to (*a*) frequency of homozygous dominant individuals (*b*) frequency of heterozygous individuals (*c*) frequency of homozygous recessive individuals (*d*) none of the above.
4. The incorporation of a mutant gene in a small population may cause the appearance of varieties having nonadaptive structures. This is an example of (*a*) spontaneous mutation (*b*) geographic isolation (*c*) biological isolation (*d*) genetic drift.
5. Which factor influences the frequency of certain genes in a population? (*a*) population size (*b*) mutation rate (*c*) biological isolation (*d*) all of the above.

Chapter Review Questions

The following questions will help you check your understanding of the material presented in the chapter.

1. Survival of organisms that show little change over time, such as the ameba and hydra, may be explained by assuming that (*a*) some forms of life have been created more recently than others (*b*) there is little tendency to vary in these species (*c*) these species are becoming extinct (*d*) the simple organization of these species is well suited to their condition of life.
2. Organic evolution could be described as (*a*) changes in species over time (*b*) Earth's five-billion-year history (*c*) causes of geographic distributions (*d*) organic compounds in living organisms.
3. The rate of evolution depends primarily on (*a*) the number of chromosomes in the species (*b*) environmental changes (*c*) the size of the organisms (*d*) random migration out of the population.
4. Which is most likely to contribute to the evolution of a new species of plant? (*a*) asexual reproduction (*b*) self-pollination (*c*) a doubling of the normal chromosome number (*d*) autotrophic nutrition during development.
5. There is a greater probability for evolution to occur in species that reproduce (*a*) sexually (*b*) by parthenogenesis (*c*) by binary fission (*d*) by budding.

6. Which statement concerning evolution probably would be attributed to Lamarck? (*a*) Organisms tend to produce more offspring than can survive. (*b*) The basis for variations in a species is provided by mutations. (*c*) Structures develop in organisms as a result of the organisms' need to survive. (*d*) DNA replication is the key to evolution.
7. Modern biologists would not agree with Lamarck's concept of (*a*) survival of the fittest (*b*) overproduction of offspring (*c*) mutation (*d*) evolution of organs because of need.

For each statement about predators in questions *8 through 11*, select a factor from the list below that is most closely related to that statement. A letter may be used more than once or not at all.

Factors

(*a*) Geographic isolation	(*d*) Struggle for existence
(*b*) Inheritance of variations	(*e*) Survival of the fittest
(*c*) Overproduction	(*f*) Noninheritable variations

8. More predators were born than the number of prey organisms could support.
9. Some of the predators had longer legs than others.
10. All predators competed for the available prey.
11. A larger percentage of the predators with longer legs lived to reproductive age.
12. Darwin's theory of evolution could not account for (*a*) finches of several different species on isolated islands (*b*) limestone fossils of extinct organisms (*c*) white-eyed fruit flies produced by red-eyed parents (*d*) white polar bears in arctic regions.

Base your answer to questions *13 and 14* on the paragraph below and on your knowledge of biology.

During the Eocene Period, there were many prehistoric horses with four toes on their front feet. Among the four-toed horses were some with larger middle toes, which enabled these horses to run faster and to better escape from predators. As a result, the horses with larger middle toes were eventually the only ones to survive and reproduce their kind.

13. The theory of evolution suggested in this paragraph was first proposed by (*a*) Mendel (*b*) Watson (*c*) Darwin (*d*) Hardy.
14. According to the modern theory of evolution, the most probable explanation for the variation in toe structure described above is (*a*) chance occurrence of the variations (*b*) need for survival of the horse (*c*) use and disuse of toes (*d*) phenotype stability of the species.
15. Most structural differences among Darwin's finches probably can be traced to (*a*) polyploidy (*b*) geographic isolation (*c*) use and disuse (*d*) the fossil record.
16. A major criticism of the original Darwin-Wallace theory of natural selection is that it (*a*) fails to explain variation (*b*) supports the theories of Mendel (*c*) does not involve use and disuse (*d*) does not recognize overproduction.
17. Strains of bacteria highly resistant to antibiotics are increasing in number. This statement can be explained by the (*a*) inheritance of acquired characteristics (*b*) law of independent assortment (*c*) heterotroph hypothesis (*d*) theory of natural selection.
18. According to the modern theory of evolution, which environmental factor would be most influential in increasing the evolutionary rate of humans? (*a*) continued use of natural grain cereals (*b*) continued use of transistor radios (*c*) increasing the quantity of pollutants in the environment (*d*) drinking of distilled water.
19. Which is the best example of species adapting to different environments? (*a*) existence of fossils in successive rock strata (*b*) finch populations on the Galápagos Islands (*c*) similarities found in vertebrate embryos (*d*) the relationship of enzymes and hormones in many organisms.
20. If an insect species lacks the variations needed to adapt to a changing environment, it probably will (*a*) acquire them through evolution (*b*) become extinct (*c*) evolve into a lower form (*d*) evolve into a higher form.
21. The current human population is much larger than it would have been according to the principle of natural selection. If this statement is correct, it is most probably because humans have (*a*) maintained the population of all other living organisms in a natural balance (*b*) permitted development of ecological climaxes (*c*) developed means of controlling the environment (*d*) strictly enforced the conservation laws.

22. Which factor would limit variation within a population? (*a*) mutation (*b*) segregation and recombination of genes (*c*) the introduction of new members of the same species (*d*) reproductive isolation.
23. On a small island off the coast of Spain, many people are polydactyl (they have more than ten fingers or ten toes). Which factor contributes most to this phenomenon? (*a*) overproduction (*b*) isolation (*c*) variation (*d*) natural selection.
24. The gene pool of a population consists of only the (*a*) heritable genes in the population (*b*) mutated genes in the population (*c*) dominant genes in the population (*d*) recessive genes in the population.
25. According to population genetics, evolution is best defined as (*a*) any change in gene frequency (*b*) variation due to environment (*c*) a tendency toward geographic isolation (*d*) the result of interbreeding.
26. In a city of 100,000 people, there are 25,000 people who exhibit a recessive characteristic. What percent of the population probably is heterozygous for the alleles that control the characteristic? (*a*) 75 percent (*b*) 50 percent (*c*) 25 percent (*d*) 4 percent.

Base your answers to questions 27 *through* 29 on the information below.

In a population genetics study, it was discovered that the ability to taste phenylthiocarbamide (PTC) is a dominant trait; inability to taste PTC was the result of recessive alleles. In a certain high school, 64 percent of the student population were tasters and 36 percent were nontasters.

27. In the Hardy-Weinberg Law, the formula $p^2 + 2pq + q^2$ is used to represent the entire population. In the above population, $2pq$ refers to (*a*) homozygous tasters (*b*) heterozygous tasters (*c*) homozygous nontasters (*d*) heterozygous nontasters.
28. The gene frequency for tasting PTC probably would change if the substance were (*a*) poisonous (*b*) pleasant-tasting to tasters (*c*) pigmented (*d*) disagreeable to tasters.
29. Which set of conditions would tend to keep the gene pool for this trait stable? (*a*) The population divides into smaller breeding units. (*b*) Mates are selected so that tasters mate with tasters and nontasters mate with nontasters. (*c*) The mutation rate increases to produce more tasters. (*d*) The population remains large and random mating continues.

30. Humans have increased the rate of evolution by (*a*) establishing game laws (*b*) preventing polyploidy (*c*) utilizing selective breeding (*d*) protecting certain species from extinction.
31. In a population, the frequency of an allele is related to the (*a*) adaptive value of the trait controlled by this allele (*b*) size of the gamete in which this allele appears (*c*) ratio of females to males in the population (*d*) evolutionary distribution of the species.
32. Inability to roll the tongue is a recessive human trait. If 36 percent of the people in a population cannot roll their tongues, what is the percentage of recessive genes in the gene pool? (*a*) 6 percent (*b*) 36 percent (*c*) 60 percent (*d*) 72 percent.

Base your answers to questions 33 *through* 35 on the information below and on your knowledge of biology.

Since the latter part of the eighteenth century, tons of soot have been settling on Earth's surface, especially near large industrial areas. This by-product of the Industrial Revolution has resulted in a general darkening of these areas and of the plant life on them. Associated with this has been the observation that of 760 light-colored species of moths in England, 70 species also showed dark or black coloration.

33. What primarily influenced the pattern of change in the moth population? (*a*) changing environmental conditions (*b*) increasing mutation rate for the darker color (*c*) migration of the darker members of the moth population (*d*) isolation of the darker-colored moths from the lighter-colored.
34. Which statement is correct regarding the genes that govern the trait for darker color? (*a*) The moths were darkened by soot settling on them, causing their genes to produce darker moths in the next generation. (*b*) Because the alleles for darker color must have had a high survival value, their frequency has increased in the gene pool. (*c*) Because of the soot in the air, the mutation rate must have been increased, producing more darker-colored moths. (*d*) The genes were unaffected and the soot settling on the moths has made them appear darker in color.
35. In 1926, the British biologist H. Harrison reported that the industrial melanism (change from light to dark) was caused by a special substance present in polluted air. He concluded that this was inherited

according to the laws of Mendel. What incorrect assumption is present in Harrison's hypothesis? (*a*) Acquired characteristics can be inherited. (*b*) There were harmful substances in the polluted air. (*c*) Mendelian laws can be applied to this type of inheritance. (*d*) The change from light to dark color was under the control of the genes.

36. In a naturally-occurring population, the mutation rate of a particular allele is predictable. The actual mutation of any single allele, however, will (*a*) always yield a recessive allele (*b*) occur by means of nondisjunction (*c*) be a completely random occurrence (*d*) result from random mating.

37. Many people with hereditary defects now have a normal life expectancy because of medical research. One long-term result of this situation probably will be (*a*) stabilization of the birth rate (*b*) changes in the gene pool (*c*) an increase of dominant alleles (*d*) a decreased frequency of defective genes.

Biology Challenge

The following questions will provide practice in answering SAT II-type questions.

Part I

Base your answers to questions *1 through 5* on the following paragraph and your knowledge of biology.

In African Americans, the gene for sickle-cell anemia has been gradually decreasing in frequency. However, among Africans still living in their ancestral homelands, the frequency for this gene has remained high.

1. In its heterozygous form, a sickle-cell anemia carrier acquires immunity to (*a*) malaria (*b*) genital herpes (*c*) sleeping sickness (*d*) typhoid (*e*) kwashiorkor.

2. Current distribution of the sickle-cell gene in the United States may be a direct result of (*a*) genetic counseling advising carriers to remain childless. (*b*) a high mortality rate among children of carriers (*c*) changing frequencies of sickle-cell alleles (*d*) susceptibility to diabetes mellitus (*e*) artificial selection.

3. Which statement is *correct*? (*a*) In Africa, the frequency of the sickle-cell gene fluctuates with the *Anopheles* mosquito population. (*b*) In its homozygous recessive form, the sickle-cell gene can cause death. (*c*) Most Black Africans are homozygous dominant for the sickle-cell gene. (*d*) Sickle-cell anemia involves the inability of red blood cells to reproduce. (*e*) The sickle-cell gene is linked with the gene that controls melanin production in the skin.

Part II

Base your answers to questions *1 through 5* on population distribution curves A and B showing phenotypic variations in wing cover colors of certain beetles. *Note*: Graph A precedes Graph B in time.

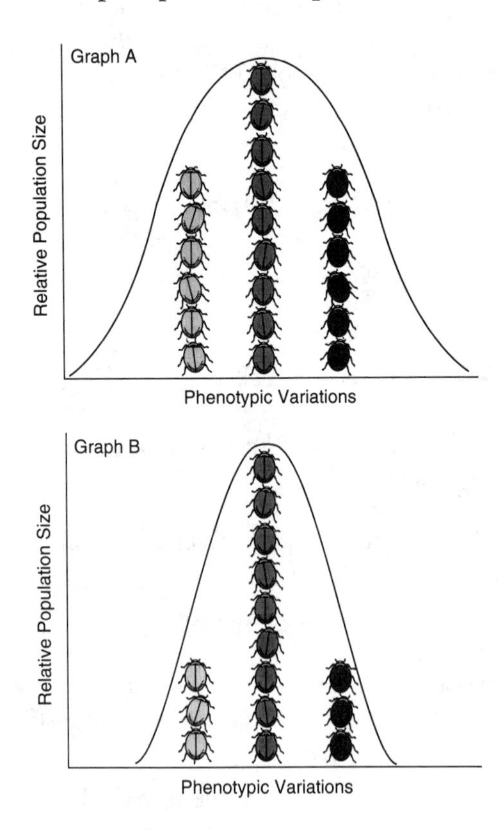

1. Which evolutionary factor is illustrated by curve B? (*a*) geographic isolation (*b*) genetic drift (*c*) selection (*d*) use and disuse (*e*) adaptation.

2. The group of beetles that tends to stabilize the species phenotype is (*a*) the group with light-colored wing covers (*b*) the group with intermediate-colored wing covers (*c*) the group with dark-colored wing covers (*d*) none of these groups (*e*) cannot be determined by studying the graph.
3. Curve B shows the effect of a certain factor on the evolutionary process during a period of time. Assume that an additional period of time affects the distribution of phenotypes in the curve. Which statement would reflect your prediction? (*a*) The most common phenotype (intermediate-colored wing covers) probably will prevail in numbers. (*b*) The group with light-colored wing covers probably increase in numbers. (*c*) The group with dark-colored wing covers probably will mate with intermediate-colored beetles to form a new species. (*d*) The time period will have no effect on the future distribution of phenotypes. (*e*) The certain factor is living space.
4. Over a period of time, the alleles of the two uncommon groups should (*a*) mutate (*b*) recombine independently (*c*) disappear from the gene pool (*d*) unite to form a fourth phenotype (*e*) increase in the gene pool.
5. Curves A and B reflect a distribution of phenotypic groups that shows (*a*) microevolution is taking place (*b*) members of a species possessing more adaptive genes can reproduce successfully (*c*) members of a species may vary in phenotypes (*d*) *a*, *b*, and *c* (*e*) *b* and *c*, only.

18

Modern Classification

Learning Objectives

When you have completed this chapter, you should be able to:

- **Describe** the method by which Aristotle classified organisms.
- **Discuss** the work of Linnaeus in developing a system of classifying organisms.
- **Contrast** old systems of classification with the modern system using five kingdoms.
- **Describe** characteristics of organisms in the Moneran, Protist, Fungus, Plant, and Animal kingdoms.
- **Discuss** the evolutionary relationships that exist among organisms in the Animal Kingdom.

OVERVIEW

The evolution of different kinds of organisms results from the inheritance of variations, or adaptations. At present, more than two million different organisms have been identified. To study this diversity of living things, it is necessary to group, or classify, living things systematically. In this chapter, you will learn how a classification scheme is used by scientists to place living things into groups in which organisms resemble each other externally and internally. A classification scheme also enables scientists to draw conclusions regarding the relationships among, and origins of, the organisms.

ARISTOTLE'S SYSTEM OF CLASSIFICATION

Aristotle (384–322 B.C.) was among the first to devise a logical system of classification for living things. However, Aristotle's system would be thought unscientific if judged by present-day standards. For example, his major basis for classifying animals was the presence or absence of red blood. He grouped fishes, snakes, and humans together in one major division because they possess red blood; he grouped insects, worms, and starfishes together in another major division because they lack red blood. Aristotle also classified animals as either land dwellers, water dwellers, or air dwellers.

Aristotle classified plants based on their size and appearance as either trees, shrubs, or herbs. For example, he classified both a moss plant and a flowering plant, like the strawberry, as herbs, because each plant grows close to the ground. Aristotle was unaware that a single plant family—for example, the rose family—may include trees (apple), shrubs (spirea), and herbs (strawberry). Modern *taxonomists,* scientists who classify organisms, know that these diverse plants of the rose family are related because of their similar flower structure (Figure 18–1). Aristotle also was unaware that a single animal group could contain diverse animals that are related because of similar internal structures. In classifying organisms, he did not consider the evolutionary relationships between the organisms.

In time, thousands of new varieties of organisms were discovered, which made Aristotle's system of classification obsolete. *Botanists,* scientists specializing in plant studies, and *zoologists*, scientists specializing in animal studies, developed more useful classification systems.

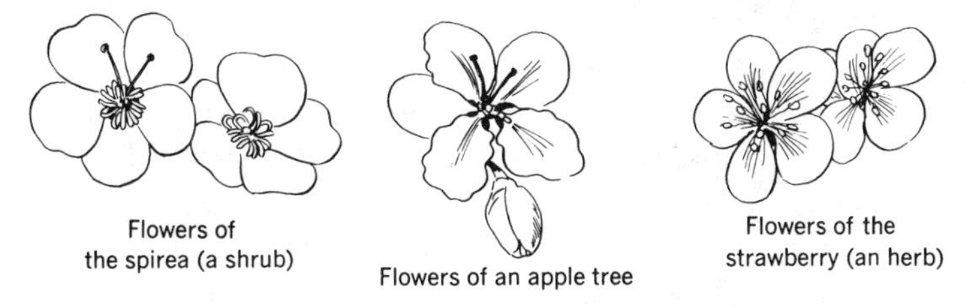

Figure 18–1 Flowers of plants in the rose family

THE SPECIES CONCEPT

Biologists have found the concept of "species" useful in developing a classification system for plants and animals. A *species* is a group of similar organisms capable of reproducing with each other, and producing fertile offspring. For example, despite the superficial differences between an African pygmy, Australian aborigine, and a northern European, all belong to the species *Homo sapiens sapiens*. All dogs—from the toy poodle to the Great Dane—belong to the species *Canis familiaris*.

LINNAEUS' SYSTEM OF NAMING ORGANISMS

Carolus Linnaeus (1707–1778) formalized classification by developing his system of *binomial nomenclature,* which uses two names derived from Latin or Greek. The first name of an organism is the *genus* (plural, genuses) name; the second is the species (plural, species) name. The genus name is written with a capital letter and both the genus and species names appear in italic type. For example, the scientific name of the wolf is *Canis lupus,* and of the sugar maple, *Acer saccharum.*

MODERN SYSTEM OF CLASSIFICATION

The modern system of classification recognizes that similarities in structure and function usually indicate a relationship between species. Moreover, this relationship has its origin in the ways organisms evolved. This system, which is based on the work of Linnaeus, is called an evolutionary or *natural system of classification.*

Taxonomy

The scientific classification of organisms is called *taxonomy.* Taxonomists classify organisms into major groups, and smaller groups within major groups, in a way that reflects structural and functional relationships among them. This is similar to the way in which merchandise in most supermarkets is arranged. A large supermarket often has numerous numbered aisles, each bearing a sign listing the products located in that aisle. Aisle 8, for example, bearing the sign "Paper Goods" includes toilet tissue, facial tissue, table napkins, and other related paper products. Likewise, modern taxonomy is a system composed of five kingdoms, or "aisles," each containing different, but related organisms. Kingdoms are divided into subgroups called phylums, classes, orders, families, genuses, and species; each subgroup is variable in size.

All subgroups, from species to phylums, are called *taxons* (singular, taxon). Thus, a taxonomist analyzes an organism's characteristics for the purpose of assigning it to a taxon. For example, taxonomists assign deer, elk, and moose to the same family, because they have features in common, such as antlers and the chewing of cud. However, deer, elk, and moose belong to different genuses because there are certain differences among them. Taxonomists classify related families together in a larger group called an *order*. Related orders are placed into a *class*, and related classes are placed into a single *phylum* (plural, phylums). Related phylums are grouped into one of five kingdoms. The following is a summary of the relationships between groups in the modern system of classification.

Each *kingdom* is divided into related *phylums*.
Each *phylum* is divided into related *classes*.
Each *class* is divided into related *orders*.
Each *order* is divided into related *families*.
Each *family* is divided into related *genuses*.
Each *genus* is divided into related *species*.
Each *species* is a subdivision of a genus and refers to a particular kind of organism.

Note: A *species* is sometimes divided further into *varieties*, for plants, and races, or *subspecies*, for animals.

Principles of modern taxonomy

The following factors are used by taxonomists to classify organisms into related groups: pattern of structure and function; development of young; similarity of DNA; and adaptations for survival.

Pattern of structure and function. Similarities in structure and function are major considerations in determining relationships among living things. For example, the flower structure of all members of the rose family illustrates that they are related.

Development of young. Organisms that appear unrelated sometimes are shown to be related based on how their young develop. For example, the early stages of development of an octopus and a clam are almost identical. In fact, it is difficult to determine which embryo will grow into an octopus and which will grow into a clam.

Similarity of DNA. Scientists can verify the placement of organisms into subgroups based on comparison of the chemical structure of their DNA. The greater the similarity in DNA structure, the closer the relationship between different organisms. For example, the DNA of a frog and of a salamander show many similarities. In contrast, the DNA of a frog and of a cat show fewer similarities.

Adaptations for survival. The adaptations of similar organisms that help them survive in a particular environment are used to identify different species. For example, a ground squirrel and a flying squirrel appear similar in many respects. However, the leg structure of the ground squirrel is adapted for jumping or running; the winglike folds of skin between the legs and the body of a flying squirrel adapt it for gliding through the air.

FIVE-KINGDOM CLASSIFICATION SYSTEM

The five-kingdom system of classification is based upon the previously mentioned principles, as well as types of cellular organization and nutrition (see Figure 18–2).

1. *Monera* were the earliest life forms on Earth; all are prokaryotes; includes all bacteria.
2. *Protista* (also called *protoctista*) includes other microbes and their descendants; all are eukaryotes. Examples are algae, seaweeds, ciliates, and water molds.

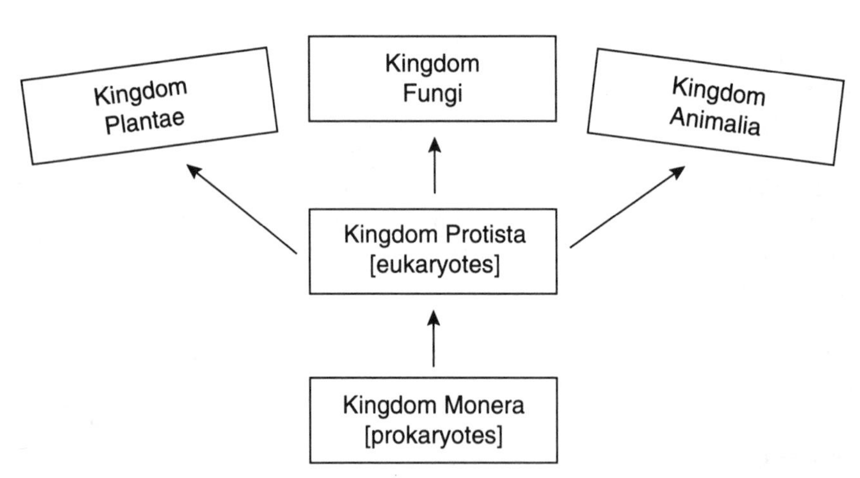

Figure 18–2 The five-kingdom system of classification

The kingdoms of monera and protista include autotrophs—organisms that synthesize food from inorganic substances using energy supplied by the sun (photosynthesis) or by chemicals (chemosynthesis). Both kingdoms also include heterotrophs—organisms that derive nourishment from ingesting dead or living organic matter. Some species may even use both types of nutrition.

3. *Fungi* are mostly *saprobes*, formerly called saprophytes. Saprobes obtain nourishment from dead matter. Fungi include yeasts, molds, and mushrooms.

4. *Plantae* includes various plants: all are multicellular, contain chlorophyll, and carry out photosynthesis. Plants (and algae) make life possible for other organisms by adding oxygen to the atmosphere. Examples of plants are mosses, ferns, trees, shrubs, and non-woody herbaceous plants.
5. *Animalia*, the most recent kingdom to evolve, contains multicellular organisms that are strictly heterotrophic.

Before the five kingdoms are each described in greater detail, let us discuss viruses, a group that still has not been classified in any taxon.

Viruses

Most scientists do not consider viruses to be living organisms for several reasons:

1. All viruses are composed of either DNA or RNA (never both) enclosed in a protein coat called a *capsid*. (See Figure 18–3, which shows a typical virus structure and some representative viruses.)
2. Viruses are ultramicroscopic, ranging from 20 to 300 nm (nanometers) in diameter. A *nanometer* is one-billionth of a meter.
3. Viruses can reproduce only within a host cell. They use the cell's living machinery for their own metabolism. Thus, viruses are parasites, unable to grow, obtain nourishment, and reproduce outside a host cell.
4. Some viruses, for example the tobacco-mosaic virus (TMV), can be crystallized, like a mineral. In a crystalline state, a virus can exist for years unchanged, until it invades a host.

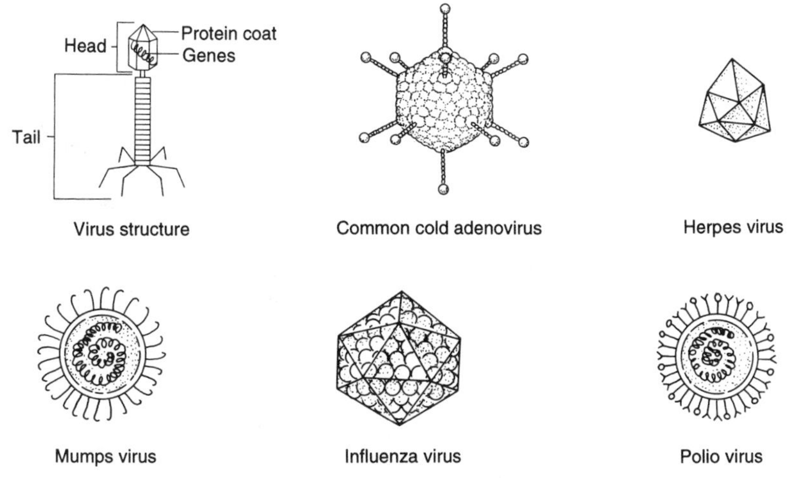

Figure 18–3 Some representative viruses

Viruses are more closely related to their hosts than they are to each other. Some scientists think that viruses originated as "loose" nucleic acids that escaped from within a cell's boundaries, and then began replicating on their own, using the cells from which they escaped as hosts. Thus, the common flu virus is more closely related to humans, and the TMV more closely related to tobacco, than the two viruses are to each other. For these reasons, taxonomists do not classify viruses in their own separate kingdom.

Kingdom Monera

All members of the Kingdom Monera are bacteria and bacterialike microbes. To date, about 5,000 species of monerans have been classified. Many scientists think that bacteria were among the first living things to evolve. For example, the oldest fossil evidence of bacteria has been dated at more than 3,000 million years, whereas the oldest fossil evidence for fungi, land plants, and animals only dates back to about 500 to 700 million years ago. Thus, the monerans were the dominant life forms on Earth for about two billion years.

Some species of bacteria can survive under the most extreme environmental conditions, including boiling, freezing, and even being submerged in hot acids. Bacteria also are found at great ocean depths and great atmospheric heights.

Major Characteristics of Monerans

1. All monerans are *prokaryotes*; their cells contain genetic material of either loose DNA threads or a grouping of DNA that lacks a nuclear membrane. Biologists call this type of unstructured genetic material a *nucleoid.* Taxonomists think that *eukaryotes*, or organisms with a membrane-enclosed nucleus, evolved from monerans.
2. Monerans are shaped like rods (*bacillus*, plural bacilli), spheres (*coccus*, plural cocci), and helical spirals (*spirillum*, plural spirilli). (See Figure 18–4.)

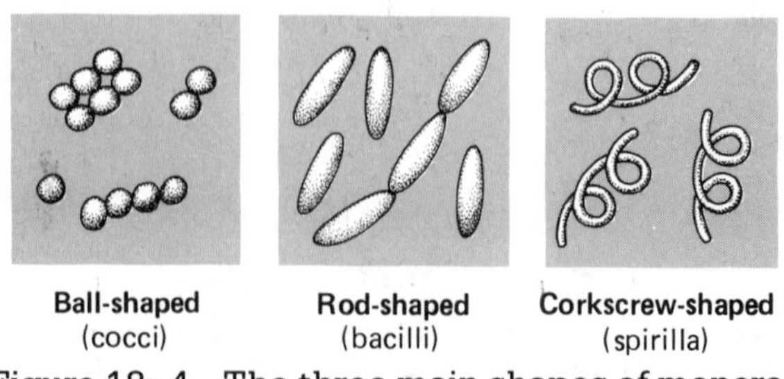

Ball-shaped (cocci) **Rod-shaped** (bacilli) **Corkscrew-shaped** (spirilla)

Figure 18–4 The three main shapes of monerans

3. Some monerans generate energy by carrying out aerobic respiration; others by anaerobic respiration. Some are *facultative anaerobes*, or organisms that ordinarily use oxygen to generate energy, but can generate energy without oxygen, if necessary, depending on environmental conditions.
4. Monerans include species that obtain nourishment by carrying out chemosynthesis; others obtain nourishment by using the energy of sunlight.
5. Monerans are microscopic-sized organisms, ranging from about 0.1 to 10.0 μm (micrometers) wide. A *micrometer* is one-millionth of a meter.
6. Monerans lack intracellular organelles, such as mitochondria, chloroplasts, and endoplasmic reticula.
7. Their complex cell walls are made up mainly of polysaccharides, peptides, and other organic compounds.
8. Some monerans are nonmotile; others move by gliding or by using flagella.
9. Monerans reproduce mainly by asexual cell division, or binary fission, without mitosis, in which equal portions of DNA are transferred to daughter cells. Some monerans transfer genetic material from donor to recipient.
10. Monerans live in all types of habitats and can obtain nourishment from all organic substances, including plastics and petroleum.

Representative Monerans

Fermenting bacteria. These anaerobic bacteria require a mixture of organic chemicals to grow and reproduce. Fermentation is a metabolic process that enables the bacteria to release energy. For example, *Lactobacillus* ferments sugars; *Clostridium denitrificans* fixes atmospheric nitrogen into organic compounds; and *Bacteroides fragilis* often inhabits the intestines of animals.

Spirochaetes. Spirochaetes, which resemble a tight coil, move by means of internal flagella. Most spirochaetes are anaerobic; others live symbiotically in the intestines of wood-eating termites and cockroaches.

Methanogenic bacteria. Each year about two billion tons of methane are produced by these monerans. Unlike other bacteria, methanogenic bacteria cannot use carbohydrates or proteins as sources of carbon and energy. Instead, they metabolize methanol (wood alcohol) and acetates (salts of acetic acid). One species, *Methanobacterium ruminantium*, is found in the rumen of cud-chewing mammals, such as cows, camels, and deer.

Cyanobacteria. These monerans were formerly classified as blue-green algae and assigned to the plant kingdom. Like algae and plants, cyanobacteria carry out photosynthesis by converting light energy, carbon dioxide, and water into food and oxygen. Also, like algae and plants, cyanobacteria possess *thylakoids* containing chlorophyll. *Anabaena* is a common filamentous form of cyanobacteria that lives in freshwater ponds and lakes, and is often studied in biology laboratory classes.

Nitrogen-fixing bacteria. Nodules on the roots of some kind of plants contain an enormous number of rod-shaped and oval-shaped cells of the nitrogen-fixing bacteria *Azotobacter* and *Rhizobium.* Nitrogen-fixing bacteria have a symbiotic relationship with leguminous plants, such as beans, peas, alfalfa, and clover.

Aerobic spore-forming bacteria. These bacteria possess an adaptation, endospore formation, which ensures successful reproduction and species survival under adverse environmental condition. For example, when subjected to extreme heat and dryness, the bacteria develop a single spore within their cells. Several species of aerobic spore-forming bacteria are abundant in soil and air; some antibiotics are derived from this type of bacteria.

Chemoautotrophic bacteria. Most organisms use preformed vitamins, sugars, and amino acids as nutrients to synthesize required organic chemicals. Chemoautotrophic bacteria, however, metabolize nitrogenous compounds, carbon dioxide, oxygen, and energy sources such as methane, ammonia, or hydrogen sulfide. They also make their own nucleic acids. Chemoautotrophic bacteria are beneficial to plants and animals by converting unusable salts and gases to usable substances. Consequently, this large group of versatile bacteria, which includes *Nitrosomonas* and *Nitrobacter* species, is essential for normal maintenance of the biosphere.

Omnibacteria. This very large and diverse group of monerans includes aerobic and anaerobic species. Some species, such as the *Klebsiella, Salmonella,* and *Pasteurella*, are well-known for the diseases they cause in humans. The genus *Neisseria* causes gonorrhea; the genus *Acetobacter* oxidizes ethyl alcohol to vinegar. The *Rickettsias* and *Chlamydians* are frequent parasites of vertebrates and invertebrates. Both types of omnibacteria cause a variety of diseases, including psittacosis of parrots and pigeons (*Chlamydians*) and Rocky Mountain spotted fever (*Rickettsias*).

Actinobacteria. Many members of this group of monerans were originally mistaken for fungi because of their funguslike structure and spore-formation. An important member is *Mycobacterium tuberculosis*, the cause of tuberculosis.

Myxobacteria. These bacteria, considered the most complex prokaryotes, are unicellular rods that often form complex colonies embedded in slime. Biologists still have not discovered how colonies of myxobacteria glide over solid surfaces. Individual cells multiply by ordinary binary fission. In the absence of nutrients or water, myxobacteria undergo a life cycle similar to that of slime molds (discussed under Protists).

Section Quiz

1. An outstanding feature of prokaryotes is that they (*a*) lack genetic material (*b*) possess genetic material enclosed in a membrane (*c*) lack a cell wall (*d*) possess a nucleoid.
2. Which is correct about viruses? (*a*) They possess both DNA and RNA. (*b*) They are usually observed with a light microscope. (*c*) Inside a host cell, they often form crystallike forms. (*d*) Inside a host cell, they replicate using the cell's living material.
3. All monerans possess (*a*) mitochondria (*b*) complex cell walls (*c*) endoplasmic reticula (*d*) flagella.
4. Which gas do facultative anaerobes normally use for respiration? (*a*) carbon dioxide (*b*) methane (*c*) hydrogen (*d*) oxygen.
5. Cyanobacteria are pigmented monerans (*a*) related to algae (*b*) formerly classified as algae (*c*) that inhabit sulfur springs (*d*) that cause disease in vertebrates and arthropods.

Kingdom Protista

This unusual kingdom includes about 27 phylums and 120,000 species. All nucleated algae, seaweeds, water molds bearing flagella, slime molds, slime nets, protozoans, and many other eukaryotic microorganisms belong to the Kingdom Protista.

Originally, taxonomists had classified only unicellular organisms, or unicells, as protists. Recent investigations, however, reveal that certain multicellular organisms, or multicells, evolved from unicellular ancestors and are more related to unicells than to other multicells. The studies had considered the genetic, reproductive, and structural similarities. The common marine kelp is an example of a complex multicell that is closely related to a tiny unicellular brown alga. Thus, both organisms have been assigned to the same kingdom. (Figure 18–5 shows some representative protists.)

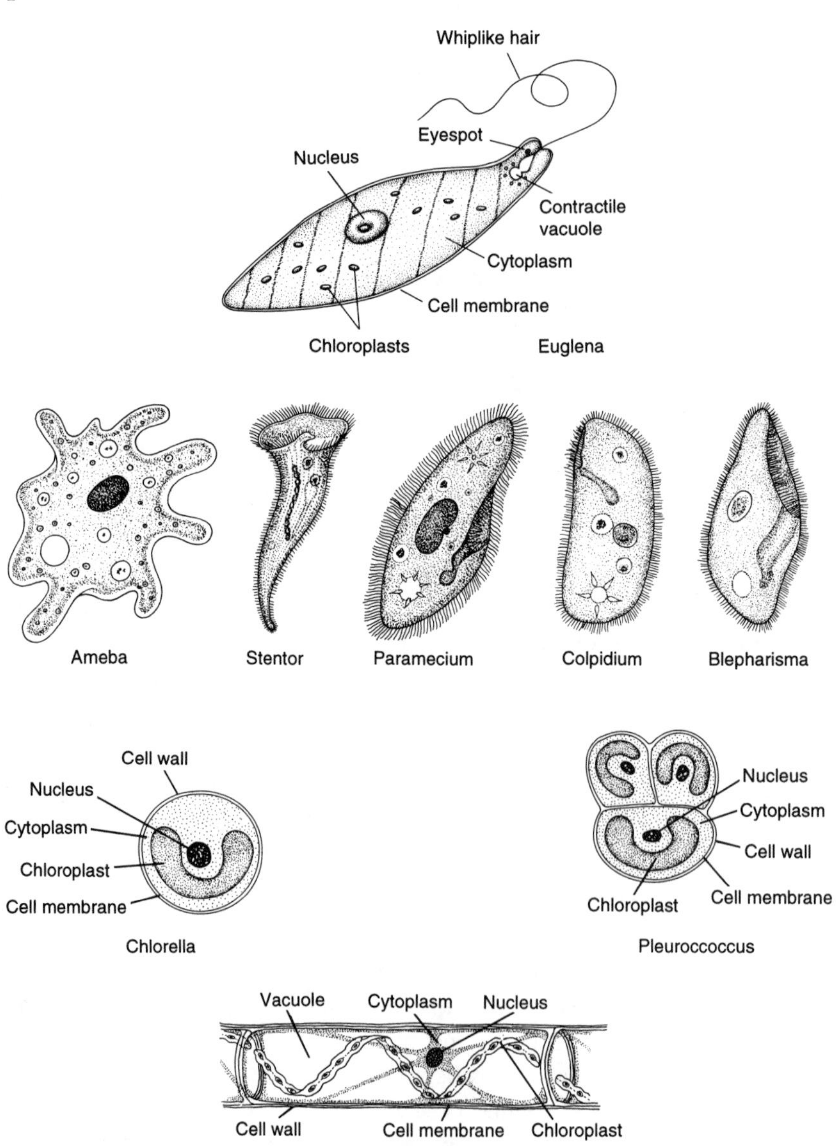

Figure 18–5 Some representative protists

Major Characteristics of Protists

1. All species are *eukaryotes;* they possess genetic material enveloped by a membrane—a true nucleus.
2. Most possess mitochondria, enabling aerobic respiration (Krebs cycle).
3. Some possess chloroplasts, enabling photosynthesis.
4. Some protists possess a cell wall composed mainly of cellulose; others lack a cell wall.
5. Most species reproduce asexually and sexually, involving meiosis and fertilization.
6. Their nutrition may be autotrophic, heterotrophic, or a combination of both.
7. Locomotion, when present, may be ameboid, or by using cilia, flagella, or internal contractile fibers.
8. Some protists are multicells, such as green, brown, and red algae. Scientists think that *differentiation,* the formation of specialized groups of cells or tissues, probably evolved in higher forms from these multicellular protists. Most taxonomists concur that early multicells were the direct ancestors of fungi, plants, and animals.

Representative Phylums of Protists

Dinoflagellates (phylum Dinoflagellata). These unicells possess two dissimilar flagella located at right angles to each other. As a result, the organisms whirl or spin as they move through water. Most dinoflagellates inhabit warm marine waters as part of the plankton. *Plankton* is a diverse biological community composed of algae, bacteria, and many other organisms floating and drifting on the surface. Plankton is a prime source of food for many aquatic animals. An eyecatching dinoflagellate is *Noctiluca,* a bioluminescent unicell that is sometimes observed illuminating ocean water. Many other species, called *zooanthellae,* live symbiotically with corals, clams, and sea anemones.

Euglenoid Flagellates (phylum Euglenophyta). Most species are unicells that inhabit fresh and stagnant water. Euglenoid flagellates usually possess a single flagellum at one end, a tough *pellicle* made up of spirally-arranged strips of specialized cytoplasm, chloroplasts, an eyespot, a contractile vacuole, and a conspicuous nucleus. They do not reproduce sexually, but reproduce asexually by mitotic cell division. *Euglena* is often studied in the biology laboratory.

Ciliates (phylum Ciliophora). A drop of stagnant water may contain dozens of these unicells, which exhibit a variety of shapes. The familiar grey *Paramecium* and pink *Blepharisma* are typical ciliates. Thousands of cilia covering the body constantly beat in rhythm, enabling ciliates to move and sweep food into the gullet. Digestion occurs in a vacuole; regulation of water balance is accomplished by one or more contractile vacuoles.

An unusual feature of ciliates is the presence of two types of nuclei. They have *micronuclei* that contain chromosomes, which are involved in *conjugation* (a form of sexual reproduction), and a single *macronucleus* that contains genes, which control normal metabolism.

Green Algae (phylum Chlorophyta). If you scrape some green algae from the bark of a tree, observe the scrapings mixed with a drop of water under a light microscope, you will observe *Chlorella,* a unicellular green alga. Other species include *Hydrodictyon, Volvox, Chlamydomonas,* and the pond weeds *Nitella* and *Chara,* all commonly studied in the biology laboratory. Like plants, green algae possess a cell wall composed mainly of cellulose and pectin. They occur as unicells, colonies, and filaments. Some species form sex cells, called *zoospores,* which possess chloroplasts and two flagella. Most taxonomists think that plants evolved from this phylum.

Plasmodial Slime Molds (phylum Myxomycota). A *plasmodium* is a moving mass of slime, composed of cytoplasm and many floating nuclei. A moving plasmodium resembles a blob of jelly. As it moves, a plasmodium digests bacteria, yeasts, and bits of organic debris. In the absence of food and water, a plasmodium stops moving and forms a mass of spore-bearing structures. When environmental conditions are favorable, the spores (which can remain dormant for years) germinate and form an amebalike mass. A plasmodium results from many masses fusing into a single blob.

Diatoms (phylum Bacillariophyta). Diatoms, among the most abundant organisms in plankton, are brownish unicells with shells called *valves* made of silica. Many people admire the valves of diatoms because of their elaborate and symmetrical patterns. Formerly classified with the golden-yellow algae, diatoms now are assigned to their own taxon due to differences in life cycle, cell structure, and reproduction. A popular abrasive is diatomaceous earth, which is composed of pulverized diatom valves.

Brown Algae (phylum Phaeophyta). Brown algae are one of the longest, fastest growing, and most photosynthetically productive of all living things. Members of the genus *Macrocystis* may grow to a length of about 100 meters—the length of a football field. The genus *Sargassum* forms immense floating masses on the Sargasso Sea, an area in the mid-Atlantic

Ocean. Taxonomists find brown algae interesting because these multicells go through an alternation of generations in their life cycle.

Amebas (phylum Rhizopoda). A tiny, shapeless mass of protoplasm that moves by means of *pseudopods* ("false feet") and that lacks cilia or flagella is called an ameba. These protists may be found in soil, in fresh water, and in salt water. Some ameba possess a shell, composed of cemented silica or calcium carbonate. One such species, *Arcella,* possesses a shell that paleontologists have recognized in the fossil record of the Paleozoic Era.

You should remember that the grouping of recognizable organisms into species, genuses, and families is usually agreed upon by most taxonomists; but criteria for grouping families into orders, classes, and phylums are often more controversial.

Section Quiz

1. Taxonomists believe that multicellularity among certain protists gave rise to (*a*) tissue differentiation (*b*) sexual reproduction (*c*) organelle formation (*d*) prokaryotic variations.
2. Dinoflagellates are characterized by their (*a*) bioluminescence (*b*) glassy valves (*c*) whirling motion (*d*) blue-green color.
3. A protist that absorbs nourishment from the environment is a (an) (*a*) autotroph (*b*) parasite (*c*) heterotroph (*d*) symbiont.
4. A protist that never relies on an autotrophic type of nutrition is a (an) (*a*) ameba (*b*) diatom (*c*) euglena (*d*) brown alga.
5. Plants are believed to have evolved from the phylum (*a*) euglenophyta (*b*) myxomycota (*c*) chlorophyta (*d*) rhizopoda.

Kingdom Fungi

There are about 100,000 known species of fungi. They are either terrestrial and free-living, parasitic, or symbiotic. Fungi range in size from microscopic organisms such as yeast to larger organisms such as the shelf fungi that can be easily seen on tree branches and trunks. Bacteria and fungi are Earth's major *decomposers*, making organic materials available for recycling and other uses by living things.

Some fungi are harmful, causing plant and animal diseases and spoiling organic matter. Other fungi are helpful when they are used to make bread, wine, beer, ethanol, soy sauce, tofu (soybean curd), citric acid, antibiotics, and steroids. The common yeast *Saccharomyces cerevisiae* is a helpful fungus that is used in the fermentation process. (Figure 18–6 shows some representative fungi.)

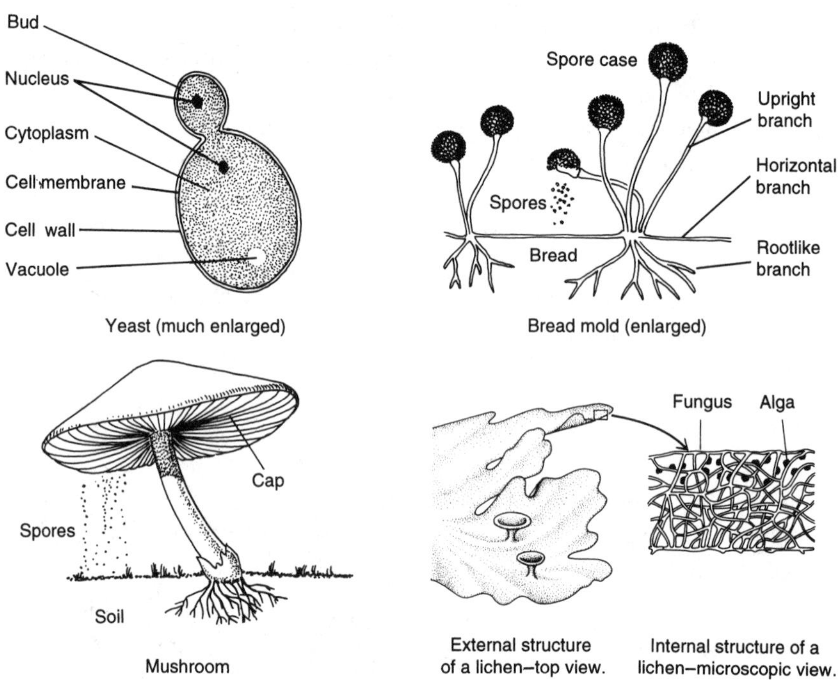

Figure 18–6 Some representative fungi and lichen

Major Characteristics of Fungi

1. Fungi are eukaryotes that can reproduce both asexually and sexually. In asexual reproduction, spores are formed; in sexual reproduction, monoploid (n) hyphae meet and fuse to form a diploid ($2n$) *zygospore*. Shortly after, meiosis occurs, resulting in monoploid nuclei. Under favorable conditions, asexual spores and zygospores germinate to form monoploid hyphae.
2. Microscopic examination of *hyphae* reveals a tubular structure and crosswalls, or *septa*. These divide hyphae into units, or "cells," that usually contain more than one nucleus.
3. The crosswalls do not completely separate one unit, or "cell," from another. As a result, the cytoplasm can flow freely through hyphae.
4. A mass of hyphae, called a *mycelium*, is the asexual form of most fungi, including mushrooms, molds, and morels.

5. Cell walls are composed of *chitin*, a compound not present in plant cell walls, but present in the exoskeleton of insects.
6. Fungi are aerobic and heterotrophic organisms. Hyphae secrete powerful enzymes that digest organic substances, dead or alive, into smaller molecules that are absorbed through the hyphae walls.
7. Fungi are considered the most resilient of eukaryotes. They can tolerate extreme heat, cold, and exposure to corrosive chemicals.

Representative Phylums of Fungi

Zygomycetes. Members of this phylum are unusual fungi. That is because zygomycetes lack crosswalls except those that separate reproductive structures from the rest of the mycelium. Many are *saprobes* (saprophytes), living on dead organic matter. Some are parasites, living on animals, plants, and sometimes even on other fungi. Asexual spores, called *conidia*, develop within *sporangia* (singular, sporangium) located on top of *sporangiophores. Rhizopus,* the common bread mold, is a typical zygomycete.

Ascomycetes. These include yeasts, certain bread molds, morels, and truffles. Ascomycetes differ from other fungi by forming *asci* (singular, ascus), tiny saclike reproductive structures that form when hyphae conjugate. Monoploid spores formed within asci are called *ascospores*. When released, the ascospores germinate on suitable materials and form mycelia.

Asexual reproduction involves the division of hyphae into huge numbers of identical *conidiospores*. Both ascospores and conidiospores are dispersed by wind, water, insects, and other animals. Some ascomycetes form mycorrhizal associations with shrubs and trees.

Basidiomycetes. These fungi include mushrooms, puffballs, stinkhorns, rusts, and smuts. Basidiomycetes differ from other fungi by possessing *basidia* (singular, basidium) located in fruiting bodies. The familiar mushroom stalk and cap, is a fruiting body. Basidia, located in gill-like filaments on the undersurface of a mushroom cap, produce large numbers of *basidiospores*. When basidiospores germinate, mycelia develop that initiate a complex life cycle that involves sexual reproduction.

Deuteromycetes. These fungi are called "imperfect" because they lack reproductive structures and cannot reproduce sexually. In a kingdom characterized by complex sexual stages, the "imperfect" fungi are considered misfits because they resemble neither the ascomycetes nor the basidiomycetes. Most deuteromycetes, including *Penicillium* (the source of penicillin) and *Aspergillus* (a fermenter used in making soy sauce and soy paste, and in flavoring Roquefort and Camembert cheese), reproduce by forming conidiospores.

Lichen. Some fungi live symbiotically with green algae. That association is a form of *mutualism* in which both organisms gain certain advantages by living together. A *lichen* is an association between a fungus and an alga. You may have observed that lichens appear to be red, yellow, green, or blue-green crusts adhering to rocks, tree bark, and soil. The different colors are due to a blending of pigments in the algae and dissolved minerals in the fungal hyphae.

Section Quiz

1. Fungi that lack sexual structures are classified as (*a*) Myxomycetes (*b*) Basidiomycetes (*c*) Ascomycetes (*d*) Deuteromycetes.
2. A basic characteristic of the Kingdom Fungi is that most members (*a*) are saprobes (*b*) possess cellulose-rich cell walls (*c*) are composed of distinct and separate cells (*d*) produce diploid conidiospores.
3. The mycelium of a fungus is composed of a mass of (*a*) basidia (*b*) septa (*c*) hyphae (*d*) asci.
4. The genetic makeup of most fungi is usually (*a*) monoploid (*b*) diploid (*c*) triploid (*d*) tetraploid.
5. Basidiomycetes are characterized by their (*a*) lack of septa (*b*) zygospore formation (*c*) pigmented mycelia (*d*) fruiting bodies.

Kingdom Plantae

Plants are multicellular, sexually reproducing eukaryotes that evolved from green algae about 400 million years ago during the Devonian Period. Plants colonized the land and became major transformers of solar energy into chemical energy. Photosynthesis, the process used by plants in this transformation, releases oxygen to the atmosphere, making terrestrial life possible for animals and all other aerobic organisms.

Taxonomists divide the plant kingdom into two major phylums: *Bryophyta*, nonvascular plants, and *Tracheophyta*, vascular plants. Vascular tissue consists of xylem and phloem. Most plants do not need a moist habitat for fertilization to occur. However, all bryophytes and certain tracheophytes, such as ferns, require a watery environment for sperm cells to

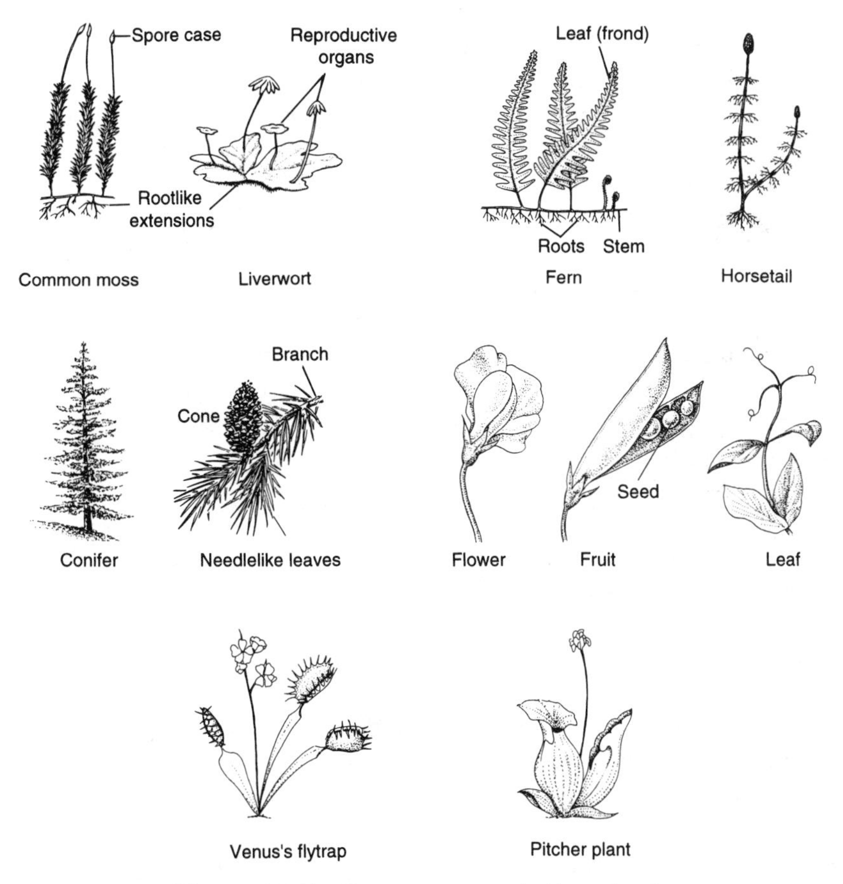

Figure 18–7 Some representative plants

reach egg cells. Plants differ from organisms in other kingdoms by going through an *alternation of generations,* or alternating sexual and asexual stages in their life cycle. (Refer back to Chapter 11, Plant Structure and Function.)

There are far fewer taxons of plants than of organisms in other kingdoms. That is because structural rather than chemical criteria are used for plant classification. (Figure 18–7 shows some representative plants.)

Major Characteristics of Bryophyta

1. *Bryophytes* usually are inconspicuous, tiny plants that inhabit watery habitats.
2. In sexual reproduction, sperm cells swim through water to reach egg cells; water also is needed for absorption of dissolved substances into the plant body.

3. As in green algae, sperm cells possess two flagella. This supports the hypothesis that plants evolved from Chlorophyta (green algae).
4. Another piece of evidence that indicates an evolutionary connection between plants and green algae is the appearance of similar chloroplasts and pigments.
5. A green, leafy plant body, or *gametophyte,* is the conspicuous stage of the bryophyte's life cycle. The gametophyte (gamete producer) is monoploid (n) and so are its sperm and egg cells. After fertilization, a nongreen, diploid ($2n$) *sporophyte* (spore producer) develops from the gametophyte and produces monoploid spores. The sporophyte, which is short-lived, depends upon the gametophyte for nourishment and growth.
6. All bryophytes lack vascular tissues.

Representative Bryophytes

Three major classes of bryophytes are: *Hepaticae* (liverworts), *Anthocerotae* (hornworts), and *Musci* (mosses).

1. *Hepaticae* species usually have a liverlike shape. The plant body adheres to a solid object or soil by rootlike *rhizoids. Marchantia,* a common liverwort, is a monoploid gametophyte. After fertilization, a diploid sporophyte develops from the gametophyte. Meiosis and mitosis produce numerous monoploid spores which, after release, germinate in a wet environment and develop into gametophytes.
2. *Anthocerotae* plants resemble Marchantia in body form and reproduction. The hornlike spikes of *Anthoceros* are sporophytes that continue to grow as long as conditions are favorable.
3. *Musci*, or mosses, is the largest and best known group of Bryophytes. *Polytrichum*, the pigeon wheat moss, is frequently found in temperate woodlands of North America. *Sphagnum*, a spongy moss, lives in acidic boglands, and is used by florists to retain water in soil. Moss reproduction is similar to that of other bryophytes.

Major Characteristics of Tracheophyta

Members of this phylum are vascular plants that have efficient conducting tissues. The plants you ordinarily see are the large and nutritionally independent sporophytes, quite different from those of the bryophytes. Other adaptations of tracheophytes include specialized leaves, stems, roots, stomates, and, in many species, flowers and seed production. *Tracheophytes* usually are divided into two groups: seedless vascular plants and seed-bearing vascular plants.

Representative Tracheophytes

Seedless Vascular Plants. These plants include the classes *Pterido-phyta*, the ferns; *Lycopodophyta*, the club mosses; and *Sphenophyta*, the horsetails.

Pteridophyta are ferns, familiar land plants, which reproduce by spores and do not produce seeds. Fern fronds resemble the leaves of higher plants, as they are compound and divided into leaflets called *pin-nae.* Sporangia, usually located on the undersurfaces of fronds, are often clustered as *sori* (singular, sorus).

The life cycles of ferns resemble those of mosses. Ferns usually have conspicuous large sporophytes; gametophytes are inconspicuous. *Os-munda cinnamomea*, the cinnamon fern, is widespread in moist and shady areas. *Polypodium virginianum*, the rock polypody, is another common fern.

Lycopodophyta include club mosses and their relatives. Most are small plants that inhabit tropical climates. Some species, however, such as *Lycopodium obscurum*, is widespread in the central and northeastern United States. This species of lycopod inhabits wooded areas under ma-ple, oak, and pine trees. Like all plants, lycopods undergo an alternation of monoploid and diploid generations.

Sphenophyta, the horsetails, are easily recognized by their jointed stems and ribbed texture. The mineral silica is concentrated in the epi-dermal cells of green horsetail stems, adding roughness to them. Horse-tails are called *scouring rushes* because they were used many years ago to scour pots. All sphenophytes belong to the genus *Equisetum*, which live on salt flats, along stream banks, and in moist wooded areas. Sphenophyte reproduction is similar to that of most bryophytes: a large diploid sporo-phyte generation and a dot-sized, green gametophyte.

Seed-Bearing Vascular Plants. Two subphylums, *Gymnosperms* and *Angiosperms*, make up this division of seed-bearing plants. Seeds are im-portant evolutionary adaptations that protect plant embryos and provide them with a source of nourishment.

1. *Gymnosperms* consist of diverse plants that belong to three major classes: *Coniferophyta*, the conifers; *Ginkgophyta*, the ginkgos or maidenhair trees; and *Cycadophyta*, the cycads.

Conifers are cone-bearing shrubs or trees, including the evergreen firs, hemlocks, pines, and spruces. The leaves of most conifers are needle-shaped and often covered with a waxy layer called a *cuticle.* Since most conifer leaves are not deciduous, they can undergo photosynthesis in winter. Conifer embryos are located on the underside of female cone scales; when mature, the embryos become winged, wind-borne seeds. *Sequoia,* the giant redwood tree in California, is a conifer.

Ginkgo trees are the only living descendants of a group that was extensive during the Mesozoic Era. Also called maidenhair trees, ginkgos have bilobed leaves resembling the fronds of the maidenhair fern. Ginkgos are pollution-resistant and decorative, so they have been planted as curbside trees in many large cities. Male ginkgos usually are planted, because the fruit produced by female trees gives off an objectionable odor. In their natural habitat, male and female ginkgo trees go through a life cycle similar to that of other gymnosperms.

Cycads range in size from shrublike plants resembling pineapples to others, such as *Cycas,* which is palmlike and about three meters tall. Unlike conifers and ginkgos, cycad leaves are fernlike and compound. Most cycads bear reproductive structures called *strobili* (singular, strobilus), which are either male or female.

2. *Angiosperms,* or *flowering plants,* are represented by nearly every familiar tree, shrub, and herb that bears flowers and seeds. There are more than 250,000 species and more than 300 families separated into two divisions: *monocots* and *dicots.* Monocots possess one cotyledon, or seed leaf; dicots possess two dicotyledons. (Refer back to Figure 11–1, which shows other differences between monocots and dicots.) Some scientists think that the dominance of angiosperms may be related to their coevolution with insects. Another indication that coevolution of insects and angiosperms may have occurred is the existence of carnivorous plants, such as the *Dionea* (Venus flytrap), *Drosera* (Sundew), and the *Saracenia* (purple pitcher plant). All of these plants trap and eat insects to obtain needed proteins and minerals.

Section Quiz

1. A distinctive adaptation of plants is that all (*a*) bear seeds (*b*) go through an alternation of generations (*c*) possess vascular tissue (*d*) possess conspicuous gametophytes.
2. Fronds are a characteristic of (*a*) Lycopodophyta (*b*) Bryophyta (*c*) Musci (*d*) Pteridophyta.

3. A plant group that requires a watery environment to complete its life cycle is (*a*) Sphenophyta (*b*) Coniferophyta (*c*) Bryophyta (*d*) Angiospermophyta.
4. A group of seed-producing vascular plants represented by only one living descendant is (*a*) Ginkgophyta (*b*) Cycadophyta (*c*) Coniferophyta (*d*) Bryophyta.
5. A major evolutionary advance of tracheophytes is (*a*) development of male and female strobili (*b*) development of different types of chlorophyll (*c*) a dominant sporophyte generation and a dependent gametophyte generation (*d*) a dominant gametophyte generation and a dependent sporophyte generation.

Kingdom Animalia

Multicellular animals were formerly called *metazoans* to distinguish them from one-celled "animals" called *protozoans.* Today, taxonomists classify protozoans as heterotrophic protists. Modern taxonomists describe animals as multicellular, diploid, heterotrophic organisms. The embryonic development of animals features continued mitotic divisions of a diploid zygote to form a *blastula,* or hollow ball of cells. This stage is followed by a *gastrula,* a three-layered hollow ball of cells with an opening at one end called a *blastopore.* Embryonic development and differentiation eventually produces an intestine with openings at both ends. At that point, the embryonic development and growth vary among different animal phylums.

Major Characteristics of Animalia

1. Most animals possess specialized tissues, organs, organ systems, a definite shape, and symmetry—either radial or bilateral.
2. Tissue cells are joined by elaborate connections, such as *desmosomes,* which provide intercellular communication and a flow of substances between cells.
3. Animals ingest food, digest food, and absorb nutrients through cells by passive and active transport.
4. Highly specialized sense organs enable animals to respond to stimuli more efficiently than do other living things. As a result, animals are more successful in adapting to changing environmental conditions than other organisms.
5. Diversity is great in the animal kingdom. The smallest and simplest animal is the ameba-sized *Trichoplax* (see Figure 18–8); the largest is

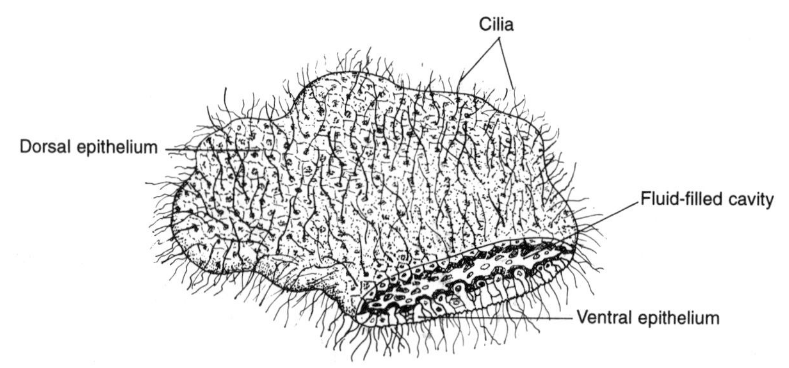

Figure 18–8 Trichoplax—the smallest animal

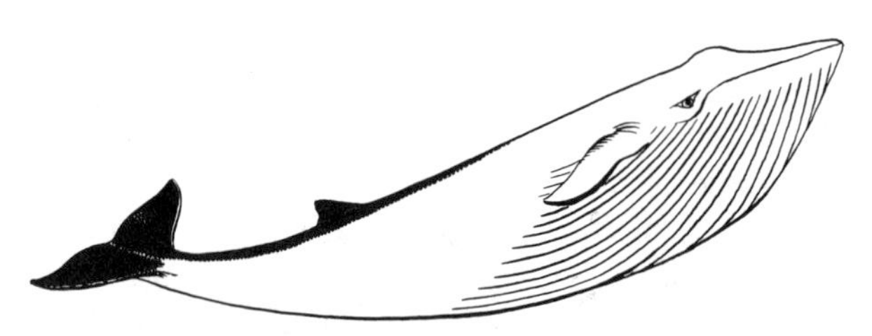

Figure 18–9 The blue whale—the largest animal

the blue whale (see Figure 18–9), which may grow to more than 30 meters. Studies have shown that the microscopic Trichoplax is an animal because it is multicellular, diploid, and heterotrophic. However, it has no tissues, no organs, nor any adaptations to ingest food.

6. Most animals require a watery environment at some point in their lives. True terrestrial animals are found in the phylums Arthropoda and Chordata.
7. Included as invertebrates are the ancestral chordates (formerly called protochordates), which belong to the subphylums *Hemichordata, Urochordata,* and *Cephalochordata.*
8. Invertebrates, or acraniates, lack a skull and brain. Chordates, or craniates, possess a skull and brain.
9. Animal phylums include radially symmetrical groups, such as *Cnidaria* (Coelenterata), and bilaterally symmetrical groups, which make up most other phylums in this kingdom.
10. Bilaterally symmetrical animals either lack a true *coelom* or body cavity; possess a body cavity, but not a true coelom; or possess a true coelom. A true coelom develops from embryonic mesoderm and is a space filled with internal organs, such as the liver and intestines.
11. Biologists think that animals evolved from protist ancestors. The recent discovery of Trichoplax provides support for this idea.

Changing Animal Classification

Traditionally, Kingdom Animalia consisted of two major divisions: *Invertebrates,* animals without backbones, and *Chordates,* or animals with backbones. Each division had its phylums, classes, orders, and other taxons based on widely accepted criteria. Thus, Chordates included ancestral aquatic forms called *Protochordates* and groups that range from Pisces (fish) to primates. Invertebrates included various phylums from Porifera to Echinodermata.

The subgroup Protochordata includes *Hemichordata* (acorn worms), *Urochordata* (tunicates), and *Cephalochordata* (lancelets). Urochordates possess a notochord only in the larval stage. Cephalochordates possess a permanent notochord. Recently, taxonomists have reclassified tunicates and lancelets as *invertebrates* because they both lack a skull and brain. (See Figure 18–10). Table 18–1 lists and compares chordate and invertebrate features.

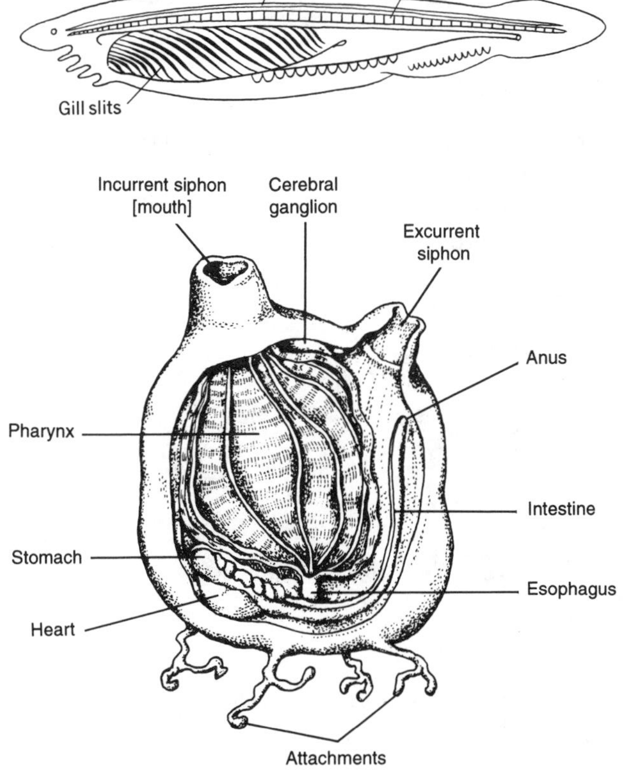

Figure 18–10 The lancelet (above) and tunicate (below)

Table 18–1 Comparison of Chordates and Invertebrates

Chordates (Craniates)	*Invertebrates (Acraniates)*
All chordates possess:	All invertebrates:
1. a skull and brain	1. lack a skull and brain
2. a hollow, dorsal nerve cord	2. possess simple nerve systems, from a nerve net to a ventral solid nerve cord
3. a vertebral column surrounding the nerve cord	3. lack an internal rigid support structure; some possess a notochord and a dorsal nerve cord sometime during their life cycle
4. a true coelom, or body cavity, derived from mesoderm	4. either lack a true coelom, possess a body cavity but not a true coelom, or have a true coelom
5. pharyngeal gill slits functional in some forms and vestigial or missing in other adult forms	5. lack pharyngeal gill slits

Representative Animals

Sponges (phylum Poriferans). Adults do not locomote, but are usually attached to an underwater object; no tissues present—organized on a cellular level; body has canals; internal skeleton of chalky, glassy, or leathery material. Examples: bath sponge, tube sponge. See Figure 18–11.

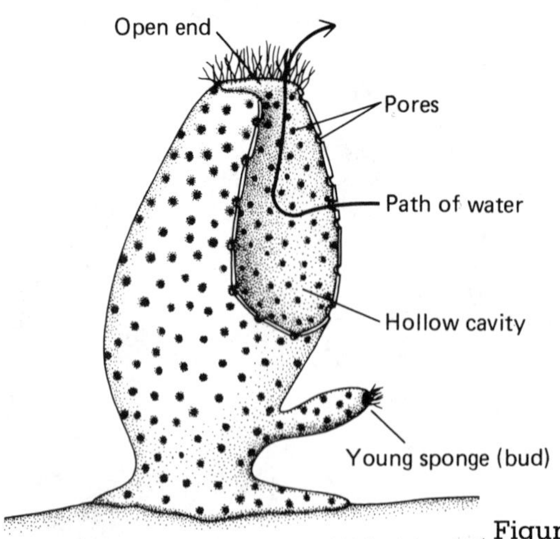

Figure 18–11 A representative sponge

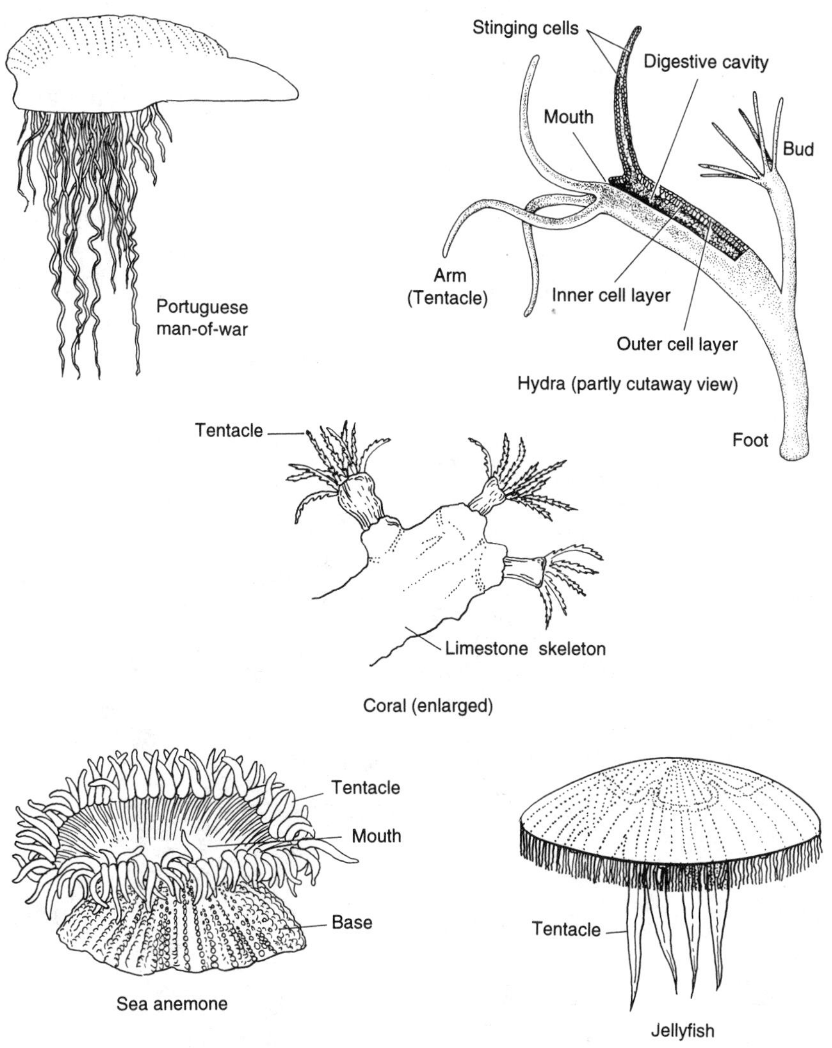

Figure 18–12 Some representative coelenterates

Hydra and its relatives (phylum Coelenterates, or Cnidaria). Two tissue layers present; organized on a tissue level; baglike body with tentacles and stinging cells; single opening of body that receives food and eliminates undigested wastes. Life cycle typically includes a medusa generation that reproduces sexually, and a polyp generation that reproduces asexually. Examples: jellyfish, hydra, coral. See Figure 18–12.

Flatworms (phylum Platyhelminthes). Three tissue layers present; organized on an organ-system level; flat body; many are parasites; digestive sac has single opening that receives food and eliminates wastes. Examples: planaria, fluke, tapeworm. See Figure 18–13.

Roundworms (phylum Nematodes). Possess an organ-system organization; digestive system with two openings (mouth and anus); round body; some are parasites. Examples: hookworm, nematode. See Figure 18–13.

Segmented worms (phylum Annelids). Cylindrical body divided into many similar segments; constructed on an organ-system level; possess complete mouth-to-anus digestive canal and digestive glands; possess complete excretory, nervous, circulatory, and reproductive systems; coelom (body cavity) present. Examples: earthworm, plumeworm, leech. The earthworm is discussed in previous chapters (see Figure 13–19).

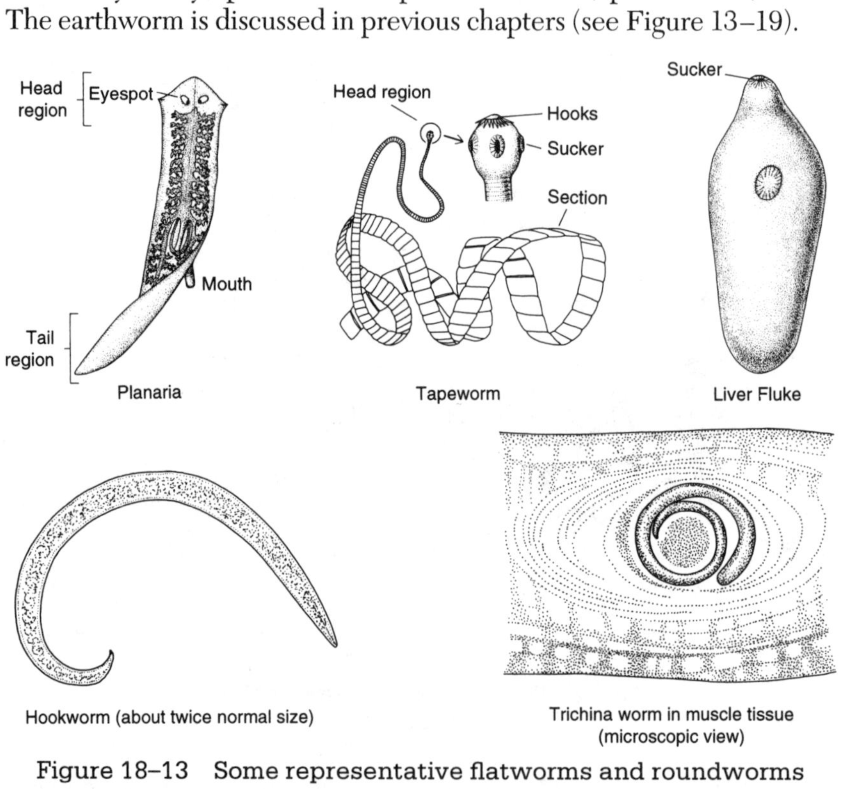

Figure 18–13 Some representative flatworms and roundworms

Phylum Rotifera. Cylindrical body; crown of cilia on anterior end; chitinlike jaws; well-developed digestive system; pseudocoelom maintains body shape. Example: rotifer. See Figure 18–14.

Hard-shelled animals (phylum Mollusks). Possess soft unsegmented body usually encased in a shell composed of calcium salts; constructed on an organ-system level; complete digestive, circulatory, respiratory, nervous, reproductive, muscular, and excretory systems; most forms are aquatic. Examples: clams, oysters, snails, octopus, slug. See Figure 18–15.

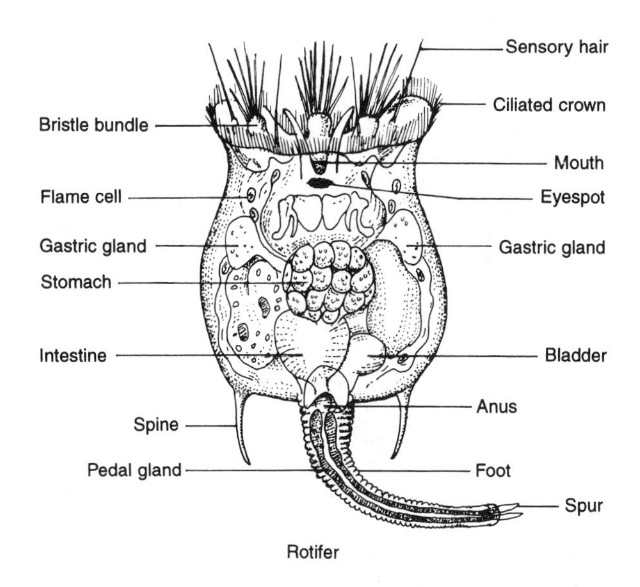

Figure 18–14 The rotifer—in a phylum by itself

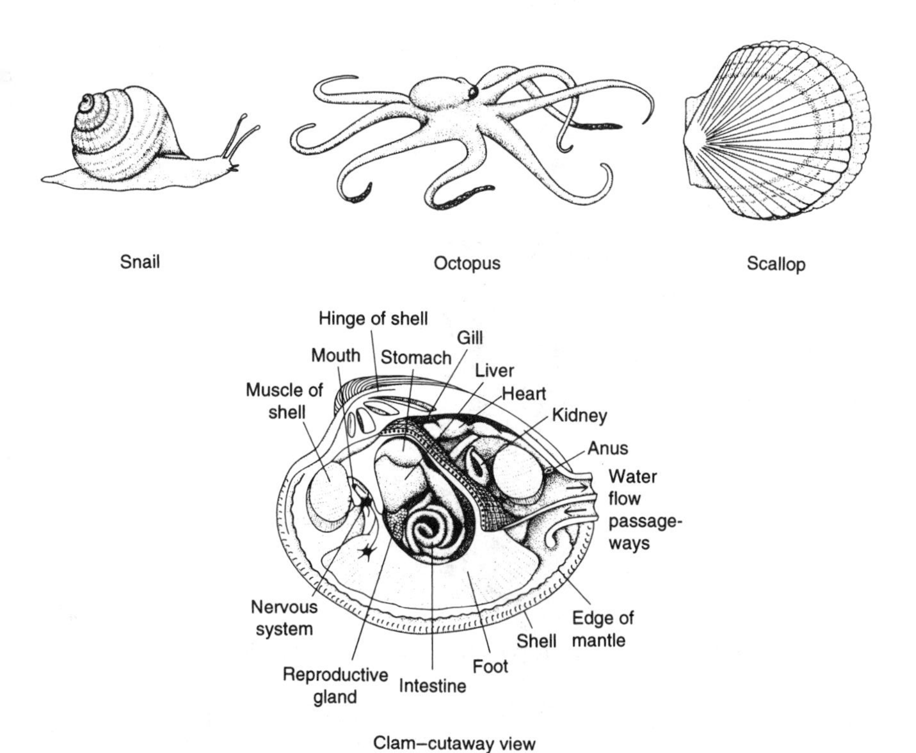

Figure 18–15 Some representative mollusks

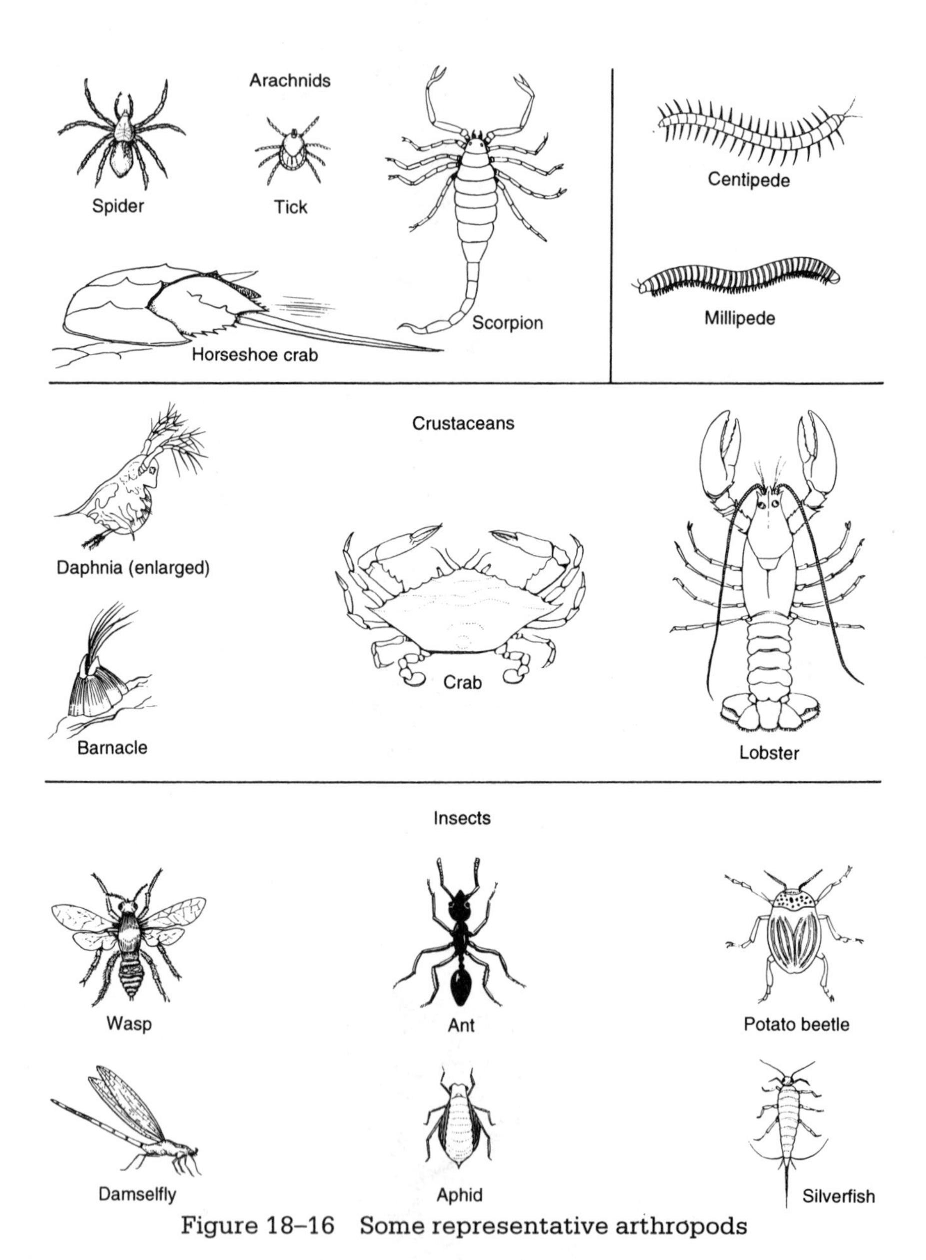

Figure 18–16 Some representative arthropods

Jointed animals (phylum Arthropods). Largest animal phylum, possess segmented, jointed body bearing many jointed appendages; body covered by a rigid exoskeleton made of chitin; movable joints provide flexibility; complete organ systems present; open circulatory system. See Figure 18–16, and the five classes of arthropods described below.

Class Crustaceans. Mostly aquatic, most live in salt water; possess exoskeleton of chitin and lime; head bears two pairs of antennae; breathe by means of gills; possess five pairs of walking legs. Examples: crab, lobster, barnacle.

Class Insects. Possess exoskeleton composed of chitin; body divided into three regions: head, thorax, and abdomen; head bears one pair of antennae; breathe by spiracles; thorax bears three pairs of legs and, usually, two pairs of wings. Examples: termite, ant, bee, fly, moth, wasp, grasshopper.

Classes Centipedes and Millipedes. Centipedes bear one pair of legs on most of the segments of their wormlike bodies; are carnivorous. Millipedes bear two pairs of legs on most of the segments of their wormlike bodies; are herbivorous.

Class Arachnids. Possess four pairs of legs; lack antennae; possess two body regions: combined head and thorax, and abdomen; lack antenna; most terrestrial. Examples: tick, spider, horseshoe crab, scorpion.

Spiny-skinned animals (phylum Echinoderms). All forms live in the sea; have a rough spiny skin; possess digestive system and a body cavity;

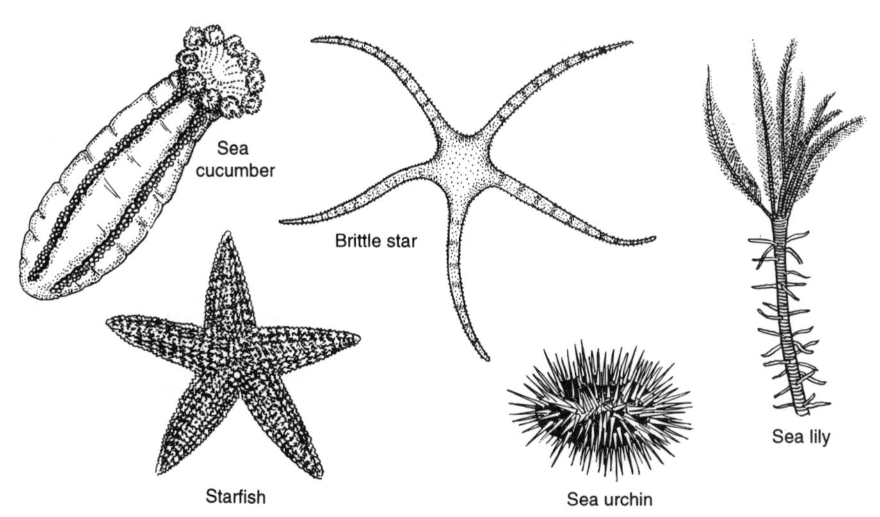

Figure 18–17 Some representative echinoderms

tube feet for locomotion in some groups; liquid circulates within the body; typically have five-part radial symmetry. Examples: sea star, sea lily, sea cucumber. See Figure 18–17.

Jawless fishes (class Agnatha). Characterized by long eellike body; simplest vertebrates; possess a shortened permanent notochord and an internal skeleton composed mainly of cartilage; usually possess fins; lack jaws, but have a circular sucking mouth and a tongue covered with horny teeth. Examples: hagfish, lamprey.

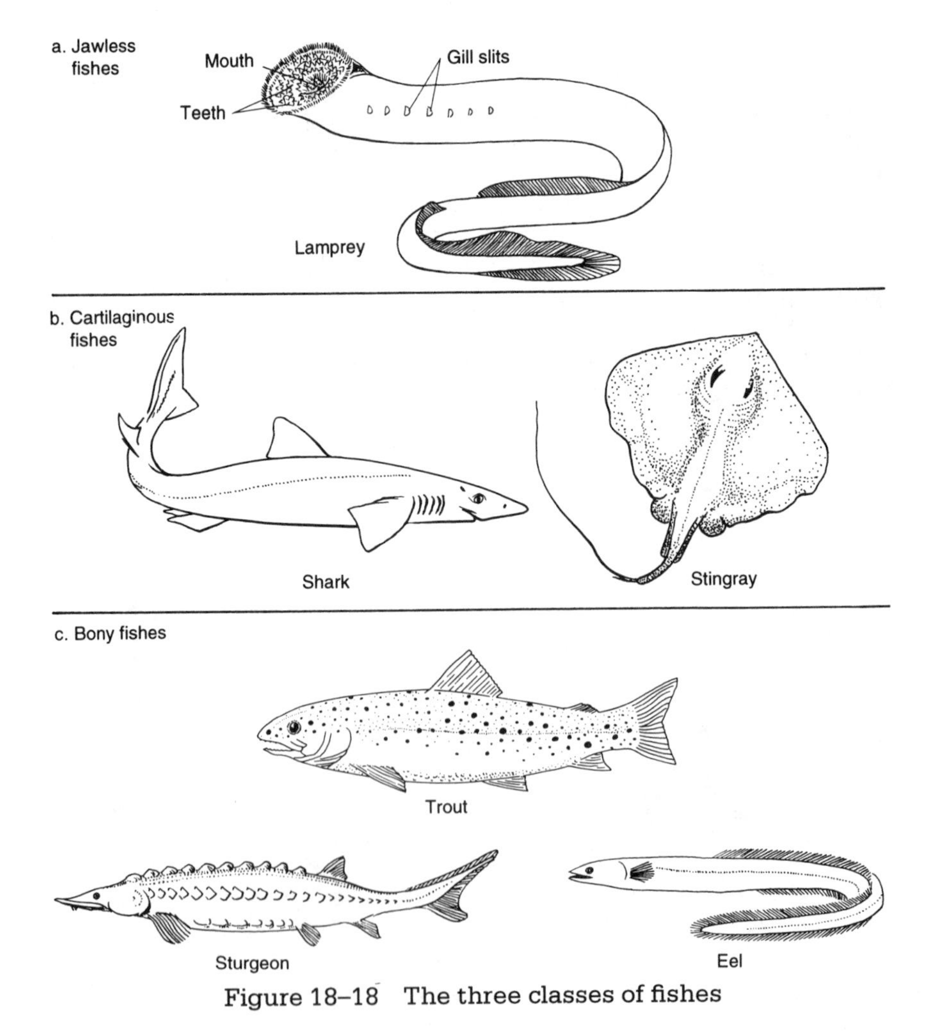

Figure 18–18 The three classes of fishes

Cartilaginous fishes (class Chondrichthyes). Possess internal skeleton of cartilage; mouth with true jaws; two pairs of fins; a skin covered with toothlike scales; and five pairs of slitlike openings that lead into gill chambers. Examples: shark, ray, skate.

Bony fishes (class Osteichthyes). Possess internal skeleton of bone; skin covered with overlapping scales; mouth has true jaws; two-chambered heart present; gills are covered; swim bladder usually present. Examples: salmon, perch, tuna. See Figure 18–18.

Class Amphibians. Adapted primarily to life in wet places; usually possess smooth, moist skin; two pairs of limbs present; three-chambered heart present; young and some adults breathe by means of gills, most adults breathe by means of lungs. Examples: salamander, frog, toad. See Figure 18–19.

Figure 18–19 Two representative amphibians

Class Reptiles. Adapted to a fully terrestrial life; possess a dry skin usually covered with scales; breathe by means of lungs only; three-chambered heart; four-chambered heart present in crocodilians. Examples: snake, turtle, lizard, alligator. See Figure 18–20.

Class Aves (Birds). Possess two pairs of limbs, of which the forelimbs are wings; body covered with feathers; scales present on legs; definite four-chambered heart; no teeth; warm-blooded. Examples: duck, ostrich, chicken, penguin. See Figure 18–21.

Class Mammals. Nervous system is most complex of all vertebrates; female has milk glands to secrete milk; diaphragm present (separates chest cavity from abdominal cavity); four-chambered heart; body covered with hair or fur at some stage of life; warm-blooded; greatest amount of parental care given to offspring. See Figure 18–22.

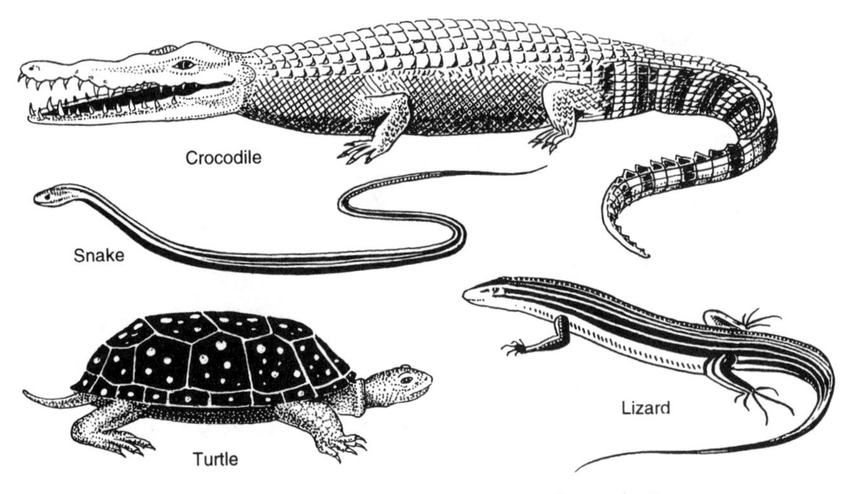

Figure 18–20 Some representative reptiles

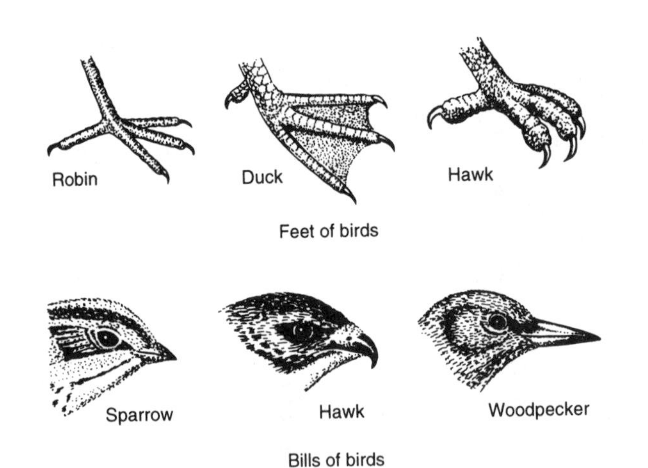

Figure 18–21 Some features of different types of birds

Figure 18–22 Some representative mammals: egg-laying, placental, and pouched (marsupial)

1. *Aplacental mammals* (Egg-laying and Pouched). Lack a placenta, a structure that connects the fetus to its mother. Examples: platypus and echidna (egg-laying); kangaroo and opossum (pouched).

2. *Placental mammals*. Possess a placenta; through this structure, food and oxygen are brought to the fetus and wastes are removed from it. Examples: human, dog, cat, elephant, bat, whale.

Evolutionary relationships of the animal kingdom. The classification of chordates—from simple to complex—further reinforces our ideas about the evolution of all living things. Some invertebrate ancestors evolved in two directions, one leading to higher invertebrates and the other leading to the chordates.

Our survey of the five kingdoms reflects the thinking of most scientists that animals probably evolved from some ancestral protists. Some protists may have evolved into forms like the multicellular algae and fungi. In time, plants appeared that possessed a tissue level of organization. Later, plants evolved with an organ-system level of organization.

Similarly, some ancestral protozoans may have evolved into simple animals possessing a tissue-level organization. In time, more complex animals appeared possessing organs and organ systems.

Many of our ideas about evolution are supported by the existence of *intermediate forms* or forms having characteristics of two groups. For example, although it is a fish, the lungfish is adapted to spending brief periods on land by using its swim bladder as a lung and stumpy fins as crawling legs. Classified as an amphibian, the tadpole of a frog breathes by means of gills and swims by means of a tail. Scientists have found remains of prehistoric reptilelike birds, and also remains of mammallike reptiles.

Section Quiz

1. An example of a radially symmetrical animal is a(an) (*a*) lobster (*b*) earthworm (*c*) sea anemone (*d*) starfish.
2. An animal that possesses a skull and brain is the (*a*) lamprey (*b*) lancelet (*c*) tunicate (*d*) acorn worm.
3. Many biologists think that animals probably evolved from a (*a*) heterotrophic moneran (*b*) heterotrophic protist (*c*) multicellular symbiont (*d*) multicellular saprobe.
4. The phylum Cnidaria includes animals such as a (*a*) hydra (*b*) comb jelly (*c*) rotifer (*d*) nematode.

5. The most simple animal is thought to be a (*a*) single-celled moneran bearing cilia (*b*) multicelled amebalike organism bearing cilia (*c*) lancelet (*d*) planarian.

Chapter Review Questions

The following questions will help you check your understanding of the material presented in the chapter.

1. In the commonly used modern system of classification, every organism is classified in one of five (*a*) phylums (*b*) kingdoms (*c*) genuses (*d*) species.
2. Which groups are arranged in correct descending order according to a modern classification system? (*a*) kingdom, genus, phylum, species (*b*) phylum, kingdom, species, genus (*c*) kingdom, phylum, genus, species (*d*) phylum, kingdom, species, kingdom.
3. Which taxon includes the other three? (*a*) genus (*b*) species (*c*) kingdom (*d*) phylum.
4. Which statement about modern classification of organisms is *incorrect*? (*a*) It reflects evolutionary similarities and differences. (*b*) It is based on the habitat of the particular organism. (*c*) There is some disagreement concerning the best classification of some organisms. (*d*) Organisms may be grouped into five kingdoms.
5. The scientific name for a dog is *Canis familiaris*, and the scientific name of a wolf is *Canis lupus*. This indicates that the dog and the wolf are in the same (*a*) species, but in different genuses (*b*) genus, but are different species (*c*) family, but in different classes (*d*) species, but in different families.
6. One-celled organisms having nuclei are classified as (*a*) algae (*b*) plants (*c*) invertebrates (*d*) protists.
7. Protists that lack true roots, stems, and leaves, but possess chlorophyll are called (*a*) algae (*b*) viruses (*c*) ciliates (*d*) mushrooms.
8. Based on nutrition, fungi are classified as (*a*) autotrophs (*b*) heterotrophs (*c*) herbivores (*d*) predators.
9. An important difference between bacteria and fungi is that bacteria lack (*a*) DNA (*b*) a well-defined nucleus (*c*) chlorophyll (*d*) a definite cell wall.

10. Which one of the following organelles is a characteristic of the ameba? (*a*) cilia (*b*) flagella (*c*) pseudopods (*d*) chloroplasts.
11. Viruses consist of an outer coat composed of protein and an inner material composed of (*a*) DNA or RNA (*b*) glucose (*c*) starch (*d*) urea.
12. According to the five-kingdom system of classification, which two are classified as plants? (*a*) ferns and maple trees (*b*) fungi and slime molds (*c*) algae and fungi (*d*) lichens and pine trees.
13. The protist kingdom includes a group of organisms that were formerly classified as (*a*) tracheophytes (*b*) multicellular heterotrophs (*c*) protozoans and algae (*d*) sponges and mollusks.
14. Needlelike or scalelike leaves are characteristic of (*a*) gymnosperms (*b*) angiosperms (*c*) ferns (*d*) liverworts.
15. An earthworm is classified as a (*a*) flatworm (*b*) roundworm (*c*) segmented worm (*d*) coelenterate.
16. Which group of organisms contains the greatest proportion of undifferentiated cells? (*a*) amphibians (*b*) fish (*c*) echinoderms (*d*) mammals.
17. Hydra, earthworms, grasshoppers, and humans belong to the same (*a*) genus (*b*) species (*c*) phylum (*d*) kingdom.
18. A two-chambered heart is characteristic of (*a*) fishes (*b*) birds (*c*) reptiles (*d*) amphibians.
19. Which one of the following mammals is a marsupial? (*a*) opossum (*b*) bat (*c*) shrew (*d*) whale.
20. A heart completely separated into four chambers is present in mammals, birds, and (*a*) toads (*b*) snakes (*c*) crocodiles (*d*) fish.
21. Although the dog, jackal, and coyote all belong to the genus *Canis*, they have been placed in different species because they (*a*) are competitive for food in the same ecological community (*b*) are natives of different parts of the world (*c*) were discovered by humans over wide intervals of time and classified separately (*d*) cannot interbreed and produce fertile offspring.
22. Endospore germination and subsequent growth depends mainly on contact with (*a*) water (*b*) water and nutrients (*c*) oxygen (*d*) none of these.
23. The body shape of a typical spirochaete resembles a (*a*) rod with square ends (*b*) tennis ball (*c*) chain of rods with flagella (*d*) wire that is a helical spiral.

24. When ciliates reproduce sexually, (*a*) partners exchange parts of their micronuclei (*b*) they undergo fission (*c*) one partner engulfs the other (*d*) zoospores unite to form a zygospore.

25. Many scientists think that chloroplasts originated from (*a*) symbiotic prokaryotes (*b*) parasitic prokaryotes (*c*) mutated cyanobacteria (*d*) chlorophyll-bearing Golgi bodies.

26. Water molds and downy mildews, once classified as fungi, are now considered protists because they (*a*) have cell walls composed of chitin (*b*) have cell walls composed mainly of cellulose (*c*) lack a nuclear membrane (*d*) lack mitochondria.

27. Rusts and smuts belong to the same phylum as a (*a*) yeast (*b*) bread mold (*c*) mushroom (*d*) lichen.

28. An important ecological function of fungi is (*a*) soil aeration (*b*) soil enrichment (*c*) nitrogen fixation (*d*) decomposition of inorganic matter.

29. Which of the following plants does *not* belong with the others? (*a*) mosses (*b*) club mosses (*c*) horsetails (*d*) ferns.

30. Strobili are structures common to (*a*) conifers (*b*) ginkgos (*c*) cycads (*d*) monocots.

31. Most taxonomists think that (*a*) plants evolved from highly differentiated brown algae (*b*) plants evolved from green algae (*c*) chemical criteria are most relevant in plant classification (*d*) structural criteria are least relevant in plant classification.

32. Evidence that supports evolution of plants from green algae is illustrated by the (*a*) flagellated sperm cells of bryophytes (*b*) dependent sporophyte generation of bryophytes (*c*) production of sori (*d*) similarity of structure and function.

33. Which statement describes most members of the Kingdom Animalia? (*a*) their life cycle features an alteration of generations (*b*) their life cycle does not include an alteration of generations (*c*) their sense organs are seldom involved in survival of the species (*d*) their coeloms are derived from endoderm.

34. The vertebral column of chordates (*a*) is composed mainly of cartilage (*b*) is a rigid support in most vertebrates (*c*) is detached from the nerve cord (*d*) protectively surrounds the nerve cord.

35. Although an octopus has a head, it is considered an acraniate because (*a*) it possesses a skull, but lacks a brain (*b*) it possesses a brain, but lacks a skull (*c*) it is radially symmetrical (*d*) it lacks a skull and a brain.
36. Pharyngeal gill slits are found in (*a*) larvae of clams (*b*) immature squids (*c*) chordate embryos (*d*) immature aquatic insects.

Biology Challenge

The following questions will provide practice in answering SAT II-type questions.

Part I

Select the letter of the word or phrase that best completes the statement or answers the question.

1. Which organism has collar cells that create water currents, feeds on organic detritus, and is structured with spicules that may be either chalky, glassy, or made of protein fibers? (*a*) hydra (*b*) sea star (*c*) starfish (*d*) sponge (*e*) coral.
2. Which of the following statements is *incorrect*? (*a*) Invertebrates and vertebrates are usually bilaterally symmetrical. (*b*) Invertebrates and vertebrates are usually radially symmetrical. (*c*) Many plants possess symmetry. (*d*) Humans are not perfectly symmetrical. (*e*) Echinoderms are radially symmetrical.
3. Different classes of mollusks possess various feeding adaptations. A class of mollusks whose members capture prey, hold and tear it apart, and then swim rapidly is (*a*) gastropods (*b*) pelecypods (*c*) cephalopods (*d*) arthropods (*e*) amphineura.
4. An arthropod considered a "living fossil" is the (*a*) horseshoe crab (*b*) trilobite (*c*) bee (*d*) flea (*e*) scorpion.
5. *Saccharomyces cerevisiae,* a fungus, is useful to humans because it (*a*) decomposes organic substances (*b*) causes athlete's foot (*c*) is edible (*d*) is a source of antibiotics (*e*) is a valuable fermenter.

Part II

The following set of questions should be answered in two parts. First, find four items in each right-hand set (1 to 5) that are related to *one* item in the corresponding left-hand set (A to C). Check this item on the left and the four related items on the right. Next, underline the checked item in the left set and the one *unchecked* (unrelated) item in the right set.

Example:	X (A) vertebrates	(1) crayfish
	(B) amphibians	X (2) toad
	(C) mammals	X (3) rabbit
		X (4) salamander
		X (5) chimpanzee

1.	(A) insects	(1) spider
	(B) annelids	(2) silverfish
	(C) mollusks	(3) ant
		(4) cockroach
		(5) aphid
2.	(A) phyllotaxy	(1) taxonomy
	(B) phenylketonuria	(2) Linnaeus
	(C) phylum	(3) abiogenesis
		(4) common evolutionary origin
		(5) genera
3.	(A) prokaryotic	(1) moneran
	(B) eukaryotic	(2) nucleoid
	(C) autotrophic	(3) omnibacterium
		(4) protist
		(5) lactobacillus
4.	(A) ameba	(1) intracellular parasite
	(B) virus	(2) phylum
	(C) yeast	(3) capsid
		(4) crystalline shape
		(5) non-cellular microbe

5. (A) bacterium	(1) hypha
(B) mold	(2) chitin
(C) alga	(3) swarm cell
	(4) rhizoid
	(5) antheridium
6. (A) cycad	(1) eukaryotic
(B) fern	(2) nonvascular
(C) liverwort	(3) biflagellate sperm cells
	(4) long-lived sporophyte
	(5) leafy gametophyte
7. (A) gymnosperm	(1) horsetail
(B) angiosperm	(2) ginkgo
(C) pteridophyte	(3) redwood
	(4) cycad
	(5) spruce
8. (A) plant	(1) desmosome
(B) fungus	(2) protozoan
(C) animal	(3) acorn worm
	(4) trichoplax
	(5) blastula
9. (A) moneran	(1) alga
(B) protist	(2) ciliate
(C) ascomycete	(3) seaweed
	(4) water mold
	(5) lichen
10. (A) Aristotle	(1) coacervate
(B) Linnaeus	(2) species concept
(C) van Leeuwenhoek	(3) binomial nomenclature
	(4) taxon
	(5) diversity

19

Ecology and Conservation

Learning Objectives

When you have completed this chapter, you should be able to:

- **Develop** a working definition of ecology terms.
- **Discuss** abiotic (nonliving) and biotic (living) factors in the environment.
- **Understand** ecological relationships organisms have with each other.
- **Describe** the flow of energy in an ecosystem.
- **Diagram** several natural cycles that are important to life on Earth.
- **Discuss** characteristics of Earth's major biomes.
- **Predict** the effects on the environment of human actions, and identify solutions for some of our environmental problems.

OVERVIEW

Today, humans face a global ecological crisis that is, in some respects, worsening as a result of past and present economic patterns and environmental degradation. Current threats to Earth's ecosystems include radioactive pollution from nuclear reactor accidents such as Chernobyl, clear-cutting of tropical rain forests, ozone depletion, and pollution from toxic wastes introduced into the atmosphere, bodies of water, and on land.

In this chapter, you will learn how the industrialization of developing countries, as well as the rapidly increasing world population, has caused pollution and resource depletion on an unprecedented scale. These problems can be effectively resolved by using educational approaches that respect the ecological integrity of natural environments.

AN ECOLOGIST'S VOCABULARY

To better understand ecology-related environmental problems, it is essential to have a working knowledge of the following basic terms:

Ecology is the study of the interaction of living things with each other and with their environment.

Environment is any factor that affects an organism. The adaptations of an organism enable it to survive in a particular environment. Environmental factors that affect organisms consist of both the *abiotic* and *biotic* factors.

A *habitat* is the location where an organism commonly is found. For example, the habitat of a sow bug may be under a stone or log; a flounder lives on the ocean bottom; and a cactus inhabits a desert.

A *niche* describes an organism's relations with the physical environment and with other members of its community. The niche has been described as an organism's "role" in a habitat.

A *population* consists of all members of the same species that inhabit a specific area. For example, a meadow may be inhabited by a population of grasshoppers. However, the same meadow may also be a habitat for other populations, such as foxtail grass, clover, and field mice.

A *community* is composed of populations of all the different species that occupy a habitat. Certain populations in a habitat can be described as a "conifer community" or a "grass community." In general, a community consists of different species that include *producers, consumers, decomposers,* and *detritivores.*

Producers are autotrophs that include plants and photosynthesizing microorganisms. Most producers manufacture organic compounds and other substances by using radiant energy from the sun.

Consumers are heterotrophs that depend on other living things to obtain organic and inorganic compounds. *Herbivores,* such as grasshoppers and cattle, eat plants; *carnivores,* such as tigers and wolves, eat other animals; *omnivores,* such as bears and humans, eat plants and animals.

Decomposers, mostly fungi and heterotrophic bacteria, obtain their nourishment by breaking down dead organic matter. Decomposers are Earth's biotic recycling agents.

Detritivores are heterotrophs, such as earthworms, that feed on small particles of organic matter—"leftovers" of the decomposers.

An *ecosystem* includes all the members of a community and their relationships with the abiotic and biotic factors that affect their lives. The flow of energy and chemical recycling are basic features of any ecosystem.

A *succession* refers to the orderly and predictable replacement of one community by another. After a forest fire, a lichen community may dominate for a while and then gradually be replaced by a ragweed community. A succession usually culminates in a *climax community*, which is a mixture of plants and animals typical of an area that remains stable year after year.

A *biome* is a large geographic area composed of a group of ecosystems in which climate is the dominant ecological factor. Aquatic biomes are far more extensive than terrestrial biomes and are made up of diverse ecosystems. Gradients in light penetration, temperature, and dissolved gases play a major role. A biome has a characteristic climax community.

THE ABIOTIC ENVIRONMENT

Major abiotic factors of the environment include *light, temperature, water, substrate, minerals,* and *oxygen.*

Light

Autotrophs use solar energy to manufacture food for themselves and, indirectly, for the heterotrophs. The intensity, duration, and wavelength of *light* determine the types of organisms that can survive in a particular region. For example, most moss plants grow in shaded areas where light is weak; daisies grow in open meadows where the light is more intense. In oceans, most of the red and yellow wavelengths of light are absorbed by the first 90 meters of water. Consequently, photosynthetic plants are found mainly in the upper water layers. These layers are inhabited by many tiny organisms that make up plankton.

The length of daylight also affects the behavior of an organism. Organisms automatically respond to changes in the number of hours of daylight. This automatic response usually is a result of hormonal fluctuations in their bodies. The time when the leaves of deciduous trees change color and fall is determined by a decrease of auxin production in leaves. As the hours of daylight get shorter in autumn, leaves produce less auxin. At the same time, a waterproof layer forms at the base of each leaf petiole. This causes each leaf to dry out, die, and eventually fall.

Lengthening spring days, and additional hours of light, cause hormonal changes within the bodies of some birds. One result is that the gonads of these birds enlarge. This change triggers other responses that cause the birds to fly northward to their breeding grounds. In the fall, the shortening days cause the gonads to get smaller and the birds fly south.

Temperature

Most life on Earth exists within a narrow *temperature* range—from 0°C to 50°C. As water freezes, it expands and destroys cells; at high temperatures, enzymes in cells are destroyed, thus stopping chemical reactions that occur within cells.

Effect of temperature on animals. Warm-blooded animals, such as birds and mammals, maintain a constant body temperature despite the variability in temperature of the physical environment. The body temperature of a person living at the equator is the same as that of a person living in Alaska. Cold-blooded animals (all invertebrates, fishes, amphibians, and reptiles) do not maintain a constant body temperature; their temperature varies with the temperature of their environment.

Regional temperatures. The temperature of a region is determined by its latitude, altitude, and geography.

1. *Latitude.* As you move north or south from the equator, temperature decreases. At higher latitudes, the sun's rays are weak because they are spread over a large area, and for most of the year the hours of daylight are shorter than at lower latitudes.
2. *Altitude.* As altitude increases, temperature decreases. High mountains located at the equator are covered with snow and ice all year. The northern slopes of mountains receive less sunlight than southern slopes. As a result, northern slopes are colder than southern slopes.
3. *Geography.* Water tends to absorb heat slowly and release heat slowly. As a result, a large body of water cools the adjacent land in summer and warms it in winter. For this reason, large bodies of water, such as an ocean, moderate the temperatures of surrounding coastal land areas.

Adaptations to regional temperatures. Different adaptations enable animals to survive in regions where the temperature is a limiting factor. For example, birds possess feathers and layers of fat under their skin to insulate them from cold. Some insects, such as the monarch butterfly, migrate to warmer regions. Mammals, such as the bear, hibernate (spend the winter in a dormant state). And, the lungfish estivates (becomes dormant during periods of high temperature when its water habitat dries out).

Effect of temperature on plants. As winter approaches in the temperate zone, leaves of perennial plants (those that live for more than one growing season) fall, and sap moves down to the roots, where it is stored. Roots are well-insulated against cold by soil, snow, and by the lowering of the freezing point of water caused by dissolved substances in sap.

Annual plants live for only one growing season in the temperate zone, and die when temperatures fall. However, species of annual plants survive the winter as a result of seeds produced during the growing season. Since seeds contain little water, they rarely freeze. In spring, the seeds germinate and re-establish the species.

Water

Because most chemical reactions in cells occur in *water*, organisms cannot survive without water.

Water relations of aquatic (water) organisms. Organisms that inhabit aquatic environments have the following advantages:

1. Water is always available for their life processes.
2. Water temperature usually remains within a narrow range. Temperature changes do not occur as often or as rapidly in water as they do in air or on land.
3. Water is a buoyant substance and can therefore support organisms (such as algae and jellyfishes) that lack supporting structures.
4. Water density increases as the temperature approaches 4°C. As a result, cooler, denser water sinks to the bottom of a pond or lake. Then, even if the surface freezes, organisms can survive in the warmer layers of water below the ice. Surface ice also acts as an insulator and prevents deeper water layers from freezing.

The following are several disadvantages to living in water:

1. Energy must be used to eliminate excess water that tends to accumulate in cells. In protozoans, for example, contractile vacuoles continually excrete excess water.
2. The tissues of some marine organisms tend to accumulate salt and lose water. These organisms prevent water loss by excreting excess salts from their tissues. In marine fishes, gills and kidneys continuously excrete excess salts.
3. The amount of oxygen dissolved in water is less than that present in air. Thus, aquatic organisms are faced with the problem of obtaining an adequate supply of oxygen. Fishes, lobsters, and clams force currents of water over their gills, which absorb dissolved oxygen.

Water relations of terrestrial (land) organisms. Terrestrial organisms have evolved adaptations that enable them to survive without being immersed in water. These adaptations include supporting tissues in plants and animals, legs or wings in animals for locomotion or dispersal, seeds as a means of dispersal for plants, and internal fertilization.

Adaptations of plants in dry, hot regions. In dry, hot regions, plants such as the cactus have developed roots that grow deep or spread out to absorb water. Cactus needles, or spines, are modified leaves that have a smaller surface area, thus reducing water loss by transpiration.

Adaptations of animals in dry, hot regions. In the desert, mammals are active mainly at night and obtain water by licking the water that condenses on plants and rocks. Insects and reptiles conserve water by excreting uric acid rather than urea. Uric acid requires less water than urea does to be excreted. Such animals also use the water produced by reactions that occur in their cells.

Substrate

The surface on which an organism lives is called its *substrate*. The water strider (an insect) spends most of its time moving across the surface of a lake or pond; the sea star lives on the ocean floor; and the lichen lives on a rock or the bark of a tree. Soil is the substrate for most terrestrial organisms. Soil is composed of varying quantities of rock particles, humus, water, air, and tiny organisms such as bacteria and roundworms. *Humus* consists mainly of decomposed organic matter (plants and animals).

Soil types. The major soil types are clay soil, sandy soil, and loam.

Clay soil consists mainly of fine particles of clay that tend to stick together and pack firmly. Clay soil is a poor substrate for plants because it is nonporous and therefore contains very little air and water.

Sandy soil consists mainly of sand grains, some clay, and some humus. Sandy soil is very porous, thus allowing water and air to seep down deeply through it. Plants with root systems long enough to reach water can grow in such soil.

Loam consists of a mixture of sand, clay, and humus in proportions that enable roots to obtain adequate amounts of water, air, and minerals. Most plants grow well in loam. Some plants, such as the azalea, grow best in acid loam; grasses grow best in a slightly alkaline loam.

Minerals

The metabolism of all living things requires *minerals* for chemical reactions in cells to proceed normally. Minerals are provided by the substrate. Plants obtain minerals from the soil or water in which they live.

Certain animals that lack sodium chloride in their food find natural outcrops of sodium chloride, called *salt licks*. Other animals, such as sheep, may die of anemia when the plants they feed upon lack the cobalt salts needed for the synthesis of vitamin B_{12} in their bodies. (Vitamin B_{12} is used to manufacture hemoglobin, a deficiency of which causes anemia.)

Oxygen

Under normal conditions, terrestrial organisms can readily obtain *oxygen*, which makes up about 20 percent of the atmosphere. Aquatic organisms have less oxygen available to them for the following reasons:

1. Only a small amount of oxygen is normally dissolved in water.
2. Only the upper layers of water are relatively rich in oxygen. The plants and algae that live in the upper layers carry on photosynthesis and provide some oxygen. Wind action and currents dissolve some atmospheric oxygen in these upper layers of water, thus adding to the supply.
3. When the temperature rises, less oxygen dissolves in water.
4. Aerobic bacteria of decay use large quantities of dissolved oxygen.

Section Quiz

1. An example of a detritivore is a (an) (*a*) earthworm (*b*) clam (*c*) crab (*d*) all of these.
2. A niche describes the (*a*) location where an organism is commonly found (*b*) members of the same species inhabiting a specific area (*c*) organism's relationship with the physical environment and with other members of its community (*d*) abiotic factors found in a particular environment.
3. Most organisms on Earth live within the following temperature range: (*a*) –20°C to 20°C (*b*) 0°C to 50°C (*c*) 20°C to 60°C (*d*) none of these.
4. An adaptation of plants that inhabit dry, hot environments is (*a*) shallow roots (*b*) large leaves (*c*) deep roots (*d*) colorful flowers.
5. The best type of soil for plants is (*a*) clay soil (*b*) sandy soil (*c*) humus (*d*) loam.

THE BIOTIC ENVIRONMENT

All organisms interact with other organisms of the same species and with organisms of other species. Thus, the *biotic factors* of the environment include mates, populations, communities, plants, animals that eat plants, animals (*predators*) that eat other animals, animals (*prey*) that are eaten by other animals, and organisms that help each other.

Habitats and Niches

The specific area of an environment in which an organism lives, or makes its home, is called its *habitat*. Although catfish and trout may live in the same lake, the habitat of a catfish is on a lake bottom and that of a trout near the surface. What the organisms of a species do to survive in a habitat and the way the habitat affects the organisms make up the *niche* of the species. Thus, a snail and a carp, which have the same habitat (the bottom of a pond), occupy different niches. The snail obtains food by scraping algae from the leaves and stems of underwater plants; the carp obtains food by poking around the bases of underwater plants where decaying material accumulates. The snail and an underwater plant occupy different niches. The snail is a heterotroph that consumes vegetation and contributes carbon dioxide and nitrogenous wastes to the pond. In contrast, the plant is an autotroph that makes food, contributes oxygen to the pond, and uses the carbon dioxide released by the snail.

If two organisms occupy the same niche but are active at different times of the day, the organisms do not compete with each other. For example, hawks and owls eat mice, but hawks hunt during the day and owls hunt at night. Competition only occurs when organisms live in the same habitat and occupy the same niche at the same time. The competing organisms with the more favorable adaptations survive and the others either die, migrate to another area, or, over time, adapt to occupy a different niche.

Symbiosis

The interactions between populations in a community are usually temporary and of short duration. The permanent relationship between two different organisms, one of which lives either on the body or in the body of the other, is called *symbiosis*. The three types of symbiosis are *mutualism*, *commensalism*, and *parasitism*.

Mutualism. In this relationship, both organisms benefit.

Termites and protozoans. The association between the wood-eating termites and the flagellate protozoans that inhabit the termites' intestines is an example of mutualism. The protozoans digest the wood into end products both organisms can use, and the termites provide the wood and a protective environment for the protozoans.

Lichens. A lichen is a permanent community consisting of unicellular green algae living closely together with threadlike fungi. The algae make food for both organisms and the fungi absorb water and minerals for both.

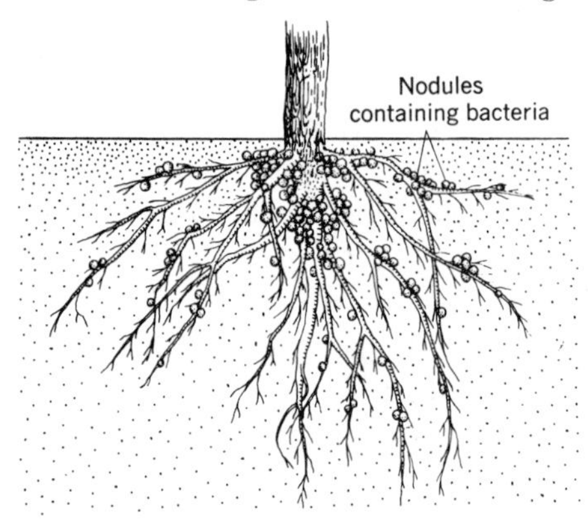

Figure 19–1 Nodules on a leguminous plant

Nitrogen-fixing bacteria and legumes. The nitrogen-fixing bacteria live in *nodules*, or swellings, on the roots of leguminous plants (clover, pea, peanut). The bacteria synthesize nitrogen compounds that are used by the plants. The plants provide moisture and nutrients for the bacteria. (See Figure 19–1.)

Commensalism. In this relationship, one species benefits; the other species neither benefits nor is harmed.

Remora. Figure 19–2 shows the relationship between the remora, or pilot fish, and the shark. The remora benefits by obtaining food scraps left by the shark, which neither benefits nor is harmed from the association.

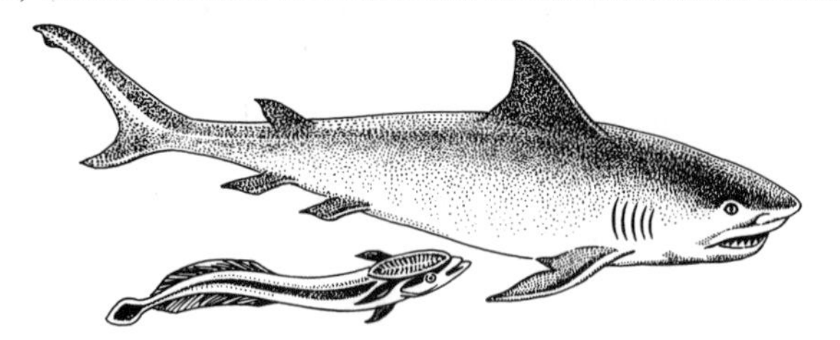

Figure 19–2 Pilotfish and shark

Spanish moss. Spanish moss is a plant that grows upon the branches of other plants. The Spanish moss benefits by being exposed to light and by obtaining water and minerals that collect in natural pockets in tree bark.

Parasitism. In this relationship, one species, the *parasite*, benefits and the other species, the *host*, is harmed.

Parasites include disease-producing bacteria, fungus that causes athlete's foot, fungi that cause diseases in plants, and tapeworms, liver flukes, and porkworms that live in the bodies of animals.

ECOSYSTEMS

The interactions of the living (biotic) and nonliving (abiotic, or physical) components of a community make up an *ecosystem*. Populations of a community not only affect each other but also affect the ecosystem's physical environment. An ecosystem remains stable as long as the abiotic factors remain constant and its populations do not fluctuate.

Food Chains and Food Webs

In each community, many *food chains* exist, each consisting of a producer, a consumer, and decomposers. A typical food chain might be represented as: wheat (producer) → mouse (consumer) → snake (consumer) → hawk (consumer). Decomposers extract energy from the dead material generated by each producer or consumer level.

In most communities, any species can be eaten by several other species. As a result, food chains are interconnected and form a network called a *food web* (Figure 19–3).

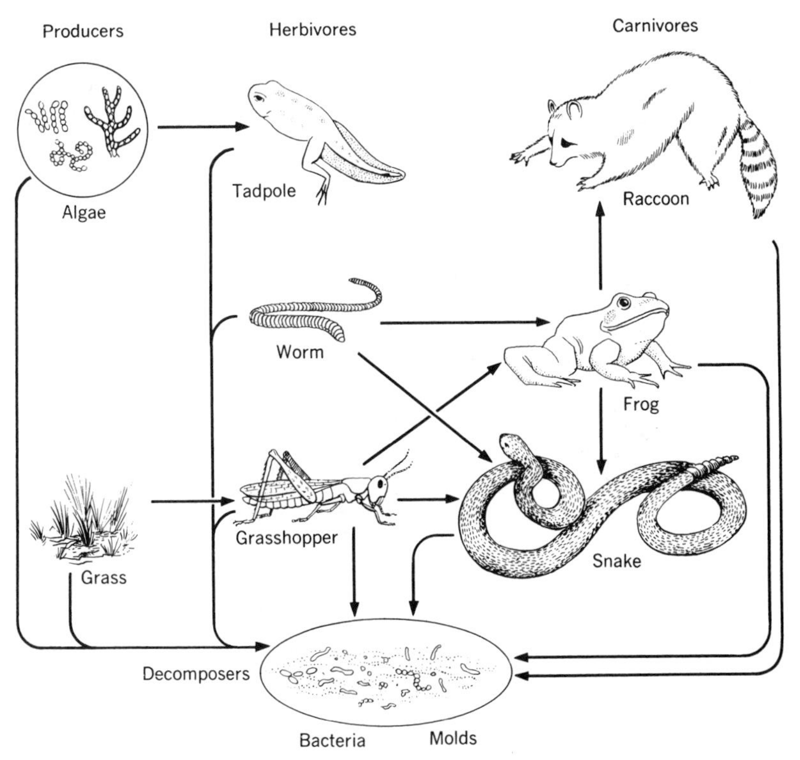

Figure 19–3 A food web

Flow of Energy in an Ecosystem

Energy is transferred from one member of a food web to another. Food eaten by one member of the web is the source of energy for another member. The flow of food or energy can be represented as a pyramid, called the *pyramid of energy* (Figure 19–4). The broad base of the pyramid represents the total energy incorporated from sunlight into grass. When a grasshopper eats grass, some of the energy in the grass is transferred to the grasshopper. However, the grass also used energy to carry on its life functions. Therefore, some of the original energy is not available to the grasshopper. When a frog eats the grasshopper, some energy is transferred to the frog. Since energy was also used by the grasshopper to carry on its life functions, more of the original energy is lost to the frog. Thus, each succeeding member of the food web receives less energy. The animal at the highest point, or apex, of the pyramid receives only a fraction of the energy represented by the broad base. This decrease in energy from the broad base to the apex of the pyramid is usually accompanied by an increase in the size of the organisms and a decrease in the number of organisms. In effect, fewer larger organisms feed on a greater number of smaller organisms. Considering the total living mass, the mass of the producers is greater than the mass of the primary consumers, which is greater than the mass of the secondary consumers, and so on to the apex of the pyramid. For these reasons, the pyramid of energy can be converted to a *pyramid of numbers* or a *pyramid of mass*.

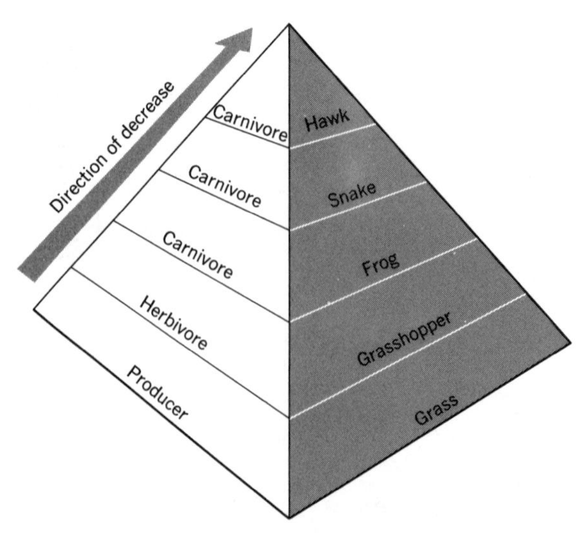

Figure 19–4 Pyramid of energy, numbers, and mass

Cycles within an Ecosystem

A stable ecosystem is usually balanced because essential materials are continually replaced by recycling. Energy, however, is not recycled but decreases in amount as food is transferred from producers to consumers. To continue, an ecosystem must have a continuous input of energy, usually supplied by the sun.

Population cycles. When environmental conditions are favorable, populations tend to increase slowly at first, and then rapidly. Predators feed on an increasing population of prey animals, reproduce, and in turn increase their numbers because the conditions for their population growth are favorable. As the population of predators grows, the prey population is reduced. Figure 19–5 shows the population cycles of the lynx and the snowshoe hare. You can see that after the hare population begins to increase, the lynx population also begins to increase. As the lynx population continues to increase, the hare population decreases. As the hare population becomes further reduced, the lynx population has less food available and consequently decreases. The decrease in the lynx population favors an increase in the hare population, and the cycle continues.

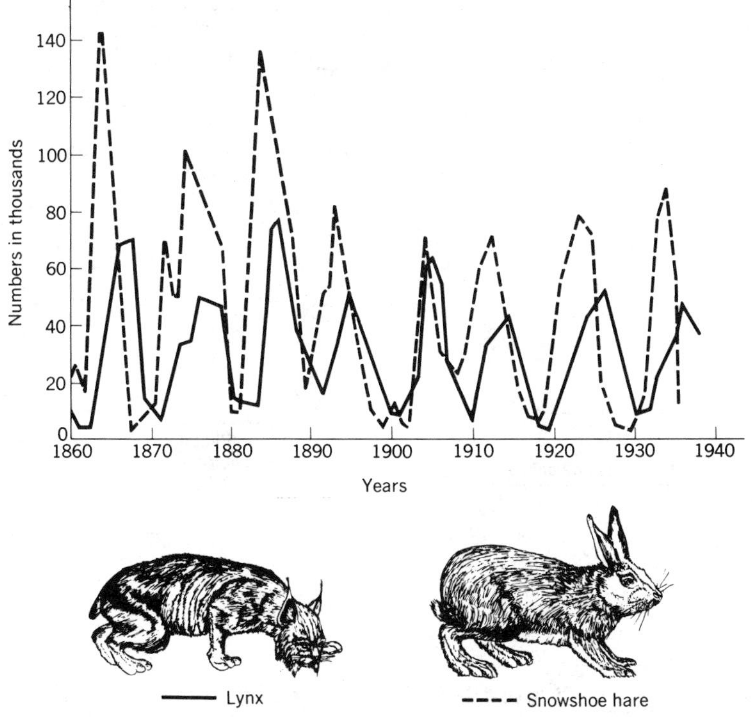

Figure 19–5 Population cycles

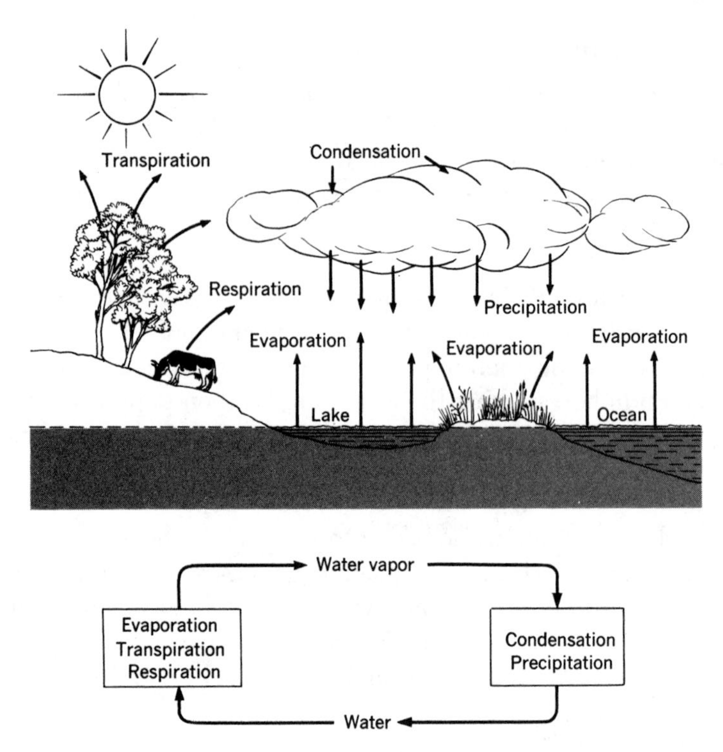

Figure 19–6 The water cycle

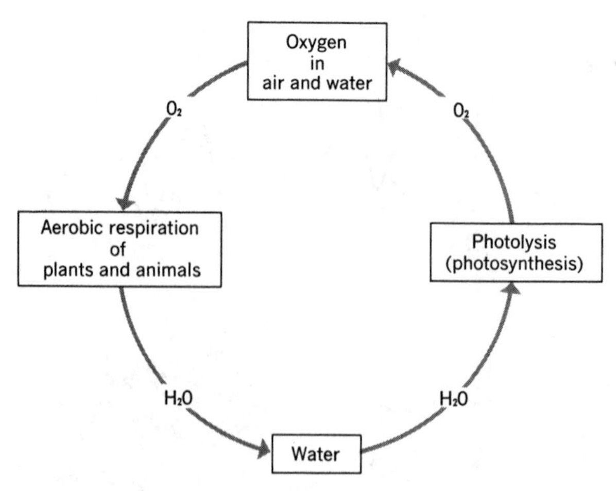

Figure 19–7 The oxygen cycle

The water cycle. The quantity of water on Earth remains constant (Figure 19–6). Water that is lost to the atmosphere as a result of evaporation, respiration, and transpiration is returned, recycled by condensation and precipitation in the form of rain, snow, sleet, or hail.

The oxygen cycle. The percentage of atmospheric oxygen also remains constant because of recycling (Figure 19–7). Oxygen that is removed from the atmosphere as a result of aerobic respiration and burning is returned to the atmosphere by photosynthesis. Humans can upset this cycle by excessive destruction of forests and grasslands, as well as by polluting lakes, rivers, and oceans.

The carbon cycle. The carbon cycle shows how carbon is recycled (Figure 19–8). The carbon in carbon dioxide is incorporated into organic compounds as a result of photosynthesis. The organic compounds, in the form of wood or fuel, are oxidized during the processes of respiration, decay, or burning. This releases energy, carbon dioxide, and water.

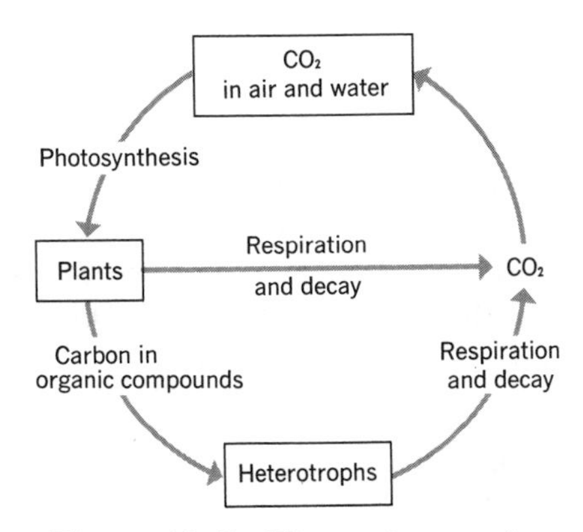

Figure 19–8 The carbon cycle

The nitrogen cycle. Refer to Figure 19–9 as you read on. About 79 percent of the atmosphere consists of gaseous nitrogen. The nitrogen, however, cannot be used by most organisms in the manufacture of proteins until nitrogen-fixing bacteria change the free nitrogen into nitrates. The nitrates are then taken in by plants, which use them in making amino acids and proteins. Animals that feed upon plants obtain amino acids from the digested plant proteins and synthesize animal protein from the amino acids.

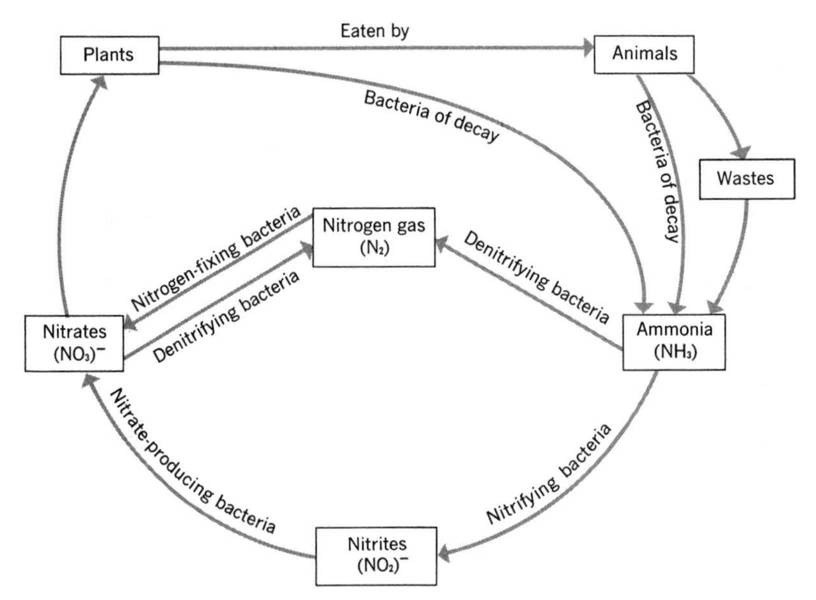

Figure 19–9 The nitrogen cycle

When plants and animals die, their bodies are decomposed into ammonia (NH_3). Certain bacteria convert the ammonia to free nitrogen or nitrites. Other bacteria convert the nitrites to nitrates, which plants can use. The free nitrogen is converted to nitrates by nitrogen-fixing bacteria in the soil and in the nodules of leguminous plants. If it were not for nitrogen-fixing bacteria, the supply of nitrogen and nitrates would soon be depleted and life could not exist.

Succession of Communities

An ecosystem is stable when its communities are in balance with the physical environment. Any major environmental change, such as a flood or drought, can upset the balance. Once upset, an ecosystem's stability can be regained by a series of gradual changes in which the same kinds of populations and communities are reestablished. In time, the same kind of ecosystem returns.

Replacement of one community by another is called *ecological succession*. Ecological succession results from changes in the abiotic and biotic factors of the environment. The final community that lives in balance with the environment and whose populations are in balance with one another is called a *climax community*.

In the central plains of the United States, grasslands are the climax community. In the northeastern United States, beech and maple forests make up the climax community. Since the animal population of a community depends upon the types of plants available, communities are named after their dominant plant types.

Succession on bare land. The first group of plants to appear in a bare region is called a *pioneer community*. On bare rock, lichens are the pioneer organisms; on an abandoned farm, annual weeds, such as foxtail grass and ragweed, are pioneer plants. In time, the pioneer organisms change the abiotic environment and establish physical conditions in which other types of plants and animals can thrive. Succession on an abandoned farm in the northeastern United States can be represented as follows: annual weeds (grasses and ragweed) → perennial weeds (goldenrod and clover) → shrubs (sumac and wild rose) → trees (beech and maple).

Each plant community is associated with a specific group of animals. The insects that feed upon the annual plants attract mice, skunks, and birds. When the perennial plants appear, snakes and rabbits also appear. When shrubs appear, cardinals (birds) and chipmunks move in. Finally, many trees in an area attract woodpeckers, raccoons, and squirrels.

Succession in a body of water. As soil is washed into a pond or lake, the body of water begins to fill up with solid material and becomes shallower. When this happens, populations of deepwater fishes, such as pike and smallmouth bass, decrease sharply, and shallow-water forms, such as sunfish and largemouth bass, increase in number. Along the shore, reeds and cattails grow, die, decay, and add more solid material to the lake. Eventually the region changes into a bog in which sumac, briers, and blueberries are the dominant plants. As the area becomes dry land, spruce, elm, and red maple make their appearance. (See Figure 19–10.)

Succession in a hay infusion. When a hay infusion is inoculated with a variety of microorganisms, a succession of populations can be observed. After one day, numerous bacteria but few protozoans are observed. Several days later, many small ciliate protozoans (colpoda) can be observed. These are later replaced by numerous larger ciliate protozoans, such as paramecium. Still later, other protozoans, such as ameba and vorticella, appear. Eventually tiny multicellular forms, such as rotifers and roundworms, dominate the infusion. Finally, all organisms in the infusion die as they are poisoned by the metabolic wastes that accumulate in the relatively small amount of water in the infusion.

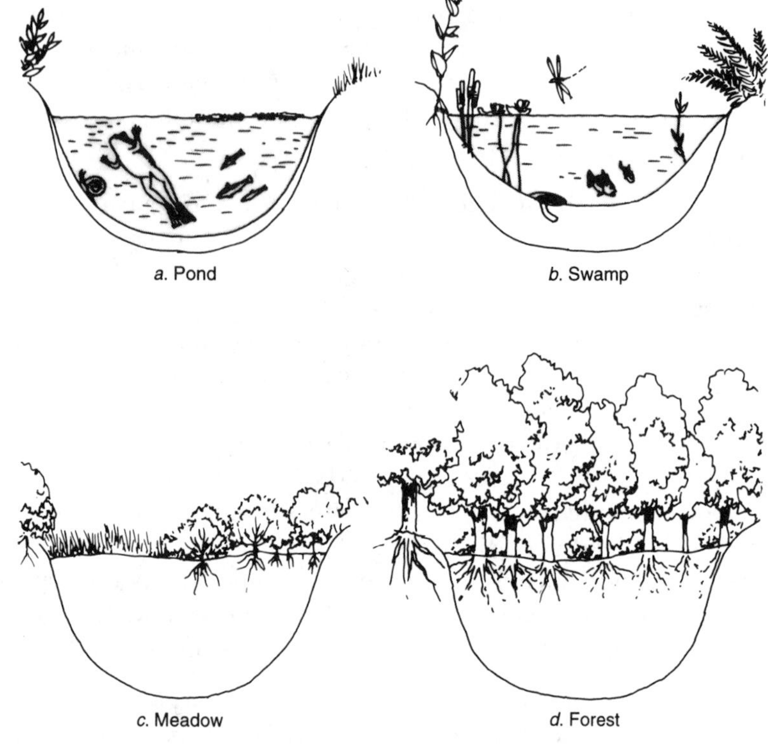

Figure 19–10 Succession in a pond

WORLD BIOMES

A *biome* is a large geographic region composed of a group of ecosystems in which climatic factors play a dominant role. Variations in abiotic factors influence the types and numbers of species that are in a particular biome. Thus, a grassland in the United States is not the same as a grassland in Brazil. The major biomes are either terrestrial or aquatic.

Terrestrial biomes

Terrestrial biomes include the tundra, taiga, temperate deciduous forest, tropical rain forest, grassland, desert, and mountain biomes. (See Table 19–1.)

Tundra. A *tundra* biome may be located just beyond a polar region or below snow-capped mountaintops. In either location, the subsoil, called *permafrost*, is permanently frozen. Topsoil thaws only during the summer. Since the meltwater cannot drain into the subsoil, many shallow ponds, bogs, and small streams are formed. In this biome, the most abun-

Table 19–1 Average Precipitation and Temperature of Terrestrial Biomes

Biome	*Yearly Precipitation (Average)*	*Yearly Temperature Range (Average)*
Tundra	Less than 25 cm	–26°C to 4°C
Coniferous forest	35 to 75 cm	–10°C to 14°C
Deciduous forest	75 to 125 cm	6°C to 28°C
Tropical rain forest	More than 200 cm	25°C to 27°C
Grassland	25 to 75 cm	0°C to 25°C
Desert	Less than 25 cm	24°C to 34°C

dant vegetation consists of lichens, grasses, and sedges (the hardiest of land plants). A few types of trees, such as birches and conifers, may grow, but these trees remain small.

During the short arctic summer, land and pond vegetation are the producers that supply food for many food chains. The consumers consist mainly of insects such as the mosquito; arctic birds such as the ptarmigan; and mammals such as the arctic hare, polar bear, caribou (reindeer), and musk oxen. Most of the animals migrate southward when winter arrives and return the next summer. Some animals, such as the ptarmigan, polar bear, and lemming, are year-round residents. The color of the ptarmigan's feathers and the hare's fur change from brown to white; the lemmings tunnel under the snow; and insects survive as eggs, which resist freezing to hatch the following summer. The annual plants survive as seeds, which resist freezing and complete their short life cycle in the summer.

Taiga. In the northern hemisphere, *taiga* (or *coniferous forest*) regions are located just south of the tundra, or just below the timberline on mountains. The ground thaws more completely in the taiga because it receives more sunlight than a tundra. The characteristic vegetation consists of a coniferous forest composed mainly of evergreens, such as spruce and fir. Lichens and mosses cover the forest floor; deer, bears, wolves, seed-eating birds, and numerous species of insects are permanent residents of the taiga. The major migratory animals are insect-eating birds, caribou, and moose that migrate southward from the tundra.

During the summer, many varieties of small birds eat insects. Herbivores include deer, moose, and porcupine. Carnivores include the wolf and lynx. During the winter, bears and squirrels sleep for long periods of time and many species of birds migrate southward.

Temperate deciduous forest. The north temperate zone lies south of the taiga. Thus, the climate of the *temperate* zone is milder than that of the taiga and tundra. There is sufficient water to support the growth of broad-leaved (deciduous) trees, which shed their leaves as the cold weather approaches. Forests of maple and beech and forests of oak and hickory make up the dominant vegetation. Shrubs, such as the rhododendron, and low-growing plants, such as ferns and Canada mayflowers, grow under the trees. Hordes of insects feed upon the vegetation and are, in turn, eaten by different species of insect-eating birds, such as the flycatcher, red-eyed vireo, and woodpeckers.

During the winter, woodchucks and snakes hibernate, insect-eating birds migrate south, trees shed their leaves, and squirrels and mice feed upon the nuts they stored during the fall. In the spring, leaves reappear on the trees, and herbaceous plants grow and flower.

Tropical rain forest. *Tropical rain forests* are located mainly along the equator. There are tropical rain forests in the Amazon basin of South America, the Congo region of Africa, the East Indies, and the southern tip of Florida. In a tropical rain forest, the average temperature is high and varies little from day to day. The combination of high temperatures and daily rains results in high humidity, which favors the growth of rich, luxuriant vegetation. Most species of trees in the tropical rain forest retain their leaves all year. The dry season here corresponds to the winter season of the temperate zone.

In the tropical rain forest, most of the trees are broad-leaved and allow little sunlight to reach the forest floor. The soil lacks minerals because the daily rains dissolve and wash away, or leach, the mineral salts. This soil condition prevents small plants from growing on the forest floor. *Epiphytes* (plants that live on other plants), such as orchids, certain ferns, and Spanish moss, obtain their water from aerial roots or from cup-shaped leaves that catch and hold water. Few animals live on the forest floor; most of them, such as monkeys, sloths, and reptiles live in the trees. The diversity of insects and birds is great. In fact, a small part of a tropical rain forest contains more species than are found in all of Europe.

Grassland. *Grasslands* receive less rainfall than forests, but enough to support the growth of grasses. Thus, grasslands of Australia, Africa, and the United States are suitable for growing wheat and corn. Grasslands supply most of the cereal grains for the world's population. In the United States, wild herbivores, such as antelopes, prairie dogs, rabbits, and grasshoppers, are the major primary consumers. In Australia, the major herbivores are kangaroos. In Africa, zebras and antelopes are major herbivores.

Depending on geographic location, carnivores of grasslands include hawks, owls, badgers, coyotes, snakes, and lions.

Desert. In *desert* regions, the rainfall is sparse (less than 25 cm a year) and irregular. Days are hot and nights are cold. Shrubs such as creosote, sagebrush, and mesquite shed their thick leaves during the dry season to conserve water normally lost by transpiration. Plants such as cactuses have deep roots that obtain water at great depths. Cactuses also have water storage cells in their stems and roots.

Desert animals are active mainly at night to avoid the heat of the day. They often obtain water by licking the dew that condenses on cold surfaces. Rodents feed upon insects; coyotes, hawks, and roadrunners (birds) feed upon the rodents.

Mountain biomes. Air temperature decreases with a rise in altitude. For this reason, biomes on a *mountain* occur in levels, or zones, according to the characteristic climate of each level. Thus, as you ascend a high mountain (Figure 19–11), you may cross a deciduous forest zone, a taiga zone, and, near the summit, a tundra zone. These mountain biomes correspond to the large biomes located in different geographical regions.

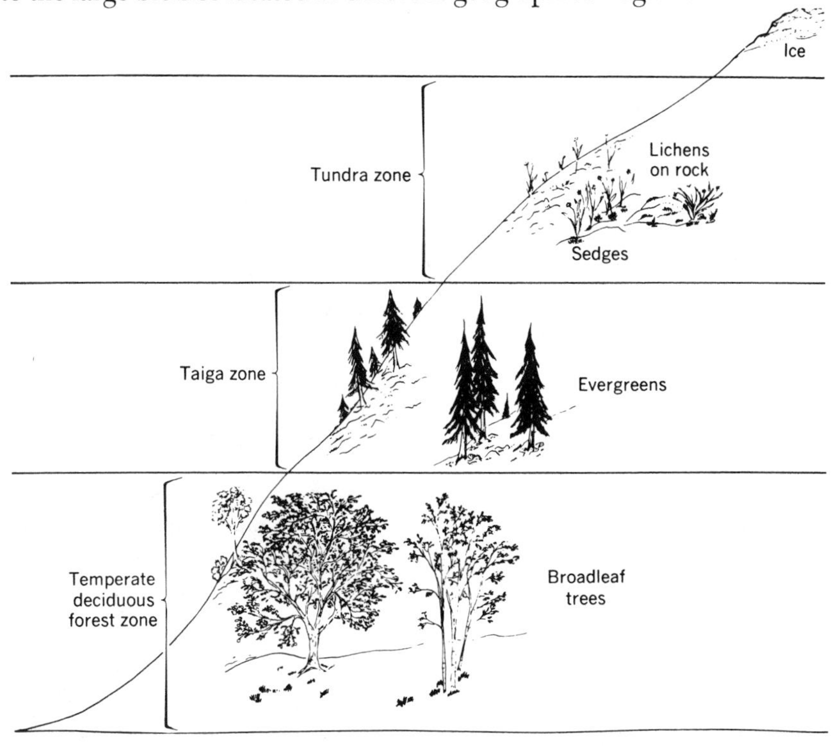

Figure 19–11 Biomes on a mountain

Aquatic Biomes

Aquatic biomes provide a relatively stable environment of diverse organisms. That is because water is not a limiting factor and the variations in temperature are not as great as they are on land. Aquatic biomes include ponds, lakes, rivers, oceans, seas, and estuaries. Major aquatic biomes are either *marine* (saltwater) or *inland* (freshwater). An *estuary* is a hybrid biome formed at a river's mouth, where it meets and flows into an ocean. Each aquatic biome features characteristic plankton of two kinds: (a) *phytoplankton*, or photosynthetic algae (producers) and (b) *zooplankton*, or small animals (consumers) that feed on phytoplankton.

Marine biomes. Figure 19–12 is a schematic drawing of ocean zones. The submerged land of a continent is called the *continental shelf*. The *intertidal zone* describes land alternately submerged and exposed as a result of changing tides. The steep drop from the continental shelf is called the *continental slope*. Canyons and gorges are often features of a continental slope. The continental slope leads to the *ocean floor* or basin, called the *abyssal plain*. The water depth above the abyssal plain is about 7000 meters; the water depth above the continental shelf is about 200 meters.

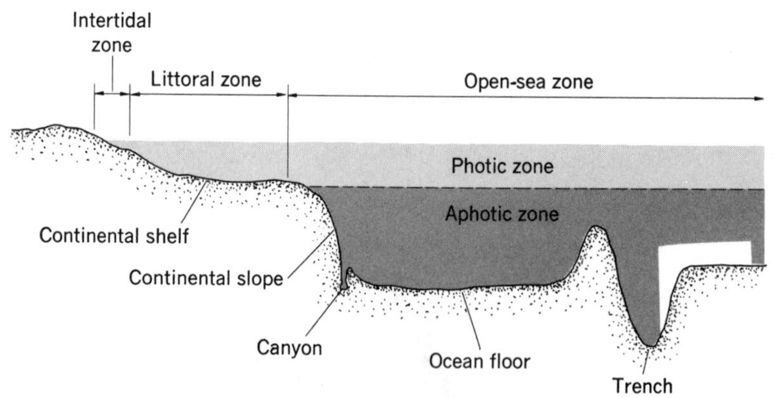

Figure 19–12 Marine biome zones

The intertidal zone. This zone is alternately submerged and exposed twice a day by changing tides. Intertidal organisms that live higher up on the shore must tolerate little food, drying out, and extremes of temperature. Organisms that live lower on the shore have access to more food, but compete for living space. At low tide, when these intertidal organisms are exposed, they face predators, such as crabs, birds, and rats; at high tide, they face hungry fishes. Sea anemones, snails, mussels, algae, and small fishes inhabit *tidal pools* found along rocky coasts of intertidal zones.

Sandy and muddy coasts usually are subjected to continuous wave action and currents that shift sediments and change the shoreline contours. As a result, few plants colonize these areas. Crabs, marine worms, and isopods are the animal inhabitants of sandy and muddy coasts.

The open ocean. The continental shelf drops steeply, beyond the intertidal zone. At the base of the continental shelf, two extensive biomes are found. The *pelagic biome* includes the entire expanse of ocean water and the *benthic biome* is the entire expanse of ocean bottom, or basin.

Pelagic biome. The neritic and oceanic zones are two divisions of the pelagic biome. The *neritic zone* is a shallow part of the ocean that overlies continental shelves. In tropical waters, coral reefs are common in this zone. A *coral reef* is built on a foundation of coral skeletons upon which living coral animals grow. Many types of coral may make up a large reef. The base of a coral reef usually is inhabited by red algae, which are eaten by many small marine animals and larvae; larger animals, such as eels, sharks, and seals, eat fish and shellfish.

The *oceanic zone* consists of great depths of water overlying ocean basins. The upper layer of the open ocean is the *photic zone*. It receives more sunlight than lower layers, which begin at a depth of about 200 meters. The photic zone receives enough sunlight for aquatic plants to carry out photosynthesis. This enables the growth and accumulation of phytoplankton which are eaten by zooplankton, such as rotifers and copepods. Zooplankton, in turn, are food for larger animals, mainly fishes.

Tropical parts of the oceanic zone have much less plankton than in higher latitudes where the surface water is cooler. In cooler climates, surface water contains more oxygen because cold water dissolves and holds more oxygen than warm water. In addition, more minerals are present because cool, denser water sinks and pushes up minerals from the ocean bottom to the surface. The circulation of nutrients is called *upwelling*. For these reasons, the Grand Banks of Newfoundland and the waters off the coast of Peru are the world's most productive fishing grounds. On sandy bottoms, flatfish, annelid worms, and some species of mollusks abound. On muddy bottoms, where seaweeds can grow, sea cucumbers, clams, and crabs are plentiful. On rocky bottoms, where water currents are strong, barnacles, mussels, sea urchins, and sea anemones are found attached to rocks. The *aphotic zone* consists of ocean layers to which light does not penetrate, thus it contains no photosynthetic organisms. Most inhabitants of the aphotic zone are scavengers and the animals that prey on them.

Benthic biome. This biome begins at the floor of the intertidal zone and extends across the ocean floor—reaching depths of more than 12,000 meters. The floor of the benthic biome is characterized by rock formations, sediments, and organic debris. Televised pictures and studies of the ooze collected and brought up from abyssal plains reveal the presence of decomposing bacteria, crustaceans, sea cucumbers, and brittle stars.

Recently, hydrothermal vent ecosystems were discovered in the deeps of the Pacific Ocean and in Lake Baikal, in Siberia. In these aphotic zones, fissures, or vents, between the Earth's crustal plates spew out heated water rich in salts of zinc, iron, copper, and calcium. These minerals are the ingredients necessary for the growth of chemosynthetic bacteria. The bacteria are primary producers for the consumer populations of clams, mussels, and worms that rely on them for nourishment.

Since Earth's human population and its need for energy resources are increasing, the marine biome is becoming a more valuable source of food and other natural resources. Today, varieties of fishes, mollusks, crustaceans, and seaweeds once overlooked as food sources are being harvested to feed growing populations. The ocean floor has yielded minerals and oil. In addition, tidal movements and wave action have been harnessed to generate electricity.

Estuary biome. Estuaries arise when fresh water, rich in nutrients, meets and mixes with ocean salt water. As a result, the salt gradient of fresh water increases and that of ocean water decreases. For this reason, organisms that inhabit an estuary must be adapted to frequent changes in salinity. Estuaries are very productive biomes. They contain communities of autotrophs that manufacture large yields of organic matter. About half of all marine fishes spend their immature stages in nutrient-rich estuaries. Phytoplankton, algae growing on plant surfaces, and grasses are primary producers. Bacteria-covered detritus, or organic debris, provides food for roundworms, crabs, snails, and small fish. Oysters, mussels, barnacles, and clams feed on organic debris suspended in water. Migratory birds, such as geese and ducks, use estuaries as temporary feeding grounds. A present threat to many estuaries are chemicals carried by rivers and streams.

Freshwater biomes. Freshwater biomes include *running water* such as rivers, and bodies of *standing water*, such as lakes and ponds.

Running water. Rivers and streams originate in mountains, and flow down slopes. The rapidly running water of mountain streams contains little, if any, plankton. Only organisms such as algae and aquatic mosses that are attached to rocks survive. The aquatic larvae of certain insects, such as the caddice fly, provide food for fishes, such as trout.

On more level ground, where streams flow more slowly, sediment is deposited on stream beds, providing a substrate for the growth of rooted plants. Bottom feeders, such as mussels, snails, crayfishes, worms, and the larvae of many insects, inhabit regions of streams where water movement is slow. Carnivores, such as the catfish, bass, and turtle, feed on small invertebrates. In marshland, where streams empty into the sea and salt water and fresh water mix, marsh grasses grow and provide a habitat for many species of waterfowl, snails, crabs, and fishes.

The water in rivers is used by farmers to irrigate crops, and by manufacturers as a coolant or to make products. Cities use rivers for domestic and commercial needs. These uses of rivers have caused pollution by industrial wastes, raw sewage, and chemical runoff—problems that must be resolved to restore clean water.

Standing water. A body of standing water that is small and shallow enough for sunlight to reach the bottom is called a *pond*. If the body of water is relatively extensive and has regions that are so deep that sunlight never reaches the bottom, it is called a *lake*.

Clear lakes that lack nutrients and do not support abundant growth of phytoplankton are called *oligotrophic*. Nutrient-rich lakes that support extensive growth of phytoplankton and plants are called *eutrophic*.

In ponds and lakes, the most active producers are algae. Consumer organisms include protozoans, rotifers, adult insects and their larvae, fishes, frogs, salamanders, turtles, birds, beavers, and weasels. (See Figure 19–13.) In deep lakes, oxygen is made available to organisms that

Figure 19–13 A pond ecosystem

inhabit deeper layers by the mixing of surface water with bottom water. Mixing occurs during the fall as the top layers of cool, denser water sink and displace the warmer, less dense water on the bottom. This sets up a vertical current of water called the *annual turnover*. As a result of the annual turnover, oxygen is circulated and mineral matter from the bottom is made available to organisms inhabiting the upper layers of water.

Excluding very large lakes, such as the Great Lakes, most lakes undergo succession. As the abiotic and biotic environments change, the succession proceeds from lake to bog to meadow to forest. (Refer back to Figure 19–10, on page 550.)

Section Quiz

1. Which statement best describes an estuary? (*a*) it has a photic and aphotic zone (*b*) it undergoes an annual upwelling (*c*) it produces large yields of organic matter (*d*) it usually contains more salt water than fresh water.
2. Tidal pools are features of a (an) (*a*) intertidal zone (*b*) estuary (*c*) neritic zone (*d*) hydrothermal vent.
3. An abyssal plain is part of a (*a*) pelagic biome (*b*) benthic biome (*c*) freshwater biome (*d*) terrestrial biome.
4. A eutrophic lake is characterized by (*a*) phytoplankton only (*b*) phytoplankton and rooted plants (*c*) rooted plants only (*d*) zooplankton and rooted plants.
5. An example of mutualism is the association between the (*a*) remora and shark (*b*) tapeworm and human (*c*) termite and flagellated protozoan (*d*) Spanish moss and tall tree.

POPULATION GROWTH

The rapid increase in the human population, as well as industrialization, has seriously affected the environment. In recent years, sometimes unknowingly, humans have created many ecological problems. These problems include the population explosion, a decrease in food-producing land, poor nutrition, and a wasting of natural resources.

The population explosion. The agricultural society in which people lived until about 1650 increased slowly in population to about 545 million people. With the conquest of many serious infectious diseases and the improvement of living conditions, the population continued to increase so that by 1990 the world's population was about 5.5 billion people.

This population explosion has caused humans to modify the environment to such an extent that many of the natural checks and balances that normally keep a community stable have been upset. For example, ranchers in the Kaibab National Forest region of Arizona killed the pumas, coyotes, and wolves that preyed upon their cattle and the deer in the national park. As a result, the deer population rose over time from 4,000 to about 100,000 in 1924. During the following year, about 90 percent of the deer died. This occurred because the forests could not support so many deer. Therefore, the deer starved to death. Population explosions of other species, such as those of English starlings and gypsy moths in the United States, are examples of organisms that have been introduced to an area by people and caused ecological problems. In general, an environment cannot indefinitely support the uncontrolled growth of a population.

Exponential growth. A population of living things grows exponentially (logarithmically) if its members obtain sufficient food and are not seriously affected by predators and disease. For example, a single bacterium may divide to form two daughter cells every 20 minutes. In 40 minutes, there are 4 bacterial cells; in 60 minutes, there are 8 bacterial cells. In effect, the size of the population doubles every 20 minutes, as long as environmental factors—food, water, temperature, lack of predators and disease—are favorable. After 10 hours (30 doublings), the bacterial count will exceed one million, provided no deaths in the bacteria have occurred.

A graph using *size of population* and *time* as coordinates produces a J-shaped growth curve, which is characteristic of populations exhibiting *exponential growth.* You can see that as the bacterial population grows larger, its size increases more rapidly (Figure 19–14). The graph in Figure 19–14 assumes that no deaths occur. Because deaths normally occur, the J-curve remains, but is skewed to the right. This indicates that more time is required to reach the same number of individuals in a population without deaths.

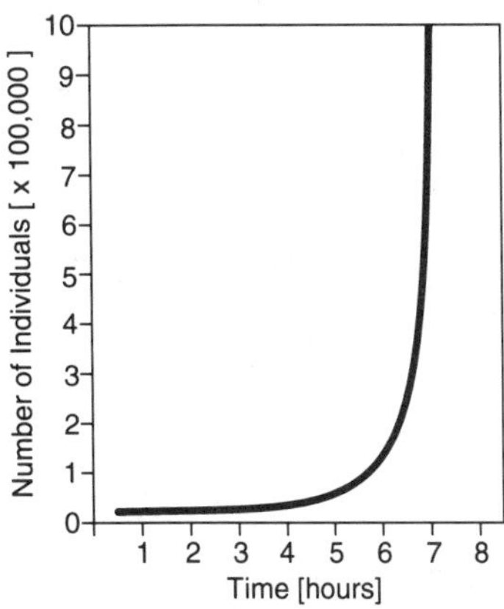

Figure 19–14 A population growth curve

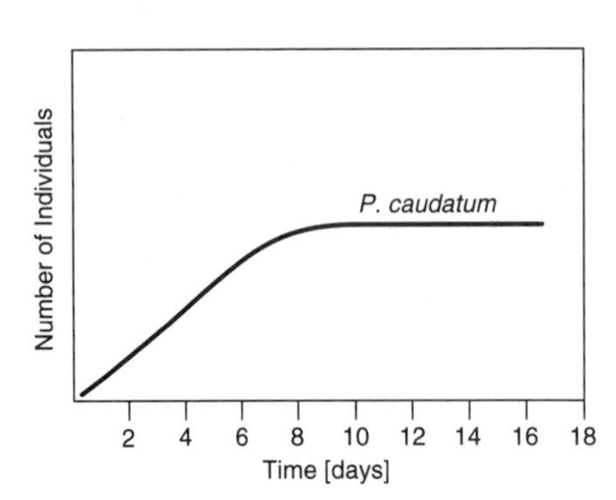

Figure 19–15 The growth curve of *Paramecium caudatum*

Carrying capacity. Most species, excluding humans, have population growth curves that tend to level off at a certain size. Figure 19–15 shows the population growth curve of a common ciliate, *Paramecium caudatum.* The growth curve levels off as the carrying capacity is reached. The environmental *carrying capacity* for a particular population includes the supply of resources provided by a habitat until the population is in equilibrium. A population is in equilibrium when births are balanced by deaths.

Carrying capacity is not fixed; it changes as the environment changes. In time, the culture will show changes in pH, oxygen and carbon dioxide concentrations, and the level of wastes. The population growth rate of parameciums levels off as the carrying capacity is reached; the quantity of biomass reaches its maximum size. *Biomass* is the dry weight of a population of organisms. Similarly, a small population of immature strawberry plants will grow, eventually reaching the environmental carrying capacity. But as the plants grow larger, competition for sunlight, water, living space, and nutrients intensifies. As a result, some strawberry plants die and the remaining plants increase in biomass.

Limiting factors. The environmental factors that limit population size are called *limiting factors*. These include the following:

- Availability of energy and raw materials.
- Production of wastes and their disposal.
- Interrelationships with other living things.

The combined effects of two or more limiting factors on population size is called *environmental resistance*.

Historically, the exponential growth of human populations has resulted from technological advances, such as the Industrial Revolution. In 1992, there were about 5.7 *billion* people compared to an estimated 5.0 million individuals in about 8000 B.C. The rate of global population growth alarms environmentalists because human growth is beginning to mirror the classic J-curve—one that cannot continue indefinitely. Thus, it is imperative that human population growth slow down to achieve equilibrium with the carrying capacity of the global environment. A reduced growth rate is especially important because humans are introducing or intensifying limiting factors such as global pollution, deforestation, ozone-layer depletion, and acid precipitation.

Resource Consumption

The *developing countries* in Asia, Africa, and Latin America are experiencing population growth mainly because the birth rate is increasing and the death rate is decreasing. As a result, many people suffer from malnutrition, poor housing, poor sanitation, and inadequate schooling. Increased consumption of environmental resources (agriculture, water, forests, land) causes another type of overpopulation, which is a characteristic of developed countries. This type of overpopulation is saddled with the consequences of pollution. In a developed country, such as the United States, each person consumes large amounts of natural resources, thereby adding to the pollution problem. In developing countries, available resources are used to capacity, and some of the resources are nonrenewable. For example, farmland depleted of its minerals and eroded by wind and water is an example of a *nonrenewable* resource. Water is a *renewable* resource because it is recycled. Nonproductive land can be restored by applying appropriate agricultural technology; the conservation of water as a renewable resource can be accomplished by wise use.

In a developed country, heavy consumption of natural resources is offset, to some degree, by employing the following practices:

- Recycling metals, paper, and water.
- Finding and using new resources, such as offshore oil and minerals from the ocean floor.
- Finding and using replacement resources, such as poorer grades of mineral ore and extracting maximum yields.

Energy. It has been estimated that by the year 2050, energy use by developing countries will outgrow that of developed countries (Figure 19–16). Energy consumption in developed countries is not increasing as rapidly as that of developing countries. In fact, energy consumption in the United States might decrease by as much as 20 percent in the near future. Renewable and nonrenewable energy sources are shown in Table 19–2.

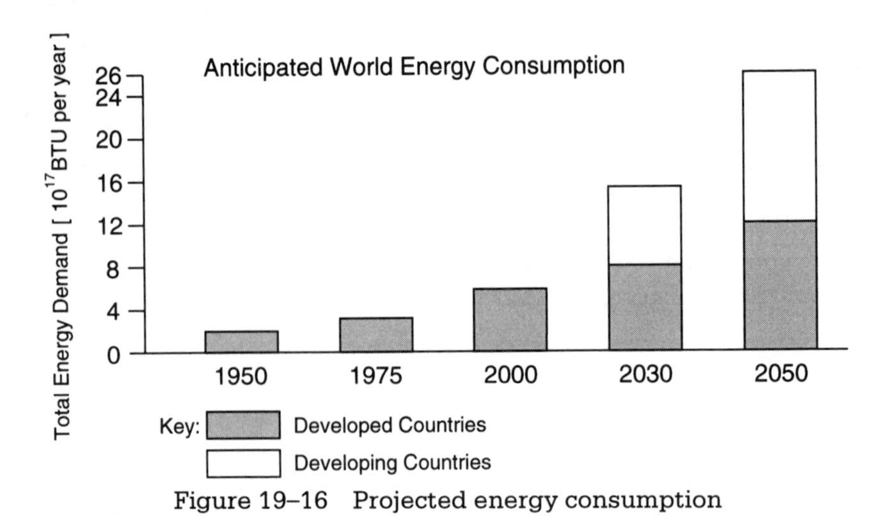

Figure 19–16 Projected energy consumption

Table 19–2 Renewable and Nonrenewable Energy Sources

Sources	*Advantages*	*Disadvantages*
Renewable	Infinite fuel supply	Technology under development
Nuclear		
Breeder	Fuel availability	Radiation pollution Thermal pollution Nuclear weapons use
Fusion	Fuel availability	Radiation pollution
Geothermal	Less pollution	Availability limited
Solar and wind	Nonpolluting Large and small scale possible	Noncompetitive cost
Ocean	Nonpolluting	Applicable only in certain areas
Biomass	Utilizes wastes	Air and water pollution
Nonrenewable	Technology well established	Finite fuel supply
Fossil fuels		
Coal, oil shale, and tar sands	Plentiful supply	Surface mining Air and water pollution
Petroleum	Cleaner burning	Limited supply
Natural gas	Cleanest burning	Limited supply

Current Problems and Solutions

As we approach the 21st century, ecological crises present serious challenges to the future of life on Earth. Increasing industrial production and an increasing human population have stressed the environment's capacity to sustain life. Resource depletion and pollution appear to be systematically linked. The ecological crises affect air, water, soil, and living things.

Air pollution. Air quality may be affected by various factors, such as the greenhouse effect, thermal inversion and smog, acid precipitation, and the ozone layer.

Greenhouse effect. "Greenhouse" gases, including methane, nitrous oxide, chlorofluorocarbons, and carbon dioxide, form a dangerous combination that threatens the quality of air. An increase of these gases, resulting from industrial processes and combination of fossil fuels, may raise Earth's temperature, creating a *greenhouse effect*. Global warming may cause more winter storms and hotter, drier summers. The ocean might rise about one meter as polar ice caps melt. This would cause the flooding of coastal areas and make hurricanes more destructive.

Thermal inversion. A thermal inversion is caused by a cool, dense layer of air being trapped by a layer of warmer air. As a result, pollutants in the cooler layer cannot be dispersed by winds or rise higher, but tend to concentrate near the ground. A complication frequently associated with thermal inversion is smog. Industrial smog appears as a gray haze over cities that burn large amounts of fossil fuels. To reduce smog, major cities in the United States and other developed countries restrict coal burning. In developing countries, such as China, industrial smog has become a serious problem. Cities with warm climates, such as Los Angeles, frequently suffer from *photochemical smog*. This type of smog is brown, which is the color of nitrogen dioxide formed when nitric oxide (from automobile exhaust) reacts with atmospheric oxygen. Thermal inversion and smog threaten the health of all organisms, especially humans.

Acid precipitation. Nonmetallic oxides, such as nitric oxide, nitrogen dioxide, sulfur dioxide, and sulfur trioxide, dissolve in water to form acids. These oxides of nitrogen and sulfur are by-products of burning fossil fuels and from applying nitrogen fertilizers. Thus, rain or snow combined with nonmetallic oxide yields acid precipitation with a relatively low pH. Acid precipitation reacts chemically with metals, marble, and plastics; more importantly, acid precipitation kills vegetation and affects the quality of natural ecosystems. In New York State, fish populations in about 200 lakes located in the Adirondack Mountains have been greatly reduced by acid precipitation.

Problems caused by acid precipitation also occur in other countries. For example, Canada generates a large volume of atmospheric sulfur dioxide, which is later deposited by wind as acid precipitation elsewhere. And, air pollutants originating in industrial regions of eastern and western Europe are carried northward to Sweden, Norway, and Finland by prevailing winds, causing acid precipitation.

Ozone layer. The ozone layer is part of the atmosphere. Life on Earth relies on the ozone layer because it screens out most ultraviolet rays from the sun. The chemical formula of ozone is O_3; it is produced when oxygen molecules react with oxygen atoms ($O_2 + O \rightarrow O_3$).

Weather satellites have shown that the ozone layer is thinning out over Antarctica and the high latitudes of North America, Europe, and Asia. Scientists think that by the year 2050, these regions may show a 25 percent decrease in the ozone layer. Some of the harmful effects of ultraviolet radiation include cataracts; skin cancers; poor immunological response to disease; disruption of food webs and ecosystems by interfering with the growth of primary producers, such as phytoplankton; and reproductive problems in some amphibian populations.

Chlorofluorocarbons, or *CFCs*, are mainly responsible for ozone depletion. CFCs are used as propellants in aerosol cans, coolants in refrigerators and air conditioners, and as reactants in the production of plastics. The ozone layer is depleted as CFC molecules rise to the stratosphere, and are subjected to ultraviolet rays. The reaction between CFC's and ultraviolet radiation cause the release of chlorine atoms. Chlorine atoms then react with ozone molecules to form oxygen and chlorine monoxide molecules. The chlorine monoxide molecules react with free oxygen atoms, yielding additional chlorine atoms, which continues the process. In 1978, some countries, including the United States, banned the use of CFCs in aerosol spray cans.

Water pollution. Lake, river, stream, estuary, and ocean ecosystems are endangered by acid rain, solid wastes, oil spills, and commercial fishing nets. Solid wastes, including plastics, have caused the deaths of about two million seabirds and 100,000 mammals yearly. Dead and dying birds are found entangled in plastic six-pack fasteners on beaches daily. Fishes, turtles, and whales ingest small plastic pellets generated as wastes by the plastic industry. Diving birds and mammals are entrapped in plastic drift nets up to 18 kilometers long. In fact, hundreds of kilometers of plastic nets are lost each season in the Pacific Ocean. Many marine animals are trapped and drowned in these lost nets.

Water is a renewable resource, but only if it is not polluted. Human sewage, animal wastes, and toxic chemicals make water unfit for human use. Runoff from farmland pollutes water with pesticides and unwanted soil nutrients, such as those that contain phosphorus compounds. Factories and industries, including power plants that use radioactive materials, also pollute water with excess heat, toxic chemicals, and radioactive wastes.

For many years, toxic wastes were dumped in hidden sites throughout the United States. In 1978, researchers discovered that the Love Canal, in Niagara Falls, New York, contained thousands of tons of industrial wastes. The Love Canal discovery motivated Congress to pass the *Comprehensive Environmental Response, Compensation and Liability Act* (*CERCLA*) to protect public health by seeking out and cleaning up toxic waste sites.

Soil. Wind and water erosion, and pollution of the soil from insecticides with long-lasting effects, threaten croplands. In the United States, about two billion tons of topsoil are lost yearly by wind and water erosion. This loss endangers about one-third of our croplands. If this continues for the next two generations, grain production will decrease to one-half the amount exported in 1980. That means millions of people in developing countries will not have enough nourishment to sustain themselves.

Soil removed from an ecosystem is normally replaced, to some degree, by humus and new soil. Intensive agriculture, however, tends to leave large areas of soil unusable, barren, and subject to greater than normal soil erosion. In some developing countries, farmers "push" the soil to the limit and clear land indiscriminately to plant food crops. In these countries, about 500 tons of topsoil per acre are lost each year!

In some parts of Asia, the Middle East, and Africa, successful crop production depends on irrigation. Water in arid lands usually contains more dissolved salts than water in other regions. Following irrigation, the evaporation of water from soil leaves mineral salt deposits behind, thereby making the soil unfit for further crop planting. The accumulation of mineral salts in soil as a result of evaporation is called *salinization*. A method, called *drip irrigation*, is used in arid parts of southern California to minimize salt accumulation in soil. Drip irrigation brings water directly to the roots of plants by pipes that drip water continuously.

Living things. Many species of wildlife have become *extinct* as the result of overhunting and habitat destruction. Some ecological studies indicate that Earth may lose about 25 percent of its present species of plants, animals, microorganisms, and fungi by the twenty-first century, unless aggressive measures are taken to protect natural ecosystems. The passenger pigeon, great auk, Carolina parakeet, and dodo already are extinct; many other species, including the whooping crane, giant panda, blue whale, and mountain gorilla are *endangered*, or threatened with extinction. (See Figure 19–17.) International agreements have helped combat threats such as habitat loss, overhunting, and pollution. However, tropical rain forests are being destroyed at a rate of 100 acres per minute, and the rate of destruction is increasing as demand increases for timber, farmland, and minerals. It is estimated that, at current rates, little will remain of Earth's tropical rain forests by the year 2040. Deforestation is not only destroying species, but also is reducing oxygen output and adding to the carbon dioxide content of the atmosphere, thus interfering with biological connections between air, water, and living things.

Figure 19–17 The dodo is extinct; the bald eagle is endangered

HUMAN SURVIVAL AND ENVIRONMENTAL PROTECTION

Ecologically sound measures must be taken in order to protect and restore environmental quality. The measures include *population control*; *increased food production*; *conservation of topsoil, water, forests, and wildlife*; and *insect control*.

Population control. Family planning is an effective way of controlling the growth of human populations and thus preventing shortages of food, living space, and water. *Family planning* means limiting the number of children in accordance with family income, available living space, general health of the parents, and amount of parental attention that can be given to each child. Thus, in a planned family, each child can receive the best opportunity to live a healthy and successful life.

Increased food production. Productivity can be increased in underdeveloped countries by introducing the use of improved strains of food plants and animals. The sea also can be used to farm certain marine vegetation and fishes (aquaculture).

Conservation of topsoil. Topsoil, which is essential for the growth of plants and the survival of decomposer organisms, is often wasted by the action of running water and wind. The erosion (wearing away of topsoil) can be reduced by the following farming methods:

Rotation of crops. Growing a particular crop, such as corn or wheat, on the same soil year after year depletes soil minerals. In time, few plants can grow in the mineral-depleted soil. As a result, soil becomes loose and easily eroded. Minerals, such as nitrates, can be restored to the soil by growing leguminous plants, such as peanuts and alfalfa (which bear nitrogen-fixing bacteria on their roots), alternately with another crop.

Adding fertilizers. Chemical fertilizers, animal wastes (manure), and *green manuring* (the plowing under of leguminous crops to add nitrates and humus to the soil) help maintain soil fertility.

Planting cover crops. Crop plants, such as clover and alfalfa, are usually planted close together. As the plants grow, they spread and cover the soil. For this reason, clover and alfalfa are called *cover crops.* Cotton and corn are called *row crops* because they are usually planted in rows separated by bare soil to facilitate weeding and harvesting. To prevent erosion after a row crop has been harvested, a cover crop should be planted on the same land soon afterward. The close-growing root systems and spreading leaves of cover-crop plants bind the soil particles together, absorb water, and shield the soil better than the roots of row crops.

Strip-cropping. In strip-cropping, parallel rows of a crop, such as corn, are alternated with parallel rows of cover crops, such as alfalfa or clover. The runoff from the corn crop is absorbed by the soil of the cover crop (Figure 19–18a). As a result, both soil and water are conserved.

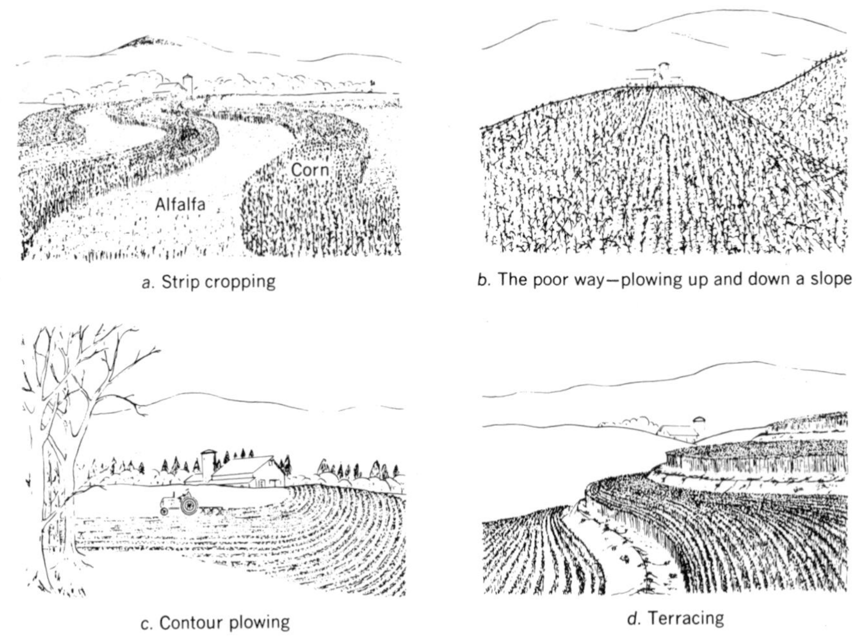

a. Strip cropping

b. The poor way—plowing up and down a slope

c. Contour plowing

d. Terracing

Figure 19–18 Prevention of soil erosion

Contour plowing. In contour plowing, the land is plowed in horizontal furrows that follow the natural contours of hills, as shown in Figure 19–18c. This method helps conserve soil and water by allowing rainwater to seep into the ground slowly.

Terracing. Terracing is used on hilly land, where the slope is too steep for contour plowing (Figure 19–18d). Rows of crops are grown on each terrace. The amount of runoff is checked because each terrace retains water, thus preventing erosion.

Soil banks. When land is allowed to remain fallow—crops are not planted on it for a year or more—nitrogen-fixing bacteria in the soil form nitrates and enrich the soil. The federal government sometimes encourages farmers to establish soil banks by paying them for not planting cash crops.

Conservation of water. The amount of usable water available to plants, animals, and humans can be increased by the following practices:

1. Construct more and better sewage disposal plants.
2. Treat industrial wastes to make them harmless before they are emptied into bodies of water or contaminate groundwater.

3. Control the spraying of chemical pesticides and herbicides. These chemicals usually enter the water supply by way of runoff. Chemical sprays should be restricted to those that kill only insects and weeds, and are changed into harmless substances in the soil or in the bodies of animals.
4. Prevent water waste by educating people to use water wisely; by fixing leaky faucets; by not using running water to wash dishes; and by watering lawns at night to avoid loss by evaporation.
5. Use proper farming methods, such as strip-cropping and contour plowing, to reduce runoff and increase the underground water supply.

Conservation of forests. To reclaim deforested land and maintain our existing forests, the following programs are recommended:

1. Establish new growth of trees in barren areas (reforestation). This will help control floods, prevent erosion, and provide wildlife habitats.
2. Reforest areas not used for farming. If this is done near the headwaters of streams, it will help prevent flooding and erosion.
3. Manage a forest like a farm. Harvest mature trees only and plant seedlings in their place. This practice can provide a constant supply of wood without having to destroy forests.
4. Use improved fire-fighting and fire-detection systems to prevent the loss of large forested regions by fire. Educate the public on ways to prevent forest fires through television, newspapers, and in schools.
5. Use biological controls, such as natural predators or natural attractants, to reduce populations of harmful insects.

Conservation of wildlife. To maintain stable communities and prevent loss of plants and animals, the following measures are recommended:

1. Enact laws that provide for closed seasons during which game animals may not be hunted; enact laws that protect birds that eat insects and rodents.
2. Protect wildlife by establishing more federal and state parks, bird sanctuaries, wildlife refuges, and fish hatcheries.
3. Require a license for hunting and fishing, to limit the catch; also, restrict the size, number, and sex of the catch.

Insect control. Insects are essential in most natural ecosystems. Many insects are economically important to humans, since the insects provide usable materials such as honey, wax, silk, and shellac. Many insects pollinate plants that provide us with food. The larvae of flies and beetles function as scavengers and thus help keep Earth's surface free from dead organisms. On the other hand, many species of insects are harmful in a variety of ways: for example, grasshoppers, potato beetles, and gypsy moths destroy crops and trees; anopheles mosquitos spread malaria; fleas spread bubonic plague; houseflies spread typhoid fever; termites destroy wooden buildings and furniture; clothes moths larvae eat fabrics; and the botfly and ox warble fly lay their eggs in the skin sores of domesticated animals.

Harmful insects can be controlled by either chemical or biological controls.

Chemical control of insects. Stomach poisons, such as lead arsenate, are used to kill insects that have biting mouth parts. Contact poisons, such as kerosene and nicotine sulfate, are used to kill insects that have sucking mouth parts. Nerve poisons, such as DDT and dieldrin, are used on a wide variety of insects. In her book *Silent Spring*, Rachel Carson described how the unrestricted use of these chemicals has upset food webs and has resulted in the death of many animals, including humans. For this reason, the use of DDT is prohibited in the United States, but not in other parts of the world.

Biological control of insects. In addition to insect control by using natural predators, biological insect control includes the use of an artificial sex attractant scent and sterilization treatment. The sex attractant scent, called *gyplure*, attracts male gypsy moths into a trap that kills them. As the number of male moths able to mate with females is reduced, the size of the next moth generation is reduced. To reduce the screwworm fly population, male flies are sterilized by irradiating them with cobalt-60 and are then released. The male flies mate with the females, but no offspring are produced.

Insect control can be made more effective by the following procedures:

1. Inspect all materials before they enter the United States to prevent the introduction of insect pests and diseased plants and animals.
2. Destroy the breeding places of insects by draining swamps, spraying swamps with oil, and burning fields after harvesting to destroy the roots and stems in which insects lay their eggs.
3. Plan the time of planting certain crops. The planting of cotton can be changed so that the flowers mature before the cotton boll weevil can lay its eggs in the cotton pod.

Many people feel that it is not their personal responsibility to improve environmental quality. Others, called environmental activists, realize how serious the global ecological crisis is, and they spearhead environmental programs to help restore and maintain natural ecosystems.

Biosocial measures. To remedy the global ecological crisis, a new approach to ecology and environmental management is reflected by pioneering environmental organizations. Their major objective is to instill in humans a sense of environmental awareness and responsibility. The involvement of concerned individuals working together on environmental projects can help resolve the global ecological crisis. Thus, environmental groups currently are lobbying against rain-forest destruction and ozone depletion, as well as ocean pollution by toxic wastes, sewage, and plastics. Community action can be effective, such as encouraging recycling and the use of non-aerosol products.

A biopolitical movement, called *Green Politics*, composed of grassroots activists, rallies for clean water, air, nontoxic foods, and against nuclear threats to the environment. *Greenpeace*, the largest international environmental organization, takes on a variety of issues, including nuclear-free seas, saving whales and seals, protesting the dumping of toxic pollutants, and preserving Antarctica. Other environmental organizations work with governmental agencies, urging them to help enact appropriate legislation to protect the environment.

Sustainable development is a socio-ecological concept in which ecologically destructive industrial activity is converted to environmentally sound production. The proponents of this concept consider soil a living thing. They claim that feeding soil, rather than vegetation, builds long-lasting fertility. Sustainable development opposes the use of pesticides and herbicides because these chemicals will degrade soil. In contrast, intensive agricultural management working in harmony with nature produces maximum yields by using compost, crop rotation, diversification, and cover crops.

Another goal of sustainable development is the restoration of ecosystems, such as prairies, wetlands, rivers, and lakes. Plant species are reintroduced into their original locations. In time, ecological relationships are reestablished between the soil, plants, insects, and other animals, recreating natural ecosystems. The sustainable development movement includes scientists, environmental organizations, political and social activists, and everyday citizens like ourselves.

On a more technical level, *ecological engineering* and *ecotechnology* deal with problems such as eutrophication of lakes, waste water management, and drinking water. *Eutrophication* of a lake is a type of pollution that results from an excess of nutrients, such as nitrates and phosphates, which cause the growth of plants to speed up. The ecotechnologist's approach to this problem is to trap the inflow of nutrient-rich water in a developed wetland, allow the water to "age," and then direct the aged, or purified, water to a lake. As a result, the lake does not undergo eutrophication prematurely.

Sludge from a waste-water treatment plant is usually incinerated, causing air pollution. In addition, slag and ash must be removed and deposited at legal sites. An ecotechnologist views sludge as an organic resource that can be used as a fertilizer. Environmental technology urges the use of ion exchange or denitrification to treat water without producing end products that are also pollutants. The ecotechnological way employs an artificial wetland designed to be a self-sustaining system, including surface vegetation. Cellulose packing under the wetland activates ion-exchange and denitrification; a second layer of sand filters out organic debris and microorganisms. Ecological engineering introduces methods of designing industrial processes that are environmentally friendly with natural ecosystems; ecotechnology advocates a symbiotic relationship between humans and nature. Looking ahead, it is possible that global environmental education, innovative technology, and a commitment by society can improve the global ecological outlook.

Section Quiz

1. Endangered peregrine falcons are bred and released in areas where pigeons are overly abundant. That procedure is an example of (*a*) species preservation and biological control (*b*) species preservation and use of parasites (*c*) overhunting and nonrenewable resource (*d*) overpopulation and use of biocides.
2. An example of a biological control against insects is the use of (*a*) pesticides (*b*) herbicides (*c*) mimicry (*d*) sex hormones.
3. A J-shaped curve reflects a population (*a*) that lacks resources such as food and water (*b*) that grows exponentially (*c*) that has encountered environmental resistance (*d*) whose death rate exceeds birth rate as limiting factors decrease.
4. When did exponential growth first occur in the human population? (*a*) at the end of World War II (*b*) at the time of the Industrial Revolution (*c*) shortly after the bubonic plague ended in Europe (*d*) after pesticides and fungicides were first used.
5. A major reason for the rapid growth of the human population is a (an) (*a*) reduction in carrying capacity (*b*) expansion in carrying

capacity (*c*) increased worldwide birth rate (*d*) increased worldwide immigration rate.

Chapter Review Questions

The following questions will help you check your understanding of the material presented in the chapter.

1. Which biotic factor is a characteristic of any natural community? (*a*) interactions among populations (*b*) climatic changes (*c*) variations in atmospheric pressure (*d*) irregularities in soil thickness.
2. Which term includes the other three? (*a*) population (*b*) species (*c*) community (*d*) ecosystem.
3. All the members of a species living in a given area constitute (*a*) a community (*b*) a biome (*c*) an ecosystem (*d*) a population.
4. "Earthworms live and reproduce in the soil. They aerate the soil and add organic material to it." These statements best describe an earthworm's (*a*) habitat (*b*) nutrition (*c*) environment (*d*) niche.
5. The prey populations of a community are strong and healthy because the predators act as a (*a*) selecting agent (*b*) source of energy (*c*) source of essential minerals (*d*) host.
6. In any natural community there is a loss of energy at each feeding level. The process that makes up for this loss and thus permits the community to continue is (*a*) decomposition (*b*) photosynthesis (*c*) nitrogen fixation (*d*) natural succession.
7. If the atmospheric carbon dioxide were reduced, the first organisms to diminish in number would most likely be the (*a*) producers (*b*) primary consumers (*c*) decomposers (*d*) scavengers.
8. Animals that obtain food by consuming only autotrophic organisms are classified as (*a*) herbivores (*b*) decomposers (*c*) scavengers (*d*) omnivores.
9. Animals that eat both plants and animals are known as (*a*) carnivores (*b*) omnivores (*c*) herbivores (*d*) saprophytes.
10. Some bacteria that live in the large intestine of humans produce B-complex vitamins that are useful to humans. This relationship between the bacteria and humans is an example of (*a*) parasitism (*b*) mutualism (*c*) commensalism (*d*) saprophytism.

11. The yucca moth pollinates the yucca flower while depositing her eggs on the flower. The moth larvae then hatch and feed on some of the seeds developing within the flower. The flower can only be pollinated by this species of moth. This relationship between moth and flower is one of (*a*) commensalism (*b*) mutualism (*c*) parasitism (*d*) saprophytism.
12. Malaria is caused by a protozoan that destroys red blood cells. The relationship between humans and the protozoan is called (*a*) parasitism (*b*) saprophytism (*c*) mutualism (*d*) commensalism.

Base your answers to questions *13 through 16* on the following information and on your knowledge of biology.

A large bottle with a layer of mud on the bottom is filled with pond water. Several fish and some green plants are then added to the bottle. The bottle is made airtight.

13. The living and nonliving contents of this bottle constitute (*a*) an ecosystem (*b*) a population (*c*) a biome (*d*) a species.
14. The fish in this bottle most likely represent (*a*) producers (*b*) decomposers (*c*) saprophytes (*d*) herbivores.
15. When one of the fish dies, the nitrogen from its body is released through the action of (*a*) viruses (*b*) bacteria (*c*) multicellular green plants (*d*) green algae.
16. To insure that the organisms in the bottle will survive for a period of time, the bottle must be (*a*) exposed to light (*b*) placed in water (*c*) kept in a cold room (*d*) placed in a well-ventilated room.
17. Protein synthesis is most closely associated with this cycle: (*a*) nitrogen (*b*) oxygen-carbon dioxide (*c*) water (*d*) phosphorus.
18. Nodules that contain nitrogen-fixing bacteria are found on the roots of members of the legume family. The relationship between these bacteria and the legume plant is best described as (*a*) mutualistic (*b*) parasitic (*c*) saprophytic (*d*) commensal.
19. Mice, insects, grasses, shrubs, owls, and trees can be found in an area consisting of a forest and a grassland. In number, the smallest population would most likely be the (*a*) insects (*b*) mice (*c*) grasses (*d*) owls.
20. A food chain includes crickets, frogs, snakes, and hawks. Which animals are the primary consumers? (*a*) crickets (*b*) frogs (*c*) snakes (*d*) hawks.

21. The natural cycling of carbon between organisms and their environment is most directly accomplished through the processes of (*a*) fermentation and oxidation (*b*) radiation and immigration (*c*) photosynthesis and respiration (*d*) isolation and dispersal.

22. The formation of nitrogen compounds from free nitrogen is called (*a*) denitrification (*b*) hydrolysis (*c*) photolysis (*d*) nitrogen fixation.

23. Some plants, when decomposed, change the pH of their environment so extensively that their offspring cannot survive but other species can survive. This is known as (*a*) symbiosis (*b*) predation (*c*) ecological succession (*d*) abiotic competition.

24. Which sequence is correctly arranged in order of decreasing average temperature? (*a*) desert, grassland, tundra, taiga (*b*) tropical forest, deciduous forest, tundra, taiga (*c*) deciduous forest, tropical forest, taiga, tundra (*d*) tropical forest, grassland, taiga, tundra.

25. The most stable environmental conditions occur in the (*a*) tropical forests (*b*) deserts (*c*) oceans (*d*) grasslands.

For each statement in questions *26 through 30*, select the letter of the land biome, chosen from the list below, that best fits that statement. A letter may be used more than once or not at all.

Land Biomes

(*a*) tundra
(*b*) desert
(*c*) grassland
(*d*) taiga
(*e*) tropical rain forest
(*f*) temperature deciduous forest

26. The primary consumers are herds of grazing animals.
27. Large mammals are not common.
28. It has the highest evaporation rate.
29. It receives the least amount of solar energy.
30. The climax vegetation is a coniferous forest.
31. The beech-maple climax forest of northern New York State is part of which world biome? (*a*) tundra (*b*) taiga (*c*) coniferous forest (*d*) deciduous forest.
32. Which zone of vegetation is found at the highest altitude? (*a*) coniferous forests (*b*) low herbs and shrubs (*c*) mosses and lichens (*d*) deciduous forests.

33. Which would probably result from the use of insecticides on flowering trees in orchards? (*a*) an increase in songbird populations (*b*) reduced fruit crops (*c*) an immediate reduction of soil minerals (*d*) an increase in the CO_2 concentration of the air.

34. Japanese beetles do relatively little damage in Japan because in Japan they (*a*) are kept in check by natural predators (*b*) are kept in check by effective insecticides (*c*) hibernate during the winter months (*d*) have gradually adapted to the environment.

35. If water pollution continues at its present rate, it may (*a*) make oxygen molecules unavailable to aquatic life (*b*) make nitrates unavailable to aquatic life (*c*) prevent precipitation (*d*) stop the water cycle.

36. Plants such as cactuses are adapted to their arid habitat because they (*a*) evolved broad leaves to reduce transpiration (*b*) evolved small leaves with recessed stomates (*c*) absorbed moisture from the atmosphere instead of from the soil (*d*) eliminated the need for water.

37. Where does the greatest amount of food production occur? (*a*) in estuaries (*b*) in continental lakes, streams, and rivers (*c*) on the floor of the oceans (*d*) in grassland biomes.

38. Most marine fish cannot survive in fresh water because of (*a*) their inability to maintain homeostatic water balance (*b*) the lower oxygen concentrations in fresh water (*c*) the higher carbon dioxide concentrations in fresh water (*d*) the lack of food plants in fresh water.

Biology Challenge

The following questions will provide practice in answering SAT II-type questions.

Part I

Select the letter of the word or phrase that best completes the statement or answers the question.

1. Which pair of animals are ecological equivalents? (*a*) rattlesnake and fox (*b*) tiger and elephant (*c*) salmon and seagull (*d*) zebra and wild horse (*e*) zooplankton and trout.

2. An example of a nonrenewable resource is (*a*) water (*b*) geothermal heat (*c*) magnesium (*d*) wind (*e*) solar energy.
3. An ecological law stating that a single environmental factor can determine the character of a biome is called the law of (*a*) minimum (*b*) Malthus (*c*) "10 percent" (*d*) limiting factors (*e*) carrying capacity.
4. Which statement is correct? (*a*) The human alimentary canal is an ecosystem. (*b*) The present increase in human population is mainly due to an increased birth rate. (*c*) Insects and humans are small food consumers. (*d*) A land biome usually is identified by its dominant animals. (*e*) Estuaries lack phytoplankton and immature organisms.
5. Demography is the study of (*a*) biotic potentials and population density (*b*) ways populations may change in the future (*c*) density-dependent factors and biomass (*d*) biogeological cycles and climate changes (*e*) all of these.

Part II

Base your answers to the following two questions on the paragraph below and your knowledge of biology.

In the course of their evolution, parasites developed special adaptations of behavior and physiology. As a result, they are better adjusted to maintain their populations in specific hosts.

It may be inferred from the statement above that:

1. (*a*) Parasites evolved from primitive ancestors that parasitized the species in a particular family. (*b*) Convergent evolution occurred. (*c*) Hosts may become extinct, but their parasites are unaffected by the carrying capacity of the environment. (*d*) All parasites occupy niches that are dependent on density-dependent factors, such as hosts. (*e*) Over time, the symbiotic relationship between host and parasite does not affect either population.
2. (*a*) Animal parasites use a secondary host when the primary host dies. (*b*) A secondary host ensures completion of the parasite's life cycle. (*c*) A secondary host does not affect species dispersal. (*d*) The anatomy and behavior of an animal parasite is different from that of a predator. (*e*) The growth curve of parasites is exponential, but skewed to the right.

GLOSSARY

absorption process by which end products of digestion move from the small intestine into the blood

acetylcholine a neurohumor secreted by the end brush of a neuron

acid a compound that dissociates in water to form hydrogen ions

ACTH (adrenocorticotropic hormone) a pituitary hormone that stimulates the secretion of several hormones by the cortex of the adrenal gland

active site the region of an enzyme molecule at which a reaction occurs with a substrate molecule

active transport the movement of substances through the plasma membrane of a cell by means of energy

actomyosin a complex combination of the proteins actin and myosin involved in muscular contraction

adaptation a characteristic of an organism that enables it to survive

adenine a nitrogen base present in nucleotides of DNA and RNA; a purine

ADP (adenosine diphosphate) a molecule that resembles ATP but lacks one high-energy phosphate group

aerobic respiration an energy-releasing process of cells that requires oxygen

agglutinin a plasma protein that reacts with a specific agglutinogen of red blood cells, causing them to clump

agglutinogen a protein of red blood cells that determines blood type

albinism a lack of pigment in the skin, eyes, and hair; a recessive trait

allantois an embryonic membrane of higher vertebrates that functions as a respiratory organ and as an excretory storage organ

allele one of a pair of contrasting genes for a given trait

allergen the protein or foreign substance that causes an allergic reaction

allergy overreaction of the immune system to a foreign substance

alveolus (*plural,* **alveoli**) one of many microscopic sacs through which the exchange of oxygen and carbon dioxide occurs in a lung

amino acid the organic building unit of polypeptides and proteins

ammonia a nitrogen compound excreted as a waste by most water-dwelling animals and animallike protists

amnion a fluid-filled sac that cushions the embryo of higher vertebrates

amylase any starch-digesting enzyme; an example is ptyalin

anaerobic respiration an energy-releasing process that does not require oxygen

angiosperm a flowering plant; may be either a dicot or a monocot

anther in a flower, the part of a stamen that produces pollen
antibiotic a substance produced by certain organisms that prevents the growth and multiplication of microorganisms
antibody a substance produced by the body that counteracts an antigen
antidiuretic hormone (ADH) a pituitary hormone that regulates the reabsorption of water in the kidneys; also called vasopressin
antigen a foreign protein that stimulates the body to form antibodies
aorta the largest artery of the body; carries oxygenated blood from the left ventricle of the heart to most body organs
arteriole a microscopic artery that joins an artery to capillaries
artery a blood vessel that carries oxygen-rich blood away from the heart
asexual reproduction reproduction by one parent only
assimilation the changing of substances into living matter
atom the smallest particle of an element that can combine with other elements
ATP (adenosine triphosphate) a molecule that stores and releases energy
autonomic nervous system the branch of the nervous system that regulates certain internal responses
autosome a chromosome other than a sex chromosome
autotroph a living thing that synthesizes its own food from inorganic substances
auxin a growth hormone secreted by certain cells of plants
axon the extension of a cyton that conducts a signal toward the end brush

bacteriophage a virus that parasitizes bacteria
base a substance that dissociates in water to form hydroxyl ions
bile a digestive juice secreted by the liver; emulsifies fats and oils
biome a very large climatic region composed of a group of ecosystems
blastula the stage of an animal embryo resembling a hollow ball of cells
bronchiole one of many branches of a bronchus leading to an air sac
bronchus (*plural,* bronchi) one of the two branches of the windpipe

calorie the amount of heat needed to raise the temperature of one gram of water one degree Celsius
cambium the permanent embryonic layer in the stem and root of a dicot; also called growing tissue or meristem
capillary small blood vessel whose thin walls allow materials to diffuse between body tissues and blood
carbohydrate an organic compound containing carbon, hydrogen, and oxygen; includes sugars, starches, glycogen, and cellulose
cardiac muscle the type of striated muscle tissue present in the heart
catalyst a substance that changes (usually speeds up) the rate of a chemical reaction without itself being permanently changed
cell membrane the living outer layer of a cell through which substances pass into and out of the cell; also called plasma membrane
cell respiration the reactions within a cell that release and store energy
cell theory the concept that the cell is the unit of structure and function of most living things

cell wall the rigid outer layer of plant cells, fungus cells, and some bacteria and protist cells
central nervous system the brain and spinal cord of vertebrates
centrioles short microtubules that aid in cell division
centrosome the structure in animal cells in which two centrioles are located; is involved in mitosis
cerebellum the part of the human brain, located behind and below the cerebrum; controls muscular coordination
cerebrum the largest part of the brain; is involved in sensation, memory, voluntary action, and intelligence
chemical bond the electrical attraction that holds atoms together; stores chemical energy
chlorophyll a complex molecule that captures light energy, which enables photosynthesis to occur
chloroplast a chlorophyll-containing organelle in green plants and algae
chordate an animal that has a notochord at one stage of its life history
chromatin a hereditary material, mainly DNA, in the nucleus of a cell
chromatography an absorption process that separates different molecules of a liquid mixture
chromosome a structure, composed of DNA, that transmits hereditary traits; contains the genes
cilia (*singular,* **cilium**) microscopic hairlike structures; their wavelike beating aids in locomotion, ingestion, and other life processes
circulation the movement of blood in major pathways to transport materials to and from cells
cleavage the rapid divisions of a fertilized egg cell
climax community the final, balanced stage in an ecological succession
coacervate a cluster of organic and inorganic compounds able to undergo respiration and reproduction; may have been the first form of life
coelenterate an invertebrate having stinging cells and a hollow body with a single opening
coenzyme a substance, usually a vitamin or a mineral, that, together with a protein (apoenzyme), forms an enzyme
color-blindness a sex-linked, inherited trait occurring mostly in males
commensalism a symbiotic relationship in which one organism benefits and the other is not harmed
community different populations of organisms living together in one area
compound two or more elements combined chemically in definite proportions by weight
conditioning using a substitute stimulus for the stimulus that originally triggered an automatic response
conifer a gymnosperm plant; usually an evergreen that bears cones
conjugation union of similar sex cells to form a zygote
consumer an organism that uses other organisms for food; a heterotroph
coronary arteries blood vessels that supply blood to the heart tissues
crossing-over the exchange of portions of chromosomes during the synapsis phase of meiosis
cyton the cell body of a neuron; contains the nucleus and cytoplasm

cytoplasm most of the material located between the nucleus and the plasma membrane of a cell; is a storehouse of chemicals
cytosine a nitrogen base present in nucleotides of DNA and RNA; a pyrimidine
cytoskeleton the tubules and filaments that make up most of a cell's structural network
cytosol the semifluid portion of the cytoplasm; 80 percent water with suspended particles

dark reaction the phase of photosynthesis that requires no light and in which glucose is produced; also called the carbon dioxide fixation phase
deamination a process in the liver that helps remove excess amino acids from the blood; forms urea
decomposer a heterotroph that obtains food from dead organic matter
dehydration synthesis a reaction that combines molecules to form a more complex molecule; it is accompanied by a loss of water molecules
dendrite an extension of a cyton forming a synapse with an end brush
diaphragm in mammals, a sheet of muscle that separates the chest cavity from the abdominal cavity; is used in breathing
dicot an angiosperm plant whose seed has two cotyledons
differentiation the transformation of similar embryonic cells into the specialized cells of different tissues
diffusion the movement of molecules from a region of greater concentration to a region of lesser concentration
digestion a chemical process that changes complex food molecules to simple food molecules; also called hydrolysis
diploid number the normal, or species, number of chromosomes characteristic of the body cells of an organism; is usually designated as $2n$
disaccharide a double sugar formed by the combination of two simple sugars by dehydration synthesis
DNA (deoxyribonucleic acid) the hereditary material in the nucleus of a cell, in chromosomes, in genes, and in certain viruses; is also present in chloroplasts
dominant trait a hereditary trait that always shows itself when present in a chromosome

ecology the study of the relationships between organisms and their environment
ecosystem the environment and all the living and nonliving things in it
ectoderm the outer tissue layer of the gastrula stage of an animal embryo
effector a muscle or gland that responds to a signal from a motor neuron
element a substance that cannot be changed into a simpler substance by ordinary means
embryo an animal in the early stages of growth and differentiation; attached to egg yolk or placenta
end brush the part of an axon at which a signal leaves a neuron
endocrine gland a gland that secretes a hormone directly into the blood; a ductless gland
endoderm the innermost layer of the gastrula stage of an animal embryo

endoplasmic reticulum the system of branching, netlike channels extending through the cytoplasm of a eukaryotic cell
endosperm a food-containing structure in seeds
enzyme an organic catalyst; lowers the activation energy of a reaction, thus speeding up the reaction
epiglottis a flaplike structure that prevents food from entering the windpipe
erythrocyte a red blood cell; lacks a nucleus and contains hemoglobin
esophagus the gullet
estrogen a female sex hormone produced by a mature follicle in an ovary
eukaryotic any cell or organism that has a membrane enclosing its genetic material
evolution the accumulation of genetic variations over time that enable living things to adapt to changes in their environment
excretion the removal of metabolic wastes from cells and from body fluids

fermentation a type of anaerobic respiration; produces ethyl alcohol
fertilization a process in sexual reproduction in which an egg cell and sperm cell unite to form a zygote
fetus a developing animal (embryo) whose major features have become visible
fibrin a protein that forms a meshwork as the basis of a blood clot
fibrinogen a protein that changes to fibrin during blood clotting
fission division of a parent cell into two or more daughter cells
flagella (*singular,* **flagellum**) microscopic hairlike structures; their whiplike motion enables movement in some cells and protozoans
follicle a pocket within an ovary that usually contains an ovum (egg cell)
food web the complex interconnecting food chains in a community
fossil the preserved remains of an ancient organism
fruit the ripened ovary of a flower
fungus (*plural,* **fungi**) a heterotrophic organism, often saprophytic; examples are molds, mildews, and mushrooms

gallbladder a bile-storing sac attached to the underpart of the liver
gamete a sex cell (usually a sperm cell or an egg cell)
gastric juice the digestive secretion of the stomach; contains hydrochloric acid and the enzyme pepsin
gastrula the stage of an animal embryo that resembles a double-walled or triple-walled cup
gene the portion of a chromosome that carries the genetic information for a specific trait; is composed of DNA
gene frequency the frequency with which an allele occurs in a population
gene pool all the allelic genes for all traits in a population
genetic drift the chance incorporation of a mutant gene into the gene pool of a small population
genotype the genetic makeup of an organism
gestation period the period of time from fertilization until birth
glomerulus (*plural,* **glomeruli**) a tuft of capillaries in the Bowman's capsule of a nephron from which a plasmalike fluid is forced
glucagon a hormone that releases glucose from the liver into the blood

glucose a six-carbon sugar used by most organisms as a source of energy
glycogen a carbohydrate formed from glucose and stored in the liver
glycolysis the anaerobic breakdown of glucose; produces molecules of ATP
goiter an enlargement of the thyroid gland
Golgi complex a cell organelle that modifies and refines proteins
gonad a reproductive gland—a testis or an ovary
granum (*plural,* **grana**) a stack of disklike bodies in a chloroplast; converts light energy to chemical energy
guanine a nitrogen base present in nucleotides of DNA and RNA; a purine
guard cells epidermal cells of a leaf that regulate stomate size
gymnosperm a nonflowering plant that bears seeds; examples are the conifers

half-life the time required for half the atoms in a radioactive specimen to change to stable end products
haploid number see *monoploid number*
Hardy-Weinberg law states that when mating is random and no external factors are involved, a population is genetically stable
hemoglobin a blood pigment that combines with oxygen and carbon dioxide
heterotroph a living thing that depends upon other organisms for food
heterotroph hypothesis the idea that the first living things were heterotrophs formed from organic compounds in the seas
heterozygous having two dissimilar genes (alleles) for a trait; hybrid
histamine a chemical compound formed by cells in response to certain antigen-antibody reactions such as allergies
homeostasis the tendency of a living system (organism) to maintain the stability of its internal environment
homologous structures structures that have the same embryonic origin but different functions
homozygous having two similar genes for a trait; pure for a given trait
hormone a secretion of an endocrine gland that enters directly into the blood
humus dead and decaying material present in soil
hybrid having a pair of contrasting genes (alleles) for a trait
hydrogen bond a weak chemical bond; connects complementary nitrogen bases in DNA chains
hydrogen transport system a series of reactions in which oxygen is used and energy is stored within the mitochondria of cells
hydrolysis a reaction that breaks up complex molecules by the addition of water molecules; digestion

immunity the ability of the body to combat foreign (pathogenic) organisms
imprinting a behavior pattern exhibited by certain animals in response to a stimulus received early in life
independent assortment Mendel's third law; states that each character for a trait operates as a unit and is inherited independently of any other character
ingestion the taking-in of food by an organism
inorganic relating to substances that were never alive; or, relating to compounds that lack carbon and hydrogen
instinct a complex form of behavior based upon an inborn series of reflexes

insulin a hormone secreted by the islets of Langerhans in the pancreas; enables cells to use and store glucose
interferon an antibodylike substance produced by a cell infected with a virus; prevents virus reproduction
interphase the period between cell divisions in which DNA replicates
invertebrate an animal without a backbone
islets of Langerhans patches of endocrine tissue in the pancreas that secrete the hormones insulin and glucagon

juvenile hormone an insect hormone that prevents metamorphosis

kinetic energy energy of motion.
Krebs cycle a series of aerobic respiratory reactions that takes place in the mitochondria; energy is released and stored in molecules of ATP

lacteal a tiny lymph vessel in the villus that carries complex fatty acids to a larger lymph vessel
larynx the voice box
leukocyte a white blood cell that defends the body against bacteria and viruses
lichen a symbiotic "organism" composed of algae and threadlike fungi; an example of mutualism
life processes activities carried out by all organisms to help them survive
light reaction the phase of photosynthesis in which water is decomposed to oxygen and hydrogen; also called photolysis
linkage the tendency of different traits to be inherited together because the genes controlling them are present in the same chromosome
lipase any enzyme that digests fats or oils
lipid any fat or oil; an organic compound composed of fatty acid molecules and glycerin molecules
loam a type of soil that consists of a mixture of sand, clay, and humus
lymph a fluid that bathes all the tissue cells of the body
lymph nodes (lymph glands) small structures located along the lymph vessels that help protect the body by producing some white blood cells and filtering out bacteria
lysosome a structure in animal cells that contains digestive enzymes

Malpighian tubule the excretory organ of the grasshopper
marsupial a pouched mammal that lacks a placenta
medulla the part of the brain connecting the brain to the spinal cord; controls breathing, the heartbeat, and reflex acts above the neck
meiosis a cell division process that reduces the diploid number ($2n$) of chromosomes to the monoploid number (n)
meristem the growing tissue of a plant
mesoderm embryonic tissue present in the gastrula from which systems such as the reproductive, skeletal, and excretory systems are derived
messenger RNA (mRNA) a single-stranded polynucleotide chain that functions on a ribosome as a template for protein synthesis

metabolism the building-up and tearing-down reactions that occur in cells

metamorphosis the change from larval to adult forms in certain animals

micrometer a unit of measure in the metric system equal to one thousandth of a millimeter (0.001mm) and symbolized μ; formerly called micron

micropyle in seed plants, an opening in an ovule through which the tip of a pollen tube passes, enabling fertilization

mitochondrion (*plural,* **mitochondria**) the cell organelle in which aerobic reactions store energy in molecules of ATP

mitosis the cell division process that duplicates nuclear material (chromosomes) and distributes the material equally between daughter cells

molecule the smallest unit of a compound

monocot an angiosperm plant whose seed has one cotyledon

monoploid number one-half the diploid, or normal, number of chromosomes; also called the haploid number, or *n*

monosaccharide a simple, or single, sugar that has six carbon atoms

motor neuron a nerve call that carries a signal away from the brain or spinal cord to a muscle or a gland

mutation a change in a gene or chromosome

mutualism a symbiotic relationship in which both organisms benefit

myelin sheath a fatty material covering the axon of a nerve cell

NADP (nicotinamide adenine dinucleotide phosphate) a compound that temporarily "holds" hydrogen during the light reaction of photosynthesis

natural selection Darwin's theory of evolution—new species arise as a result of the survival of organisms that inherit favorable variations

nematocysts stinging cells on the tentacles of hydra and other coelenterates

nephridium (*plural,* **nephridia**) the excretory organ of the earthworm

nephron one of millions of kidney tubules that remove metabolic wastes from the blood and form urine

nerve net the simple nervous system of the hydra

nerve signal an electrochemical wave that travels along a nerve cell

neurohumor a hormonelike substance secreted at the end brush of an axon

niche the interaction between an organism and its environment

nitrogen fixation the formation of nitrates from free nitrogen by certain bacteria

nondisjunction the failure of homologous chromosomes to separate during meiosis

notochord an internal supporting rod (not a spinal column) present in one stage of the life history of all chordates

nucleic acids DNA and RNA (composed of nucleotides); they both control heredity and protein synthesis

nucleoid a coiled mass (organelle) of DNA found in bacteria

nucleolus (*plural,* **nucleoli**) a small spherical structure located within the nucleus of a cell; contains RNA

nucleotide a complex molecule composed of a nitrogen base, a five-carbon sugar group, and a phosphate group

nucleus the densest organelle of a cell; it carries the hereditary material
nutrients the usable organic compounds found in foods

organ several tissues working together to perform a function
organelle any cell structure
organic relating to compounds containing carbon and hydrogen; or, relating to substances produced in or by living things
organ system several organs working together
osmosis the movement of water molecules from a region of greater concentration to a region of lesser concentration
ovary the female reproductive gland of plants or animals
ovulation the release of an egg cell from an ovary
ovule a structure within the ovary of a seed plant in which an egg nucleus is located; after fertilization, the ovule develops into a seed
oxidation the chemical union of oxygen with a substance; a loss of electrons
oxyhemoglobin hemoglobin chemically combined with oxygen

parasitism a symbiotic relationship in which one organism (the parasite) benefits and the other (the host) is harmed
parenchyma general storage tissue in plants
parthenogenesis the development of an unfertilized egg cell into an embryo
passive transport the simple passage of substances across a plasma membrane without an energy boost
pepsin a protein-splitting enzyme in gastric juice
peptide bond a bond that joins amino acids to form a protein in the process of dehydration synthesis
peristalsis a wave of rhythmic, involuntary muscular contractions that move the contents of the alimentary canal and blood vessels
permafrost the permanently frozen subsoil of the tundra region
permeability the extent to which a membrane allows different molecules to pass through it
pH a measure of the acidity of a solution; a pH of 7 is neutral, less than 7 is acidic, and greater than 7 is basic
phagocyte a white blood cell that engulfs and ingests foreign matter
phagocytosis the ingestion of foreign matter by white blood cells
pharynx a region in the back of the mouth that leads into the openings of the gullet and windpipe; the throat
phenotype the inherited physical appearance of an organism
phenylalanine an essential amino acid; inherited defects in its metabolism are the cause of albinism and phenylketonuria
phenylketonuria (PKU) an inherited disease characterized by brain damage in infants; it is caused by faulty metabolism of phenylalanine
phloem the conducting (vascular) tissue in plants that transports manufactured materials from the leaves to the stem and roots.
phosphate a chemical group, or ion, containing phosphorus; usually functions as an energy carrier in cells
phospholipids the molecules that make up a cell's plasma membrane
photic zone the upper layer of the open sea that receives the most sunlight

photolysis the light phase of photosynthesis; water is broken down to hydrogen and oxygen
photosynthesis the series of complex reactions in chloroplasts resulting in glucose production and oxygen release
pinocytosis a type of active transport in which large molecules enter a cell
pistil the female reproductive organ of a flower
placenta an embryonic organ in the uterus of mammals that nourishes the embryo and removes its wastes
plasma the liquid portion of blood; contains water and dissolved substances
plasma membrane see *cell membrane*
platelets see *thrombocytes*
pollen grain a structure produced by the anther of a flower; develops into a tiny male sexual plant
pollination the transfer of pollen grains from an anther to a stigma
polyploid more than the normal (diploid) number of chromosomes
population a group of organisms of the same species living in an area
potential energy stored energy
producer an autotroph (algae or plant); an organism that manufactures complex organic molecules (food) from simple inorganic molecules
progesterone a hormone that builds up the epithelium of the uterus and stimulates the growth of blood vessels in the uterus
prokaryotic any cell or organism that lacks a membrane enclosing its genetic material
prolactin a pituitary hormone that helps in the development of the mammary glands of a pregnant female and, after birth, stimulates the production of milk
protease any protein-digesting enzyme
protein a complex organic molecule composed of a chain of amino acids
protist a single-celled organism that has characteristics of plants or animals or both; algae and protozoans
pulmonary circulation circulation of blood between the heart and lungs
purine a nitrogen base in DNA and RNA; examples are cytosine and thymine; uracil occurs in place of thymine in all types of RNA
pyruvic acid a compound resulting from the glycolysis of glucose; in complex organisms, pyruvic acid enters the Krebs cycle

recessive trait a hereditary trait that does not appear if present with a dominant gene for the same trait
reduction the removal of oxygen from a compound; a gain of electrons by an atom
reflex action an inborn automatic response to a stimulus
renal circulation the circulation of blood between the dorsal aorta, the kidneys, and the heart
replication the duplication of DNA at cell division
reproduction the life activity by which organisms produce offspring
respiration (cellular) the process that releases and stores energy in cells
retina a complex layer of light-sensitive nerve cells in the eye
Rh factor a blood protein occurring in about 85 percent of the population

ribosome the cell organelle at which protein synthesis occurs; is located along the edges of the endoplasmic reticulum; possesses RNA (rRNA)
RNA (ribonucleic acid) a single chain of nucleotides patterned from a DNA template; the base uracil is present instead of thymine
root pressure osmotic pressure within the cells of a root; helps water rise in a plant

salt an inorganic compound formed from the union of an acid with a base
saprophyte an organism that obtains preformed organic molecules (food) from dead organic matter; examples are bacteria, molds, and mushrooms
sclerenchyma the tough supporting tissue of a plant
secretin a hormone secreted by the lining of the duodenum that stimulates the secretion of bile and pancreatic juice
secretion the formation of useful substances by an organism
seed a ripened ovule; contains an embryo plant and stored food
segregation Mendel's second law of inheritance; states that traits separate (during meiosis) and later recombine (at fertilization)
semicircular canals structures located in each inner ear that help detect changes in body movement and maintain balance
sensitivity the ability of an organism to respond to changes in its environment
sensory neuron a nerve cell that carries signals to the spinal cord and the brain
sex-linked trait a trait carried by a gene in the X chromosome
sickle-cell anemia an inherited disease in which abnormal hemoglobin molecules in red blood cells cause these cells to become misshapen when subjected to a lack of oxygen
solute a substance that dissolve in a solvent to form a solution
solution a homogeneous mixture formed when one substance dissolves in another
solvent a substance in which a solute dissolves to form a solution
specialization division of labor among cells, tissues, or organs
species a group of related organisms capable of mating with one another and producing fertile offspring
spindle fibers threadlike structures appearing in a cell during mitosis; may be involved in separating chromosomes
spiracle an opening in the body of an insect that allows air to enter and leave the body of the insect
spore an asexual cell that can resist unfavorable conditions and is capable of producing a new organism
stamen the male reproductive organ of a flower; consists of the filament and the anther, which produces pollen grains
stimulus any change in an organism's internal or external environment
stomate a pore in the epidermal layer of a leaf through which gases, including water vapor, move into and out of the leaf
substrate a molecule upon which an enzyme works
succession the replacement of one ecological community by another
symbiosis a permanent relationship between two different organisms living together; one organism lives either on or in the body of the other

synapse the space between the end brush of one neuron and the dendrities of another across which a nerve impulse (signal) passes
systemic circulation the circulation of the blood through all parts of the body except the lungs

taiga the land biome south of the tundra; is characterized by coniferous forests composed largely of evergreens, lichens, and mosses
taxis (*plural,* **taxes**) an inborn automatic response in which an animal's entire body moves either toward or away from a stimulus
testcross a mating process to determine whether an organism showing a dominant trait is homozygous for the trait
testis (*plural,* **testes**) the male reproductive gland of an animal
testosterone a male sex hormone secreted in the testes
thoracic duct a large lymph vessel that receives lymph from smaller lymph vessels; empties its contents into a large vein near the base of the neck
thorax in insects and chordates, the region of the body between the head and abdomen; in humans, the chest region
thrombocytes tiny blood cell fragments that help the blood clot normally
thymine a nitrogen base present in nucleotides of DNA; a pyrimidine
tissue a group of similar cells carrying out a specialized activity
toxoid weakened bacterial toxin used to produce active immunity
trachea the air tube through which air passes from the pharynx into the lungs; the windpipe; also, one of many air tubes of an insect
tracheophytes higher plants possessing specialized conducting tissues
transcription the process in which RNA is made from DNA
transduction the transfer of DNA from one organism to another by a virus
transfer RNA (tRNA) a single-chain polynucleotide patterned from a DNA template; carries a specific amino acid to a ribosome in protein synthesis
translation the process by which the DNA code in RNA is used to assemble the amino acid sequence that makes up a protein
transpiration the loss of water vapor from the stomates of leaves
transport all of the processes that move usable substances and wastes to and from cells
triplet a grouping of three nitrogen bases in DNA and RNA molecules
tropism an automatic response of a plant or part of a plant to an environmental stimulus such as sunlight or gravity
trypsin a protein-digesting enzyme in pancreatic juice
tundra the land biome located north of the taiga biome; is characterized by permanently frozen soil

ultramicroscopic too small to be observed with an ordinary light microscope
umbilical cord a structure in mammals that connects an embryo with a placenta
universal donor a person with blood type O; can give blood to a person having any blood type
universal recipient a person with blood type AB; can receive blood from a person having any blood type
uracil a nitrogen base in nucleotides of RNA; a pyrimidine

urea a nitrogen compound derived from excess amino acids and excreted as a waste
ureter a tube passing from a kidney to the urinary bladder
urethra a tube passing from the urinary bladder to the outside of the body
uterus a muscular organ in female mammals in which an embryo develops

vacuoles the cell organelles that store water and dissolved substances
variation an inherited difference between parents and offspring or between different offspring
vascular tissue the conducting tissue of plants; the xylem and phloem
vegetative propagation a process in which a nonreproductive (nutritive) structure of a plant develops into an entire new plant
vein a blood vessel that carries blood to the heart
venule a tiny vein that leads from capillaries to a larger vein
vertebra (*plural,* **vertebrae**) one of the 33 bones of the spinal column
vertebrate a chordate with a backbone
vestigial structure a structure, reduced in size, that has no use anymore in the body of an organism
villus (*plural,* **villi**) one of millions of fingerlike projections in the small intestine through which digested food passes into the blood
virus a particle showing some characteristics of life and visible only with the electron microscope; survives only as a parasite in a host cell
vitamin a nutrient not manufactured by the body and usually obtained from an outside source; is required for the synthesis of many enzymes

water cycle the movement of water from the atmosphere to the ground, through organisms, and back to the atmosphere
white blood cell a blood cell that possesses a nucleus; one type produces antibodies, another type ingests bacteria

xylem conducting (vascular) tissue in plants; carries water and dissolved minerals upward from the roots

yolk stored food material in an egg cell

zygote a fertilized egg cell; has the diploid number ($2n$) of chromosomes

INDEX